KU-162-190

Multivariable Feedback Control

Multivariable Feedback Control
Control
Analysis and Design

Sigurd Skogestad
Norwegian University of Science and Technology

Ian Postlethwaite
University of Leicester

JOHN WILEY & SONS
Chichester · New York · Brisbane · Toronto · Singapore

Copyright © 1996 by John Wiley & Sons Ltd,
    Baffins Lane, Chichester,
    West Sussex PO19 1UD, England

        National   01243 779777
        International (+44) 1243 779777
        e-mail (for orders and customer service enquiries): cs-books@wiley.co.uk.
        Visit our Home Page on http://www.wiley.co.uk
            or http://www.wiley.com

Reprinted July 1996

All rights reserved. No part of this book may be reproduced, stored in a retrieval system,
or transmitted, in any form or by any means, electronic, mechanical, photocopying,
recording or otherwise, except under the terms of the Copyright Designs and Patents Act
1988 or under the terms of a licence issued by the Copyright Licensing Agency,
90 Tottenham Court Road, London, UK W1P 9HE, without the permission in writing of the
publisher.

Other Wiley Editorial Offices

John Wiley & Sons, Inc., 605 Third Avenue,
New York, NY 10158-0012, USA

Jacaranda Wiley Ltd, 33 Park Road, Milton,
Queensland 4064, Australia

John Wiley & Sons (Canada) Ltd, 22 Worcester Road,
Rexdale, Ontario M9W 1L1, Canada

John Wiley & Sons (Asia) Pte Ltd, 2 Clementi Loop #02-01,
Jin Xing Distripark, Singapore 0512

British Library Cataloguing in Publication Data

A catalogue record for this book is available from the British Library

ISBN 0 471 94277 4; 0 471 94330 4 (pbk.)

Produced from camera-ready copy supplied by the authors
Printed and bound in Great Britain by Biddles Ltd, Guildford and King's Lynn
This book is printed on acid-free paper responsibly manufactured from sustainable forestation,
for which at least two trees are planted for each one used for paper production.

CONTENTS

PREFACE

This is a book on practical feedback control and not on system theory generally. Feedback is used in control systems to change the dynamics of the system (usually to make the response stable and sufficiently fast), and to reduce the sensitivity of the system to signal uncertainty (disturbances) and model uncertainty. Important topics covered in the book, include

- classical frequency-domain methods
- analysis of directions in multivariable systems using the singular value decomposition
- input-output controllability (inherent control limitations in the plant)
- model uncertainty and robustness
- performance requirements
- methods for controller design and model reduction
- control structure selection and decentralized control

The treatment is for linear systems. The theory is then much simpler and more well developed, and a large amount of practical experience tells us that in many cases linear controllers designed using linear methods provide satisfactory performance when applied to real nonlinear plants.

We have attempted to keep the mathematics at a reasonably simple level, and we emphasize results that enhance *insight* and *intuition*. The design methods currently available for linear systems are well developed, and with associated software it is relatively straightforward to design controllers for most multivariable plants. However, without insight and intuition it is difficult to judge a solution, and to know how to proceed (e.g. how to change weights) in order to improve a design.

The book is appropriate for use as a text for an introductory graduate course in multivariable control or for an advanced undergraduate course. We also think it will be useful for engineers who want to understand multivariable control, its limitations, and how it can be applied in practice. There are numerous worked examples, exercises and case studies which make frequent use of MATLAB™ [1].

---

[1] MATLAB is a registered trademark of The MathWorks, Inc.

The prerequisites for reading this book are an introductory course in classical single-input single-output (SISO) control and some elementary knowledge of matrices and linear algebra. Parts of the book can be studied alone, and provide an appropriate background for a number of linear control courses at both undergraduate and graduate levels: classical loop-shaping control, an introduction to multivariable control, advanced multivariable control, robust control, controller design, control structure design and controllability analysis.

The book is partly based on a graduate multivariable control course given by the first author in the Cybernetics Department at the Norwegian University of Science and Technology in Trondheim. About 10 students from Electrical, Chemical and Mechanical Engineering have taken the course each year since 1989. The course has usually consisted of 3 lectures a week for 12 weeks. In addition to regular assignments, the students have been required to complete a 50 hour design project using MATLAB. In Appendix B, a project outline is given together with a sample exam.

Examples and internet

Most of the numerical examples have been solved using MATLAB. Some sample files are included in the text to illustrate the steps involved. Most of these files use the μ-toolbox, and some the Robust Control toolbox, but in most cases the problems could have been solved easily using other software packages.

The following are available over the internet from Trondheim[2] and Leicester:

- MATLAB files for examples and figures
- Solutions to selected exercises
- Linear state-space models for plants used in the case studies
- Corrections, comments to chapters, extra exercises

This information can be accessed from the authors' home pages:

- http://www.kjemi.unit.no/~skoge
- http://www.engg.le.ac.uk/staff/Ian.Postlethwaite

Comments and questions

Please send questions, errors and any comments you may have to the authors. Their email addresses are:

- Sigurd.Skogestad@kjemi.unit.no
- ixp@le.ac.uk

---

[2] The internet site name in Trondheim will change from unit to ntnu during 1996.

Acknowledgements

The contents of the book are strongly influenced by the ideas and courses of Professors John Doyle and Manfred Morari from the first author's time as a graduate student at Caltech during the period 1983-1986, and by the formative years, 1975-1981, the second author spent at Cambridge University with Professor Alistair MacFarlane. We thank the organizers of the 1993 European Control Conference for inviting us to present a short course on applied $\mathcal{H}_\infty$ control, which was the starting point for our collaboration. The final manuscript began to take shape in 1994-95 during a stay the authors had at the University of California at Berkeley – thanks to Andy Packard, Kameshwar Poolla, Masayoshi Tomizuka and others at the BCCI-lab, and to the stimulating coffee at *Brewed Awakening*.

We are grateful for the numerous technical and editorial contributions of Yi Cao, Kjetil Havre, Ghassan Murad and Ying Zhao. The computations for Example 4.5 were performed by Roy S. Smith who shared an office with the authors at Berkeley. Helpful comments and corrections were provided by Richard Braatz, Atle C. Christiansen, Wankyun Chung, Bjørn Glemmestad, John Morten Godhavn, Finn Are Michelsen and Per Johan Nicklasson. A number of people have assisted in editing and typing various versions of the manuscript, including Zi-Qin Wang, Yongjiang Yu, Greg Becker, Fen Wu, Regina Raag and Anneli Laur. We also acknowledge the contributions from our graduate students, notably Neale Foster, Morten Hovd, Elling W. Jacobsen, Petter Lundström, John Morud, Raza Samar and Erik A. Wolff.

The aero-engine model (Chapters 11 and 12) and the helicopter model (Chapter 12) are provided with the kind permission of Rolls-Royce Military Aero Engines Ltd, and the UK Ministry of Defence, DRA Bedford, respectively.

Finally, thanks to colleagues and former colleagues at Trondheim and Caltech from the first author, and at Leicester, Oxford and Cambridge from the second author.

We have made use of material from several books. In particular, we recommend Zhou, Doyle and Glover (1996) as an excellent reference on system theory and $\mathcal{H}_\infty$ control. Of the others we would like to acknowledge, and recommend for further reading, the following: Rosenbrock (1970), Rosenbrock (1974), Kwakernaak and Sivan (1972), Kailath (1980), Chen (1984), Francis (1987), Anderson and Moore (1989), Maciejowski (1989), Morari and Zafiriou (1989), Boyd and Barratt (1991), Doyle et al. (1992), Green and Limebeer (1995), and the MATLAB toolbox manuals of Balas et al. (1993) and Chiang and Safonov (1992).

BORGHEIM, an engineer:

Herregud, en kan da ikke gjøre noe bedre enn leke i denne velsignede verden. Jeg synes hele livet er som en lek, jeg!

Good heavens, one can't do anything better than play in this blessed world. The whole of life seems like playing to me!

Act one, LITTLE EYOLF, Henrik Ibsen.

1

INTRODUCTION

In this chapter, we begin with a brief outline of the design process for control systems. We then discuss linear models and transfer functions which are the basic building blocks for the analysis and design techniques presented in this book. The scaling of variables is critical in applications and so we provide a simple procedure for this. An example is given to show how to derive a linear model in terms of deviation variables for a practical application. Finally, we summarize the most important notation used in the book.

1.1   The process of control system design

The process of designing a control system usually makes many demands of the engineer or engineering team. These demands often emerge in a step by step design procedure as follows:

1. Study the system (plant) to be controlled and obtain initial information about the control objectives.
2. Model the system and simplify the model, if necessary.
3. Analyze the resulting model; determine its properties.
4. Decide which variables are to be controlled (controlled outputs).
5. Decide on the measurements and manipulated variables: what sensors and actuators will be used and where will they be placed?
6. Select the control configuration.
7. Decide on the type of controller to be used.
8. Decide on performance specifications, based on the overall control objectives.
9. Design a controller.
10. Analyze the resulting controlled system to see if the specifications are satisfied; and if they are not satisfied modify the specifications or the type of controller.
11. Simulate the resulting controlled system, either on a computer or a pilot plant.
12. Repeat from step 2, if necessary.
13. Choose hardware and software and implement the controller.
14. Test and validate the control system, and tune the controller on-line, if necessary.

Control courses and text books usually focus on steps 9 and 10 in the above procedure; that is, on methods for controller design and control system analysis. Interestingly, many real control systems are designed without any consideration of these two steps. For example, even for complex systems with many inputs and outputs, it may be possible to design workable control systems, often based on a hierarchy of cascaded control loops, using only on-line tuning (involving steps 1, 4 5, 6, 7, 13 and 14). However, in this case a suitable control structure may not be known at the outset, and there is a need for systematic tools and insights to assist the designer with steps 4, 5 and 6. A special feature of this book is the provision of tools for *input-output controllability analysis* (step 3) and for *control structure design* (steps 4, 5, 6 and 7).

Input-output controllability is the ability to achieve acceptable control performance. It is affected by the location of sensors and actuators, but otherwise it cannot be changed by the control engineer. Simply stated, "even the best control system cannot make a Ferrari out of a Volkswagen". Therefore, the process of control system design should in some cases also include a step 0, involving the design of the process equipment itself. The idea of looking at process equipment design and control system design as an integrated whole is not new, as is clear from the following quote taken from a paper by Ziegler and Nichols (1943):

> In the application of automatic controllers, it is important to realize that controller and process form a unit; credit or discredit for results obtained are attributable to one as much as the other. A poor controller is often able to perform acceptably on a process which is easily controlled. The finest controller made, when applied to a miserably designed process, may not deliver the desired performance. True, on badly designed processes, advanced controllers are able to eke out better results than older models, but on these processes, there is a definite end point which can be approached by instrumentation and it falls short of perfection.

Ziegler and Nichols then proceed to observe that there is a factor in equipment design that is neglected, and state that

> ...the missing characteristic can be called the "controllability", the ability of the process to achieve and maintain the desired equilibrium value.

To derive simple tools with which to quantify the inherent input-output controllability of a plant is the goal of Chapters 5 and 6.

1.2   The control problem

The objective of a control system is to make the output y behave in a desired way by manipulating the plant input u. The *regulator problem* is to manipulate u to counteract the effect of a disturbance d. The *servo problem* is to manipulate u to keep the output close to a given reference input r. Thus, in both cases we want the *control error* $e = y - r$ to be small. The algorithm for adjusting u based on the available information is the controller K. To arrive at a good design for K we need *a priori* information about the expected disturbances and reference inputs, and of the plant model (G) and disturbance model (G_d). In this book we make use of linear models of the form

$$y = Gu + G_d d \qquad (1.1)$$

A major source of difficulty is that the models (G, G_d) may be inaccurate or may change with time. In particular, inaccuracy in G may cause problems because the plant will be part of a feedback loop. To deal with such a problem we will make use of the concept of model uncertainty. For example, instead of a single model G we may study the behaviour of a class of models, $G_p = G + E$, where the "uncertainty" or "perturbation" E is bounded, but otherwise unknown. In most cases weighting functions, $w(s)$, are used to express $E = w\Delta$ in terms of normalized perturbations, Δ, where the magnitude (norm) of Δ is less than or equal to 1. The following terms are useful:

Nominal stability (NS). The system is stable with no model uncertainty.

Nominal Performance (NP). The system satisfies the performance specifications with no model uncertainty.

Robust stability (RS). The system is stable for all perturbed plants about the nominal model up to the worst-case model uncertainty.

Robust performance (RP). The system satisfies the performance specifications for all perturbed plants about the nominal model up to the worst-case model uncertainty.

1.3 Transfer functions

The book makes extensive use of transfer functions, $G(s)$, and of the frequency domain, which are very useful in applications for the following reasons:

- Invaluable insights are obtained from simple frequency-dependent plots.
- Important concepts for feedback such as bandwidth and peaks of closed-loop transfer functions may be defined.
- $G(j\omega)$ gives the response to a sinusoidal input of frequency ω.
- A series interconnection of systems corresponds in the frequency domain to multiplication of the individual system transfer functions, whereas in the time domain the evaluation of complicated convolution integrals is required.
- Poles and zeros appear explicitly in factorized scalar transfer functions.

• Uncertainty is more easily handled in the frequency domain. This is related to the fact that two systems can be described as close (i.e. have similar behaviour) if their frequency responses are similar. On the other hand, a small change in a parameter in a state-space description can result in an entirely different system response.

We consider linear, time-invariant systems whose input-output responses are governed by linear ordinary differential equations with constant coefficients. An example of such a system is

$$
\begin{aligned}
\dot{x}_1(t) &= -a_1 x_1(t) + x_2(t) + \beta_1 u(t) \\
\dot{x}_2(t) &= -a_0 x_1(t) + \beta_0 u(t) \\
y(t) &= x_1(t)
\end{aligned}
\tag{1.2}
$$

where $\dot{x}(t) \equiv dx/dt$. Here $u(t)$ represents the input signal, $x_1(t)$ and $x_2(t)$ the states, and $y(t)$ the output signal. The system is time-invariant since the coefficients a_1, a_0, β_1 and β_0 are independent of time. If we apply the Laplace transform to (1.2) we obtain

$$
\begin{aligned}
s\bar{x}_1(s) - x_1(t=0) &= -a_1 \bar{x}_1(s) + \bar{x}_2(s) + \beta_1 \bar{u}(s) \\
s\bar{x}_2(s) - x_2(t=0) &= -a_0 \bar{x}_1(s) + \beta_0 \bar{u}(s) \\
\bar{y}(s) &= \bar{x}_1(s)
\end{aligned}
\tag{1.3}
$$

where $\bar{y}(s)$ denotes the Laplace transform of $y(t)$, and so on. To simplify our presentation we will make the usual abuse of notation and replace $\bar{y}(s)$ by $y(s)$, etc.. In addition, we will omit the independent variables s and t when the meaning is clear.

If $u(t), x_1(t), x_2(t)$ and $y(t)$ represent deviation variables away from a nominal operating point or trajectory, then we can assume $x_1(t=0) = x_2(t=0) = 0$. The elimination of $x_1(s)$ and $x_2(s)$ from (1.3) then yields the transfer function

$$
\frac{y(s)}{u(s)} = G(s) = \frac{\beta_1 s + \beta_0}{s^2 + a_1 s + a_0}
\tag{1.4}
$$

Importantly, for linear systems, the transfer function is independent of the input signal (forcing function). Notice that the transfer function in (1.4) may also represent the following system

$$
\ddot{y}(t) + a_1 \dot{y}(t) + a_0 y(t) = \beta_1 \dot{u}(t) + \beta_0 u(t)
\tag{1.5}
$$

with input $u(t)$ and output $y(t)$.

Transfer functions, such as $G(s)$ in (1.4), will be used throughout the book to model systems and their components. More generally, we consider rational transfer functions of the form

$$
G(s) = \frac{\beta_{n_z} s^{n_z} + \cdots + \beta_1 s + \beta_0}{s^n + a_{n-1} s^{n-1} + \cdots + a_1 s + a_0}
\tag{1.6}
$$

For multivariable systems, $G(s)$ is a matrix of transfer functions. In (1.6) n is the order of the denominator (or pole polynomial) and is also called the *order of the system*, and n_z is the order of the numerator (or zero polynomial). Then $n - n_z$ is referred to as the pole excess or *relative order*.

Definition 1.1

- *A system $G(s)$ is* strictly proper *if $G(s) \to 0$ as $s \to \infty$.*
- *A system $G(s)$ is* semi-proper *or* bi-proper *if $G(s) \to D \neq 0$ as $s \to \infty$.*
- *A system $G(s)$ which is strictly proper or semi-proper is* proper.
- *A system $G(s)$ is* improper *if $G(s) \to \infty$ as $s \to \infty$.*

For a proper system, with $n \geq n_z$, we may realize (1.6) by a state-space description, $\dot{x} = Ax + Bu$, $y = Cx + Du$, similar to (1.2). The transfer function may then be written as

$$G(s) = C(sI - A)^{-1}B + D \tag{1.7}$$

Remark. All practical systems will have zero gain at a sufficiently high frequency, and are therefore strictly proper. It is often convenient, however, to model high frequency effects by a non-zero D-term, and hence semi-proper models are frequently used. Furthermore, certain derived transfer functions, such as $S = (I + GK)^{-1}$, are semi-proper.

Usually we use $G(s)$ to represent the effect of the inputs u on the outputs y, whereas $G_d(s)$ represents the effect on y of the disturbances d. We then have the following linear process model in terms of deviation variables

$$y(s) = G(s)u(s) + G_d(s)d(s) \tag{1.8}$$

We have made use of the superposition principle for linear systems, which implies that a change in a dependent variable (here y) can simply be found by adding together the separate effects resulting from changes in the independent variables (here u and d) considered one at a time.

All the signals $u(s)$, $d(s)$ and $y(s)$ are deviation variables. This is sometimes shown explicitly, for example, by use of the notation $\delta u(s)$, but since we always use deviation variables when we consider Laplace transforms, the δ is normally omitted.

1.4 Scaling

Scaling is very important in practical applications as it makes model analysis and controller design (weight selection) much simpler. It requires the engineer to make a judgement at the start of the design process about the required performance of the system. To do this, decisions are made on the expected magnitudes of disturbances and reference changes, on the allowed magnitude of each input signal, and on the allowed deviation of each output.

Let the unscaled (or originally scaled) linear model of the process in deviation variables be

$$\widehat{y} = \widehat{G}\widehat{u} + \widehat{G}_d\widehat{d}; \quad \widehat{e} = \widehat{y} - \widehat{r} \tag{1.9}$$

where a hat ($\widehat{}$) is used to show that the variables are in their unscaled units. A useful approach for scaling is to make the variables less than one in magnitude. This is done by *dividing each variable by its maximum expected or allowed change*. For disturbances and manipulated inputs, we use the scaled variables

$$d = \widehat{d}/\widehat{d}_{\max}, \quad u = \widehat{u}/\widehat{u}_{\max} \tag{1.10}$$

where:

- $\widehat{d}_{\max}$ — largest expected change in disturbance
- $\widehat{u}_{\max}$ — largest allowed input change

The maximum deviation from a nominal value should be chosen by thinking of the maximum value one can expect, or allow, as a function of time.

The variables $\widehat{y}$, $\widehat{e}$ and $\widehat{r}$ are in the same units, so the same scaling factor should be applied to each. Two alternatives are possible:

- $\widehat{e}_{\max}$ — largest allowed control error
- $\widehat{r}_{\max}$ — largest expected change in reference value

Since a major objective of control is to minimize the control error $\widehat{e}$, we here usually choose to scale with respect to the maximum control error:

$$y = \widehat{y}/\widehat{e}_{\max}, \quad r = \widehat{r}/\widehat{e}_{\max}, \quad e = \widehat{e}/\widehat{e}_{\max} \tag{1.11}$$

To formalize the scaling procedure, introduce the scaling factors

$$D_e = \widehat{e}_{\max}, \ D_u = \widehat{u}_{\max}, \ D_d = \widehat{d}_{\max}, \ D_r = \widehat{r}_{\max} \tag{1.12}$$

For MIMO systems each variable in the vectors $\widehat{d}$, $\widehat{r}$, $\widehat{u}$ and $\widehat{e}$ may have a different maximum value, in which case D_e, D_u, D_d and D_r become diagonal scaling *matrices*. This ensures, for example, that all errors (outputs) are of about equal importance in terms of their magnitude.

The corresponding scaled variables to use for control purposes are then

$$d = D_d^{-1}\widehat{d}, \ u = D_u^{-1}\widehat{u}, \ y = D_e^{-1}\widehat{y}, \ e = D_e^{-1}\widehat{e}, \ r = D_e^{-1}\widehat{r} \tag{1.13}$$

On substituting (1.13) into (1.9) we get

$$D_e y = \widehat{G}D_u u + \widehat{G}_d D_d d; \quad D_e e = D_e y - D_e r$$

and introducing the scaled transfer functions

$$G = D_e^{-1}\widehat{G}D_u, \quad G_d = D_e^{-1}\widehat{G}_d D_d \tag{1.14}$$

then yields the following model in terms of scaled variables

$$y = Gu + G_dd; \quad e = y - r \tag{1.15}$$

Here u and d should be less than 1 in magnitude, and it is useful in some cases to introduce a scaled reference $\tilde{r}$, which is less than 1 in magnitude. This is done by dividing the reference by the maximum expected reference change

$$\tilde{r} = \hat{r}/\hat{r}_{\max} = D_r^{-1}\hat{r} \tag{1.16}$$

We then have that

$$r = R\tilde{r} \quad \text{where} \quad R \triangleq D_e^{-1}D_r = \hat{r}_{\max}/\hat{e}_{\max} \tag{1.17}$$

Here R is the largest expected change in reference relative to the allowed control

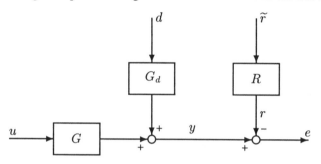

Figure 1.1: Model in terms of scaled variables

error (typically, $R \geq 1$). The block diagram for the system in scaled variables may then be written as in Figure 1.1, for which the following control objective is relevant:

- In terms of scaled variables we have that $|d(t)| \leq 1$ and $|\tilde{r}(t)| \leq 1$, and our control objective is to manipulate u with $|u(t)| \leq 1$ such that $|e(t)| = |y(t) - r(t)| \leq 1$ (at least most of the time).

Remark 1 A number of the interpretations used in the book depend critically on a correct scaling. In particular, this applies to the input-output controllability analysis presented in Chapters 5 and 6. Furthermore, for a MIMO system one cannot correctly make use of the sensitivity function $S = (I + GK)^{-1}$ unless the output errors are of comparable magnitude.

Remark 2 With the above scalings, the worst-case behaviour of a system is analyzed by considering disturbances d of magnitude 1, and references $\tilde{r}$ of magnitude 1.

Remark 3 The control error is

$$e = y - r = Gu + G_dd - R\tilde{r} \tag{1.18}$$

and we see that a normalized reference change $\tilde{r}$ may be viewed as a special case of a disturbance with $G_d = -R$, where R is usually a constant diagonal matrix. We will sometimes use this to unify our treatment of disturbances and references.

Remark 4 The scaling of the outputs in (1.11) in terms of the control error is used when analyzing a given plant. However, if the issue is to *select* which outputs to control, see Section 10.3, then one may choose to scale the outputs with respect to their expected variation (which is usually similar to $\hat{r}_{max}$).

Remark 5 If the expected or allowed variation of a variable about 0 (its nominal value) is not symmetric, then the largest variation should be used for $\hat{d}_{max}$ and the smallest variation for $\hat{u}_{max}$ and $\hat{e}_{max}$. For example, if the disturbance is $-5 \leq d \leq 10$ then $\hat{d}_{max} = 10$, and if the manipulated input is $-5 \leq \hat{u} \leq 10$ then $\hat{u}_{max} = 5$. This approach may be conservative (in terms of allowing too large disturbances etc.) when the variations for *several* variables are not symmetric.

A further discussion on scaling and performance is given in Chapter 5 on page 161.

1.5   Deriving linear models

Linear models may be obtained from physical "first-principle" models, from analyzing input-output data, or from a combination of these two approaches. Although modelling and system identification are not covered in this book, it is always important for a control engineer to have a good understanding of a model's origin. The following steps are usually taken when deriving a linear model for controller design based on a first-principle approach:

1. Formulate a nonlinear state-space model based on physical knowledge.
2. Determine the steady-state operating point (or trajectory) about which to linearize.
3. Introduce deviation variables and linearize the model. There are essentially three parts to this step:

   (a) Linearize the equations using a Taylor expansion where second and higher order terms are omitted.
   (b) Introduce the deviation variables, e.g. $\delta x(t)$ defined by

   $$\delta x(t) = x(t) - x^*$$

   where the superscript * denotes the steady-state operating point or trajectory along which we are linearizing.
   (c) Subtract the steady-state to eliminate the terms involving only steady-state quantities.

   These parts are usually accomplished together. For example, for a nonlinear state-space model of the form

$$\frac{dx}{dt} = f(x, u) \tag{1.19}$$

the linearized model in deviation variables $(\delta x, \delta u)$ is

$$\frac{d\delta x(t)}{dt} = \underbrace{\left(\frac{\partial f}{\partial x}\right)^*}_{A} \delta x(t) + \underbrace{\left(\frac{\partial f}{\partial u}\right)^*}_{B} \delta u(t) \tag{1.20}$$

Here x and u may be vectors, in which case the Jacobians A and B are matrices.
4. Scale the variables to obtain scaled models which are more suitable for control purposes.

In most cases steps 2 and 3 are performed numerically based on the model obtained in step 1. Also, since (1.20) is in terms of deviation variables, its Laplace transform becomes $s\delta x(s) = A\delta x(s) + B\delta u(s)$, or

$$\delta x(s) = (sI - A)^{-1} B\delta u(s) \tag{1.21}$$

Example 1.1 Physical model of a room heating process.

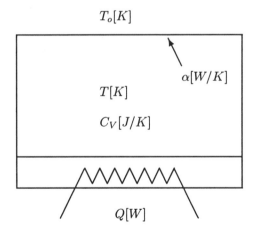

$T_o[K]$

$\alpha[W/K]$

$T[K]$

$C_V[J/K]$

$Q[W]$

Figure 1.2: Room heating process

The above steps for deriving a linear model will be illustrated on the simple example depicted in Figure 1.2, where the control problem is to adjust the heat input Q to maintain constant room temperature T (within ± 1 K). The outdoor temperature T_o is the main disturbance. Units are shown in square brackets.
1. Physical model. *An energy balance for the room requires that the change in energy in the room must equal the net inflow of energy to the room (per unit of time). This yields the following state-space model*

$$\frac{d}{dt}(C_V T) = Q + \alpha(T_o - T) \tag{1.22}$$

where T [K] is the room temperature, C_V [J/K] is the heat capacity of the room, Q [W] is the heat input (from some heat source), and the term $\alpha(T_o - T)$ [W] represents the net heat loss due to exchange of air and heat conduction through the walls.

2. Operating point. Consider a case where the heat input Q^* is 2000 W and the difference between indoor and outdoor temperatures $T^* - T_o^*$ is 20 K. Then the steady-state energy balance yields $\alpha^* = 2000/20 = 100$ W/K. We assume the room heat capacity is constant, $C_V = 100$ kJ/K. (This value corresponds approximately to the heat capacity of air in a room of about 100 m^3; thus we neglect heat accumulation in the walls.)

3. Linear model in deviation variables. If we assume α is constant the model in (1.22) is already linear. Then introducing deviation variables

$$\delta T(t) = T(t) - T^*(t), \ \delta Q(t) = Q(t) - Q^*(t), \ \delta T_o(t) = T_o(t) - T_o^*(t)$$

yields

$$C_V \frac{d}{dt}\delta T(t) = \delta Q(t) + \alpha(\delta T_o(t) - \delta T(t)) \tag{1.23}$$

Remark. If α depended on the state variable (T in this example), or on one of the independent variables of interest (Q or T_o in this example), then one would have to include an extra term $(T^* - T_o^*)\delta\alpha(t)$ on the right hand side of Equation (1.23).

On taking Laplace transforms in (1.23), assuming $\delta T(t) = 0$ at $t = 0$, and rearranging we get

$$\delta T(s) = \frac{1}{\tau s + 1}\left(\frac{1}{\alpha}\delta Q(s) + \delta T_o(s)\right); \quad \tau = \frac{C_V}{\alpha} \tag{1.24}$$

The time constant for this example is $\tau = 100 \cdot 10^3/100 = 1000$ s ≈ 17 min which is reasonable. It means that for a step increase in heat input it will take about 17min for the temperature to reach 63% of its steady-state increase.

4. Linear model in scaled variables. Introduce the following scaled variables

$$y(s) = \frac{\delta T(s)}{\delta T_{\max}}; \quad u(s) = \frac{\delta Q(s)}{\delta Q_{\max}}; \quad d(s) = \frac{\delta T_o(s)}{\delta T_{o,max}} \tag{1.25}$$

In our case the acceptable variations in room temperature T are ± 1 K, i.e. $\delta T_{\max} = \delta e_{\max} = 1$ K. Furthermore, the heat input can vary between 0 W and 6000 W, and since its nominal value is 2000 W we have $\delta Q_{\max} = 2000$ W (see Remark 5 on page 8). Finally, the expected variations in outdoor temperature are ± 10 K, i.e. $\delta T_{o,max} = 10$ K. The model in terms of scaled variables then becomes

$$G(s) = \frac{1}{\tau s + 1}\frac{\delta Q_{\max}}{\delta T_{\max}}\frac{1}{\alpha} = \frac{20}{1000s + 1}$$

$$G_d(s) = \frac{1}{\tau s + 1}\frac{\delta T_{o,max}}{\delta T_{\max}} = \frac{10}{1000s + 1} \tag{1.26}$$

Note that the static gain for the input is $k = 20$, whereas the static gain for the disturbance is $k_d = 10$. The fact that $|k_d| > 1$ means that we need some control (feedback or feedforward) to keep the output within its allowed bound ($|e| \leq 1$) when there is a disturbance of magnitude $|d| = 1$. The fact that $|k| > |k_d|$ means that we have enough "power" in the inputs to reject the disturbance at steady state, that is, we can, using an input of magnitude $|u| \leq 1$, have perfect disturbance rejection ($e = 0$) for the maximum disturbance ($|d| = 1$). We will return with a detailed discussion of this in Section 5.16.2 where we analyze the input-output controllability of the room heating process.

1.6   Notation

There is no standard notation to cover all of the topics covered in this book. We have tried to use the most familiar notation from the literature whenever possible, but an overriding concern has been to be consistent within the book, to ensure that the reader can follow the ideas and techniques through from one chapter to another.

The most important notation is summarized in Figure 1.3, which shows a one degree-of-freedom control configuration with negative feedback, a two degrees-of-freedom control configuration, and a general control configuration. The latter can be used to represent a wide class of controllers, including the one and two degrees-of-freedom configurations, as well as feedforward and estimation schemes and many others; and, as we will see, it can also be used to formulate optimization problems for controller design. The symbols used in Figure 1.3 are defined in Table 1.1. Apart from the use of v to represent the controller inputs for the general configuration, this notation is reasonably standard.

Lower-case letters are used for vectors and signals (e.g. u, y, n), and capital letters for matrices, transfer functions and systems (e.g. G, K). Matrix elements are usually denoted by lower-case letters, so g_{ij} is the ij'th element in the matrix G. However, sometimes we use upper-case letters G_{ij}, for example if G is partitioned so that G_{ij} is itself a matrix, or to avoid conflicts in notation. The Laplace variable s is often omitted for simplicity, so we often write G when we mean $G(s)$.

For state-space realizations we use the standard (A, B, C, D)-notation. That is, a system G with a state-space realization (A, B, C, D) has a transfer function $G(s) = C(sI - A)^{-1}B + D$. We sometimes write

$$G(s) \stackrel{s}{=} \left[\begin{array}{c|c} A & B \\ \hline C & D \end{array} \right] \tag{1.27}$$

to mean that the transfer function $G(s)$ has a state-space realization given by the quadruple (A, B, C, D).

For closed-loop transfer functions we use S to denote the sensitivity at the plant output, and $T = I - S$ to denote the complementary sensitivity. With negative feedback, $S = (I + L)^{-1}$ and $T = L(I + L)^{-1}$, where L is the transfer function around the loop as seen from the output. In most cases $L = GK$, but if we also include measurement dynamics ($y_m = G_m y + n$) then $L = GKG_m$. The corresponding transfer functions as seen from the input of the plant are $L_I = KG$ (or $L_I = KG_m G$), $S_I = (I + L_I)^{-1}$ and $T_I = L_I(I + L_I)^{-1}$.

To represent uncertainty we use perturbations E (not normalized) or perturbations Δ (normalized such that their magnitude (norm) is less than or equal to one). The nominal plant model is G, whereas the perturbed model with uncertainty is denoted G_p (usually for a set of possible perturbed plants) or G' (usually for a particular perturbed plant). For example, with additive uncertainty we may have $G_p = G + E_A = G + w_A \Delta_A$, where w_A is a weight representing the magnitude of the uncertainty.

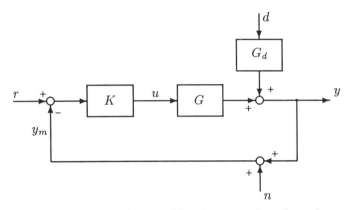

(a) One degree-of-freedom control configuration

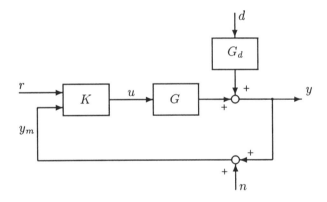

(b) Two degrees-of-freedom control configuration

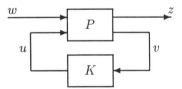

(c) General control configuration

Figure 1.3: Control configurations

Table 1.1: Nomenclature

K	controller, in whatever configuration. Sometimes the controller is broken down into its constituent parts. For example, in the two degrees-of-freedom controller in Figure 1.3(b), $K = \begin{bmatrix} K_r \\ K_y \end{bmatrix}$ where K_r is a prefilter and K_y is the feedback controller.

For the conventional control configurations (Figure 1.3(a) and (b)):

G	plant model
G_d	disturbance model
r	reference inputs (commands, setpoints)
d	disturbances (process noise)
n	measurement noise
y	plant outputs. These signals include the variables to be controlled ("primary" outputs with reference values r) and possibly some additional "secondary" measurements to improve control. Usually the signals y are measurable.
y_m	measured y
u	control signals (manipulated plant inputs)

For the general control configuration (Figure 1.3(c)):

P	generalized plant model. It will include G and G_d and the interconnection structure between the plant and the controller. In addition, if P is being used to formulate a design problem, then it will also include weighting functions.
w	exogenous inputs: commands, disturbances and noise
z	exogenous outputs; "error" signals to be minimized, e.g. $y - r$
v	controller inputs for the general configuration, e.g. commands, measured plant outputs, measured disturbances, etc. For the special case of a one degree-of-freedom controller with perfect measurements we have $v = r - y$.
u	control signals

By the right-half plane (RHP) we mean the closed right half of the complex plane, including the imaginary axis ($j\omega$-axis). The left-half plane (LHP) is the open left half of the complex plane, excluding the imaginary axis. A RHP-pole (unstable pole) is a pole located in the right-half plane, and thus includes poles on the imaginary axis. Similarly, a RHP-zero ("unstable" zero) is a zero located in the right-half plane.

We use A^T to denote the transpose of a matrix A, and A^H to represent its complex conjugate transpose.

Mathematical terminology

The symbol $\triangleq$ is used to denote *equal by definition*, $\overset{\text{def}}{\Leftrightarrow}$ is used to denote equivalent by definition, and $A \equiv B$ means that A is identically equal to B.

Let A and B be logic statements. Then the following expressions are equivalent:

A ⇐ B

A if B, or: If B then A

A is necessary for B

B ⇒ A, or: B implies A

B is sufficient for A

B only if A

not A ⇒ not B

The remaining notation, special terminology and abbreviations will be defined in the text.

2

CLASSICAL FEEDBACK CONTROL

In this chapter, we review the classical frequency-response techniques for the analysis and design of single-loop (single-input single-output, SISO) feedback control systems. These loop-shaping techniques have been successfully used by industrial control engineers for decades, and have proved to be indispensable when it comes to providing insight into the benefits, limitations and problems of feedback control. During the 1980's the classical methods were extended to a more formal method based on shaping closed-loop transfer functions, for example, by considering the $\mathcal{H}_\infty$ norm of the weighted sensitivity function. We introduce this method at the end of the chapter.

The same underlying ideas and techniques will recur throughout the book as we present practical procedures for the analysis and design of multivariable (multi-input multi-output, MIMO) control systems.

2.1 Frequency response

On replacing s by $j\omega$ in a transfer function model $G(s)$ we get the so-called frequency response description. Frequency responses can be used to describe: 1) a system's response to sinusoids of varying frequency, 2) the frequency content of a deterministic signal via the Fourier transform, and 3) the frequency distribution of a stochastic signal via the power spectral density function.

In this book we use the first interpretation, namely that of frequency-by-frequency sinusoidal response. This interpretation has the advantage of being directly linked to the time domain, and at each frequency ω the complex number $G(j\omega)$ (or complex matrix for a MIMO system) has a clear physical interpretation. It gives the response to an input sinusoid of frequency ω. This will be explained in more detail below. For the other two interpretations we cannot assign a clear physical meaning to $G(j\omega)$ or $y(j\omega)$ at a particular frequency – it is the distribution relative to other frequencies which matters then.

One important advantage of a frequency response analysis of a system is that it provides insight into the benefits and trade-offs of feedback control. Although this insight may be obtained by viewing the frequency response in terms of its relationship between power spectral densities, as is evident from the excellent treatment by Kwakernaak and Sivan (1972), we believe that the frequency-by-frequency sinusoidal response interpretation is the most transparent and useful.

Frequency-by-frequency sinusoids

We now want to give a physical picture of frequency response in terms of a system's response to persistent sinusoids. It is important that the reader has this picture in mind when reading the rest of the book. For example, it is needed to understand the response of a multivariable system in terms of its singular value decomposition. A physical interpretation of the frequency response for a stable linear system $y = G(s)u$ is a follows. Apply a sinusoidal input signal with frequency ω [rad/s] and magnitude u_0, such that

$$u(t) = u_0 \sin(\omega t + \alpha)$$

This input signal is persistent, that is, it has been applied since $t = -\infty$. Then the output signal is also a persistent sinusoid of the same frequency, namely

$$y(t) = y_0 \sin(\omega t + \beta)$$

Here u_0 and y_0 represent magnitudes and are therefore both non-negative. Note that the output sinusoid has a different amplitude y_0 and is also shifted in phase from the input by

$$\phi \triangleq \beta - \alpha$$

Importantly, it can be shown that y_0/u_0 and ϕ can be obtained directly from the Laplace transform $G(s)$ after inserting the imaginary number $s = j\omega$ and evaluating the magnitude and phase of the resulting complex number $G(j\omega)$. We have

$$y_0/u_0 = |G(j\omega)|; \quad \phi = \angle G(j\omega) \text{ [rad]} \qquad (2.1)$$

For example, let $G(j\omega) = a + jb$, with real part $a = \operatorname{Re} G(j\omega)$ and imaginary part $b = \operatorname{Im} G(j\omega)$, then

$$|G(j\omega)| = \sqrt{a^2 + b^2}; \quad \angle G(j\omega) = \arctan(b/a) \qquad (2.2)$$

In words, (2.1) says that *after sending a sinusoidal signal through a system $G(s)$, the signal's magnitude is amplified by a factor $|G(j\omega)|$ and its phase is shifted by $\angle G(j\omega)$*. In Figure 2.1, this statement is illustrated for the following first-order delay system (time in seconds),

$$G(s) = \frac{ke^{-\theta s}}{\tau s + 1}; \quad k = 5, \theta = 2, \tau = 10 \qquad (2.3)$$

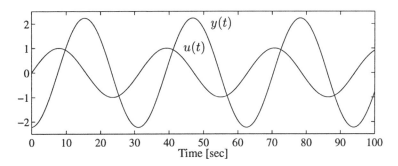

Figure 2.1: Sinusoidal response for system $G(s) = 5e^{-2s}/(10s + 1)$ at frequency $\omega = 0.2$ rad/s

At frequency $\omega = 0.2$ rad/s, we see that the output y lags behind the input by about a quarter of a period and that the amplitude of the output is approximately twice that of the input. More accurately, the amplification is

$$|G(j\omega)| = k/\sqrt{(\tau\omega)^2 + 1} = 5/\sqrt{(10\omega)^2 + 1} = 2.24$$

and the phase shift is

$$\phi = \angle G(j\omega) = -\arctan(\tau\omega) - \theta\omega = -\arctan(10\omega) - 2\omega = -1.51\,\text{rad} = -86.5°$$

$G(j\omega)$ is called the *frequency response* of the system $G(s)$. It describes how the system responds to persistent sinusoidal inputs of frequency ω. The magnitude of the frequency response, $|G(j\omega)|$, being equal to $|y_0(\omega)|/|u_0(\omega)|$, is also referred to as the *system gain*. Sometimes the gain is given in units of dB (decibel) defined as

$$A\,[\text{dB}] = 20\log_{10} A \tag{2.4}$$

For example, $A = 2$ corresponds to $A = 6.02$ dB, and $A = \sqrt{2}$ corresponds to $A = 3.01$ dB, and $A = 1$ corresponds to $A = 0$ dB.

Both $|G(j\omega)|$ and $\angle G(j\omega)$ depend on the frequency ω. This dependency may be plotted explicitly in Bode plots (with ω as independent variable) or somewhat implicitly in a Nyquist plot (phase plane plot). In Bode plots we usually employ a log-scale for frequency and gain, and a linear scale for the phase.

In Figure 2.2, the Bode plots are shown for the system in (2.3). We note that in this case both the gain and phase fall monotonically with frequency. This is quite common for process control applications. The delay θ only shifts the sinusoid in time, and thus affects the phase but not the gain. The system gain $|G(j\omega)|$ is equal to k at low frequencies; this is the steady-state gain and is obtained by setting $s = 0$ (or $\omega = 0$). The gain remains relatively constant up to the break frequency $1/\tau$ where it starts falling sharply. Physically, the system responds too slowly to let high-frequency

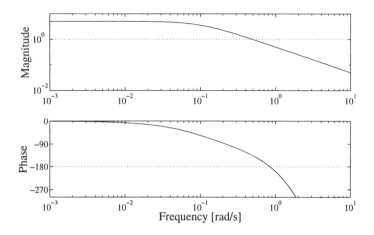

Figure 2.2: Frequency response (Bode plots) of $G(s) = 5e^{-2s}/(10s + 1)$

("fast") inputs have much effect on the outputs, and sinusoidal inputs with $\omega > 1/\tau$ are attenuated by the system dynamics.

The frequency response is also useful for an *unstable plant* $G(s)$, which by itself has no steady-state response. Let $G(s)$ be stabilized by feedback control, and consider applying a sinusoidal forcing signal to the stabilized system. In this case all signals within the system are persistent sinusoids with the same frequency ω, and $G(j\omega)$ yields as before the sinusoidal response from the input to the output of $G(s)$.

Phasor notation. From Euler's formula for complex numbers we have that $e^{jz} = \cos z + j \sin z$. It then follows that $\sin(\omega t)$ is equal to the imaginary part of the complex function $e^{j\omega t}$, and we can write the time domain sinusoidal response in complex form as follows:

$$u(t) = u_0 \text{Im } e^{j(\omega t + \alpha)} \text{ gives as } t \to \infty \quad y(t) = y_0 \text{Im } e^{j(\omega t + \beta)} \quad (2.5)$$

where

$$y_0 = |G(j\omega)|u_0, \quad \beta = \angle G(j\omega) + \alpha \quad (2.6)$$

and $|G(j\omega)|$ and $\angle G(j\omega)$ are defined in (2.2). Now introduce the complex numbers

$$u(\omega) \triangleq u_0 e^{j\alpha}, \quad y(\omega) \triangleq y_0 e^{j\beta} \quad (2.7)$$

where we have used ω as an argument because y_0 and β depend on frequency, and in some cases so may u_0 and α. Note that $u(\omega)$ is *not* equal to $u(s)$ evaluated at $s = \omega$ nor is it equal to $u(t)$ evaluated at $t = \omega$. Since $G(j\omega) = |G(j\omega)|\, e^{j\angle G(j\omega)}$ the sinusoidal response in (2.5) and (2.6) can then be written on complex form as follows

$$y(\omega)e^{j\omega t} = G(j\omega)u(\omega)e^{j\omega t} \quad (2.8)$$

or because the term $e^{j\omega t}$ appears on both sides

$$\boxed{y(\omega) = G(j\omega)u(\omega)} \qquad (2.9)$$

which we refer to as the phasor notation. At each frequency, $u(\omega)$, $y(\omega)$ and $G(j\omega)$ are complex numbers, and the usual rules for multiplying complex numbers apply. We will use this phasor notation throughout the book. Thus *whenever we use notation such as $u(\omega)$ (with ω and not $j\omega$ as an argument), the reader should interpret this as a (complex) sinusoidal signal, $u(\omega)e^{j\omega t}$.* (2.9) also applies to MIMO systems where $u(\omega)$ and $y(\omega)$ are complex vectors representing the sinusoidal signal in each channel and $G(j\omega)$ is a complex matrix.

Minimum phase systems. For stable systems which are minimum phase (no time delays or right-half plane (RHP) zeros) there is a unique relationship between the gain and phase of the frequency response. This may be quantified by the Bode gain-phase relationship which gives the phase of G (normalized[1] such that $G(0) > 0$) at a given frequency ω_0 as a function of $|G(j\omega)|$ over the entire frequency range:

$$\angle G(j\omega_0) = \frac{1}{\pi} \int_{-\infty}^{\infty} \underbrace{\frac{d\ln|G(j\omega)|}{d\ln\omega}}_{N(\omega)} \ln\left|\frac{\omega + \omega_0}{\omega - \omega_0}\right| \cdot \frac{d\omega}{\omega} \qquad (2.10)$$

The name *minimum phase* refers to the fact that such a system has the minimum possible phase lag for the given magnitude response $|G(j\omega)|$. The term $N(\omega)$ is the slope of the magnitude in log-variables at frequency ω. In particular, the local slope at frequency ω_0 is

$$N(\omega_0) = \left(\frac{d\ln|G(j\omega)|}{d\ln\omega}\right)_{\omega=\omega_0}$$

The term $\ln\left|\frac{\omega+\omega_0}{\omega-\omega_0}\right|$ in (2.10) is infinite at $\omega = \omega_0$, so it follows that $\angle G(j\omega_0)$ is primarily determined by the local slope $N(\omega_0)$. Also $\int_{-\infty}^{\infty} \ln\left|\frac{\omega+\omega_0}{\omega-\omega_0}\right| \cdot \frac{d\omega}{\omega} = \frac{\pi^2}{2}$ which justifies the commonly used approximation for stable minimum phase systems

$$\angle G(j\omega_0) \approx \frac{\pi}{2}N(\omega_0) \text{ [rad]} = 90° \cdot N(\omega_0) \qquad (2.11)$$

The approximation is exact for the system $G(s) = 1/s^n$ (where $N(\omega) = -n$), and it is good for stable minimum phase systems except at frequencies close to those of resonance (complex) poles or zeros.

RHP-zeros and time delays contribute additional phase lag to a system when compared to that of a minimum phase system with the same gain (hence the term *non-minimum phase* system). For example, the system $G(s) = \frac{-s+a}{s+a}$ with a RHP-zero at

---

[1] The normalization of $G(s)$ is necessary to handle systems such as $\frac{1}{s+2}$ and $\frac{-1}{s+2}$, which have equal gain, are stable and minimum phase, but their phases differ by $180°$. Systems with integrators may be treated by replacing $\frac{1}{s}$ by $\frac{1}{s+\epsilon}$ where ϵ is a small positive number.

$s = a$ has a constant gain of 1, but its phase is $-2 \arctan(\omega/a)$ [rad] (and not 0 [rad] as it would be for the minimum phase system $G(s) = 1$ of the same gain). Similarly, the time delay system $e^{-\theta s}$ has a constant gain of 1, but its phase is $-\omega\theta$ [rad].

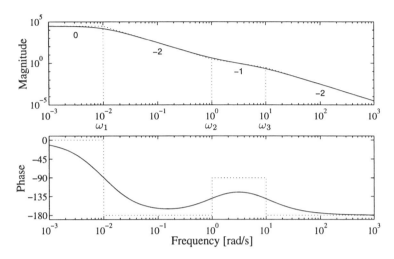

Figure 2.3: Bode plots of transfer function $L_1 = 30 \frac{s+1}{(s+0.01)^2(s+10)}$. The asymptotes are given by dotted lines. The vertical dotted lines on the upper plot indicate the break frequencies ω_1, ω_2 and ω_3.

Straight-line approximations (asymptotes). For the design methods used in this book it is useful to be able to sketch Bode plots quickly, and in particular the magnitude (gain) diagram. The reader is therefore advised to become familiar with asymptotic Bode plots (straight-line approximations). For example, for a transfer function

$$G(s) = k\frac{(s + z_1)(s + z_2)\cdots}{(s + p_1)(s + p_2)\cdots} \tag{2.12}$$

the asymptotic Bode plots of $G(j\omega)$ are obtained by using for each term $s + a$ the approximation $j\omega + a \approx a$ for $\omega < a$ and by $j\omega + a \approx j\omega$ for $\omega > a$. These approximations yield straight lines on a log-log plot which meet at the so-called break point frequency $\omega = a$. In (2.12) therefore, the frequencies $z_1, z_2, \ldots, p_1, p_2, \ldots$ are the break points where the asymptotes meet. For complex poles or zeros, the term $s^2 + 2\zeta s\omega_0 + \omega_0^2$ (where $|\zeta| < 1$) is approximated by ω_0^2 for $\omega < \omega_0$ and by $s^2 = (j\omega)^2 = -\omega^2$ for $\omega > \omega_0$. The magnitude of a transfer function is usually close to its asymptotic value, and the only case when there is significant deviation is around the resonance frequency ω_0 for complex poles or zeros with a damping $|\zeta|$ of about 0.3 or less. In Figure 2.3, the Bode plots are shown for

$$L_1(s) = 30\frac{(s + 1)}{(s + 0.01)^2(s + 10)} \tag{2.13}$$

The asymptotes (straight-line approximations) are shown by dotted lines. We note that the magnitude follows the asymptotes closely, whereas the phase does not. In this example the asymptotic slope of L_1 is 0 up to the first break frequency at $\omega_1 = 0.01$ rad/s where we have two poles and then the slope changes to $N = -2$. Then at $\omega_2 = 1$ rad/s there is a zero and the slope changes to $N = -1$. Finally, there is a break frequency corresponding to a pole at $\omega_3 = 10$ rad/s and so the slope is $N = -2$ at this and higher frequencies.

2.2 Feedback control

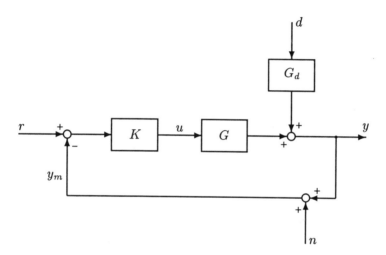

Figure 2.4: Block diagram of one degree-of-freedom feedback control system

2.2.1 One degree-of-freedom controller

In most of this chapter, we examine the simple one degree-of-freedom negative feedback structure shown in Figure 2.4. The input to the controller $K(s)$ is $r - y_m$ where $y_m = y + n$ is the measured output and n is the measurement noise. Thus, the input to the plant is

$$u = K(s)(r - y - n) \qquad (2.14)$$

The objective of control is to manipulate u (design K) such that the control error e remains small in spite of disturbances d. The control error e is defined as

$$e = y - r \qquad (2.15)$$

where r denotes the reference value (setpoint) for the output.

Remark. In the literature, the control error is frequently defined as $r - y_m$ which is often the controller input. However, this is not a good definition of an error variable. First, the error is normally defined as the actual value (here y) minus the desired value (here r). Second, the error should involve the actual value (y) and not the measured value (y_m).

Note that we do not define e as the controller input $r - y_m$ which is frequently done.

2.2.2   Closed-loop transfer functions

The plant model is written as

$$y = G(s)u + G_d(s)d \qquad (2.16)$$

and for a one degree-of-freedom controller the substitution of (2.14) into (2.16) yields

$$y = GK(r - y - n) + G_d d$$

or

$$(I + GK)y = GKr + G_d d - GKn \qquad (2.17)$$

and hence the closed-loop response is

$$y = \underbrace{(I + GK)^{-1}GK}_{T}\, r + \underbrace{(I + GK)^{-1}}_{S}\, G_d d - \underbrace{(I + GK)^{-1}GK}_{T}\, n \qquad (2.18)$$

The control error is

$$e = y - r = -Sr + SG_d d - Tn \qquad (2.19)$$

where we have used the fact $T - I = -S$. The corresponding plant input signal is

$$u = KSr - KSG_d d - KSn \qquad (2.20)$$

The following notation and terminology are used

$$L = GK \quad \text{loop transfer function}$$
$$S = (I + GK)^{-1} = (I + L)^{-1} \quad \text{sensitivity function}$$
$$T = (I + GK)^{-1}GK = (I + L)^{-1}L \quad \text{complementary sensitivity function}$$

We see that S is the closed-loop transfer function from the output disturbances to the outputs, while T is the closed-loop transfer function from the reference signals to the outputs. The term complementary sensitivity for T follows from the identity:

$$S + T = I \qquad (2.21)$$

To derive (2.21), write $S + T = (I + L)^{-1} + (I + L)^{-1}L$ and factor out the term $(I + L)^{-1}$. The term sensitivity function is natural because S gives the sensitivity reduction afforded by feedback. To see this, consider the "open-loop" case i.e. with no feedback. Then

$$y = GKr + G_d d + 0 \cdot n \tag{2.22}$$

and a comparison with (2.18) shows that, with the exception of noise, the response with feedback is obtained by premultiplying the right hand side by S.

Remark 1 Actually, the above is not the original reason for the name "sensitivity". Bode first called S sensitivity because it gives the relative sensitivity of the closed-loop transfer function T to the relative plant model error. In particular, at a given frequency ω we have for a SISO plant, by straightforward differentiation of T, that

$$\frac{dT/T}{dG/G} = S \tag{2.23}$$

Remark 2 Equations (2.14)-(2.22) are written in matrix form because they also apply to MIMO systems. Of course, for SISO systems we may write $S + T = 1$, $S = \frac{1}{1+L}$, $T = \frac{L}{1+L}$ and so on.

Remark 3 In general, closed-loop transfer functions for SISO systems with *negative* feedback may be obtained from the rule

$$\text{OUTPUT} = \frac{\text{``direct''}}{1 + \text{``loop''}} \cdot \text{INPUT} \tag{2.24}$$

where "direct" represents the transfer function for the direct effect of the input on the output (with the feedback path open) and "loop" is the transfer function around the loop (denoted $L(s)$). In the above case $L = GK$. If there is also a measurement device, $G_m(s)$, in the loop, then $L(s) = GKG_m$. The rule in (2.24) is easily derived by generalizing (2.17). In Section 3.2, we present a more general form of this rule which also applies to multivariable systems.

2.2.3 Why feedback?

At this point it is pertinent to ask why we should use feedback control at all — rather than simply using feedforward control. A "perfect" feedforward controller is obtained by removing the feedback signal and using the controller

$$K_r(s) = G^{-1}(s) \tag{2.25}$$

(we assume for now that it is possible to obtain and physically realize such an inverse, although this may of course not be true). We assume that the plant and controller are both stable and that all the disturbances are known, that is, we know $G_d d$, the effect of the disturbances on the outputs. Then with $r - G_d d$ as the controller input, this feedforward controller would yield perfect control:

$$y = Gu + G_d d = GK(r - G_d d) + G_d d = r$$

Unfortunately, G is never an exact model, and the disturbances are never known exactly. *The fundamental reasons for using feedback control are therefore the presence of*

1. Signal uncertainty – Unknown disturbance
2. Model uncertainty
3. An unstable plant

The third reason follows because unstable plants can only be stabilized by feedback (see internal stability in Chapter 4). The ability of feedback to reduce the effect of model uncertainty is of crucial importance in controller design.

2.3   Closed-loop stability

One of the main issues in designing feedback controllers is stability. If the feedback gain is too large, then the controller may "overreact" and the closed-loop system becomes unstable. This is illustrated next by a simple example.

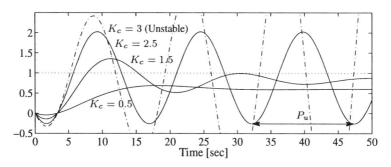

Figure 2.5: Effect of proportional gain K_c on the closed-loop response $y(t)$ of the inverse response process

Example 2.1 Inverse response process. *Consider the plant (time in seconds)*

$$G(s) = \frac{3(-2s + 1)}{(5s + 1)(10s + 1)} \tag{2.26}$$

This is one of two main example processes used in this chapter to illustrate the techniques of classical control. The model has a right-half plane (RHP) zero at $s = 0.5$ rad/s. This imposes a fundamental limitation on control, and high controller gains will induce closed-loop instability.

This is illustrated for a proportional (P) controller $K(s) = K_c$ in Figure 2.5, where the response $y = Tr = GK_c(1+GK_c)^{-1}r$ to a step change in the reference ($r(t) = 1$ for $t > 0$) is shown for four different values of K_c. The system is seen to be stable for $K_c < 2.5$, and unstable for $K_c > 2.5$. The controller gain at the limit of instability, $K_u = 2.5$, is sometimes called the ultimate gain and for this value the system is seen to cycle continuously with a period $P_u = 15.2$ s, corresponding to the frequency $\omega_u \triangleq 2\pi/P_u = 0.42$ rad/s.

Two methods are commonly used to determine closed-loop stability:

1. The poles of the closed-loop system are evaluated. That is, the roots of $1 + L(s) = 0$ are found, where L is the transfer function around the loop. The system is stable *if and only if* all the closed-loop poles are in the open left-half plane (LHP) (that is, poles on the imaginary axis are considered "unstable"). The poles are also equal to the eigenvalues of the state-space A-matrix, and this is usually how the poles are computed numerically.

2. The frequency response (including negative frequencies) of $L(j\omega)$ is plotted in the complex plane and the number of encirclements it makes of the critical point -1 is counted. By Nyquist's stability criterion (for which a detailed statement is given in Theorem 4.7) closed-loop stability is inferred by equating the number of encirclements to the number of open-loop unstable poles (RHP-poles).

   For open-loop stable systems where $\angle L(j\omega)$ falls with frequency such that $\angle L(j\omega)$ crosses $-180°$ only once (from above at frequency ω_{180}), one may equivalently use *Bode's stability condition* which says that the closed-loop system is stable if and only if the loop gain $|L|$ is less than 1 at this frequency, that is

$$\text{Stability} \quad \Leftrightarrow \quad |L(j\omega_{180})| < 1 \tag{2.27}$$

where ω_{180} is the phase crossover frequency defined by $\angle L(j\omega_{180}) = -180°$.

Method 1, which involves computing the poles, is best suited for numerical calculations. However, time delays must first be approximated as rational transfer functions, e.g. Padé approximations. Method 2, which is based on the frequency response, has a nice graphical interpretation, and may also be used for systems with time delays. Furthermore, it provides useful measures of relative stability and forms the basis for several of the robustness tests used later in this book.

Example 2.2 Stability of inverse response process with proportional control. *Let us determine the condition for closed-loop stability of the plant G in (2.26) with proportional control, that is, with $K(s) = K_c$ and $L(s) = K_c G(s)$.*

1. The system is stable if and only if all the closed-loop poles are in the LHP. The poles are solutions to $1 + L(s) = 0$ or equivalently the roots of

$$(5s + 1)(10s + 1) + K_c 3(-2s + 1) = 0$$

$$\Leftrightarrow \quad 50s^2 + (15 - 6K_c)s + (1 + 3K_c) = 0 \tag{2.28}$$

But since we are only interested in the half plane location of the poles, it is not necessary to solve (2.28). Rather, one may consider the coefficients a_i of the characteristic equation $a_n s^n + \cdots a_1 s + a_0 = 0$ in (2.28), and use the Routh-Hurwitz test to check for stability. For second order systems, this test says that we have stability if and only if all the coefficients have the same sign. This yields the following stability conditions

$$(15 - 6K_c) > 0; \quad (1 + 3K_c) > 0$$

or equivalently $-1/3 < K_c < 2.5$. With negative feedback ($K_c \geq 0$) only the upper bound is of practical interest, and we find that the maximum allowed gain ("ultimate gain") *is $K_u = 2.5$ which agrees with the simulation in Figure 2.5. The poles at the onset of instability may be found by substituting $K_c = K_u = 2.5$ into (2.28) to get $50s^2 + 8.5 = 0$, i.e. $s = \pm j\sqrt{8.5/50} = \pm j0.412$. Thus, at the onset of instability we have two poles on the imaginary axis, and the system will be continuously cycling with a frequency $\omega = 0.412$ rad/s corresponding to a period $P_u = 2\pi/\omega = 15.2$ s. This agrees with the simulation results in Figure 2.5.*

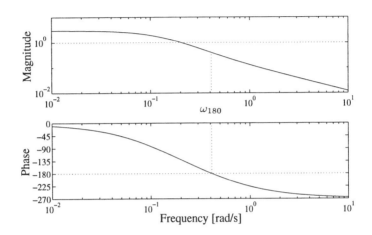

Figure 2.6: Bode plots for $L(s) = K_c \frac{3(-2s+1)}{(10s+1)(5s+1)}$ with $K_c = 1$

2. *Stability may also be evaluated from the frequency response of $L(s)$. A graphical evaluation is most enlightening. The Bode plots of the plant (i.e. $L(s)$ with $K_c = 1$) are shown in Figure 2.6. From these one finds the frequency ω_{180} where $\angle L$ is $-180°$ and then reads off the corresponding gain. This yields $|L(j\omega_{180})| = K_c|G(j\omega_{180})| = 0.4K_c$, and we get from (2.27) that the system is stable if and only if $|L(j\omega_{180})| < 1 \Leftrightarrow K_c < 2.5$ (as found above). Alternatively, the phase crossover frequency may be obtained analytically from:*

$$\angle L(j\omega_{180}) = -\arctan(2\omega_{180}) - \arctan(5\omega_{180}) - \arctan(10\omega_{180}) = -180°$$

which gives $\omega_{180} = 0.412$ rad/s as found in the pole calculation above. The loop gain at this frequency is

$$|L(j\omega_{180})| = K_c \frac{3 \cdot \sqrt{(2\omega_{180})^2 + 1}}{\sqrt{(5\omega_{180})^2 + 1} \cdot \sqrt{(10\omega_{180})^2 + 1}} = 0.4K_c$$

which is the same as found from the graph in Figure 2.6. The stability condition $|L(j\omega_{180})| < 1$ then yields $K_c < 2.5$ as expected.

2.4 Evaluating closed-loop performance

Although closed-loop stability is an important issue, the real objective of control is to improve performance, that is, to make the output $y(t)$ behave in a more desirable manner. Actually, the possibility of inducing instability is one of the disadvantages of feedback control which has to be traded off against performance improvement. The objective of this section is to discuss ways of evaluating closed-loop performance.

2.4.1 Typical closed-loop responses

The following example which considers proportional plus integral (PI) control of the inverse response process in (2.26), illustrates what type of closed-loop performance one might expect.

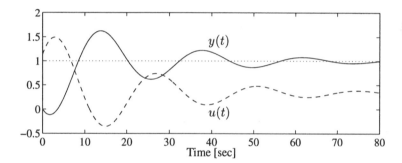

Figure 2.7: Closed-loop response to a step change in reference for the inverse response process with PI-control

Example 2.3 PI-control of the inverse response process. *We have already studied the use of a proportional controller for the process in (2.26). We found that a controller gain of $K_c = 1.5$ gave a reasonably good response, except for a steady-state offset (see Figure 2.5). The reason for this offset is the non-zero steady-state sensitivity function, $S(0) = \frac{1}{1+K_c G(0)} = 0.18$ (where $G(0) = 3$ is the steady-state gain of the plant). From $e = -Sr$ it follows that for $r = 1$ the steady-state control error is -0.18 (as is confirmed by the simulation in Figure 2.5). To remove the steady-state offset we add integral action in the form of a PI-controller*

$$K(s) = K_c \left(1 + \frac{1}{\tau_I s}\right) \tag{2.29}$$

The settings for K_c and τ_I can be determined from the classical tuning rules of Ziegler and Nichols (1942):

$$K_c = K_u/2.2, \quad \tau_I = P_u/1.2 \tag{2.30}$$

where K_u is the maximum (ultimate) P-controller gain and P_u is the corresponding period of oscillations. In our case $K_u = 2.5$ and $P_u = 15.2$ s (as observed from the simulation in

Figure 2.5), and we get $K_c = 1.14$ and $\tau_I = 12.7$ s. Alternatively, K_u and P_u can be obtained from the model $G(s)$,

$$K_u = 1/|G(j\omega_u)|, \quad P_u = 2\pi/\omega_u \tag{2.31}$$

where ω_u is defined by $\angle G(j\omega_u) = -180°$.

The closed-loop response, with PI-control, to a step change in reference is shown in *Figure 2.7. The output $y(t)$ has an initial inverse response due to the RHP-zero, but it then rises quickly and $y(t) = 0.9$ at $t = 8.0$ s (the rise time). The response is quite oscillatory and it does not settle to within $\pm 5\%$ of the final value until after $t = 65$ s (the settling time). The overshoot (height of peak relative to the final value) is about 62% which is much larger than one would normally like for reference tracking. The decay ratio, which is the ratio between subsequent peaks, is about 0.35 which is also a bit large. However, for disturbance rejection the controller settings may be more reasonable, and one can always add a prefilter to improve the response for reference tracking, resulting in a two degrees-of-freedom controller.*

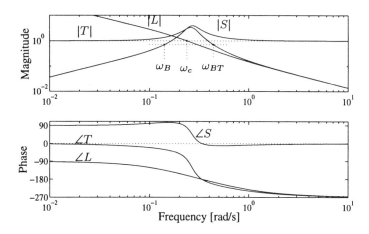

Figure 2.8: Bode magnitude and phase plots of $L = GK$, S and T when
$G(s) = \frac{3(-2s+1)}{(5s+1)(10s+1)}$, and $K(s) = 1.136(1 + \frac{1}{12.7s})$ (a Ziegler-Nichols PI controller)

The corresponding Bode plots for L, S and T are shown in Figure 2.8. Later, in Section 2.4.3, we define stability margins, and from the plot of $L(j\omega)$, repeated in Figure 2.11, we find that the phase margin (PM) is 0.34 rad $= 19.4°$ and the gain margin (GM) is 1.63. These margins are too small according to common rules of thumb. The peak value of $|S|$ is $M_S = 3.92$, and the peak value of $|T|$ is $M_T = 3.35$ which again are high according to normal design rules.

Exercise 2.1 *Use (2.31) to compute K_u and P_u for the process in (2.26).*

In summary, for this example, the Ziegler-Nichols' PI-tunings are somewhat "aggressive" and give a closed-loop system with smaller stability margins and a more oscillatory response than would normally be regarded as acceptable.

2.4.2   Time domain performance

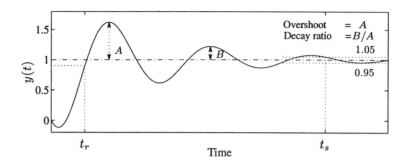

Figure 2.9: Characteristics of closed-loop response to step in reference

Step response analysis. The above example illustrates the approach often taken by engineers when evaluating the performance of a control system. That is, one simulates the response to a step in the reference input, and considers the following characteristics (see Figure 2.9):

- *Rise time:* (t_r) the time it takes for the output to first reach 90% of its final value, which is usually required to be small.
- *Settling time:* (t_s) the time after which the output remains within ±5% of its final value, which is usually required to be small.
- *Overshoot:* the peak value divided by the final value, which should typically be 1.2 (20%) or less.
- *Decay ratio:* the ratio of the second and first peaks, which should typically be 0.3 or less.
- *Steady-state offset:* the difference between the final value and the desired final value, which is usually required to be small.

The rise time and settling time are measures of the *speed of the response*, whereas the overshoot, decay ratio and steady-state offset are related to the *quality of the response*. Another measure of the quality of the response is:

- *Excess variation:*   the total variation (TV) divided by the overall change at steady state, which should be as close to 1 as possible.

The total variation is the total movement of the output as illustrated in Figure 2.10. For the cases considered here the overall change is 1, so the excess variation is equal to the total variation.

The above measures address the output response, $y(t)$. In addition, one should consider the magnitude of the manipulated input (control signal, u), which usually should be as small and smooth as possible. If there are important disturbances, then

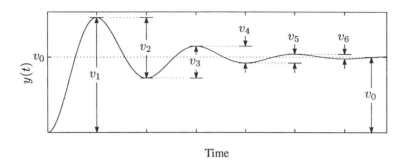

Figure 2.10: Total variation is TV $= \sum_i v_i$, and Excess variation is TV$/v_0$

the response to these should also be considered. Finally, one may investigate in simulation how the controller works if the plant model parameters are different from their nominal values.

Remark 1 Another way of quantifying time domain performance is in terms of some norm of the error signal $e(t) = y(t) - r(t)$. For example, one might use the integral squared error (ISE), or its square root which is the 2-norm of the error signal, $\|e(t)\|_2 = \sqrt{\int_0^\infty |e(\tau)|^2 d\tau}$. Note that in this case the various objectives related to both the speed and quality of response are combined into one number. Actually, in most cases minimizing the 2-norm seems to give a reasonable trade-off between the various objectives listed above. Another advantage of the 2-norm is that the resulting optimization problems (such as minimizing ISE) are numerically easy to solve. One can also take input magnitudes into account by considering, for example, $J = \sqrt{\int_0^\infty (Q|e(t)|^2 + R|u(t)|^2)dt}$ where Q and R are positive constants. This is similar to linear quadratic (LQ) optimal control, but in LQ-control one normally considers an impulse rather than a step change in $r(t)$.

Remark 2 The step response is equal to the integral of the corresponding impulse response, e.g. set $u(\tau) = 1$ in (4.11). Some thought then reveals that one can compute the total variation as the integrated absolute area (1-norm) of the corresponding impulse response (Boyd and Barratt, 1991, p. 98). That is, let $y = Tr$, then the total variation in y for a step change in r is

$$\text{TV} = \int_0^\infty |g_T(\tau)|d\tau \triangleq \|g_T(t)\|_1 \tag{2.32}$$

where $g_T(t)$ is the impulse response of T, i.e. $y(t)$ resulting from an impulse change in $r(t)$.

2.4.3 Frequency domain performance

The frequency-response of the loop transfer function, $L(j\omega)$, or of various closed-loop transfer functions, may also be used to characterize closed-loop performance. Typical Bode plots of L, T and S are shown in Figure 2.8. One advantage of the frequency domain compared to a step response analysis, is that it considers a broader class of signals (sinusoids of any frequency). This makes it easier to characterize feedback properties, and in particular system behaviour in the crossover (bandwidth) region. We will now describe some of the important frequency-domain measures used to assess performance e.g. gain and phase margins, the maximum peaks of S and T, and the various definitions of crossover and bandwidth frequencies used to characterize speed of response.

Gain and phase margins

Let $L(s)$ denote the loop transfer function of a system which is closed-loop stable under negative feedback. A typical Bode plot and a typical Nyquist plot of $L(j\omega)$ illustrating the gain margin (GM) and phase margin (PM) are given in Figures 2.11 and 2.12, respectively.

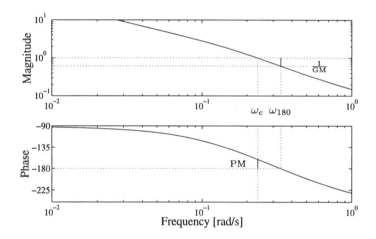

Figure 2.11: Typical Bode plot of $L(j\omega)$ with PM and GM indicated

The *gain margin* is defined as

$$GM = 1/|L(j\omega_{180})| \tag{2.33}$$

where the *phase crossover frequency* ω_{180} is where the Nyquist curve of $L(j\omega)$ crosses the negative real axis between -1 and 0, that is

$$\angle L(j\omega_{180}) = -180° \tag{2.34}$$

If there is more than one crossing the largest value of $|L(j\omega_{180})|$ is taken. On a Bode plot with a logarithmic axis for $|L|$, we have that GM (in logarithms, e.g. in dB) is the vertical distance from the unit magnitude line down to $|L(j\omega_{180})|$, see Figure 2.11. The GM is the factor by which the loop gain $|L(j\omega)|$ may be increased

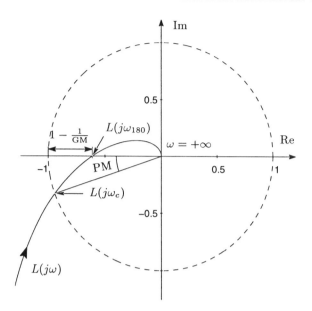

Figure 2.12: Typical Nyquist plot of $L(j\omega)$ for stable plant with PM and GM indicated. Closed-loop instability occurs if $L(j\omega)$ encircles the critical point -1

before the closed-loop system becomes unstable. The GM is thus a direct safeguard against steady-state gain uncertainty (error). Typically we require GM > 2. If the Nyquist plot of L crosses the negative real axis between -1 and $-\infty$ then a *gain reduction margin* can be similarly defined from the smallest value of $|L(j\omega_{180})|$ of such crossings.

The *phase margin* is defined as

$$\text{PM} = \angle L(j\omega_c) + 180° \qquad (2.35)$$

where the *gain crossover frequency* ω_c is where $|L(j\omega)|$ first crosses 1 from above, that is

$$|L(j\omega_c)| = 1 \qquad (2.36)$$

The phase margin tells how much negative phase (phase lag) we can add to $L(s)$ at frequency ω_c before the phase at this frequency becomes $-180°$ which corresponds to closed-loop instability (see Figure 2.12). Typically, we require PM larger than $30°$ or more. The PM is a direct safeguard against time delay uncertainty; the system becomes unstable if we add a time delay of

$$\theta_{\max} = \text{PM}/\omega_c \qquad (2.37)$$

Note that the units must be consistent, and so if w_c is in [rad/s] then PM must be in radians. It is also important to note that by decreasing the value of w_c (lowering the closed-loop bandwidth, resulting in a slower response) the system can tolerate larger time delay errors.

Example 2.4 *For the PI-controlled inverse response process example we have* PM $=$ 19.4° $=$ 19.4/57.3 *rad* $=$ 0.34 *rad and* w_c $=$ 0.236 *rad/s. The allowed time delay error is then* $\theta_{\max}$ $=$ 0.34 *rad*/0.236 *rad/s* $=$ 1.44 *s.*

From the above arguments we see that gain and phase margins provide *stability margins* for gain and delay uncertainty. However, as we show below the gain and phase margins are closely related to the peak values of $|S(j\omega)|$ and $|T(j\omega)|$ and are therefore also useful in terms of *performance*. In short, the gain and phase margins are used to provide the appropriate trade-off between performance and stability.

Exercise 2.2 *Prove that the maximum additional delay for which closed-loop stability is maintained is given by (2.37).*

Exercise 2.3 *Derive the approximation for K_u $=$ $1/|G(j\omega_u)|$ given in (5.73) for a first-order delay system.*

Maximum peak criteria

The maximum peaks of the sensitivity and complementary sensitivity functions are defined as

$$M_S = \max_\omega |S(j\omega)|; \quad M_T = \max_\omega |T(j\omega)| \tag{2.38}$$

(Note that M_S $=$ $\|S\|_\infty$ and M_T $=$ $\|T\|_\infty$ in terms of the $\mathcal{H}_\infty$ norm introduced later.) Typically, it is required that M_S is less than about 2 (6 dB) and M_T is less than about 1.25 (2 dB). A large value of M_S or M_T (larger than about 4) indicates poor performance as well as poor robustness. Since $S + T = 1$ it follows that at any frequency

$$\big| |S| - |T| \big| \leq |S + T| = 1$$

so M_S and M_T differ at most by 1. A large value of M_S therefore occurs if and only if M_T is large. For stable plants we usually have $M_S > M_T$, but this is not a general rule. An upper bound on M_T has been a common design specification in classical control and the reader may be familiar with the use of M-circles on a Nyquist plot or a Nichols chart used to determine M_T from $L(j\omega)$.

We now give some justification for why we may want to bound the value of M_S. Without control (u $=$ 0), we have $e = y - r = G_d d - r$, and with feedback control $e = S(G_d d - r)$. Thus, feedback control improves performance in terms of reducing $|e|$ at all frequencies where $|S| < 1$. Usually, $|S|$ is small at low frequencies, for example, $|S(0)| = 0$ for systems with integral action. But because all real systems are strictly proper we must at high frequencies have that $L \to 0$ or equivalently $S \to 1$. At intermediate frequencies one cannot avoid in practice a peak value, M_S, larger than 1 (e.g. see the remark below). Thus, there is an intermediate frequency range where feedback control degrades performance, and the value of M_S is a measure of the worst-case performance degradation. One may also view M_S as a robustness measure, as is now explained. To maintain closed-loop stability the number of encirclements of the critical point -1 by $L(j\omega)$ must not change; so we want L to stay away from this point. The smallest distance between $L(j\omega)$ and the -1 point is M_S^{-1}, and therefore for robustness, the smaller M_S, the better. In summary, both for stability and performance we want M_S close to 1.

There is a close relationship between these maximum peaks and the gain and phase margins. Specifically, for a given M_S we are guaranteed

$$\mathrm{GM} \geq \frac{M_S}{M_S - 1}; \qquad \mathrm{PM} \geq 2 \arcsin \left(\frac{1}{2M_S} \right) \geq \frac{1}{M_S} \text{ [rad]} \qquad (2.39)$$

For example, with $M_S = 2$ we are guaranteed $\mathrm{GM} \geq 2$ and $\mathrm{PM} \geq 29.0°$. Similarly, for a given value of M_T we are guaranteed

$$\mathrm{GM} \geq 1 + \frac{1}{M_T}; \qquad \mathrm{PM} \geq 2 \arcsin \left(\frac{1}{2M_T} \right) > \frac{1}{M_T} \text{ [rad]} \qquad (2.40)$$

and therefore with $M_T = 2$ we have $\mathrm{GM} \geq 1.5$ and $\mathrm{PM} \geq 29.0°$.

Proof of (2.39) and (2.40): To derive the GM-inequalities notice that $L(j\omega_{180}) = -1/\mathrm{GM}$ (since $\mathrm{GM} = 1/|L(j\omega_{180})|$ and L is real and negative at ω_{180}), from which we get

$$T(j\omega_{180}) = \frac{-1}{\mathrm{GM} - 1}; \qquad S(j\omega_{180}) = \frac{1}{1 - \frac{1}{\mathrm{GM}}} \qquad (2.41)$$

and the GM-results follow. To derive the PM-inequalities in (2.39) and (2.40) consider Figure 2.13 where we have $|S(j\omega_c)| = 1/|1 + L(j\omega_c)| = 1/|-1 - L(j\omega_c)|$ and we obtain

$$|S(j\omega_c)| = |T(j\omega_c)| = \frac{1}{2 \sin(\mathrm{PM}/2)} \qquad (2.42)$$

and the inequalities follow. Alternative formulas, which are sometimes used, follow from the identity $2 \sin(\mathrm{PM}/2) = \sqrt{2(1 - \cos(\mathrm{PM}))}$.   □

Remark. We note with interest that (2.41) requires $|S|$ to be larger than 1 at frequency ω_{180}. This means that provided ω_{180} exists, that is, $L(j\omega)$ has more than $-180°$ phase lag at some frequency (which is the case for any real system), then the peak of $|S(j\omega)|$ must exceed 1.

In conclusion, we see that specifications on the peaks of $|S(j\omega)|$ or $|T(j\omega)|$ (M_S or M_T), can make specifications on the gain and phase margins unnecessary. For instance, requiring $M_S < 2$ implies the common rules of thumb $\mathrm{GM} > 2$ and $\mathrm{PM} > 30°$.

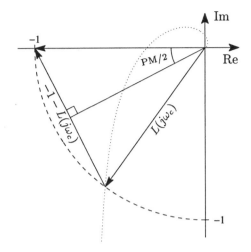

Figure 2.13: At frequency ω_c we see from the figure that $|1 + L(j\omega_c)| = 2\sin(\text{PM}/2)$

2.4.4   Relationship between time and frequency domain peaks

For a change in reference r, the output is $y(s) = T(s)r(s)$. Is there any relationship between the frequency domain peak of $T(j\omega)$, M_T, and any characteristic of the time domain step response, for example the overshoot or the total variation? To answer this consider a prototype second-order system with complementary sensitivity function

$$T(s) = \frac{1}{\tau^2 s^2 + 2\tau\zeta s + 1} \tag{2.43}$$

For underdamped systems with $\zeta < 1$ the poles are complex and yield oscillatory step responses. With $r(t) = 1$ (a unit step change) the values of the overshoot and total variation for $y(t)$ are given, together with M_T and M_S, as a function of ζ in Table 2.1. From Table 2.1, we see that the total variation TV correlates quite well with M_T. This is further confirmed by (A.95) and (2.32) which together yield the following general bounds

$$M_T \leq \text{TV} \leq (2n + 1)M_T \tag{2.44}$$

Here n is the order of $T(s)$, which is 2 for our prototype system in (2.43). Given that the response of many systems can be crudely approximated by fairly low-order systems, the bound in (2.44) suggests that M_T may provide a reasonable approximation to the total variation. This provides some justification for the use of M_T in classical control to evaluate the quality of the response.

Table 2.1: Peak values and total variation of prototype second-order system

	Time domain		Frequency domain	
ζ	Overshoot	Total variation	M_T	M_S
2.0	1	1	1	1.05
1.5	1	1	1	1.08
1.0	1	1	1	1.15
0.8	1.02	1.03	1	1.22
0.6	1.09	1.21	1.04	1.35
0.4	1.25	1.68	1.36	1.66
0.2	1.53	3.22	2.55	2.73
0.1	1.73	6.39	5.03	5.12
0.01	1.97	63.7	50.0	50.0

```
% MATLAB code (Mu toolbox) to generate Table:
tau=1;zeta=0.1;t=0:0.01:100;
T = nd2sys(1,[tau*tau 2*tau*zeta 1]); S = msub(1,T);
[A,B,C,D]=unpck(T); y1 = step(A,B,C,D,1,t);
overshoot=max(y1),tv=sum(abs(diff(y1)))
Mt=hinfnorm(T,1.e-4),Ms=hinfnorm(S,1.e-4)
```

2.4.5   Bandwidth and crossover frequency

The concept of bandwidth is very important in understanding the benefits and trade-offs involved when applying feedback control. Above we considered peaks of closed-loop transfer functions, M_S and M_T, which are related to the quality of the response. However, for performance we must also consider the speed of the response, and this leads to considering the bandwidth frequency of the system. In general, a large bandwidth corresponds to a faster rise time, since high frequency signals are more easily passed on to the outputs. A high bandwidth also indicates a system which is sensitive to noise and to parameter variations. Conversely, if the bandwidth is small, the time response will generally be slow, and the system will usually be more robust.

Loosely speaking, *bandwidth* may be defined as the frequency range $[\omega_1, \omega_2]$ over which control is effective. In most cases we require tight control at steady-state so $\omega_1 = 0$, and we then simply call $\omega_2 = \omega_B$ the bandwidth.

The word "effective" may be interpreted in different ways, and this may give rise to different definitions of bandwidth. The interpretation we use is that control is *effective* if we obtain some *benefit* in terms of performance. For tracking performance the error is $e = y - r = -Sr$ and we get that feedback is effective (in terms of improving performance) as long as the relative error $e/r = -S$ is reasonably small, which we may define to be less than 0.707 in magnitude. We then get the following definition:

Definition 2.1 *The (closed-loop) bandwidth, ω_B, is the frequency where $|S(j\omega)|$ first*

crosses $1/\sqrt{2} = 0.707(\approx -3 \, dB)$ *from below.*

Another interpretation is to say that control is *effective* if it significantly *changes* the output response. For tracking performance, the output is $y = Tr$ and since without control $y = 0$, we may say that control is effective as long as T is reasonably large, which we may define to be larger than 0.707. This leads to an alternative definition which has been traditionally used to define the bandwidth of a control system: *The bandwidth in terms of* T, ω_{BT}, *is the highest frequency at which* $|T(j\omega)|$ *crosses* $1/\sqrt{2} = 0.707(\approx -3 \, dB)$ *from above.*

Remark 1 The definition of bandwidth in terms of ω_{BT} has the advantage of being closer to how the term is used in other fields, for example, in defining the frequency range of an amplifier in an audio system.

Remark 2 In most cases, the two definitions in terms of S and T yield similar values for the bandwidth. In cases where ω_B and ω_{BT} differ, the situation is generally as follows. Up to the frequency ω_B, $|S|$ is less than 0.7, and control is effective in terms of improving performance. In the frequency range $[\omega_B, \omega_{BT}]$ control still affects the response, but does not improve performance — in most cases we find that in this frequency range $|S|$ is larger than 1 and control degrades performance. Finally, at frequencies higher than ω_{BT} we have $S \approx 1$ and control has no significant effect on the response. The situation just described is illustrated in Example 2.5 below (see Figure 2.15).

The *gain crossover frequency*, ω_c, defined as the frequency where $|L(j\omega_c)|$ first crosses 1 from above, is also sometimes used to define closed-loop bandwidth. It has the advantage of being simple to compute and usually gives a value between ω_B and ω_{BT}. Specifically, for systems with PM < 90° we have

$$\omega_B < \omega_c < \omega_{BT} \tag{2.45}$$

Proof of (2.45): Note that $|L(j\omega_c)| = 1$ so $|S(j\omega_c)| = |T(j\omega_c)|$. Thus, when PM = 90° we get $|S(j\omega_c)| = |T(j\omega_c)| = 0.707$ (see (2.42)), and we have $\omega_B = \omega_c = \omega_{BT}$. For PM < 90° we get $|S(j\omega_c)| = |T(j\omega_c)| > 0.707$, and since ω_B is the frequency where $|S(j\omega)|$ crosses 0.707 from below we must have $\omega_B < \omega_c$. Similarly, since ω_{BT} is the frequency where $|T(j\omega)|$ crosses 0.707 from above, we must have $\omega_{BT} > \omega_c$. □

Another important frequency is the *phase crossover frequency*, ω_{180}, defined as the first frequency where the Nyquist curve of $L(j\omega)$ crosses the negative real axis between -1 and 0.

Remark. From (2.41) we get that $\omega_{180} > \omega_{BT}$ for GM > 2.414, and $\omega_{180} < \omega_{BT}$ for GM < 2.414, and since in many cases the gain margin is about 2.4 we conclude that ω_{180} is usually close to ω_{BT}. It is also interesting to note from (2.41) that at ω_{180} the phase of T (and of L) is $-180°$, so from $y = Tr$ we conclude that at frequency ω_{180} the tracking response is completely out of phase. Since as just noted ω_{BT} is often close to ω_{180}, this further illustrates that ω_{BT} may be a poor indicator of closed-loop performance.

Example 2.5 Comparison of ω_B and ω_{BT} as indicators of performance. *An example where ω_{BT} is a poor indicator of performance is the following:*

$$L = \frac{-s + z}{s(\tau s + \tau z + 2)}; \quad T = \frac{-s + z}{s + z}\frac{1}{\tau s + 1}; \quad z = 0.1, \ \tau = 1 \qquad (2.46)$$

For this system, both L and T have a RHP-zero at $z = 0.1$, and we have GM $= 2.1$, *PM $= 60.1°$, $M_S = 1.93$ and $M_T = 1$. We find that $\omega_B = 0.036$ and $\omega_c = 0.054$ are both less than $z = 0.1$ (as one should expect because speed of response is limited by the presence of RHP-zeros), whereas $\omega_{BT} = 1/\tau = 1.0$ is ten times larger than z. The closed-loop response to a unit step change in the reference is shown in Figure 2.14. The rise time is 31.0 s, which is close to $1/\omega_B = 28.0$ s, but very different from $1/\omega_{BT} = 1.0$ s, illustrating that ω_B is a better indicator of closed-loop performance than ω_{BT}.*

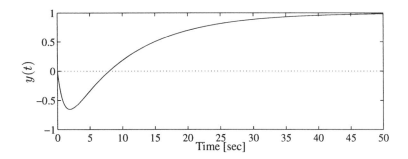

Figure 2.14: Step response for system $T = \frac{-s+0.1}{s+0.1}\frac{1}{s+1}$

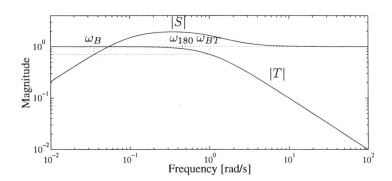

Figure 2.15: Plots of $|S|$ and $|T|$ for system $T = \frac{-s+0.1}{s+0.1}\frac{1}{s+1}$

The magnitude Bode plots of S and T are shown in Figure 2.15. We see that $|T| \approx 1$ up to about ω_{BT}. However, in the frequency range from ω_B to ω_{BT} the phase of T (not shown) drops from about $-40°$ to about $-220°$, so in practice tracking is poor in this frequency range. For *example, at frequency $\omega_{180} = 0.46$ we have $T \approx -0.9$, and the response to a sinusoidally varying reference $r(t) = \sin \omega_{180} t$ is completely out of phase, i.e. $y(t) \approx -0.9r(t)$.*

In conclusion, ω_B (which is defined in terms of $|S|$) and also ω_c (in terms of $|L|$) are good indicators of closed-loop performance, while ω_{BT} (in terms of $|T|$) may be misleading in some cases. The reason is that we want $T \approx 1$ in order to have good performance, and it is not sufficient that $|T| \approx 1$; we must also consider its phase. On the other hand, for for good performance we want S close to 0, and this will be the case if $|S| \approx 0$ irrespective of the phase of S.

2.5 Controller design

We have considered ways of evaluating performance, but one also needs methods for controller design. The Ziegler-Nichols' method used earlier is well suited for on-line tuning, but most other methods involve minimizing some cost function. The overall design process is iterative between controller design and performance (or cost) evaluation. If performance is not satisfactory then one must either adjust the controller parameters directly (for example, by reducing K_c from the value obtained by the Ziegler-Nichols' rules) or adjust some weighting factor in an objective function used to synthesize the controller.

There exist a large number of methods for controller design and some of these will be discussed in Chapter 9. In addition to heuristic rules and on-line tuning we can distinguish between three main approaches to controller design:

1. **Shaping of transfer functions.** In this approach the designer specifies the *magnitude* of some transfer function(s) as a function of frequency, and then finds a controller which gives the desired shape(s).

   (a) **Loop shaping.** This is the classical approach in which the magnitude of the open-loop transfer function, $L(j\omega)$, is shaped. Usually no optimization is involved and the designer aims to obtain $|L(j\omega)|$ with desired bandwidth, slopes etc. We will look at this approach in detail later in this chapter. However, classical loop shaping is difficult to apply for complicated systems, and one may then instead use the Glover-McFarlane $\mathcal{H}_\infty$ loop-shaping design presented in Chapter 9. The method consists of a second step where optimization is used to make an initial loop-shaping design more robust.

   (b) **Shaping of closed-loop transfer functions, such as S, T and KS.** Optimization is usually used, resulting in various $\mathcal{H}_\infty$ optimal control problems such as mixed weighted sensitivity; more on this later.

2. **The signal-based approach.** This involves time domain problem formulations resulting in the minimization of a norm of a transfer function. Here one considers a particular disturbance or reference change and then one tries to optimize the closed-loop response. The "modern" state-space methods from the 1960's, such

as Linear Quadratic Gaussian (LQG) control, are based on this signal-oriented approach. In LQG the input signals are assumed to be stochastic (or alternatively impulses in a deterministic setting) and the expected value of the output variance (or the 2-norm) is minimized. These methods may be generalized to include frequency dependent weights on the signals leading to what is called the Wiener-Hopf (or $\mathcal{H}_2$-norm) design method.

By considering sinusoidal signals, frequency-by-frequency, a signal-based $\mathcal{H}_\infty$ optimal control methodology can be derived in which the $\mathcal{H}_\infty$ norm of a combination of closed-loop transfer functions is minimized. This approach has attracted significant interest, and may be combined with model uncertainty representations, to yield quite complex robust performance problems requiring μ-synthesis; an important topic which will be addressed in later chapters.

3. **Numerical optimization.** This often involves multi-objective optimization where one attempts to optimize directly the true objectives, such as rise times, stability margins, etc. Computationally, such optimization problems may be difficult to solve, especially if one does not have convexity in the controller parameters. Also, by effectively including performance evaluation and controller design in a single step procedure, the problem formulation is far more critical than in iterative two-step approaches. The numerical optimization approach may also be performed on-line, which might be useful when dealing with cases with constraints on the inputs and outputs. On-line optimization approaches such as model predictive control are likely to become more popular as faster computers and more efficient and reliable computational algorithms are developed.

2.6   Loop shaping

In the classical loop-shaping approach to controller design, "loop shape" refers to the magnitude of the loop transfer function $L = GK$ as a function of frequency. An understanding of how K can be selected to shape this loop gain provides invaluable insight into the multivariable techniques and concepts which will be presented later in the book, and so we will discuss loop shaping in some detail in the next two sections.

2.6.1   Trade-offs in terms of L

Recall equation (2.19), which yields the closed-loop response in terms of the control error $e = y - r$:

$$e = -\underbrace{(I + L)^{-1}}_{S} r + \underbrace{(I + L)^{-1}}_{S} G_d d - \underbrace{(I + L)^{-1} L}_{T} n \qquad (2.47)$$

For "perfect control" we want $e = y - r = 0$; that is, we would like

$$e \approx 0 \cdot d + 0 \cdot r + 0 \cdot n$$

The first two requirements in this equation, namely disturbance rejection and command tracking, are obtained with $S \approx 0$, or equivalently, $T \approx I$. Since $S = (I + L)^{-1}$, this implies that the loop transfer function L must be large in magnitude. On the other hand, the requirement for zero noise transmission implies that $T \approx 0$, or equivalently, $S \approx I$, which is obtained with $L \approx 0$. This illustrates the fundamental nature of feedback design which always involves a trade-off between conflicting objectives; in this case between large loop gains for disturbance rejection and tracking, and small loop gains to reduce the effect of noise.

It is also important to consider the magnitude of the control action u (which is the input to the plant). We want u small because this causes less wear and saves input energy, and also because u is often a disturbance to other parts of the system (e.g. consider opening a window in your office to adjust your comfort and the undesirable disturbance this will impose on the air conditioning system for the building). In particular, we usually want to avoid fast changes in u. The control action is given by $u = K(r - y_m)$ and we find as expected that a small u corresponds to small controller gains and a small $L = GK$.

The most important design objectives which necessitate trade-offs in feedback control are summarized below:

1. Performance, good disturbance rejection: needs large controller gains, i.e. L large.
2. Performance, good command following: L large.
3. Stabilization of unstable plant: L large.
4. Mitigation of measurement noise on plant outputs: L small.
5. Small magnitude of input signals: K small and L small.
6. Physical controller must be strictly proper: $K \to 0$ and $L \to 0$ at high frequencies.
7. Nominal stability (stable plant): L small (because of RHP-zeros and time delays).
8. Robust stability (stable plant): L small (because of uncertain or neglected dynamics).

Fortunately, the conflicting design objectives mentioned above are generally in different frequency ranges, and we can meet most of the objectives by using a large loop gain ($|L| > 1$) at low frequencies below crossover, and a small gain ($|L| < 1$) at high frequencies above crossover.

2.6.2   Fundamentals of loop-shaping design

By *loop shaping* we mean a design procedure that involves explicitly shaping the magnitude of the loop transfer function, $|L(j\omega)|$. Here $L(s) = G(s)K(s)$ where $K(s)$ is the feedback controller to be designed and $G(s)$ is the product of all other transfer functions around the loop, including the plant, the actuator and the measurement device. Essentially, to get the benefits of feedback control we want the loop gain, $|L(j\omega)|$, to be as large as possible within the bandwidth region. However, due to time delays, RHP-zeros, unmodelled high-frequency dynamics and limitations on the allowed manipulated inputs, the loop gain has to drop below one at and above some frequency which we call the crossover frequency ω_c. Thus, disregarding stability for the moment, it is desirable that $|L(j\omega)|$ falls sharply with frequency. To measure how $|L|$ falls with frequency we consider the logarithmic slope $N = d\ln|L|/d\ln\omega$. For example, a slope $N = -1$ implies that $|L|$ drops by a factor of 10 when ω increases by a factor of 10. If the gain is measured in decibels (dB) then a slope of $N = -1$ corresponds to -20 dB/ decade. The value of $-N$ at high frequencies is often called the *roll-off* rate.

The design of $L(s)$ is most crucial and difficult in the crossover region between ω_c (where $|L| = 1$) and ω_{180} (where $\angle L = -180°$). For stability, we at least need the loop gain to be less than 1 at frequency ω_{180}, i.e. $|L(j\omega_{180})| < 1$. Thus, to get a high bandwidth (fast response) we want ω_c and therefore ω_{180} large, that is, we want the phase lag in L to be small. Unfortunately, this is not consistent with the desire that $|L(j\omega)|$ should fall sharply. For example, the loop transfer function $L = 1/s^n$ (which has a slope $N = -n$ on a log-log plot) has a phase $\angle L = -n \cdot 90°$. Thus, to have a phase margin of $45°$ we need $\angle L > -135°$, and the slope of $|L|$ cannot exceed $N = -1.5$.

In addition, if the slope is made steeper at lower or higher frequencies, then this will add unwanted phase lag at intermediate frequencies. As an example, consider $L_1(s)$ given in (2.13) with the Bode plot shown in Figure 2.3. Here the slope of the asymptote of $|L|$ is -1 at the gain crossover frequency (where $|L_1(j\omega_c)| = 1$), which by itself gives $-90°$ phase lag. However, due to the influence of the steeper slopes of -2 at lower and higher frequencies, there is a "penalty" of about $-35°$ at crossover, so the actual phase of L_1 at ω_c is approximately $-125°$.

The situation becomes even worse for cases with delays or RHP-zeros in $L(s)$ which add undesirable phase lag to L without contributing to a desirable negative slope in L. At the gain crossover frequency ω_c, the additional phase lag from delays and RHP-zeros may in practice be $-30°$ or more.

In summary, a desired loop shape for $|L(j\omega)|$ typically has a slope of about -1 in the crossover region, and a slope of -2 or higher beyond this frequency, that is, the roll-off is 2 or larger. Also, with a proper controller, which is required for any real system, we must have that $L = GK$ rolls off at least as fast as G. At low frequencies, the desired shape of $|L|$ depends on what disturbances and references we are designing for. For example, if we are considering step changes in the references or disturbances which affect the outputs as steps, then a slope for $|L|$ of -1 at low frequencies is acceptable. If the references or disturbances require the outputs to change in a ramp-like fashion then a slope of -2 is required. In practice, integrators are included in the controller to get the desired low-frequency performance, and for offset-free reference tracking the rule is that

- $L(s)$ *must contain at least one integrator for each integrator in* $r(s)$.

Proof: Let $L(s) = \widehat{L}(s)/s^{n_I}$ where $\widehat{L}(0)$ is non-zero and finite and n_I is the number of integrators in $L(s)$ — sometimes n_I is called the *system type*. Consider a reference signal of the form $r(s) = 1/s^{n_r}$. For example, if $r(t)$ is a unit step, then $r(s) = 1/s$ $(n_r = 1)$, and if

$r(t)$ is a ramp then $r(s) = 1/s^2$ ($n_r = 2$). The final value theorem for Laplace transforms is

$$\lim_{t \to \infty} e(t) = \lim_{s \to 0} se(s) \tag{2.48}$$

In our case, the control error is

$$e(s) = -\frac{1}{1 + L(s)} r(s) = -\frac{s^{n_I - n_r}}{s^{n_I} + \widehat{L}(s)} \tag{2.49}$$

and to get zero offset (i.e. $e(t \to \infty) = 0$) we must from (2.48) require $n_I \geq n_r$, and the rule follows. $\square$

In conclusion, one can define the desired loop transfer function in terms of the following specifications:

1. The gain crossover frequency, ω_c, where $|L(j\omega_c)| = 1$.
2. The shape of $L(j\omega)$, e.g. in terms of the slope of $|L(j\omega)|$ in certain frequency ranges. Typically, we desire a slope of about $N = -1$ around crossover, and a larger roll-off at higher frequencies. The desired slope at lower frequencies depends on the nature of the disturbance or reference signal.
3. The system type, defined as the number of pure integrators in $L(s)$.

In Section 2.6.4, we discuss how to specify the loop shape when disturbance rejection is the primary objective of control. Loop-shaping design is typically an iterative procedure where the designer shapes and reshapes $|L(j\omega)|$ after computing the phase and gain margins, the peaks of closed-loop frequency responses (M_T and M_S), selected closed-loop time responses, the magnitude of the input signal, etc. The procedure is illustrated next by an example.

Example 2.6 Loop-shaping design for the inverse response process.
We will now design a loop-shaping controller for the example process in (2.26) which has a RHP-zero at $s = 0.5$. The RHP-zero limits the achievable bandwidth and so the crossover region (defined as the frequencies between ω_c and ω_{180}) will be at about 0.5 rad/s. We require the system to have one integrator (type 1 system), and therefore a reasonable approach is to let the loop transfer function have a slope of -1 at low frequencies, and then to roll off with a higher slope at frequencies beyond 0.5 rad/s. The plant and our choice for the loop-shape is

$$G(s) = \frac{3(-2s + 1)}{(5s + 1)(10s + 1)}; \quad L(s) = 3K_c \frac{(-2s + 1)}{s(2s + 1)(0.33s + 1)} \tag{2.50}$$

The frequency response (Bode plots) of L is shown in Figure 2.16 for $K_c = 0.05$. The controller gain K_c was selected to get a reasonable stability margins (PM and GM). The asymptotic slope of $|L|$ is -1 up to 3 rad/s where it changes to -2. The controller corresponding to the loop-shape in (2.50) is

$$K(s) = K_c \frac{(10s + 1)(5s + 1)}{s(2s + 1)(0.33s + 1)}, \quad K_c = 0.05 \tag{2.51}$$

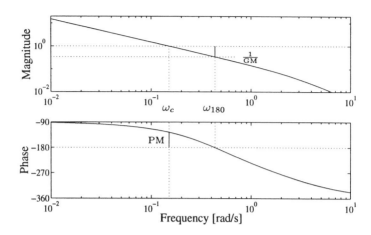

Figure 2.16: Frequency response of $L(s)$ in (2.50) for loop-shaping design with $K_c = 0.05$
(GM = 2.92, PM = 54°, $\omega_c = 0.15$, $\omega_{180} = 0.43$, $M_S = 1.75$, $M_T = 1.11$)

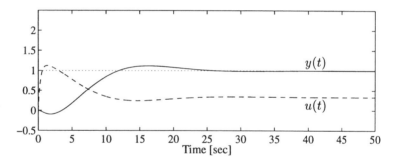

Figure 2.17: Response to step in reference for loop-shaping design

*The controller has zeros at the locations of the plant poles. This is desired in this case because
we do not want the slope of the loop shape to drop at the break frequencies $1/10 = 0.1$ rad/s
and $1/5 = 0.2$ rad/s just before crossover. The phase of L is $-90°$ at low frequency, and
at $\omega = 0.5$ rad/s the additional contribution from the term $\frac{-2s+1}{2s+1}$ in (2.50) is $-90°$, so for
stability we need $\omega_c < 0.5$ rad/s. The choice $K_c = 0.05$ yields $\omega_c = 0.15$ rad/s corresponding
to GM = 2.92 and PM=54°. The corresponding time response is shown in Figure 2.17. It is
seen to be much better than the responses with either the simple PI-controller in Figure 2.7 or
with the P-controller in Figure 2.5. Figure 2.17 also shows that the magnitude of the input
signal remains less than about 1 in magnitude. This means that the controller gain is not
too large at high frequencies. The magnitude Bode plot for the controller (2.51) is shown in
Figure 2.18. It is interesting to note that in the crossover region around $\omega = 0.5$ rad/s the
controller gain is quite constant, around 1 in magnitude, which is similar to the "best" gain
found using a P-controller (see Figure 2.5).*

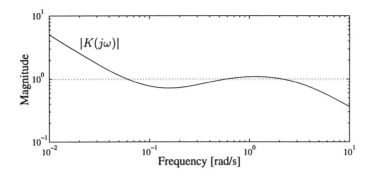

Figure 2.18: Magnitude Bode plot of controller (2.51) for loop-shaping design

Limitations imposed by RHP-zeros and time delays

Based on the above loop-shaping arguments we can now examine how the presence of delays and RHP-zeros limit the achievable control performance. We have already argued that if we want the loop shape to have a slope of -1 around crossover (ω_c), with preferably a steeper slope before and after crossover, then the phase lag of L at ω_c will necessarily be at least $-90°$, even when there are no RHP-zeros or delays. Therefore, if we assume that for performance and robustness we want a phase margin of about $35°$ or more, then the additional phase contribution from any delays and RHP-zeros at frequency ω_c cannot exceed about $-55°$.

First consider a time delay θ. It yields an additional phase contribution of $-\theta\omega$, which at frequency $\omega = 1/\theta$ is -1 rad $= -57°$ (which is more than $-55°$). Thus, for acceptable control performance we need $\omega_c < 1/\theta$, approximately.

Next consider a real RHP-zero at $s = z$. To avoid an increase in slope caused by this zero we place a pole at $s = -z$ such that the loop transfer function contains the term $\frac{-s+z}{s+z}$, the form of which is referred to as all-pass since its magnitude equals 1 at all frequencies. The phase contribution from the all-pass term at $\omega = z/2$ is $-2\arctan(0.5) = -53°$ (which is close to $-55°$), so for acceptable control performance we need $\omega_c < z/2$, approximately.

2.6.3   Inverse-based controller design

In Example 2.6, we made sure that $L(s)$ contained the RHP-zero of $G(s)$, but otherwise the specified $L(s)$ was independent of $G(s)$. This suggests the following possible approach for a minimum-phase plant (i.e. one with no RHP-zeros or time delays). Select a loop shape which has a slope of -1 throughout the frequency range, namely

$$L(s) = \frac{\omega_c}{s} \tag{2.52}$$

where ω_c is the desired gain crossover frequency. This loop shape yields a phase margin of 90° and an infinite gain margin since the phase of $L(j\omega)$ never reaches $-180°$. The controller corresponding to (2.52) is

$$K(s) = \frac{\omega_c}{s}G^{-1}(s) \tag{2.53}$$

That is, the controller inverts the plant and adds an integrator $(1/s)$. This is an old idea, and is also the essential part of the internal model control (IMC) design procedure (Morari and Zafiriou, 1989) which has proved successful in many applications. However, there are at least two good reasons for why this inverse-based controller may not be a good choice:

1. The controller will not be realizable if $G(s)$ has a pole excess of two or larger, and may in any case yield large input signals. These problems may be partly fixed by adding high-frequency dynamics to the controller.
2. The loop shape resulting from (2.52) and (2.53) is *not* generally desirable, unless the references and disturbances affect the outputs as steps. This is illustrated by the following example.

Example 2.7 Disturbance process. *We now introduce our second SISO example control problem in which disturbance rejection is an important objective in addition to command tracking. We assume that the plant has been appropriately scaled as outlined in Section 1.4.*
Problem formulation. *Consider the disturbance process described by*

$$\boxed{G(s) = \frac{200}{10s + 1}\frac{1}{(0.05s + 1)^2}, \quad G_d(s) = \frac{100}{10s + 1}} \tag{2.54}$$

with time in seconds (a block diagram is shown in Figure 2.20). The control objectives are:

1. *Command tracking: The rise time (to reach 90% of the final value) should be less than 0.3 s and the overshoot should be less than 5%.*
2. *Disturbance rejection: The output in response to a unit step disturbance should remain within the range $[-1, 1]$ at all times, and it should return to 0 as quickly as possible ($|y(t)|$ should at least be less than 0.1 after 3 s).*
3. *Input constraints: $u(t)$ should remain within the range $[-1, 1]$ at all times to avoid input saturation (this is easily satisfied for most designs).*

Analysis. *Since $G_d(0) = 100$ we have that without control the output response to a unit disturbance ($d = 1$) will be 100 times larger than what is deemed to be acceptable. The magnitude $|G_d(j\omega)|$ is lower at higher frequencies, but it remains larger than 1 up to $\omega_d \approx 10$ rad/s (where $|G_d(j\omega_d)| = 1$). Thus, feedback control is needed up to frequency ω_d, so we need ω_c to be approximately equal to 10 rad/s for disturbance rejection. On the other hand, we do not want ω_c to be larger than necessary because of sensitivity to noise and stability problems associated with high gain feedback. We will thus aim at a design with $\underline{\omega_c \approx 10 \text{ rad/s}}$.*
Inverse-based controller design. *We will consider the inverse-based design as given by (2.52) and (2.53) with $\omega_c = 10$. Since $G(s)$ has a pole excess of three this yields an unrealizable controller, and therefore we choose to approximate the plant term $(0.05s + 1)^2$*

by $(0.1s + 1)$ and then in the controller we let this term be effective over one decade, i.e. we use $(0.1s + 1)/(0.01s + 1)$ to give the realizable design

$$K_0(s) = \frac{\omega_c}{s} \frac{10s + 1}{200} \frac{0.1s + 1}{0.01s + 1}, \quad L_0(s) = \frac{\omega_c}{s} \frac{0.1s + 1}{(0.05s + 1)^2(0.01s + 1)}, \quad \omega_c = 10 \quad (2.55)$$

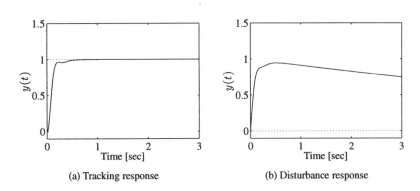

(a) Tracking response   (b) Disturbance response

Figure 2.19: Responses with "inverse-based" controller $K_0(s)$ for the disturbance process

The response to a step reference is excellent as shown in Figure 2.19(a). The rise time is about 0.16 s and there is no overshoot so the specifications are more than satisfied. However, the response to a step disturbance (Figure 2.19(b)) is much too sluggish. Although the output stays within the range $[-1, 1]$, it is still 0.75 at $t = 3$ s (whereas it should be less than 0.1). Because of the integral action the output does eventually return to zero, but it does not drop below 0.1 until after 23 s.

The above example illustrates that the simple inverse-based design method where L has a slope of about $N = -1$ at all frequencies, does not always yield satisfactory designs. In the example, reference tracking was excellent, but disturbance rejection was poor. The objective of the next section is to understand why the disturbance response was so poor, and to propose a more desirable loop shape for disturbance rejection.

2.6.4 Loop shaping for disturbance rejection

At the outset we assume that the disturbance has been scaled such that at each frequency $|d(\omega)| \leq 1$, and the main control objective is to achieve $|e(\omega)| < 1$. With feedback control we have $e = y = SG_d d$, so to achieve $|e(\omega)| \leq 1$ for $|d(\omega)| = 1$ (the worst-case disturbance) we require $|SG_d(j\omega)| < 1, \forall \omega$, or equivalently,

$$|1 + L| \geq |G_d| \quad \forall \omega \quad (2.56)$$

At frequencies where $|G_d| > 1$, this is approximately the same as requiring $|L| > |G_d|$. However, in order to minimize the input signals, thereby reducing the sensitivity to noise and avoiding stability problems, we do not want to use larger loop gains than necessary (at least at frequencies around crossover). A reasonable initial loop shape $L_{\min}(s)$ is then one that just satisfies the condition

$$|L_{\min}| \approx |G_d| \tag{2.57}$$

where the subscript min signifies that $L_{\min}$ is the smallest loop gain to satisfy $|e(\omega)| \leq 1$. Since $L = GK$ the corresponding controller with the minimum gain satisfies

$$|K_{\min}| \approx |G^{-1}G_d| \tag{2.58}$$

In addition, to improve low-frequency performance (e.g. to get zero steady-state offset), we often add integral action at low frequencies, and use

$$|K| = |\frac{s + \omega_I}{s}||G^{-1}G_d| \tag{2.59}$$

This can be summarized as follows:

- For disturbance rejection a good choice for the controller is one which contains the dynamics (G_d) of the disturbance and inverts the dynamics (G) of the inputs (at least at frequencies just before crossover).
- For disturbances entering directly at the plant output, $G_d = 1$, we get $|K_{\min}| = |G^{-1}|$, so an inverse-based design provides the best trade-off between performance (disturbance rejection) and minimum use of feedback.
- For disturbances entering directly at the plant input (which is a common situation in practice – often referred to as a load disturbance), we have $G_d = G$ and we get $|K_{\min}| = 1$, so a simple proportional controller with unit gain yields a good trade-off between output performance and input usage.
- Notice that a reference change may be viewed as a disturbance directly affecting the output. This follows from (1.18), from which we get that a maximum reference change $r = R$ may be viewed as a disturbance $d = 1$ with $G_d(s) = -R$ where R is usually a constant. This explains why selecting K to be like G^{-1} (an inverse-based controller) yields good responses to step changes in the reference.

In addition to satisfying $|L| \approx |G_d|$ (eq. 2.57) at frequencies around crossover, the desired loop-shape $L(s)$ may be modified as follows:

1. Around crossover make the slope N of $|L|$ to be about -1. This is to achieve good transient behaviour with acceptable gain and phase margins.
2. Increase the loop gain at low frequencies as illustrated in (2.59) to improve the settling time and to reduce the steady-state offset. Adding an integrator yields zero steady-state offset to a step disturbance.
3. Let $L(s)$ roll off faster at higher frequencies (beyond the bandwidth) in order to reduce the use of manipulated inputs, to make the controller realizable and to reduce the effects of noise.

The above requirements are concerned with the magnitude, $|L(j\omega)|$. In addition, the dynamics (phase) of $L(s)$ must be selected such that the closed-loop system is stable. When selecting $L(s)$ to satisfy $|L| \approx |G_d|$ one should replace $G_d(s)$ by the corresponding minimum-phase transfer function with the same magnitude, that is, time delays and RHP-zeros in $G_d(s)$ should not be included in $L(s)$ as this will impose undesirable limitations on feedback. On the other hand, any time delays or RHP-zeros in $G(s)$ must be included in $L = GK$ because RHP pole-zero cancellations between $G(s)$ and $K(s)$ yield internal instability; see Chapter 4.

Remark. The idea of including a disturbance model in the controller is well known and is more rigorously presented in, for example, research on the internal model principle (Wonham, 1974), or the internal model control design for disturbances (Morari and Zafiriou, 1989). However, our development is simple, and sufficient for gaining the insight needed for later chapters.

Example 2.8 Loop-shaping design for the disturbance process. *Consider again the plant described by (2.54). The plant can be represented by the block diagram in Figure 2.20, and we see that the disturbance enters at the plant input in the sense that G and G_d share the same dominating dynamics as represented by the term $200/(10s + 1)$.*

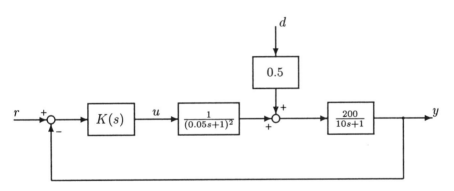

Figure 2.20: Block diagram representation of the disturbance process in (2.54)

Step 1. Initial design. *From (2.57) we know that a good initial loop shape looks like $|L_{min}| = |G_d| = \left|\frac{100}{10s+1}\right|$ at frequencies up to crossover. The corresponding controller is $K(s) = G^{-1}L_{min} = 0.5(0.05s + 1)^2$. This controller is not proper (i.e. it has more zeros than poles), but since the term $(0.05s + 1)^2$ only comes into effect at $1/0.05 = 20$ rad/s, which is beyond the desired gain crossover frequency $\omega_c = 10$ rad/s, we may replace it by a constant gain of 1 resulting in a proportional controller*

$$K_1(s) = 0.5 \qquad\qquad (2.60)$$

The magnitude of the corresponding loop transfer function, $|L_1(j\omega)|$, and the response ($y_1(t)$) to a step change in the disturbance are shown in Figure 2.21. This simple controller works *surprisingly well, and for $t < 3$ s the response to a step change in the disturbance is not much different from that with the more complicated inverse-based controller $K_0(s)$ of (2.55) as shown earlier in Figure 2.19. However, there is no integral action and $y_1(t) \rightarrow 1$ as $t \rightarrow \infty$.*

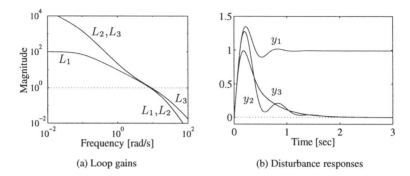

| (a) Loop gains | (b) Disturbance responses |

Figure 2.21: Loop shapes and disturbance responses for controllers K_1, K_2 and K_3 for the disturbance process

Step 2. More gain at low frequency. *To get integral action we multiply the controller by the term $\frac{s+\omega_I}{s}$, see (2.59), where ω_I is the frequency up to which the term is effective (the asymptotic value of the term is 1 for $\omega > \omega_I$). For performance we want large gains at low frequencies, so we want ω_I to be large, but in order to maintain an acceptable phase margin (which is $44.7°$ for controller K_1) the term should not add too much negative phase at frequency ω_c, so ω_I should not be too large. A reasonable value is $\omega_I = 0.2\omega_c$ for which the phase contribution from $\frac{s+\omega_I}{s}$ is $\arctan(1/0.2) - 90° = -11°$ at ω_c. In our case $\omega_c \approx 10$ rad/s, so we select the following controller*

$$K_2(s) = 0.5\frac{s+2}{s} \tag{2.61}$$

The resulting disturbance response (y_2) shown in Figure 2.21(b) satisfies the requirement that $|y(t)| < 0.1$ at time $t = 3$ s, but $y(t)$ exceeds 1 for a short time. Also, the response is slightly oscillatory as might be expected since the phase margin is only $31°$ and the peak values for $|S|$ and $|T|$ are $M_S = 2.28$ and $M_T = 1.89$.

Step 3. High-frequency correction. *To increase the phase margin and improve the transient response we supplement the controller with "derivative action" by multiplying $K_2(s)$ by a lead-lag term which is effective over one decade starting at 20 rad/s:*

$$K_3(s) = 0.5\frac{s+2}{s}\frac{0.05s+1}{0.005s+1} \tag{2.62}$$

This gives a phase margin of $51°$, and peak values $M_S = 1.43$ and $M_T = 1.23$. From Figure 2.21(b), it is seen that the controller $K_3(s)$ reacts quicker than $K_2(s)$ and the disturbance response $y_3(t)$ stays below 1.

Table 2.2 summarizes the results for the four loop-shaping designs; the inverse-based design K_0 for reference tracking and the three designs K_1, K_2 and K_3 for disturbance rejection. Although controller K_3 satisfies the requirements for disturbance rejection, it is not

Table 2.2: Alternative loop-shaping designs for the disturbance process

	GM	PM	ω_c	M_S	M_T	Reference		Disturbance	
						t_r	y_{max}	y_{max}	$y(t = 3)$
Spec.→			≈ 10			$\leq .3$	≤ 1.05	≤ 1	≤ 0.1
K_0	9.95	72.9°	11.4	1.34	1	0.16	1.00	0.95	0.75
K_1	4.04	44.7°	8.48	1.83	1.33	0.21	1.24	1.35	0.99
K_2	3.24	30.9°	8.65	2.28	1.89	0.19	1.51	1.27	0.001
K_3	19.7	50.9°	9.27	1.43	1.23	0.16	1.24	0.99	0.001

satisfactory for reference tracking; the overshoot is 24% which is significantly higher than the maximum value of 5%. On the other hand, the inverse-based controller K_0 inverts the term $1/(10s + 1)$ which is also in the disturbance model, and therefore yields a very sluggish response to disturbances (the output is still 0.75 at $t = 3$ s whereas it should be less than 0.1).

In summary, for this process none of the controller designs meet all the objectives for both reference tracking and disturbance rejection. The solution is to use a two degrees-of-freedom controller as is discussed next.

2.6.5   Two degrees-of-freedom design

For reference tracking we typically want the controller to look like $\frac{1}{s}G^{-1}$, see (2.53), whereas for disturbance rejection we want the controller to look like $\frac{1}{s}G^{-1}G_d$, see (2.59). We cannot achieve both of these simultaneously with a single (feedback) controller.

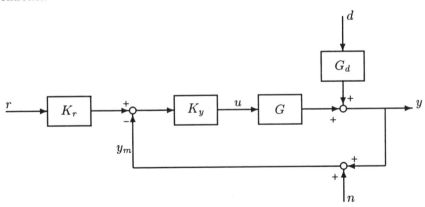

Figure 2.22: Two degrees-of-freedom controller

The solution is to use a two degrees-of-freedom controller where the reference signal r and output measurement y_m are independently treated by the controller, rather than operating on their difference $r - y_m$. There exist several alternative implementations of a two degrees-of-freedom controller. The most general form is shown in Figure 1.3(b) on page 12 where the controller has two inputs (r and y_m) and one output (u). However, the controller is often split into two separate blocks as shown in Figure 2.22 where K_y denotes the feedback part of the controller and K_r a reference prefilter. The feedback controller K_y is used to reduce the effect of uncertainty (disturbances and model error) whereas the prefilter K_r shapes the commands r to improve tracking performance. In general, it is optimal to design the combined two degrees-of-freedom controller K in one step. However, in practice K_y is often designed first for disturbance rejection, and then K_r is designed to improve reference tracking. This is the approach taken here.

Let $T = L(1 + L)^{-1}$ (with $L = GK_y$) denote the complementary sensitivity function for the feedback system. Then for a one degree-of-freedom controller $y = Tr$, whereas for a two degrees-of-freedom controller $y = TK_r r$. If the desired transfer function for reference tracking (often denoted the reference model) is T_{ref}, then the corresponding ideal reference prefilter K_r satisfies $TK_r = T_{\text{ref}}$, or

$$K_r(s) = T^{-1}(s)T_{\text{ref}}(s) \tag{2.63}$$

Thus, in theory we may design $K_r(s)$ to get any desired tracking response $T_{\text{ref}}(s)$. However, in practice it is not so simple because the resulting $K_r(s)$ may be unstable (if $G(s)$ has RHP-zeros) or unrealizable, and also $TK_r \neq T_{\text{ref}}$ if $T(s)$ is not known exactly.

Remark. A convenient practical choice of prefilter is the lead-lag network

$$K_r(s) = \frac{\tau_{\text{lead}}s + 1}{\tau_{\text{lag}}s + 1} \tag{2.64}$$

Here we select $\tau_{\text{lead}} > \tau_{\text{lag}}$ if we want to speed up the response, and $\tau_{\text{lead}} < \tau_{\text{lag}}$ if we want to slow down the response. If one does not require fast reference tracking, which is the case in many process control applications, a simple lag is often used (with $\tau_{\text{lead}} = 0$).

Example 2.9 Two degrees-of-freedom design for the disturbance process. *In Example 2.8 we designed a loop-shaping controller $K_3(s)$ for the plant in (2.54) which gave good performance with respect to disturbances. However, the command tracking performance was not quite acceptable as is shown by y_3 in Figure 2.23. The rise time is 0.16 s which is better than the required value of 0.3s, but the overshoot is 24% which is significantly higher than the maximum value of 5%. To improve upon this we can use a two degrees-of-freedom controller with $K_y = K_3$, and we design $K_r(s)$ based on (2.63) with reference model $T_{\text{ref}} = 1/(0.1s + 1)$ (a first-order response with no overshoot). To get a low-order $K_r(s)$, we may either use the actual $T(s)$ and then use a low-order approximation of $K_r(s)$, or we may start with a low-order approximation of $T(s)$. We will do the latter. From the step response y_3 in Figure 2.23 we approximate the response by two parts; a fast response with time constant 0.1 s and gain 1.5, and a slower response with time constant 0.5 s and gain −0.5 (the sum of the gains is 1). Thus we use $T(s) \approx \frac{1.5}{0.1s+1} - \frac{0.5}{0.5s+1} = \frac{(0.7s+1)}{(0.1s+1)(0.5s+1)}$, from which (2.63) yields*

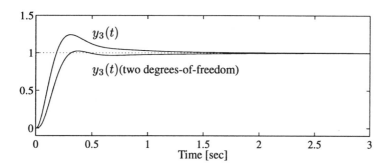

Figure 2.23: Tracking responses with the one degree-of-freedom controller (K_3) and the two degrees-of-freedom controller (K_3, K_{r3}) for the disturbance process

$K_r(s) = \frac{0.5s+1}{0.7s+1}$. *Following closed-loop simulations we modified this slightly to arrive at the design*

$$K_{r3}(s) = \frac{0.5s+1}{0.65s+1} \cdot \frac{1}{0.03s+1} \qquad (2.65)$$

where the term $1/(0.03s+1)$ was included to avoid the initial peaking of the input signal $u(t)$ above 1. The tracking response with this two degrees-of-freedom controller is shown in Figure 2.23. The rise time is 0.25 s which is better than the requirement of 0.3 s, and the overshoot is only 2.3% which is better than the requirement of 5%. The disturbance response is the same as curve y_3 in Figure 2.21. In conclusion, we are able to satisfy all specifications using a two degrees-of-freedom controller.

Loop shaping applied to a flexible structure

The following example shows how the loop-shaping procedure for disturbance rejection, can be used to design a one degree-of-freedom controller for a very different kind of plant.

Example 2.10 Loop shaping for a flexible structure. *Consider the following model of a flexible structure with a disturbance occurring at the plant input*

$$G(s) = G_d(s) = \frac{2.5s(s^2+1)}{(s^2+0.5^2)(s^2+2^2)} \qquad (2.66)$$

From the Bode magnitude plot in Figure 2.24(a) we see that $|G_d(j\omega)| \gg 1$ around the resonance frequencies of 0.5 and 2 rad/s, so control is needed at these frequencies. The dashed line in Figure 2.24(b) shows the open-loop response to a unit step disturbance. The output is seen to cycle between -2 and 2 (outside the allowed range -1 to 1), which confirms that control is needed. From (2.58) a controller which meets the specification $|y(\omega)| \leq 1$ for $|d(\omega)| = 1$ is given by $|K_{min}(j\omega)| = |G^{-1}G_d| = 1$. Indeed the controller

$$K(s) = 1 \qquad (2.67)$$

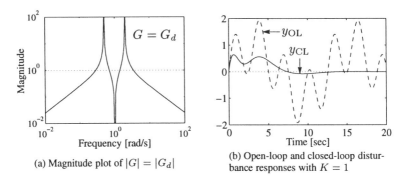

(a) Magnitude plot of $|G| = |G_d|$

(b) Open-loop and closed-loop disturbance responses with $K = 1$

Figure 2.24: Flexible structure in (2.66)

*turns out to be a good choice as is verified by the closed-loop disturbance response (solid line)
in Figure 2.24(b); the output goes up to about 0.5 and then returns to zero. The fact that the
choice $L(s) = G(s)$ gives closed-loop stability is not immediately obvious since $|G|$ has 4
gain crossover frequencies. However, instability cannot occur because the plant is "passive"
with $\angle G > -180°$ at all frequencies.*

2.6.6   Conclusions on loop shaping

The loop-shaping procedure outlined and illustrated by the examples above is well
suited for relatively simple problems, as might arise for stable plants where $L(s)$
crosses the negative real axis only once. Although the procedure may be extended to
more complicated systems the effort required by the engineer is considerably greater.
In particular, it may be very difficult to achieve stability.

Fortunately, there exist alternative methods where the burden on the engineer is
much less. One such approach is the Glover-McFarlane $\mathcal{H}_\infty$ loop-shaping procedure
which is discussed in detail in Chapter 9. It is essentially a two-step procedure, where
in the first step the engineer, as outlined in this section, decides on a loop shape, $|L|$
(denoted the "shaped plant" G_s), and in the second step an optimization provides the
necessary phase corrections to get a stable and robust design. The method is applied
to the disturbance process in Example 9.3 on page 381.

Another design philosophy which deals directly with shaping both the gain and
phase of $L(s)$ is the quantitative feedback theory (QFT) of Horowitz (1991).

2.7   Shaping closed-loop transfer functions

In this section, we introduce the reader to the shaping of the magnitudes of closed-
loop transfer functions, where we synthesize a controller by minimizing an $\mathcal{H}_\infty$

performance objective. The topic is discussed further in Section 3.4.6 and in more detail in Chapter 9.

Specifications directly on the *open-loop transfer function* $L = GK$, as in the loop-shaping design procedures of the previous section, make the design process transparent as it is clear how changes in $L(s)$ affect the controller $K(s)$ and *vice versa*. An apparent problem with this approach, however, is that it does not consider directly the *closed-loop transfer functions*, such as S and T, which determine the final response. The following approximations apply

$$
\begin{aligned}
|L(j\omega)| \gg 1 &\quad \Rightarrow \quad S \approx L^{-1}; \quad T \approx 1 \\
|L(j\omega)| \ll 1 &\quad \Rightarrow \quad S \approx 1; \qquad T \approx L
\end{aligned}
$$

but in the crossover region where $|L(j\omega)|$ is close to 1, one cannot infer anything about S and T from the magnitude of the loop shape, $|L(j\omega)|$. For example, $|S|$ and $|T|$ may experience large peaks if $L(j\omega)$ is close to -1, i.e. the phase of $L(j\omega)$ is crucial in this frequency range.

An alternative design strategy is to directly shape the magnitudes of closed-loop transfer functions, such as $S(s)$ and $T(s)$. Such a design strategy can be formulated as an $\mathcal{H}_\infty$ optimal control problem, thus automating the actual controller design and leaving the engineer with the task of selecting reasonable bounds ("weights") on the desired closed-loop transfer functions. Before explaining how this may be done in practice, we discuss the terms $\mathcal{H}_\infty$ and $\mathcal{H}_2$.

2.7.1   The terms $\mathcal{H}_\infty$ and $\mathcal{H}_2$

The $\mathcal{H}_\infty$ norm of a stable scalar transfer function $f(s)$ is simply the peak value of $|f(j\omega)|$ as a function of frequency, that is,

$$
\|f(s)\|_\infty \triangleq \max_\omega |f(j\omega)| \tag{2.68}
$$

Remark. Strictly speaking, we should here replace "max" (the maximum value) by "sup" (the supremum, the least upper bound). This is because the maximum may only be approached as $w \to \infty$ and may therefore not actually be achieved. However, for engineering purposes there is no difference between "sup" and "max".

The terms $\mathcal{H}_\infty$ norm and $\mathcal{H}_\infty$ control are intimidating at first, and a name conveying the engineering significance of $\mathcal{H}_\infty$ would have been better. After all, we are simply talking about a design method which aims to press down the peak(s) of one or more selected transfer functions. However, the term $\mathcal{H}_\infty$, which is purely mathematical, has now established itself in the control community. To make the term less forbidding, an explanation of its background may help. First, the symbol ∞ comes from the fact that the maximum magnitude over frequency may be written as

$$
\max_\omega |f(j\omega)| = \lim_{p \to \infty} \left(\int_{-\infty}^{\infty} |f(j\omega)|^p d\omega \right)^{1/p}
$$

Essentially, by raising $|f|$ to an infinite power we pick out its peak value. Next, the symbol $\mathcal{H}$ stands for "Hardy space", and $\mathcal{H}_\infty$ in the context of this book is the set of transfer functions with bounded ∞-norm, which is simply the set of *stable and proper* transfer functions.

Similarly, the symbol $\mathcal{H}_2$ stands for the Hardy space of transfer functions with bounded 2-norm, which is the set of *stable and strictly proper* transfer functions. The $\mathcal{H}_2$ norm of a strictly proper stable scalar transfer function is defined as

$$\|f(s)\|_2 \triangleq \left(\frac{1}{2\pi} \int_{-\infty}^{\infty} |f(j\omega)|^2 d\omega \right)^{1/2} \tag{2.69}$$

The factor $1/\sqrt{2\pi}$ is introduced to get consistency with the 2-norm of the corresponding impulse response; see (4.117). Note that the $\mathcal{H}_2$ norm of a semi-proper (or bi-proper) transfer function (where $\lim_{s\to\infty} f(s)$ is a non-zero constant) is infinite, whereas its $\mathcal{H}_\infty$ norm is finite. An example of a semi-proper transfer function (with an infinite $\mathcal{H}_2$ norm) is the sensitivity function $S = (I + GK)^{-1}$.

2.7.2 Weighted sensitivity

As already discussed, the sensitivity function S is a very good indicator of closed-loop performance, both for SISO and MIMO systems. The main advantage of considering S is that because we ideally want S small, it is sufficient to consider just its magnitude $|S|$; that is, we need not worry about its phase. Typical specifications in terms of S include:

1. Minimum bandwidth frequency ω_B^* (defined as the frequency where $|S(j\omega)|$ crosses 0.707 from below).
2. Maximum tracking error at selected frequencies.
3. System type, or alternatively the maximum steady-state tracking error, A.
4. Shape of S over selected frequency ranges.
5. Maximum peak magnitude of S, $\|S(j\omega)\|_\infty \leq M$.

The peak specification prevents amplification of noise at high frequencies, and also introduces a margin of robustness; typically we select $M = 2$. Mathematically, these specifications may be captured by an upper bound, $1/|w_P(s)|$, on the magnitude of S, where $w_P(s)$ is a weight selected by the designer. The subscript P stands for *performance* since S is mainly used as a performance indicator, and the performance requirement becomes

$$|S(j\omega)| < 1/|w_P(j\omega)|, \ \forall\omega \tag{2.70}$$

$$\Leftrightarrow \quad |w_P S| < 1, \ \forall\omega \quad \Leftrightarrow \quad \boxed{\|w_P S\|_\infty < 1} \tag{2.71}$$

The last equivalence follows from the definition of the $\mathcal{H}_\infty$ norm, and in words the performance requirement is that the $\mathcal{H}_\infty$ norm of the weighted sensitivity, $w_P S$,

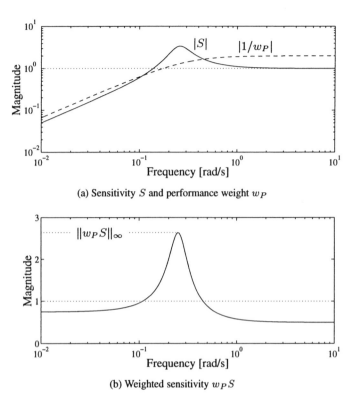

(a) Sensitivity S and performance weight w_P

(b) Weighted sensitivity $w_P S$

Figure 2.25: Case where $|S|$ exceeds its bound $1/|w_P|$, resulting in $\|w_P S\|_\infty > 1$

must be less than one. In Figure 2.25(a), an example is shown where the sensitivity, $|S|$, exceeds its upper bound, $1/|w_P|$, at some frequencies. The resulting weighted sensitivity, $|w_P S|$ therefore exceeds 1 at the same frequencies as is illustrated in Figure 2.25(b). Note that we usually do not use a log-scale for the magnitude when plotting weighted transfer functions, such as $|w_P S|$.

Weight selection. An asymptotic plot of a typical upper bound, $1/|w_P|$, is shown in Figure 2.26. The weight illustrated may be represented by

$$w_P(s) = \frac{s/M + \omega_B^*}{s + \omega_B^* A} \qquad (2.72)$$

and we see that $1/|w_P(j\omega)|$ (the upper bound on $|S|$) is equal to $A \leq 1$ at low frequencies, is equal to $M \geq 1$ at high frequencies, and the asymptote crosses 1 at the frequency ω_B^*, which is approximately the bandwidth requirement.

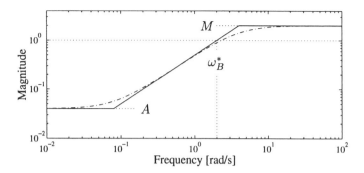

Figure 2.26: Inverse of performance weight. Exact and asymptotic plot of $1/|w_P(j\omega)|$ in (2.72)

Remark. For this weight the loop shape $L = \omega_B^*/s$ yields an S which exactly matches the bound (2.71) at frequencies below the bandwidth and easily satisfies (by a factor M) the bound at higher frequencies. This L has a slope in the frequency range below crossover of $N = -1$.

In some cases, in order to improve performance, we may want a steeper slope for L (and S) below the bandwidth, and then a higher-order weight may be selected. A weight which asks for a slope of -2 for L in a range of frequencies below crossover is

$$w_P(s) = \frac{(s/M^{1/2} + \omega_B^*)^2}{(s + \omega_B^* A^{1/2})^2} \qquad (2.73)$$

The insights gained in the previous section on loop-shaping design are very useful for selecting weights. For example, for disturbance rejection we must satisfy $|SG_d(j\omega)| < 1$ at all frequencies (assuming the variables have been scaled to be less than 1 in magnitude). It then follows that a good initial choice for the performance weight is to let $|w_P(j\omega)|$ look like $|G_d(j\omega)|$ at frequencies where $|G_d| > 1$.

Exercise 2.4 *Make an asymptotic plot of $1/|w_P|$ in (2.73) and compare with the asymptotic plot of $1/|w_P|$ in (2.72).*

2.7.3   Stacked requirements: mixed sensitivity

The specification $\|w_P S\|_\infty < 1$ puts a lower bound on the bandwidth, but not an upper one, and nor does it allow us to specify the roll-off of $L(s)$ above the bandwidth. To do this one can make demands on another closed-loop transfer function, for example, on the complementary sensitivity $T = I - S = GKS$. For instance, one might specify an upper bound $1/|w_T|$ on the magnitude of T to make sure that L rolls off sufficiently fast at high frequencies. Also, to achieve robustness or to restrict the magnitude of the input signals, $u = KS(r - G_d d)$, one may place an upper bound, $1/|w_u|$, on the magnitude of KS. To combine these "mixed sensitivity" specifications, a "stacking approach" is usually used, resulting in the following overall specification:

$$\|N\|_\infty = \max_\omega \bar{\sigma}(N(j\omega)) < 1; \quad N = \begin{bmatrix} w_P S \\ w_T T \\ w_u KS \end{bmatrix} \quad (2.74)$$

We here use the maximum singular value, $\bar{\sigma}(N(j\omega))$, to measure the size of the matrix N at each frequency. For SISO systems, N is a vector and $\bar{\sigma}(N)$ is the usual Euclidean vector norm:

$$\bar{\sigma}(N) = \sqrt{|w_P S|^2 + |w_T T|^2 + |w_u KS|^2} \quad (2.75)$$

After selecting the form of N and the weights, the $\mathcal{H}_\infty$ optimal controller is obtained by solving the problem

$$\min_K \|N(K)\|_\infty \quad (2.76)$$

where K is a stabilizing controller. A good tutorial introduction to $\mathcal{H}_\infty$ control is given by Kwakernaak (1993).

Remark 1 The stacking procedure is selected for mathematical convenience as it does not allow us to exactly specify the bounds on the individual transfer functions as described above. For example, assume that $\phi_1(K)$ and $\phi_2(K)$ are two functions of K (which might represent $\phi_1(K) = w_P S$ and $\phi_2(K) = w_T T$) and that we want to achieve

$$|\phi_1| < 1 \quad \text{and} \quad |\phi_2| < 1 \quad (2.77)$$

This is similar to, but not quite the same as the stacked requirement

$$\bar{\sigma} \begin{bmatrix} \phi_1 \\ \phi_2 \end{bmatrix} = \sqrt{|\phi_1|^2 + |\phi_2|^2} < 1 \quad (2.78)$$

Objectives (2.77) and (2.78) are very similar when either $|\phi_1|$ or $|\phi_2|$ is small, but in the worst case when $|\phi_1| = |\phi_2|$, we get from (2.78) that $|\phi_1| \le 0.707$ and $|\phi_2| \le 0.707$. That is, there is a possible "error" in each specification equal to at most a factor $\sqrt{2} \approx 3$ dB. In general, with n stacked requirements the resulting error is at most $\sqrt{n}$. This inaccuracy in the specifications is something we are probably willing to sacrifice in the interests of mathematical convenience. In any case, the specifications are in general rather rough, and are effectively knobs for the engineer to select and adjust until a satisfactory design is reached.

Remark 2 Let $\gamma_0 = \min_K \|N(K)\|_\infty$ denote the optimal $\mathcal{H}_\infty$ norm. An important property of $\mathcal{H}_\infty$ optimal controllers is that they yield a flat frequency response, that is, $\bar{\sigma}(N(j\omega)) = \gamma_0$ at all frequencies. The practical implication is that, except for at most a factor $\sqrt{n}$, the transfer functions resulting from a solution to (2.76) will be close to γ_0 times the bounds selected by the designer. This gives the designer a mechanism for directly shaping the magnitudes of $\bar{\sigma}(S)$, $\bar{\sigma}(T)$, $\bar{\sigma}(KS)$, and so on.

Example 2.11 $\mathcal{H}_\infty$ **mixed sensitivity design for the disturbance process.** *Consider again the plant in (2.54), and consider an* $\mathcal{H}_\infty$ *mixed sensitivity* S/KS *design in which*

$$N = \begin{bmatrix} w_P S \\ w_u K S \end{bmatrix} \tag{2.79}$$

Appropriate scaling of the plant has been performed so that the inputs should be about 1 or less in magnitude, and we therefore select a simple input weight $w_u = 1$. *The performance weight is chosen, in the form of (2.72), as*

$$w_{P1}(s) = \frac{s/M + \omega_B^*}{s + \omega_B^* A}; \quad M = 1.5, \ \omega_B^* = 10, \quad A = 10^{-4} \tag{2.80}$$

A value of $A = 0$ *would ask for integral action in the controller, but to get a stable weight and to prevent numerical problems in the algorithm used to synthesize the controller, we have moved the integrator slightly by using a small non-zero value for* A. *This has no practical significance in terms of control performance. The value* $\omega_B^* = 10$ *has been selected to achieve approximately the desired crossover frequency* ω_c *of 10 rad/s. The* $\mathcal{H}_\infty$ *problem is solved with the* μ-*toolbox in MATLAB using the commands in Table 2.3.*

Table 2.3: MATLAB program to synthesize an $\mathcal{H}_\infty$ controller

```
% Uses the Mu-toolbox
G=nd2sys(1,conv([10 1],conv([0.05 1],[0.05 1])),200);         % Plant is G.
M=1.5; wb=10; A=1.e-4; Wp = nd2sys([1/M wb], [1 wb*A]); Wu = 1;  % Weights.
%
% Generalized plant P is found with function sysic:
% (see Section 3.8 for more details)
%
systemnames = 'G Wp Wu';
inputvar = '[ r(1); u(1)]';
outputvar = '[Wp; Wu; r-G]';
input_to_G = '[u]';
input_to_Wp = '[r-G]';
input_to_Wu = '[u]';
sysoutname = 'P';
cleanupsysic = 'yes';
sysic;
%
% Find H-infinity optimal controller:
%
nmeas=1; nu=1; gmn=0.5; gmx=20; tol=0.001;
[khinf,ghinf,gopt] = hinfsyn(P,nmeas,nu,gmn,gmx,tol);
```

For this problem, we achieved an optimal $\mathcal{H}_\infty$ *norm of 1.37, so the weighted sensitivity requirements are not quite satisfied (see design 1 in Figure 2.27). Nevertheless, the design seems good with* $\|S\|_\infty = M_S = 1.30$, $\|T\|_\infty = M_T = 1.0$, GM = 8.04, PM = 71.2° *and* $\omega_c = 7.22$ *rad/s, and the tracking response is very good as shown by curve* y_1 *in Figure 2.28(a). The design is actually very similar to the loop-shaping design for references,* K_0, *which was an inverse-based controller.*

However, we see from curve y_1 *in Figure 2.28(b) that the disturbance response is very sluggish. If disturbance rejection is the main concern, then from our earlier discussion in Section 2.6.4 this motivates the need for a performance weight that specifies higher gains at*

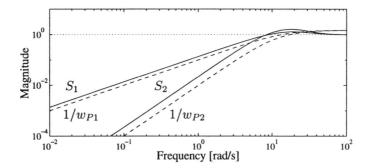

Figure 2.27: Inverse of performance weight (dashed line) and resulting sensitivity function (solid line) for two $\mathcal{H}_\infty$ designs (1 and 2) for the disturbance process

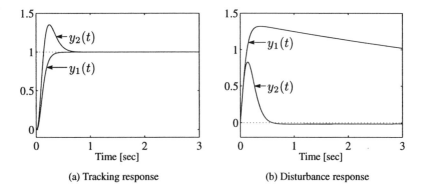

(a) Tracking response   (b) Disturbance response

Figure 2.28: Closed-loop step responses for two alternative $\mathcal{H}_\infty$ designs (1 and 2) for the disturbance process

low frequencies. We therefore try

$$w_{P2}(s) = \frac{(s/M^{1/2} + \omega_B^*)^2}{(s + \omega_B^* A^{1/2})^2}, \quad M = 1.5, \omega_B^* = 10, A = 10^{-4} \quad (2.81)$$

The inverse of this weight is shown in Figure 2.27, and is seen from the dashed line to cross 1 in magnitude at about the same frequency as weight w_{P1}, but it specifies tighter control at lower frequencies. With the weight w_{P2}, we get a design with an optimal $\mathcal{H}_\infty$ norm of 2.21, yielding $M_S = 1.63$, $M_T = 1.43$, GM = 4.76, PM = 43.3° and $\omega_c = 11.34$ rad/s. The design is actually very similar to the loop-shaping design for disturbances, K_3. The disturbance response is very good, whereas the tracking response has a somewhat high overshoot; see curve y_2 in Figure 2.28(a).

In conclusion, design 1 is best for reference tracking whereas design 2 is best for disturbance rejection. To get a design with both good tracking and good disturbance rejection we need a two degrees-of-freedom controller, as was discussed in Example 2.9.

2.8   Conclusion

The main purpose of this chapter has been to present the classical ideas and techniques of feedback control. We have concentrated on SISO systems so that insights into the necessary design trade-offs, and the design approaches available, can be properly developed before MIMO systems are considered. We also introduced the $\mathcal{H}_\infty$ problem based on weighted sensitivity, for which typical performance weights are given in (2.72) and (2.73).

3

INTRODUCTION TO MULTIVARIABLE CONTROL

In this chapter, we introduce the reader to multi-input multi-output (MIMO) systems. We discuss the singular value decomposition (SVD), multivariable control, and multivariable right-half plane (RHP) zeros. The need for a careful analysis of the effect of uncertainty in MIMO systems is motivated by two examples. Finally we describe a general control configuration that can be used to formulate control problems. Many of these important topics are considered again in greater detail later in the book. The chapter should be accessible to readers who have attended a classical SISO control course.

3.1 Introduction

We consider a multi-input multi-output (MIMO) plant with m inputs and l outputs. Thus, the basic transfer function model is $y(s) = G(s)u(s)$, where y is an $l \times 1$ vector, u is an $m \times 1$ vector and $G(s)$ is an $l \times m$ transfer function matrix.

If we make a change in the first input, u_1, then this will generally affect all the outputs, $y_1, y_2, \ldots, y_l$, that is, there is *interaction* between the inputs and outputs. A non-interacting plant would result if u_1 only affects y_1, u_2 only affects y_2, and so on.

The main difference between a scalar (SISO) system and a MIMO system is the presence of *directions* in the latter. Directions are relevant for vectors and matrices, but not for scalars. However, despite the complicating factor of directions, most of the ideas and techniques presented in the previous chapter on SISO systems may be extended to MIMO systems. The singular value decomposition (SVD) provides a useful way of quantifying multivariable directionality, and we will see that most SISO results involving the absolute value (magnitude) may be generalized to multivariable systems by considering the maximum singular value. An exception to this is Bode's stability condition which has no generalization in terms of singular values. This is related to the fact that it is difficult to find a good measure of phase for MIMO transfer functions.

The chapter is organized as follows. We start by presenting some rules for determining multivariable transfer functions from block diagrams. Although most of the formulas for scalar systems apply, we must exercise some care since matrix multiplication is not commutative, that is, in general $GK \neq KG$. Then we introduce the singular value decomposition and show how it may be used to study directions in multivariable systems. We also give a brief introduction to multivariable control and decoupling. We then consider a simple plant with a multivariable RHP-zero and show how the effect of this zero may be shifted from one output channel to another. After this we discuss robustness, and study two example plants, each 2×2, which demonstrate that the simple gain and phase margins used for SISO systems do not generalize easily to MIMO systems. Finally, we consider a general control problem formulation.

At this point, you may find it useful to browse through Appendix A where some important mathematical tools are described. Exercises to test your understanding of this mathematics are given at the end of this chapter.

3.2   Transfer functions for MIMO systems

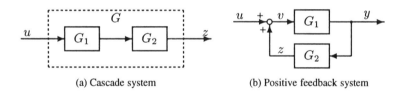

(a) Cascade system   (b) Positive feedback system

Figure 3.1: Block diagrams for the cascade rule and the feedback rule

The following three rules are useful when evaluating transfer functions for MIMO systems.

1. **Cascade rule.**   *For the cascade (series) interconnection of G_1 and G_2 in Figure 3.1(a), the overall transfer function matrix is $G = G_2 G_1$.*

Remark. The order of the transfer function matrices in $G = G_2 G_1$ (from left to right) is the reverse of the order in which they appear in the block diagram of Figure 3.1(a) (from left to right). This has led some authors to use block diagrams in which the inputs enter at the right hand side. However, in this case the order of the transfer function blocks in a feedback path will be reversed compared with their order in the formula, so no fundamental benefit is obtained.

2. **Feedback rule.**   *With reference to the positive feedback system in Figure 3.1(b), we have $v = (I - L)^{-1} u$ where $L = G_2 G_1$ is the transfer function around the loop.*

3. **Push-through rule.** *For matrices of appropriate dimensions*

$$G_1(I - G_2G_1)^{-1} = (I - G_1G_2)^{-1}G_1 \tag{3.1}$$

Proof: Equation (3.1) is verified by pre-multiplying both sides by $(I - G_1G_2)$ and post-multiplying both sides by $(I - G_2G_1)$. □

Exercise 3.1 *Derive the cascade and feedback rules.*

The cascade and feedback rules can be combined into the following MIMO rule for evaluating closed-loop transfer functions from block diagrams.

MIMO Rule: *Start from the output and write down the blocks as you meet them when moving backwards (against the signal flow), taking the most direct path towards the input. If you exit from a feedback loop then include a term $(I-L)^{-1}$ for positive feedback (or $(I + L)^{-1}$ for negative feedback) where L is the transfer function around that loop (evaluated against the signal flow starting at the point of exit from the loop).*

Care should be taken when applying this rule to systems with nested loops. For such systems it is probably safer to write down the signal equations and eliminate internal variables to get the transfer function of interest. The rule is best understood by considering an example.

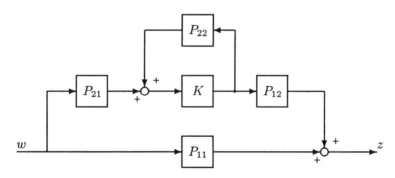

Figure 3.2: Block diagram corresponding to (3.2)

Example 3.1 *The transfer function for the block diagram in Figure 3.2 is given by*

$$z = (P_{11} + P_{12}K(I - P_{22}K)^{-1}P_{21})w \tag{3.2}$$

To derive this from the MIMO rule above we start at the output z and move backwards towards w. There are two branches, one of which gives the term P_{11} directly. In the other branch we move backwards and meet P_{12} and then K. We then exit from a feedback loop and get a term $(I - L)^{-1}$ (positive feedback) with $L = P_{22}K$, and finally we meet P_{21}.

Exercise 3.2 *Use the MIMO rule to derive the transfer functions from u to y and from u to z in Figure 3.1(b). Use the push-through rule to rewrite the two transfer functions.*

Exercise 3.3 *Use the MIMO rule to show that (2.18) corresponds to the negative feedback system in Figure 2.4.*

Negative feedback control systems

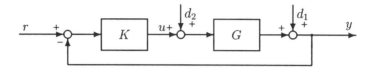

Figure 3.3: Conventional negative feedback control system

For the negative feedback system in Figure 3.3, we define L to be the loop transfer function as seen when breaking the loop at the *output* of the plant. Thus, for the case where the loop consists of a plant G and a feedback controller K we have

$$L = GK \tag{3.3}$$

The sensitivity and complementary sensitivity are then defined as

$$S \triangleq (I + L)^{-1}; \quad T \triangleq I - S = L(I + L)^{-1} \tag{3.4}$$

In Figure 3.3, T is the transfer function from r to y, and S is the transfer function from d_1 to y; also see equations (2.16) to (2.20) which apply to MIMO systems.

S and T are sometimes called the *output sensitivity* and *output complementary sensitivity*, respectively, and to make this explicit one may use the notation $L_O \equiv L$, $S_O \equiv S$ and $T_O \equiv T$. This is to distinguish them from the corresponding transfer functions evaluated at the *input* to the plant.

We define L_I to be the loop transfer function as seen when breaking the loop at the *input* to the plant with negative feedback assumed. In Figure 3.3

$$L_I = KG \tag{3.5}$$

The *input* sensitivity and *input* complementary sensitivity functions are then defined as

$$S_I \triangleq (I + L_I)^{-1}; \quad T_I \triangleq I - S_I = L_I(I + L_I)^{-1} \tag{3.6}$$

In Figure 3.3, $-T_I$ is the transfer function from d_2 to u. Of course, for SISO systems $L_I = L$, $S_I = S$, and $T_I = T$.

Exercise 3.4 *In Figure 3.3, what transfer function does S_I represent? Evaluate the transfer functions from d_1 and d_2 to $r - y$.*

The following relationships are useful:

$$(I + L)^{-1} + L(I + L)^{-1} = S + T = I \tag{3.7}$$

$$G(I + KG)^{-1} = (I + GK)^{-1}G \tag{3.8}$$

$$GK(I + GK)^{-1} = G(I + KG)^{-1}K = (I + GK)^{-1}GK \tag{3.9}$$

$$T = L(I + L)^{-1} = (I + (L)^{-1})^{-1} \tag{3.10}$$

Note that the matrices G and K in (3.7)-(3.10) need not be square whereas $L = GK$ is square. (3.7) follows trivially by factorizing out the term $(I + L)^{-1}$ from the right. (3.8) says that $GS_I = SG$ and follows from the push-through rule. (3.9) also follows from the push-through rule. (3.10) can be derived from the identity $M_1^{-1}M_2^{-1} = (M_2 M_1)^{-1}$.

Similar relationships, but with G and K interchanged, apply for the transfer functions evaluated at the plant input. To assist in remembering (3.7)-(3.10) note that G comes first (because the transfer function is evaluated at the output) and then G and K alternate in sequence. A given transfer matrix never occurs twice in sequence. For example, the closed-loop transfer function $G(I + GK)^{-1}$ does *not* exist (unless G is repeated in the block diagram, but then these G's would actually represent two different physical entities).

Remark 1 The above identities are clearly useful when deriving transfer functions analytically, but they are also useful for numerical calculations involving state-space realizations, e.g. $L(s) = C(sI - A)^{-1}B + D$. For example, assume we have been given a state-space realization for $L = GK$ with n states (so A is a $n \times n$ matrix) and we want to find the state space realization of T. Then we can first form $S = (I + L)^{-1}$ with n states, and then multiply it by L to obtain $T = SL$ with $2n$ states. However, a minimal realization of T has only n states. This may be obtained numerically using model reduction, but it is preferable to find it directly using $T = I - S$, see (3.7).

Remark 2 Note also that the right identity in (3.10) can only be used to compute the state-space realization of T if that of L^{-1} exists, so L must be semi-proper with $D \neq 0$ (which is rarely the case in practice). On the other hand, since L is square, we can always compute the frequency response of $L(j\omega)^{-1}$ (except at frequencies where $L(s)$ has $j\omega$-axis poles), and then obtain $T(j\omega)$ from (3.10).

Remark 3 In Appendix A.6 we present some factorizations of the sensitivity function which will be useful in later applications. For example, (A.139) relates the sensitivity of a perturbed plant, $S' = (I + G'K)^{-1}$, to that of the nominal plant, $S = (I + GK)^{-1}$. We have

$$S' = S(I + E_O T)^{-1}, \quad E_O \triangleq (G' - G)G^{-1} \tag{3.11}$$

where E_O is an output multiplicative perturbation representing the difference between G and G', and T is the nominal complementary sensitivity function.

3.3 Multivariable frequency response analysis

The transfer function $G(s)$ is a function of the Laplace variable s and can be used to represent a dynamic system. However, if we fix $s = s_0$ then we may view $G(s_0)$ simply as a complex matrix, which can be analyzed using standard tools in matrix algebra. In particular, the choice $s_0 = j\omega$ is of interest since $G(j\omega)$ represents the response to a sinusoidal signal of frequency ω.

3.3.1 Obtaining the frequency response from $G(s)$

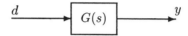

Figure 3.4: System $G(s)$ with input d and output y

The frequency domain is ideal for studying directions in multivariable systems at any given frequency. Consider the system $G(s)$ in Figure 3.4 with input $d(s)$ and output $y(s)$:

$$y(s) = G(s)d(s) \tag{3.12}$$

(We here denote the input by d rather than by u to avoid confusion with the matrix U used below in the singular value decomposition). In Section 2.1 we considered the sinusoidal response of scalar systems. These results may be directly generalized to multivariable systems by considering the elements g_{ij} of the matrix G. We have

• $g_{ij}(j\omega)$ represents the sinusoidal response from input j to output i.

To be more specific, apply to input channel j a scalar sinusoidal signal given by

$$d_j(t) = d_{j0} \sin(\omega t + \alpha_j) \tag{3.13}$$

This input signal is persistent, that is, it has been applied since $t = -\infty$. Then the corresponding persistent output signal in channel i is also a sinusoid with the same frequency

$$y_i(t) = y_{i0} \sin(\omega t + \beta_i) \tag{3.14}$$

where the amplification (gain) and phase shift may be obtained from the complex number $g_{ij}(j\omega)$ as follows

$$\frac{y_{io}}{d_{jo}} = |g_{ij}(j\omega)|, \quad \beta_i - \alpha_j = \angle g_{ij}(j\omega) \tag{3.15}$$

In phasor notation, see (2.7) and (2.9), we may compactly represent the sinusoidal time response described in (3.13)-(3.15) by

$$y_i(\omega) = g_{ij}(j\omega)d_j(\omega) \tag{3.16}$$

where

$$d_j(\omega) = d_{jo}e^{j\alpha_j}, \quad y_i(\omega) = y_{io}e^{j\beta_i} \tag{3.17}$$

Here the use of ω (and not $j\omega$) as the argument of $d_j(\omega)$ and $y_i(\omega)$ implies that these are complex numbers, representing at each frequency ω the magnitude and phase of the sinusoidal signals in (3.13) and (3.14).

The overall response to simultaneous input signals of the same frequency in several input channels is, by the superposition principle for linear systems, equal to the sum of the individual responses, and we have from (3.16)

$$y_i(\omega) = g_{i1}(j\omega)d_1(\omega) + g_{i2}(j\omega)d_2(\omega) + \cdots = \sum_j g_{ij}(j\omega)d_j(\omega) \tag{3.18}$$

or in matrix form

$$\boxed{y(\omega) = G(j\omega)d(\omega)} \tag{3.19}$$

where

$$d(\omega) = \begin{bmatrix} d_1(\omega) \\ d_2(\omega) \\ \vdots \\ d_m(\omega) \end{bmatrix} \quad \text{and} \quad y(\omega) = \begin{bmatrix} y_1(\omega) \\ y_2(\omega) \\ \vdots \\ y_l(\omega) \end{bmatrix} \tag{3.20}$$

represent the vectors of sinusoidal input and output signals.

Example 3.2 *Consider a 2×2 multivariable system where we simultaneously apply sinusoidal signals of the same frequency ω to the two input channels:*

$$d(t) = \begin{bmatrix} d_1(t) \\ d_2(t) \end{bmatrix} = \begin{bmatrix} d_{10}\sin(\omega t + \alpha_1) \\ d_{20}\sin(\omega t + \alpha_2) \end{bmatrix} \tag{3.21}$$

The corresponding output signal is

$$y(t) = \begin{bmatrix} y_1(t) \\ y_2(t) \end{bmatrix} = \begin{bmatrix} y_{10}\sin(\omega t + \beta_1) \\ y_{20}\sin(\omega t + \beta_2) \end{bmatrix} \tag{3.22}$$

which can be computed by multiplying the complex matrix $G(j\omega)$ by the complex vector $d(\omega)$:

$$y(\omega) = G(j\omega)d(\omega); \quad y(\omega) = \begin{bmatrix} y_{10}e^{j\beta_1} \\ y_{20}e^{j\beta_2} \end{bmatrix}, \ d(\omega) = \begin{bmatrix} d_{10}e^{j\alpha_1} \\ d_{20}e^{j\alpha_2} \end{bmatrix} \tag{3.23}$$

3.3.2 Directions in multivariable systems

For a SISO system, $y = Gd$, the gain at a given frequency is simply

$$\frac{|y(\omega)|}{|d(\omega)|} = \frac{|G(j\omega)d(\omega)|}{|d(\omega)|} = |G(j\omega)|$$

The gain depends on the frequency ω, but since the system is linear it is independent of the input magnitude $|d(\omega)|$.

Things are not quite as simple for MIMO systems where the input and output signals are both vectors, and we need to "sum up" the magnitudes of the elements in each vector by use of some norm, as discussed in Appendix A.5.1. If we select the vector 2-norm, the usual measure of length, then at a given frequency ω the magnitude of the vector input signal is

$$\|d(\omega)\|_2 = \sqrt{\sum_j |d_j(\omega)|^2} = \sqrt{d_{10}^2 + d_{20}^2 + \cdots} \qquad (3.24)$$

and the magnitude of the vector output signal is

$$\|y(\omega)\|_2 = \sqrt{\sum_i |y_i(\omega)|^2} = \sqrt{y_{10}^2 + y_{20}^2 + \cdots} \qquad (3.25)$$

The *gain* of the system $G(s)$ for a particular input signal $d(\omega)$ is then given by the ratio

$$\frac{\|y(\omega)\|_2}{\|d(\omega)\|_2} = \frac{\|G(j\omega)d(\omega)\|_2}{\|d(\omega)\|_2} = \frac{\sqrt{y_{10}^2 + y_{20}^2 + \cdots}}{\sqrt{d_{10}^2 + d_{20}^2 + \cdots}} \qquad (3.26)$$

Again the gain depends on the frequency ω, and again it is independent of the input magnitude $\|d(\omega)\|_2$. However, for a MIMO system there are additional degrees of freedom and the gain depends also on the *direction* of the input d.

Example 3.3 *For a system with two inputs, $d = \begin{bmatrix} d_{10} \\ d_{20} \end{bmatrix}$, the gain is in general different for the following five inputs:*

$$d_1 = \begin{bmatrix} 1 \\ 0 \end{bmatrix}, \; d_2 = \begin{bmatrix} 0 \\ 1 \end{bmatrix}, \; d_3 = \begin{bmatrix} 0.707 \\ 0.707 \end{bmatrix}, \; d_4 = \begin{bmatrix} 0.707 \\ -0.707 \end{bmatrix}, \; d_5 = \begin{bmatrix} 0.6 \\ -0.8 \end{bmatrix}$$

(which all have the same magnitude $\|d\|_2 = 1$ but are in different directions). For example, for the 2×2 system

$$G_1 = \begin{bmatrix} 5 & 4 \\ 3 & 2 \end{bmatrix} \qquad (3.27)$$

(a constant matrix) we compute for the five inputs d_j the following output vectors

$$y_1 = \begin{bmatrix} 5 \\ 3 \end{bmatrix}, \; y_2 = \begin{bmatrix} 4 \\ 2 \end{bmatrix}, \; y_3 = \begin{bmatrix} 6.36 \\ 3.54 \end{bmatrix}, \; y_4 = \begin{bmatrix} 0.707 \\ 0.707 \end{bmatrix}, \; y_5 = \begin{bmatrix} -0.2 \\ 0.2 \end{bmatrix}$$

and the 2-norms of these five outputs (i.e. the gains for the five inputs) are

$$\|y_1\|_2 = 5.83, \; \|y_2\|_2 = 4.47, \; \|y_3\|_2 = 7.30, \; \|y_4\|_2 = 1.00, \; \|y_5\|_2 = 0.28$$

This dependency of the gain on the input direction is illustrated graphically in Figure 3.5 where we have used the ratio d_{20}/d_{10} as an independent variable to represent the input direction. We see that, depending on the ratio d_{20}/d_{10}, the gain varies between 0.27 and 7.34.

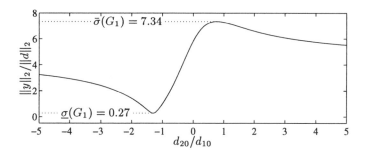

Figure 3.5: Gain $\|G_1 d\|_2 / \|d\|_2$ as a function of d_{20}/d_{10} for G_1 in (3.27)

The maximum value of the gain in (3.26) as the direction of the input is varied is the maximum singular value of G,

$$\max_{d \neq 0} \frac{\|Gd\|_2}{\|d\|_2} = \max_{\|d\|_2=1} \|Gd\|_2 = \bar{\sigma}(G) \qquad (3.28)$$

whereas the minimum gain is the minimum singular value of G,

$$\min_{d \neq 0} \frac{\|Gd\|_2}{\|d\|_2} = \min_{\|d\|_2=1} \|Gd\|_2 = \underline{\sigma}(G) \qquad (3.29)$$

We will discuss this in detail below. The first identities in (3.28) and (3.29) follow because the gain is independent of the input magnitude for a linear system.

3.3.3   Eigenvalues are a poor measure of gain

Before discussing the singular values we want to demonstrate that the magnitudes of the eigenvalues of a transfer function matrix, e.g. $|\lambda_i(G(j\omega))|$, do *not* provide a useful means of generalizing the SISO gain, $|G(j\omega)|$. First of all, eigenvalues can only be computed for square systems, and even then they can be very misleading. To see this, consider the system $y = Gd$ with

$$G = \begin{bmatrix} 0 & 100 \\ 0 & 0 \end{bmatrix} \qquad (3.30)$$

which has both eigenvalues λ_i equal to zero. However, to conclude from the eigenvalues that the system gain is zero is clearly misleading. For example, with an input vector $d = [0 \ 1]^T$ we get an output vector $y = [100 \ 0]^T$.

The "problem" is that the eigenvalues measure the gain for the special case when the inputs and the outputs are in the same direction, namely in the direction of the eigenvectors. To see this let t_i be an eigenvector of G and consider an input $d = t_i$. Then the output is $y = Gt_i = \lambda_i t_i$ where λ_i is the corresponding eigenvalue. We get

$$\|y\|/\|d\| = \|\lambda_i t_i\|/\|t_i\| = |\lambda_i|$$

so $|\lambda_i|$ measures the gain in the direction t_i. This may be useful for stability analysis, but not for performance.

To find useful generalizations of $|G|$ for the case when G is a matrix, we need the concept of a *matrix norm*, denoted $\|G\|$. Two important properties which must be satisfied for a matrix norm are the *triangle inequality*

$$\|G_1 + G_2\| \leq \|G_1\| + \|G_2\| \tag{3.31}$$

and the multiplicative property

$$\|G_1 G_2\| \leq \|G_1\| \cdot \|G_2\| \tag{3.32}$$

(see Appendix A.5 for more details). As we may expect, the magnitude of the largest eigenvalue, $\rho(G) \triangleq |\lambda_{max}(G)|$ (the spectral radius), does *not* satisfy the properties of a matrix norm; also see (A.115).

In Appendix A.5.2 we introduce several matrix norms, such as the Frobenius norm $\|G\|_F$, the sum norm $\|G\|_{\text{sum}}$, the maximum column sum $\|G\|_{i1}$, the maximum row sum $\|G\|_{i\infty}$, and the maximum singular value $\|G\|_{i2} = \bar{\sigma}(G)$ (the latter three norms are induced by a vector norm, e.g. see (3.28); this is the reason for the subscript i). We will use all of these norms in this book, each depending on the situation. However, in this chapter we will mainly use the induced 2-norm, $\bar{\sigma}(G)$. Notice that $\bar{\sigma}(G) = 100$ for the matrix in (3.30).

Exercise 3.5 *Compute the spectral radius and the five matrix norms mentioned above for the matrices in (3.27) and (3.30).*

3.3.4   Singular value decomposition

The singular value decomposition (SVD) is defined in Appendix A.3. Here we are interested in its physical interpretation when applied to the frequency response of a MIMO system $G(s)$ with m inputs and l outputs.

Consider a fixed frequency ω where $G(j\omega)$ is a constant $l \times m$ complex matrix, and denote $G(j\omega)$ by G for simplicity. Any matrix G may be decomposed into its singular value decomposition, and we write

$$G = U\Sigma V^H \tag{3.33}$$

where

Σ is an $l \times m$ matrix with $k = \min\{l, m\}$ non-negative singular values, σ_i, arranged in descending order along its main diagonal; the other entries are zero. The singular values are the positive square roots of the eigenvalues of $G^H G$, where G^H is the complex conjugate transpose of G.

$$\sigma_i(G) = \sqrt{\lambda_i(G^H G)} \tag{3.34}$$

U is an $l \times l$ unitary matrix of output singular vectors, u_i,

V is an $m \times m$ unitary matrix of input singular vectors, v_i,

This is illustrated by the SVD of a real 2×2 matrix which can always be written in the form

$$G = \underbrace{\begin{bmatrix} \cos\theta_1 & -\sin\theta_1 \\ \sin\theta_1 & \cos\theta_1 \end{bmatrix}}_{U} \underbrace{\begin{bmatrix} \sigma_1 & 0 \\ 0 & \sigma_2 \end{bmatrix}}_{\Sigma} \underbrace{\begin{bmatrix} \cos\theta_2 & \pm\sin\theta_2 \\ -\sin\theta_2 & \pm\cos\theta_2 \end{bmatrix}^T}_{V^T} \tag{3.35}$$

where the angles θ_1 and θ_2 depend on the given matrix. From (3.35) we see that the matrices U and V involve rotations and that their columns are orthonormal.

The singular values are sometimes called the principal values or principal gains, and the associated directions are called principal directions. In general, the singular values must be computed numerically. For 2×2 matrices however, analytic expressions for the singular values are given in (A.36).

Caution. It is standard notation to use the symbol U to denote the matrix of *output* singular vectors. This is unfortunate as it is also standard notation to use u (lower case) to represent the *input* signal. The reader should be careful not to confuse these two.

Input and output directions. The column vectors of U, denoted u_i, represent the *output directions* of the plant. They are orthogonal and of unit length (orthonormal), that is

$$\|u_i\|_2 = \sqrt{|u_{i1}|^2 + |u_{i2}|^2 + \ldots + |u_{il}|^2} = 1 \tag{3.36}$$

$$u_i^H u_i = 1, \quad u_i^H u_j = 0, \quad i \neq j \tag{3.37}$$

Likewise, the column vectors of V, denoted v_i, are orthogonal and of unit length, and represent the *input directions*. These input and output directions are related through the singular values. To see this, note that since V is unitary we have $V^H V = I$, so (3.33) may be written as $GV = U\Sigma$, which for column i becomes

$$Gv_i = \sigma_i u_i \tag{3.38}$$

where v_i and u_i are vectors, whereas σ_i is a scalar. That is, if we consider an *input* in the direction v_i, then the *output* is in the direction u_i. Furthermore, since $\|v_i\|_2 = 1$ and $\|u_i\|_2 = 1$ we see that the i'th singular value σ_i gives directly the gain of the matrix G in this direction. In other words

$$\sigma_i(G) = \|Gv_i\|_2 = \frac{\|Gv_i\|_2}{\|v_i\|_2} \tag{3.39}$$

Some advantages of the SVD over the eigenvalue decomposition for analyzing gains and directionality of multivariable plants are:

1. The singular values give better information about the gains of the plant.

2. The plant directions obtained from the SVD are orthogonal.
3. The SVD also applies directly to non-square plants.

Maximum and minimum singular values. As already stated, it can be shown that
the largest gain for *any* input direction is equal to the maximum singular value

$$\bar{\sigma}(G) \equiv \sigma_1(G) = \max_{d \neq 0} \frac{\|Gd\|_2}{\|d\|_2} = \frac{\|Gv_1\|_2}{\|v_1\|_2} \tag{3.40}$$

and that the smallest gain for any input direction is equal to the minimum singular
value

$$\underline{\sigma}(G) \equiv \sigma_k(G) = \min_{d \neq 0} \frac{\|Gd\|_2}{\|d\|_2} = \frac{\|Gv_k\|_2}{\|v_k\|_2} \tag{3.41}$$

where $k = \min\{l, m\}$. Thus, for any vector d we have that

$$\underline{\sigma}(G) \leq \frac{\|Gd\|_2}{\|d\|_2} \leq \bar{\sigma}(G) \tag{3.42}$$

Define $u_1 = \bar{u}, v_1 = \bar{v}, u_k = \underline{u}$ and $v_k = \underline{v}$. Then it follows that

$$G\bar{v} = \bar{\sigma}\bar{u}, \qquad G\underline{v} = \underline{\sigma}\,\underline{u} \tag{3.43}$$

The vector $\bar{v}$ corresponds to the input direction with largest amplification, and $\bar{u}$ is the
corresponding output direction in which the inputs are most effective. The directions
involving $\bar{v}$ and $\bar{u}$ are sometimes referred to as the "strongest", "high-gain" or "most
important" directions. The next most important directions are associated with v_2 and
u_2, and so on (see Appendix A.3.5) until the "least important", "weak" or "low-gain"
directions which are associated with $\underline{v}$ and $\underline{u}$.

Example 3.4 *Consider again the system (3.27) in Example 3.3,*

$$G_1 = \begin{bmatrix} 5 & 4 \\ 3 & 2 \end{bmatrix} \tag{3.44}$$

The singular value decomposition of G_1 is

$$G_1 = \underbrace{\begin{bmatrix} 0.872 & 0.490 \\ 0.490 & -0.872 \end{bmatrix}}_{U} \underbrace{\begin{bmatrix} 7.343 & 0 \\ 0 & 0.272 \end{bmatrix}}_{\Sigma} \underbrace{\begin{bmatrix} 0.794 & -0.608 \\ 0.608 & 0.794 \end{bmatrix}^H}_{V^H}$$

The largest gain of 7.343 is for an input in the direction $\bar{v} = \begin{bmatrix} 0.794 \\ 0.608 \end{bmatrix}$, and the smallest gain of

0.272 is for an input in the direction $\underline{v} = \begin{bmatrix} -0.608 \\ 0.794 \end{bmatrix}$. This confirms the findings in Example 3.3.

Since in (3.44) both inputs affect both outputs, we say that the system is *interactive*.
This follows from the relatively large off-diagonal elements in G_1. Furthermore, the
system is *ill-conditioned*, that is, some combinations of the inputs have a strong effect
on the outputs, whereas other combinations have a weak effect on the outputs. This
may be quantified by the *condition number*; the ratio between the gains in the strong
and weak directions; which for the system in (3.44) is $\bar{\sigma}/\underline{\sigma} = 7.343/0.272 = 27.0$.

Example 3.5 Shopping cart. *Consider a shopping cart (supermarket trolley) with fixed wheels which we may want to move in three directions; forwards, sideways and upwards. This is a simple illustrative example where we can easily figure out the principal directions from experience. The strongest direction, corresponding to the largest singular value, will clearly be in the forwards direction. The next direction, corresponding to the second singular value, will be sideways. Finally, the most "difficult" or "weak" direction, corresponding to the smallest singular value, will be upwards (lifting up the cart).*

For the shopping cart the gain depends strongly on the input direction, i.e. the plant is ill-conditioned. Control of ill-conditioned plants is sometimes difficult, and the control problem associated with the shopping cart can be described as follows: Assume we want to push the shopping cart sideways (maybe we are blocking someone). This is rather difficult (the plant has low gain in this direction) so a strong force is needed. However, if there is any uncertainty in our knowledge about the direction the cart is pointing, then some of our applied force will be directed forwards (where the plant gain is large) and the cart will suddenly move forward with an undesired large speed. We thus see that the control of an ill-conditioned plant may be especially difficult if there is input uncertainty which can cause the input signal to "spread" from one input direction to another. We will discuss this in more detail later.

Example 3.6 Distillation process. *Consider the following steady-state model of a distillation column*

$$G = \begin{bmatrix} 87.8 & -86.4 \\ 108.2 & -109.6 \end{bmatrix} \tag{3.45}$$

The variables have been scaled as discussed in Section 1.4. Thus, since the elements are much larger than 1 in magnitude this suggests that there will be no problems with input constraints. However, this is somewhat misleading as the gain in the low-gain direction (corresponding to the smallest singular value) is actually only just above 1. To see this consider the SVD of G:

$$G = \underbrace{\begin{bmatrix} 0.625 & -0.781 \\ 0.781 & 0.625 \end{bmatrix}}_{U} \underbrace{\begin{bmatrix} 197.2 & 0 \\ 0 & 1.39 \end{bmatrix}}_{\Sigma} \underbrace{\begin{bmatrix} 0.707 & -0.708 \\ -0.708 & -0.707 \end{bmatrix}^H}_{V^H} \tag{3.46}$$

From the first input singular vector, $\bar{v} = \begin{bmatrix} 0.707 & -0.708 \end{bmatrix}^T$, we see that the gain is 197.2 when we increase one input and decrease the other input by a similar amount. On the other hand, from the second input singular vector, $\underline{v} = \begin{bmatrix} -0.708 & -0.707 \end{bmatrix}^T$, we see that if we increase both inputs by the same amount then the gain is only 1.39. The reason for this is that the plant is such that the two inputs counteract each other. Thus, the distillation process is ill-conditioned, at least at steady-state, and the condition number is $197.2/1.39 = 141.7$. The physics of this example is discussed in more detail below, and later in this chapter we will consider a simple controller design (see Motivating robustness example No. 2 in Section 3.7.2).

Example 3.7 Physics of the distillation process. *The model in (3.45) represents two-point (dual) composition control of a distillation column, where the top composition is to be controlled at $y_D = 0.99$ (output y_1) and the bottom composition at $x_B = 0.01$ (output y_2), using reflux L (input u_1) and boilup V (input u_2) as manipulated inputs (see Figure 10.6 on page 426). Note that we have here returned to the convention of using u_1 and u_2 to denote the manipulated inputs; the output singular vectors will be denoted by $\bar{u}$ and $\underline{u}$.*

The 1, 1-element of the gain matrix G is 87.8. Thus an increase in u_1 by 1 (with u_2 constant) yields a large steady-state change in y_1 of 87.8, that is, the outputs are very sensitive to changes *in u_1. Similarly, an increase in u_2 by 1 (with u_1 constant) yields $y_1 = -86.4$. Again, this is a very large change, but in the opposite direction of that for the increase in u_1. We therefore see that changes in u_1 and u_2 counteract each other, and if we increase u_1 and u_2 simultaneously by 1, then the overall steady-state change in y_1 is only $87.8 - 86.4 = 1.4$.*

Physically, the reason for this small change is that the compositions in the distillation column are only weakly dependent on changes in the internal flows *(i.e. simultaneous changes in the internal flows L and V). This can also be seen from the smallest singular value, $\underline{\sigma}(G) = 1.39$, which is obtained for inputs in the direction $\underline{v} = \begin{bmatrix} -0.708 \\ -0.707 \end{bmatrix}$. From the output singular vector $\underline{u} = \begin{bmatrix} -0.781 \\ 0.625 \end{bmatrix}$ we see that the effect is to move the outputs in different directions, that is, to change $y_1 - y_2$. Therefore, it takes a large control action to move the compositions in different directions, that is, to make both products purer simultaneously. This makes sense from a physical point of view.*

On the other hand, the distillation column is very sensitive to changes in external flows *(i.e. increase $u_1 - u_2 = L - V$). This can be seen from the input singular vector $\bar{v} = \begin{bmatrix} 0.707 \\ -0.708 \end{bmatrix}$ associated with the largest singular value, and is a general property of distillation columns where both products are of high purity. The reason for this is that the external distillate flow (which varies as $V - L$) has to be about equal to the amount of light component in the feed, and even a small imbalance leads to large changes in the product compositions.*

For dynamic systems the singular values and their associated directions vary with frequency, and for control purposes it is usually the frequency range corresponding to the closed-loop bandwidth which is of main interest. The singular values are usually plotted as a function of frequency in a Bode magnitude plot with a log-scale for frequency and magnitude. Typical plots are shown in Figure 3.6.

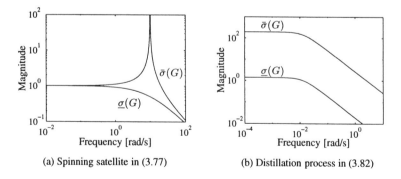

(a) Spinning satellite in (3.77)   (b) Distillation process in (3.82)

Figure 3.6: Typical plots of singular values

Non-Square plants

The SVD is also useful for non-square plants. For example, consider a plant with 2 inputs and 3 outputs. In this case the third output singular vector, u_3, tells us in which output direction the plant cannot be controlled. Similarly, for a plant with more inputs than outputs, the additional input singular vectors tell us in which directions the input will have no effect.

Exercise 3.6 *For a system with m inputs and 1 output, what is the interpretation of the singular values and the associated input directions (V)? What is U in this case?*

Use of the minimum singular value of the plant

The minimum singular value of the plant, $\underline{\sigma}(G(j\omega))$, evaluated as a function of frequency, is a useful measure for evaluating the feasibility of achieving acceptable control. If the inputs and outputs have been scaled as outlined in Section 1.4, then with a manipulated input of unit magnitude (measured by the 2-norm), we can achieve an output magnitude of at least $\underline{\sigma}(G)$ in *any* output direction. We generally want $\underline{\sigma}(G)$ as large as possible.

Remark. The requirement $\underline{\sigma}(G) > 1$, to avoid input saturation, is discussed in Section 6.9. In Section 10.3, it is shown that it may be desirable to have $\underline{\sigma}(G(j\omega))$ large even when input saturation is not a concern. The minimum singular value of the plant and its use is also discussed by Morari (1983), and Yu and Luyben (1986) call $\underline{\sigma}(G(j\omega))$ the "Morari resilience index".

3.3.5   Singular values for performance

So far we have used the SVD primarily to gain insight into the directionality of MIMO systems. But the maximum singular value is also very useful in terms of frequency-domain performance and robustness. We here consider performance.

For SISO systems we earlier found that $|S(j\omega)|$ evaluated as a function of frequency gives useful information about the effectiveness of feedback control. For example, it is the gain from a sinusoidal reference input (or output disturbance) to the control error, $|e(\omega)|/|r(\omega)| = |S(j\omega)|$.

For MIMO systems a useful generalization results if we consider the ratio $\|e(\omega)\|_2/\|r(\omega)\|_2$, where r is the vector of reference inputs, e is the vector of control errors, and $\|\cdot\|_2$ is the vector 2-norm. As explained above, this gain depends on the *direction* of $r(\omega)$ and we have from (3.42) that it is bounded by the maximum and minimum singular value of S,

$$\underline{\sigma}(S(j\omega)) \leq \frac{\|e(\omega)\|_2}{\|r(\omega)\|_2} \leq \bar{\sigma}(S(j\omega)) \tag{3.47}$$

In terms of *performance*, it is reasonable to require that the gain $\|e(\omega)\|_2/\|r(\omega)\|_2$ remains small for any direction of $r(\omega)$, including the "worst-case" direction which gives a gain of $\bar{\sigma}(S(j\omega))$. Let $1/|w_P(j\omega)|$ (the inverse of the performance weight) represent the maximum allowed magnitude of $\|e\|_2/\|r\|_2$ at each frequency. This results in the following performance requirement:

$$\bar{\sigma}(S(j\omega)) < 1/|w_P(j\omega)|, \ \forall\omega \quad \Leftrightarrow \quad \bar{\sigma}(w_P S) < 1, \ \forall\omega$$

$$\Leftrightarrow \quad \|w_P S\|_\infty < 1 \tag{3.48}$$

where the $\mathcal{H}_\infty$ norm (see also page 55) is defined as the peak of the maximum singular value of the frequency response

$$\|M(s)\|_\infty \triangleq \max_\omega \bar{\sigma}(M(j\omega)) \tag{3.49}$$

Typical performance weights $w_P(s)$ are given in Section 2.7.2, which should be studied carefully.

The singular values of $S(j\omega)$ may be plotted as functions of frequency, as illustrated later in Figure 3.10(a). Typically, they are small at low frequencies where feedback is effective, and they approach 1 at high frequencies because any real system is strictly proper:

$$\omega \to \infty: \quad L(j\omega) \to 0 \quad \Rightarrow \quad S(j\omega) \to I \tag{3.50}$$

The maximum singular value, $\bar{\sigma}(S(j\omega))$, usually has a peak larger than 1 around the crossover frequencies. This peak is undesirable, but it is unavoidable for real systems.

As for SISO systems we define the bandwidth as the frequency up to which feedback is effective. For MIMO systems the bandwidth will depend on directions, and we have a *bandwidth region* between a lower frequency where the maximum singular value, $\bar{\sigma}(S)$, reaches 0.7 (the low-gain or worst-case direction), and a higher frequency where the minimum singular value, $\underline{\sigma}(S)$, reaches 0.7 (the high-gain or best direction). If we want to associate a single bandwidth frequency for a multivariable system, then we consider the worst-case (low-gain) direction, and define

- *Bandwidth, ω_B*: Frequency where $\bar{\sigma}(S)$ crosses $\frac{1}{\sqrt{2}} = 0.7$ from below.

It is then understood that the bandwidth is at least ω_B for any direction of the input (reference or disturbance) signal. Since $S = (I + L)^{-1}$, (A.52) yields

$$\underline{\sigma}(L) - 1 \leq \frac{1}{\bar{\sigma}(S)} \leq \underline{\sigma}(L) + 1 \tag{3.51}$$

Thus at frequencies where feedback is effective (namely where $\underline{\sigma}(L) \gg 1$) we have $\bar{\sigma}(S) \approx 1/\underline{\sigma}(L)$, and at the bandwidth frequency (where $1/\bar{\sigma}(S(j\omega_B)) = \sqrt{2} = 1.41$) we have that $\underline{\sigma}(L(j\omega_B))$ is between 0.41 and 2.41. Thus, the bandwidth is approximately where $\underline{\sigma}(L)$ crosses 1. Finally, at higher frequencies where for any real system $\underline{\sigma}(L)$ (and $\bar{\sigma}(L)$) is small we have that $\bar{\sigma}(S) \approx 1$.

3.4 Control of multivariable plants

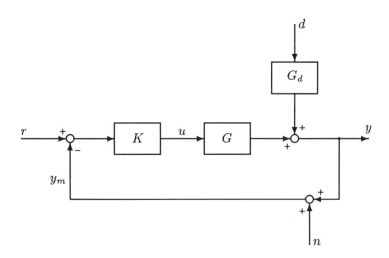

Figure 3.7: One degree-of-freedom feedback control configuration

Consider the simple feedback system in Figure 3.7. A conceptually simple approach to multivariable control is given by a two-step procedure in which we first design a "compensator" to deal with the interactions in G, and then design a *diagonal* controller using methods similar to those for SISO systems. This approach is discussed below.

The most common approach is to use a pre-compensator, $W_1(s)$, which counteracts the interactions in the plant and results in a "new" shaped plant:

$$G_s(s) = G(s)W_1(s) \qquad (3.52)$$

which is more diagonal and easier to control than the original plant $G(s)$. After finding a suitable $W_1(s)$ we can design a *diagonal* controller $K_s(s)$ for the shaped plant $G_s(s)$. The overall controller is then

$$K(s) = W_1(s)K_s(s) \qquad (3.53)$$

In many cases effective compensators may be derived on physical grounds and may include nonlinear elements such as ratios.

Remark 1 Some design approaches in this spirit are the Nyquist Array technique of Rosenbrock (1974) and the characteristic loci technique of MacFarlane and Kouvaritakis (1977).

Remark 2 The $\mathcal{H}_\infty$ loop-shaping design procedure, described in detail in Section 9.4, is similar in that a pre-compensator is first chosen to yield a shaped plant, $G_s = GW_1$, with desirable properties, and then a controller $K_s(s)$ is designed. The main difference is that in $\mathcal{H}_\infty$ loop shaping, $K_s(s)$ is a full multivariable controller, designed based on optimization (to optimize $\mathcal{H}_\infty$ robust stability).

3.4.1 Decoupling

Decoupling control results when the compensator is chosen such that G_s in (3.52) is diagonal at a selected frequency. The following different cases are possible:

1. *Dynamic decoupling:* $G_s(s)$ is diagonal (at all frequencies). For example, with $G_s(s) = I$ and a square plant, we get $W_1 = G^{-1}(s)$ (disregarding the possible problems involved in realizing $G^{-1}(s)$). If we then select $K_s(s) = l(s)I$ (e.g. with $l(s) = k/s$), the overall controller is

$$K(s) = K_{\text{inv}}(s) \triangleq l(s)G^{-1}(s) \tag{3.54}$$

We will later refer to (3.54) as an *inverse-based* controller. It results in a decoupled nominal system with identical loops, i.e. $L(s) = l(s)I$, $S(s) = \frac{1}{1+l(s)}I$ and $T(s) = \frac{l(s)}{1+l(s)}I$.

Remark. In some cases we may want to keep the diagonal elements in the shaped plant unchanged by selecting $W_1 = G^{-1}G_{diag}$. In other cases we may want the diagonal elements in W_1 to be 1. This may be obtained by selecting $W_1 = G^{-1}((G^{-1})_{diag})^{-1}$, and the off-diagonal elements of W_1 are then called "decoupling elements".

2. *Steady-state decoupling:* $G_s(0)$ is diagonal. This may be obtained by selecting a constant pre-compensator $W_1 = G^{-1}(0)$ (and for a non-square plant we may use the pseudo-inverse provided $G(0)$ has full row (output) rank).

3. *Approximate decoupling at frequency w_o:* $G_s(j\omega_o)$ is as diagonal as possible. This is usually obtained by choosing a constant pre-compensator $W_1 = G_o^{-1}$ where G_o is a real approximation of $G(j\omega_o)$. G_o may be obtained, for example, using the align algorithm of Kouvaritakis (1974). The bandwidth frequency is a good selection for ω_o because the effect on performance of reducing interaction is normally greatest at this frequency.

The idea of decoupling control is appealing, but there are several difficulties:

1. As one might expect, decoupling may be very sensitive to modelling errors and uncertainties. This is illustrated below in Section 3.7.2.
2. The requirement of decoupling and the use of an inverse-based controller may not be desirable for disturbance rejection. The reasons are similar to those given for SISO systems in Section 2.6.4, and are discussed further below; see (3.58).
3. If the plant has RHP-zeros then the requirement of decoupling generally introduces extra RHP-zeros into the closed-loop system (see Section 6.5.1).

Even though decoupling controllers may not always be desirable in practice, they are of interest from a theoretical point of view. They also yield insights into the limitations imposed by the multivariable interactions on achievable performance. One popular design method, which essentially yields a decoupling controller is the internal model control (IMC) approach (Morari and Zafiriou, 1989).

Another common strategy, which avoids most of the problems just mentioned, is to use *partial (one-way) decoupling* where $G_s(s)$ in (3.52) is upper or lower triangular.

3.4.2 Pre- and post-compensators and the SVD-controller

The above pre-compensator approach may be extended by introducing a post-compensator $W_2(s)$, as shown in Figure 3.8. One then designs a *diagonal* controller

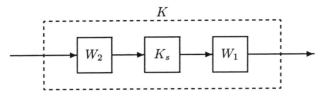

Figure 3.8: Pre- and post-compensators, W_1 and W_2. K_s is diagonal

K_s for the shaped plant W_2GW_1. The overall controller is then

$$K(s) = W_1 K_s W_2 \qquad (3.55)$$

The *SVD-controller* is a special case of a pre- and post-compensator design. Here

$$W_1 = V_o \quad \text{and} \quad W_2 = U_o^T \qquad (3.56)$$

where V_o and U_o are obtained from a singular value decomposition of $G_o = U_o \Sigma_o V_o^T$, where G_o is a real approximation of $G(j\omega_o)$ at a given frequency w_o (often around the bandwidth). SVD-controllers are studied by Hung and MacFarlane (1982), and by Hovd et al. (1994) who found that the SVD controller structure is optimal in some cases, e.g. for plants consisting of symmetrically interconnected subsystems.

In summary, the SVD-controller provides a useful class of controllers. By selecting $K_s = l(s)\Sigma_o^{-1}$ a decoupling design is achieved, and by selecting a diagonal K_s with a low condition number ($\gamma(K_s)$ small) generally results in a robust controller (see Section 6.10).

3.4.3 Diagonal controller (decentralized control)

Another simple approach to multivariable controller design is to use a diagonal or block-diagonal controller $K(s)$. This is often referred to as decentralized control. Clearly, this works well if $G(s)$ is close to diagonal, because then the plant to be controlled is essentially a collection of independent sub-plants, and each element in $K(s)$ may be designed independently. However, if off-diagonal elements in $G(s)$ are large, then the performance with decentralized diagonal control may be poor because no attempt is made to counteract the interactions.

3.4.4 What is the shape of the "best" feedback controller?

Consider the problem of disturbance rejection. The closed-loop disturbance response is $y = SG_d d$. Suppose we have scaled the system (see Section 1.4) such that at each frequency the disturbances are of magnitude 1, $\|d\|_2 \leq 1$, and our performance requirement is that $\|y\|_2 \leq 1$. This is equivalent to requiring $\bar{\sigma}(SG_d) \leq 1$. In many cases there is a trade-off between input usage and performance, such that the controller that minimizes the input magnitude is one that yields all singular values of SG_d equal to 1, i.e. $\sigma_i(SG_d) = 1, \forall \omega$. This corresponds to

$$S_{\min}G_d = U_1 \tag{3.57}$$

where $U_1(s)$ is some all-pass transfer function (which at each frequency has all its singular values equal to 1). The subscript min refers to the use of the smallest loop gain that satisfies the performance objective. For simplicity, we assume that G_d is square so $U_1(j\omega)$ is a unitary matrix. At frequencies where feedback is effective we have $S = (I + L)^{-1} \approx L^{-1}$, and (3.57) yields $L_{\min} = GK_{\min} \approx G_dU_1^{-1}$. In conclusion, the controller and loop shape with the minimum gain will often look like

$$K_{\min} \approx G^{-1}G_dU_2, \quad L_{\min} \approx G_dU_2 \tag{3.58}$$

where $U_2 = U_1^{-1}$ is some all-pass transfer function matrix. This provides a generalization of $|K_{\min}| \approx |G^{-1}G_d|$ which was derived in (2.58) for SISO systems, and the summary following (2.58) on page 48 therefore also applies to MIMO systems. For example, we see that for disturbances entering at the plant inputs, $G_d = G$, we get $K_{\min} = U_2$, so a simple constant unit gain controller yields a good trade-off between output performance and input usage. We also note with interest that it is generally not possible to select a unitary matrix U_2 such that $L_{\min} = G_dU_2$ is diagonal, so a decoupling design is generally not optimal for disturbance rejection. These insights can be used as a basis for a loop-shaping design; see more on $\mathcal{H}_\infty$ loop-shaping in Chapter 9.

3.4.5   Multivariable controller synthesis

The above design methods are based on a two-step procedure in which we first design a pre-compensator (for decoupling control) or we make an input-output pairing selection (for decentralized control) and then we design a diagonal controller $K_s(s)$. Invariably this two-step procedure results in a suboptimal design.

The alternative is to synthesize directly a multivariable controller $K(s)$ based on minimizing some objective function (norm). We here use the word *synthesize* rather than *design* to stress that this is a more formalized approach. Optimization in controller design became prominent in the 1960's with "optimal control theory" based on minimizing the expected value of the output variance in the face of stochastic disturbances. Later, other approaches and norms were introduced, such as $\mathcal{H}_\infty$ optimal control.

3.4.6 Summary of mixed-sensitivity $\mathcal{H}_\infty$ design (S/KS)

We here provide a brief summary of the S/KS and other mixed-sensitivity $\mathcal{H}_\infty$ design methods which are used in later examples. In the S/KS problem, the objective is to minimize the $\mathcal{H}_\infty$ norm of

$$N = \begin{bmatrix} W_P S \\ W_u KS \end{bmatrix} \tag{3.59}$$

This problem was discussed earlier for SISO systems, and another look at Section 2.7.3 would be useful now. A sample MATLAB file is provided in Example 2.11, page 60.

The following issues and guidelines are relevant when selecting the weights W_P and W_u:

1. KS is the transfer function from r to u in Figure 3.7, so for a system which has been scaled as in Section 1.4, a reasonable initial choice for the input weight is $W_u = I$.
2. S is the transfer function from r to $-e = r - y$. A common choice for the performance weight is $W_P = \text{diag}\{w_{Pi}\}$ with

$$w_{Pi} = \frac{s/M_i + \omega_{Bi}^*}{s + \omega_{Bi}^* A_i}, \quad A_i \ll 1 \tag{3.60}$$

(see also Figure 2.26 on page 58). Selecting $A_i \ll 1$ ensures approximate integral action with $S(0) \approx 0$. Often we select M_i about 2 for all outputs, whereas ω_{Bi}^* may be different for each output. A large value of ω_{Bi}^* yields a faster response for output i.
3. To find a reasonable initial choice for the weight W_P, one can first obtain a controller with some other design method, plot the magnitude of the resulting diagonal elements of S as a function of frequency, and select $w_{Pi}(s)$ as a rational approximation of $1/|S_{ii}|$.
4. For disturbance rejection, we may in some cases want a steeper slope for $w_{Pi}(s)$ at low frequencies than that given in (3.60), e.g. as see the weight in (2.73). However, it may be better to consider the disturbances explicitly by considering the $\mathcal{H}_\infty$ norm of

$$N = \begin{bmatrix} W_P S & W_P S G_d \\ W_u KS & W_u KS G_d \end{bmatrix} \tag{3.61}$$

or equivalently

$$N = \begin{bmatrix} W_P S W_d \\ W_u KS W_d \end{bmatrix} \quad \text{with } W_d = [\, I \quad G_d \,] \tag{3.62}$$

where N represents the transfer function from $\begin{bmatrix} r \\ d \end{bmatrix}$ to the weighted outputs $\begin{bmatrix} W_P e \\ W_u u \end{bmatrix}$. In some situations we may want to adjust W_P or G_d in order to satisfy better our original objectives. The helicopter case study in Section 12.2 illustrates this by introducing a scalar parameter α to adjust the magnitude of G_d.

5. T is the transfer function from $-n$ to y. To reduce sensitivity to noise and uncertainty, we want T small at high frequencies, and so we may want additional roll-off in L. This can be achieved in several ways. One approach is to add $W_T T$ to the stack for N in (3.59), where $W_T = \text{diag}\{w_{Ti}\}$ and $|w_{Ti}|$ is smaller than 1 at low frequencies and large at high frequencies. A more direct approach is to add high-frequency dynamics, $W_1(s)$, to the plant model to ensure that the resulting shaped plant, $G_s = GW_1$, rolls off with the desired slope. We then obtain an $\mathcal{H}_\infty$ optimal controller K_s for this shaped plant, and finally include $W_1(s)$ in the controller, $K = W_1 K_s$.

More details about $\mathcal{H}_\infty$ design are given in Chapter 9.

3.5   Introduction to multivariable RHP-zeros

By means of an example, we now give the reader an appreciation of the fact that MIMO systems have zeros even though their presence may not be obvious from the elements of $G(s)$. As for SISO systems, we find that RHP-zeros impose fundamental limitations on control.

The zeros z of MIMO systems are defined as the values $s = z$ where $G(s)$ loses rank, and we can find the *direction* of a zero by looking at the direction in which the matrix $G(z)$ has zero gain. For square systems we essentially have that the poles and zeros of $G(s)$ are the poles and zeros of $\det G(s)$. However, this crude method may fail in some cases, as it may incorrectly cancel poles and zeros with the same location but different directions (see Sections 4.5 and 4.6.1 for more details).

Example 3.8 *Consider the following plant*

$$G(s) = \frac{1}{(0.2s+1)(s+1)} \begin{bmatrix} 1 & 1 \\ 1+2s & 2 \end{bmatrix} \tag{3.63}$$

The responses to a step in each individual input are shown in Figure 3.9(a) and (b). We see that the plant is interactive, but for these two inputs there is no inverse response to indicate the presence of a RHP-zero. Nevertheless, the plant does have a multivariable RHP-zero at $z = 0.5$, that is, $G(s)$ loses rank at $s = 0.5$, and $\det G(0.5) = 0$. The singular value decomposition of $G(0.5)$ is

$$G(0.5) = \frac{1}{1.65} \begin{bmatrix} 1 & 1 \\ 2 & 2 \end{bmatrix} = \underbrace{\begin{bmatrix} 0.45 & 0.89 \\ 0.89 & -0.45 \end{bmatrix}}_{U} \underbrace{\begin{bmatrix} 1.92 & 0 \\ 0 & 0 \end{bmatrix}}_{\Sigma} \underbrace{\begin{bmatrix} 0.71 & -0.71 \\ 0.71 & 0.71 \end{bmatrix}^H}_{V^H} \tag{3.64}$$

and we have as expected $\underline{\sigma}(G(0.5)) = 0$. The input and output directions corresponding to the RHP-zero are $\underline{v} = \begin{bmatrix} -0.71 \\ 0.71 \end{bmatrix}$ and $\underline{u} = \begin{bmatrix} 0.89 \\ -0.45 \end{bmatrix}$. Thus, the RHP-zero is associated with

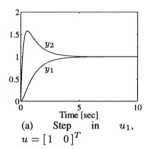 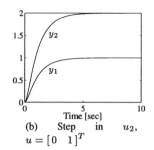 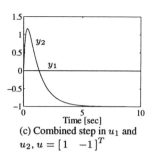

(a)   Step   in u_1,   (b)   Step   in u_2,   (c) Combined step in u_1 and
$u = \begin{bmatrix} 1 & 0 \end{bmatrix}^T$ $u = \begin{bmatrix} 0 & 1 \end{bmatrix}^T$ $u_2, u = \begin{bmatrix} 1 & -1 \end{bmatrix}^T$

Figure 3.9: Open-loop response for $G(s)$ in (3.63)

both inputs and with both outputs. The presence of the multivariable RHP-zero is also observed from the time response in Figure 3.9(c), which is for a simultaneous input change in opposite directions, $u = \begin{bmatrix} 1 \\ -1 \end{bmatrix}^T$. We see that y_2 displays an inverse response whereas y_1 happens to remain at zero for this particular input change.

To see how the RHP-zero affects the closed-loop response, we design a controller which minimizes the $\mathcal{H}_\infty$ norm of the weighted S/KS matrix

$$N = \begin{bmatrix} W_P S \\ W_u KS \end{bmatrix} \qquad (3.65)$$

with weights

$$W_u = I, \ W_P = \begin{bmatrix} w_{P1} & 0 \\ 0 & w_{P2} \end{bmatrix}, w_{Pi} = \frac{s/M_i + \omega^*_{Bi}}{s + w^*_{Bi} A_i}, \quad A_i = 10^{-4} \qquad (3.66)$$

The MATLAB file for the design is the same as in Table 2.3 on page 60, except that we now have a 2 × 2 system. Since there is a RHP-zero at $z = 0.5$ we expect that this will somehow limit the bandwidth of the closed-loop system.

Design 1. *We weight the two outputs equally and select*

Design 1 : $M_1 = M_2 = 1.5;$ $\omega^*_{B1} = \omega^*_{B2} = z/2 = 0.25$

This yields an $\mathcal{H}_\infty$ norm for N of 2.80 and the resulting singular values of S are shown by the solid lines in Figure 3.10(a). The closed-loop response to a reference change $r = \begin{bmatrix} 1 & -1 \end{bmatrix}^T$ is shown by the solid lines in Figure 3.10(b). We note that both outputs behave rather poorly and both display an inverse response.

Design 2. *For MIMO plants, one can often move most of the deteriorating effect (e.g. inverse response) of a RHP-zero to a particular output channel. To illustrate this, we change the weight w_{P2} so that more emphasis is placed on output 2. We do this by increasing the bandwidth requirement in output channel 2 by a factor of 100:*

Design 2 : $M_1 = M_2 = 1.5;$ $\omega^*_{B1} = 0.25, \ \omega^*_{B2} = 25$

This yields an $\mathcal{H}_\infty$ norm for N of 2.92. In this case we see from the dashed line in Figure 3.10(b) that the response for output 2 (y_2) is excellent with no inverse response.

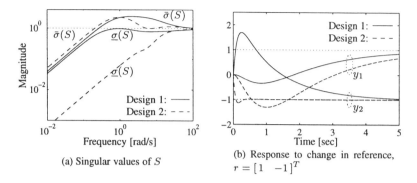

(a) Singular values of S

(b) Response to change in reference,
$r = \begin{bmatrix} 1 & -1 \end{bmatrix}^T$

Figure 3.10: Alternative designs for 2×2 plant (3.63) with RHP-zero

However, this comes at the expense of output 1 (y_1) where the response is somewhat poorer than for Design 1.

Design 3. *We can also interchange the weights w_{P1} and w_{P2} to stress output 1 rather than output 2. In this case (not shown) we get an excellent response in output 1 with no inverse response, but output 2 responds very poorly (much poorer than output 1 for Design 2). Furthermore, the $\mathcal{H}_\infty$ norm for N is 6.73, whereas it was only 2.92 for Design 2.*

Thus, we see that it is easier, for this example, to get tight control of output 2 than of output 1. This may be expected from the output direction of the RHP-zero, $\underline{u} = \begin{bmatrix} 0.89 \\ -0.45 \end{bmatrix}$, which is mostly in the direction of output 1. We will discuss this in more detail in Section 6.5.1.

Remark 1 We find from this example that we can direct the effect of the RHP-zero to either of the two outputs. This is typical of multivariable RHP-zeros, but there are cases where the RHP-zero is associated with a particular output channel and it is *not* possible to move its effect to another channel. The zero is then called a "pinned zero" (see Section 4.6.2).

Remark 2 It is observed from the plot of the singular values in Figure 3.10(a), that we were able to obtain by Design 2 a very large improvement in the "good" direction (corresponding to $\underline{\sigma}(S)$) at the expense of only a minor deterioration in the "bad" direction (corresponding to $\bar{\sigma}(S)$). Thus Design 1 demonstrates a shortcoming of the $\mathcal{H}_\infty$ norm: only the worst direction (maximum singular value) contributes to the $\mathcal{H}_\infty$ norm and it may not always be easy to get a good trade-off between the various directions.

3.6   Condition number and RGA

Two measures which are used to quantify the degree of directionality and The level of (two-way) interactions in MIMO systems, are the condition number and the relative gain array (RGA), respectively. We here define the two measures and present an overview of their *practical use*. We do not give detailed proofs, but refer to other places in the book for further details.

3.6.1 Condition number

We define the *condition number* of a matrix as the ratio between the maximum and minimum singular values,

$$\gamma(G) \triangleq \bar{\sigma}(G)/\underline{\sigma}(G) \tag{3.67}$$

A matrix with a large condition number is said to be *ill-conditioned*. For a non-singular (square) matrix $\underline{\sigma}(G) = 1/\bar{\sigma}(G^{-1})$, so $\gamma(G) = \bar{\sigma}(G)\bar{\sigma}(G^{-1})$. It then follows from (A.119) that the condition number is large if both G and G^{-1} have large elements.

The condition number depends strongly on the scaling of the inputs and outputs. To be more specific, if D_1 and D_2 are diagonal scaling matrices, then the condition numbers of the matrices G and $D_1 G D_2$ may be arbitrarily far apart. In general, the matrix G should be scaled on physical grounds, for example, by dividing each input and output by its largest expected or desired value as discussed in Section 1.4.

One might also consider minimizing the condition number over all possible scalings. This results in the *minimized or optimal condition number* which is defined by

$$\gamma^*(G) = \min_{D_1, D_2} \gamma(D_1 G D_2) \tag{3.68}$$

and can be computed using (A.73).

The condition number has been used as an input-output controllability measure, and in particular it has been postulated that a large condition number indicates sensitivity to uncertainty. This is not true in general, but the reverse holds; if the condition number is small, then the multivariable effects of uncertainty are not likely to be serious (see (6.72)).

If the condition number is large (say, larger than 10), then this may *indicate* control problems:

1. A large condition number $\gamma(G) = \bar{\sigma}(G)/\underline{\sigma}(G)$ may be caused by a small value of $\underline{\sigma}(G)$, which is generally undesirable (on the other hand, a large value of $\bar{\sigma}(G)$ need not necessarily be a problem).
2. A large condition number may mean that the plant has a large minimized condition number, or equivalently, it has large RGA-elements which indicate fundamental control problems; see below.
3. A large condition number *does* imply that the system is sensitive to "unstructured" (full-block) input uncertainty (e.g. with an inverse-based controller, see (8.135)), but this kind of uncertainty often does not occur in practice. We therefore *cannot* generally conclude that a plant with a large condition number is sensitive to uncertainty, e.g. see the diagonal plant in Example 3.9.

3.6.2 Relative Gain Array (RGA)

The relative gain array (RGA) of a non-singular square matrix G is a square matrix defined as

$$\text{RGA}(G) = \Lambda(G) \triangleq G \times (G^{-1})^T \qquad (3.69)$$

where $\times$ denotes element-by-element multiplication (the Hadamard or Schur product). For a 2×2 matrix with elements g_{ij} the RGA is

$$\Lambda(G) = \begin{bmatrix} \lambda_{11} & \lambda_{12} \\ \lambda_{21} & \lambda_{22} \end{bmatrix} = \begin{bmatrix} \lambda_{11} & 1 - \lambda_{11} \\ 1 - \lambda_{11} & \lambda_{11} \end{bmatrix}; \quad \lambda_{11} = \frac{1}{1 - \frac{g_{12}g_{21}}{g_{11}g_{22}}} \qquad (3.70)$$

Bristol (1966) originally introduced the RGA as a steady-state measure of interactions for decentralized control. Unfortunately, based on the original definition, many people have dismissed the RGA as being "only meaningful at $\omega = 0$". To the contrary, in most cases it is the value of the RGA at frequencies close to crossover which is most important.

The RGA has a number of interesting *algebraic properties*, of which the most important are (see Appendix A.4 for more details):

1. It is independent of input and output scaling.
2. Its rows and columns sum to one.
3. The sum-norm of the RGA, $\|\Lambda\|_{\text{sum}}$, is very close to the minimized condition number γ^*; see (A.78). This means that plants with large RGA-elements are always ill-conditioned (with a large value of $\gamma(G)$), but the reverse may not hold (i.e. a plant with a large $\gamma(G)$ may have small RGA-elements).
4. A relative change in an element of G equal to the negative inverse of its corresponding RGA-element yields singularity.
5. The RGA is the identity matrix if G is upper or lower triangular.

From the last property it follows that the RGA (or more precisely $\Lambda - I$) provides a measure of *two-way interaction*. The definition of the RGA may be generalized to non-square matrices by using the pseudo inverse; see Appendix A.4.2.

In addition to the algebraic properties listed above, the RGA has a surprising number of useful *control properties*:

1. The RGA is a good indicator of sensitivity to uncertainty:

    (a) *Uncertainty in the input channels (diagonal input uncertainty)*. Plants with large RGA-elements around the crossover frequency are fundamentally difficult to control because of sensitivity to input uncertainty (e.g. caused by uncertain or neglected actuator dynamics). In particular, decouplers or other inverse-based controllers should not be used for plants with large RGA-elements (see page 244).

    (b) *Element uncertainty*. As implied by algebraic property no. 4 above, large RGA-elements imply sensitivity to element-by-element uncertainty. However, this kind of uncertainty may not occur in practice due to physical couplings between the transfer function elements. Therefore, diagonal input uncertainty (which is always present) is usually of more concern for plants with large RGA-elements.

2. *RGA and RHP-zeros.* If the sign of an RGA-element changes from $s = 0$ to $s = \infty$, then there is a RHP-zero in G or in some subsystem of G (see Theorem 10.5).
3. *Non-square plants.* Extra inputs: If the sum of the elements in a column of RGA is small ($\ll 1$), then one may consider deleting the corresponding input. Extra outputs: If all elements in a row of RGA are small ($\ll 1$), then the corresponding output cannot be controlled (see Section 10.4).
4. *Diagonal dominance.* The RGA can be used to measure diagonal dominance, by the simple quantity

$$\text{RGA-number} = \|\Lambda(G) - I\|_{\text{sum}} \qquad (3.71)$$

For decentralized control we prefer pairings for which the RGA-number at crossover frequencies is close to 1 (see pairing rule 1 on page 435). Similarly, for certain multivariable design methods, shaping, it is simpler to choose the weights and shape the plant if we first rearrange the inputs and outputs to make the plant diagonally dominant with a small RGA-number.

5. *RGA and decentralized control.*

   (a) *Integrity:* For stable plants avoid input-output pairing on negative steady-state RGA-elements. Otherwise, if the sub-controllers are designed independently each with integral action, then the interactions will cause instability either when all of the loops are closed, or when the loop corresponding to the negative relative gain becomes inactive (e.g. because of saturation) (see Theorem 10.4 page 439). Interestingly, this is the only use of the RGA directly related to Bristol's original definition.
   (b) *Stability:* Prefer pairings corresponding to an RGA-number close to 0 at crossover frequencies (see page 435).

Remark. An iterative evaluation of the RGA, $\Lambda^2(G) = \Lambda(\Lambda(G))$ etc., has in applications proved to be useful for choosing pairings for large systems. Wolff (1994) found numerically that

$$\Lambda^\infty \triangleq \lim_{k \to \infty} \Lambda^k(G) \qquad (3.72)$$

is a permuted identity matrix (with the exception of "borderline" cases, the result is proved for a positive definite Hermitian matrix G by Johnson and Shapiro (1986)). Typically, Λ^k approaches Λ^∞ for k between 4 and 8. This permuted identity matrix may then be used as a candidate pairing choice. For example, for $G = \begin{bmatrix} 1 & 2 \\ -1 & 1 \end{bmatrix}$ we get $\Lambda = \begin{bmatrix} 0.33 & 0.67 \\ 0.67 & 0.33 \end{bmatrix}$, $\Lambda^2 = \begin{bmatrix} -0.33 & 1.33 \\ 1.33 & -0.33 \end{bmatrix}$, $\Lambda^3 = \begin{bmatrix} -0.07 & 1.07 \\ 1.07 & -0.07 \end{bmatrix}$ and $\Lambda^4 = \begin{bmatrix} 0.00 & 1.00 \\ 1.00 & 0.00 \end{bmatrix}$, which indicates that the off-diagonal pairing should be considered. Note that Λ^∞ may sometimes "recommend" a pairing on negative RGA-elements, even if a positive pairing is possible.

Example 3.9 *Consider a diagonal plant and compute the RGA and condition number,*

$$G = \begin{bmatrix} 100 & 0 \\ 0 & 1 \end{bmatrix}, \; \Lambda(G) = I, \; \gamma(G) = \frac{\bar{\sigma}(G)}{\underline{\sigma}(G)} = \frac{100}{1} = 100, \; \gamma^*(G) = 1 \qquad (3.73)$$

Here the condition number is large which means that the plant gain depends strongly on the input direction. However, since the plant is diagonal there are no interactions so $\Lambda(G) = I$ and $\gamma^*(G) = 1$, and no sensitivity to uncertainty (or other control problems) is normally expected.

Remark. *An exception would be if there was uncertainty caused by unmodelled or neglected off-diagonal elements in G. This would couple the high-gain and low-gain directions, and the large condition number implies sensitivity to this off-diagonal ("unstructured") uncertainty.*

Example 3.10 *Consider a triangular plant G for which we get*

$$G = \begin{bmatrix} 1 & 2 \\ 0 & 1 \end{bmatrix}, \; G^{-1} = \begin{bmatrix} 1 & -2 \\ 0 & 1 \end{bmatrix}, \; \Lambda(G) = I, \; \gamma(G) = \frac{2.41}{0.41} = 5.83, \; \gamma^*(G) = 1 \; (3.74)$$

Note that for a triangular matrix, the RGA is always the identity matrix and $\gamma^*(G)$ is always 1.

Example 3.11 *Consider again the distillation process for which we have at steady-state*

$$G = \begin{bmatrix} 87.8 & -86.4 \\ 108.2 & -109.6 \end{bmatrix}, \; G^{-1} = \begin{bmatrix} 0.399 & -0.315 \\ 0.394 & -0.320 \end{bmatrix}, \; \Lambda(G) = \begin{bmatrix} 35.1 & -34.1 \\ -34.1 & 35.1 \end{bmatrix}$$
$$(3.75)$$

In this case $\gamma(G) = 197.2/1.391 = 141.7$ is only slightly larger than $\gamma^*(G) = 138.268$. The magnitude sum of the elements in the RGA-matrix is $\|\Lambda\|_{sum} = 138.275$. This confirms (A.79) which states that, for 2×2 systems, $\|\Lambda(G)\|_{sum} \approx \gamma^*(G)$ when $\gamma^*(G)$ is large. The condition number is large, but since the minimum singular value $\underline{\sigma}(G) = 1.391$ is larger than 1 this does not by itself imply a control problem. However, the large RGA-elements indicate control problems, and fundamental control problems are expected if analysis shows that $G(j\omega)$ has large RGA-elements also in the crossover frequency range. (Indeed, the idealized dynamic model (3.82) used below has large RGA-elements at all frequencies, and we will confirm in simulations that there is a strong sensitivity to input channel uncertainty with an inverse-based controller).

Example 3.12 *Consider a 3×3 plant for which we have*

$$G = \begin{bmatrix} 16.8 & 30.5 & 4.30 \\ -16.7 & 31.0 & -1.41 \\ 1.27 & 54.1 & 5.40 \end{bmatrix}, \; \Lambda(G) = \begin{bmatrix} 1.50 & 0.99 & -1.48 \\ -0.41 & 0.97 & 0.45 \\ -0.08 & -0.95 & 2.03 \end{bmatrix} \qquad (3.76)$$

and $\gamma = 69.6/1.63 = 42.6$ and $\gamma^* = 7.80$. The magnitude sum of the elements in the RGA is $\|\Lambda\|_{sum} = 8.86$ which is close to γ^* as expected from (A.78). Note that the rows and the columns of Λ sum to 1. Since $\underline{\sigma}(G)$ is larger than 1 and the RGA-elements are relatively small, this steady-state analysis does not indicate any particular control problems for the plant.

Remark. *The plant in (3.76) represents the steady-state model of a fluid catalytic cracking (FCC) process. A dynamic model of the FCC process in (3.76) is given in Exercise 6.16.*

For a detailed analysis of achievable performance of the plant (input-output controllability analysis), one must also consider the singular values, RGA and condition number as functions of frequency. In particular, the crossover frequency range is important. In addition, disturbances and the presence of unstable (RHP) plant poles and zeros must be considered. All these issues are discussed in much more detail in Chapters 5 and 6 where we discuss achievable performance and input-output controllability analysis for SISO and MIMO plants, respectively.

3.7   Introduction to MIMO robustness

To motivate the need for a deeper understanding of robustness, we present two examples which illustrate that MIMO systems can display a sensitivity to uncertainty not found in SISO systems. We focus our attention on diagonal input uncertainty, which is present in any real system and often limits achievable performance because it enters between the controller and the plant.

3.7.1   Motivating robustness example no. 1: Spinning Satellite

Consider the following plant (Doyle, 1986; Packard et al., 1993) which can itself be motivated by considering the angular velocity control of a satellite spinning about one of its principal axes:

$$G(s) = \frac{1}{s^2 + a^2} \begin{bmatrix} s - a^2 & a(s+1) \\ -a(s+1) & s - a^2 \end{bmatrix}; \quad a = 10 \qquad (3.77)$$

A minimal, state-space realization, $G = C(sI - A)^{-1}B + D$, is

$$\left[\begin{array}{c|c} A & B \\ \hline C & D \end{array} \right] = \left[\begin{array}{cc|cc} 0 & a & 1 & 0 \\ -a & 0 & 0 & 1 \\ \hline 1 & a & 0 & 0 \\ -a & 1 & 0 & 0 \end{array} \right] \qquad (3.78)$$

The plant has a pair of $j\omega$-axis poles at $s = \pm ja$ so it needs to be stabilized. Let us apply negative feedback and try the simple diagonal constant controller

$$K = I$$

The complementary sensitivity function is

$$T(s) = GK(I + GK)^{-1} = \frac{1}{s+1} \begin{bmatrix} 1 & a \\ -a & 1 \end{bmatrix} \qquad (3.79)$$

Nominal stability (NS). The closed-loop system has two poles at $s = -1$ and so it is stable. This can be verified by evaluating the closed-loop state matrix

$$A_{cl} = A - BKC = \begin{bmatrix} 0 & a \\ -a & 0 \end{bmatrix} - \begin{bmatrix} 1 & a \\ -a & 1 \end{bmatrix} = \begin{bmatrix} -1 & 0 \\ 0 & -1 \end{bmatrix}$$

(To derive A_{cl} use $\dot{x} = Ax + Bu$, $y = Cx$ and $u = -Ky$).

Nominal performance (NP). The singular values of $L = GK = G$ are shown in Figure 3.6(a), page 76. We see that $\underline{\sigma}(L) = 1$ at low frequencies and starts dropping off at about $\omega = 10$. Since $\underline{\sigma}(L)$ never exceeds 1, we do not have tight control in the low-gain direction for this plant (recall the discussion following (3.51)), so we expect poor closed-loop performance. This is confirmed by considering S and T. For example, at steady-state $\bar{\sigma}(T) = 10.05$ and $\bar{\sigma}(S) = 10$. Furthermore, the large off-diagonal elements in $T(s)$ in (3.79) show that we have strong interactions in the closed-loop system. (For reference tracking, however, this may be counteracted by use of a two degrees-of-freedom controller).

Robust stability (RS). Now let us consider stability robustness. In order to determine stability margins with respect to perturbations in each input channel, one may consider Figure 3.11 where we have broken the loop at the first input. The loop transfer function at this point (the transfer function from w_1 to z_1) is $L_1(s) = 1/s$ (which can be derived from $t_{11}(s) = \frac{1}{1+s} = \frac{L_1(s)}{1+L_1(s)}$). This corresponds to an infinite gain margin and a phase margin of $90°$. On breaking the loop at the second input we get the same result. This suggests good robustness properties irrespective of the value of a. However, the design is far from robust as a further analysis shows. Consider

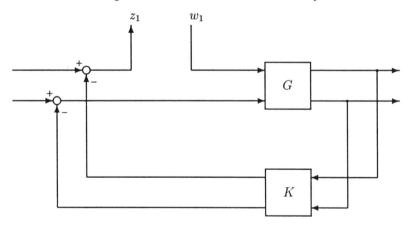

Figure 3.11: Checking stability margins "one-loop-at-a-time"

input gain uncertainty, and let ϵ_1 and ϵ_2 denote the relative error in the gain in each input channel. Then

$$u_1' = (1 + \epsilon_1)u_1, \quad u_2' = (1 + \epsilon_2)u_2 \tag{3.80}$$

where u_1' and u_2' are the actual changes in the manipulated inputs, while u_1 and u_2 are the desired changes as computed by the controller. It is important to stress that this diagonal input uncertainty, which stems from our inability to know the exact values of the manipulated inputs, is *always* present. In terms of a state-space description, (3.80) may be represented by replacing B by

$$B' = \begin{bmatrix} 1 + \epsilon_1 & 0 \\ 0 & 1 + \epsilon_2 \end{bmatrix}$$

The corresponding closed-loop state matrix is

$$A'_{cl} = A - B'KC = \begin{bmatrix} 0 & a \\ -a & 0 \end{bmatrix} - \begin{bmatrix} 1 + \epsilon_1 & 0 \\ 0 & 1 + \epsilon_2 \end{bmatrix} \begin{bmatrix} 1 & a \\ -a & 1 \end{bmatrix}$$

which has a characteristic polynomial given by

$$\det(sI - A'_{cl}) = s^2 + \underbrace{(2 + \epsilon_1 + \epsilon_2)}_{a_1} s + \underbrace{1 + \epsilon_1 + \epsilon_2 + (a^2 + 1)\epsilon_1\epsilon_2}_{a_0} \quad (3.81)$$

The perturbed system is stable if and only if both the coefficients a_0 and a_1 are positive. We therefore see that *the system is always stable if we consider uncertainty in only one channel at a time* (at least as long as the channel gain is positive). More precisely, we have stability for $(-1 < \epsilon_1 < \infty, \epsilon_2 = 0)$ and $(\epsilon_1 = 0, -1 < \epsilon_2 < \infty)$. This confirms the infinite gain margin seen earlier. However, the system can only tolerate *small simultaneous changes* in the two channels. For example, let $\epsilon_1 = -\epsilon_2$, then the system is unstable ($a_0 < 0$) for

$$|\epsilon_1| > \frac{1}{\sqrt{a^2 + 1}} \approx 0.1$$

In summary, we have found that checking single-loop margins is inadequate for MIMO problems. We have also observed that large values of $\bar{\sigma}(T)$ or $\bar{\sigma}(S)$ indicate robustness problems. We will return to this in Chapter 8, where we show that with input uncertainty of magnitude $|\epsilon_i| < 1/\bar{\sigma}(T)$, we are guaranteed robust stability (even for "full-block complex perturbations").

In the next example we find that there can be sensitivity to diagonal input uncertainty even in cases where $\bar{\sigma}(T)$ and $\bar{\sigma}(S)$ have no large peaks. This can not happen for a diagonal controller, see (6.77), but it will happen if we use an inverse-based controller for a plant with large RGA-elements, see (6.78).

3.7.2 Motivating robustness example no. 2: Distillation Process

The following is an idealized dynamic model of a distillation column,

$$G(s) = \frac{1}{75s + 1} \begin{bmatrix} 87.8 & -86.4 \\ 108.2 & -109.6 \end{bmatrix} \quad (3.82)$$

(time is in minutes). The physics of this example was discussed in Example 3.7. The plant is ill-conditioned with condition number $\gamma(G) = 141.7$ at all frequencies. The plant is also strongly two-way interactive and the RGA-matrix at all frequencies is

$$\text{RGA}(G) = \begin{bmatrix} 35.1 & -34.1 \\ -34.1 & 35.1 \end{bmatrix} \qquad (3.83)$$

The large elements in this matrix indicate that this process is fundamentally difficult to control.

Remark. (3.82) is admittedly a very crude model of a real distillation column; there should be a high-order lag in the transfer function from input 1 to output 2 to represent the liquid flow down to the column, and higher-order composition dynamics should also be included. Nevertheless, the model is simple and displays important features of distillation column behaviour. It should be noted that with a more detailed model, the RGA-elements would approach 1 at frequencies around 1 rad/min, indicating less of a control problem.

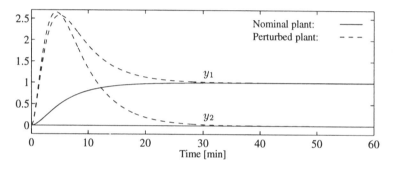

Figure 3.12: Response with decoupling controller to filtered reference input $r_1 = 1/(5s + 1)$. The perturbed plant has 20% gain uncertainty as given by (3.86).

We consider the following inverse-based controller, which may also be looked upon as a steady-state decoupler with a PI controller:

$$K_{\text{inv}}(s) = \frac{k_1}{s}G^{-1}(s) = \frac{k_1(1 + 75s)}{s} \begin{bmatrix} 0.3994 & -0.3149 \\ 0.3943 & -0.3200 \end{bmatrix}, \quad k_1 = 0.7 \quad (3.84)$$

Nominal performance (NP). We have $GK_{\text{inv}} = K_{\text{inv}}G = \frac{0.7}{s}I$. With no model error this controller should counteract all the interactions in the plant and give rise to two decoupled first-order responses each with a time constant of $1/0.7 = 1.43$ min. This is confirmed by the solid line in Figure 3.12 which shows the simulated response to a reference change in y_1. The responses are clearly acceptable, and we conclude that *nominal performance (NP) is achieved with the decoupling controller*.

Robust stability (RS). The resulting sensitivity and complementary sensitivity functions with this controller are

$$S = S_I = \frac{s}{s + 0.7}I; \quad T = T_I = \frac{1}{1.43s + 1}I \qquad (3.85)$$

Thus, $\bar{\sigma}(S)$ and $\bar{\sigma}(T)$ are both less than 1 at all frequencies, so there are no peaks which would indicate robustness problems. We also find that this controller gives an infinite gain margin (GM) and a phase margin (PM) of $90°$ in each channel. Thus, use of the traditional margins and the peak values of S and T indicate no robustness problems. However, from the large RGA-elements there is cause for concern, and this is confirmed in the following.

We consider again the input gain uncertainty (3.80) as in the previous example, and we select $\epsilon_1 = 0.2$ and $\epsilon_2 = -0.2$. We then have

$$u'_1 = 1.2u_1, \quad u'_2 = 0.8u_2 \tag{3.86}$$

Note that the uncertainty is on the *change* in the inputs (flow rates), and not on their absolute values. A 20% error is typical for process control applications (see Remark 2 on page 300). The uncertainty in (3.86) does not by itself yield instability. This is verified by computing the closed-loop poles, which, assuming no cancellations, are solutions to $\det(I + L(s)) = \det(I + L_I(s)) = 0$ (see (4.102) and (A.12)). In our case

$$L'_I(s) = K_{\text{inv}}G' = K_{\text{inv}}G \begin{bmatrix} 1+\epsilon_1 & 0 \\ 0 & 1+\epsilon_2 \end{bmatrix} = \frac{0.7}{s} \begin{bmatrix} 1+\epsilon_1 & 0 \\ 0 & 1+\epsilon_2 \end{bmatrix}$$

so the perturbed closed-loop poles are

$$s_1 = -0.7(1+\epsilon_1), \quad s_2 = -0.7(1+\epsilon_2) \tag{3.87}$$

and we have closed-loop stability as long as the input gains $1 + \epsilon_1$ and $1 + \epsilon_2$ remain positive, so we can have up to 100% error in each input channel. We thus conclude that *we have robust stability (RS) with respect to input gain errors for the decoupling controller.*

Robust performance (RP). For SISO systems we generally have that nominal performance (NP) and robust stability (RS) imply robust performance (RP), but this is not the case for MIMO systems. This is clearly seen from the dotted lines in Figure 3.12 which show the closed-loop response of the perturbed system. It differs drastically from the nominal response represented by the solid line, and even though it is stable, the response is clearly not acceptable; it is no longer decoupled, and $y_1(t)$ and $y_2(t)$ reach a value of about 2.5 before settling at their desired values of 1 and 0. *Thus RP is not achieved by the decoupling controller.*

Remark 1 There is a simple reason for the observed poor response to the reference change in y_1. To accomplish this change, which occurs mostly in the direction corresponding to the low plant gain, the inverse-based controller generates relatively *large* inputs u_1 and u_2, while trying to keep $u_1 - u_2$ very *small*. However, the input uncertainty makes this impossible – the result is an undesired *large* change in the actual value of $u'_1 - u'_2$, which subsequently results in large changes in y_1 and y_2 because of the large plant gain ($\bar{\sigma}(G) = 197.2$) in this direction, as seen from (3.46).

Remark 2 The system remains stable for gain uncertainty up to 100% because the uncertainty occurs only at one side of the plant (at the input). If we also consider uncertainty at the output then we find that the decoupling controller yields instability for relatively small errors in the input and output gains. This is illustrated in Exercise 3.8 below.

Remark 3 It is also difficult to get a robust controller with other standard design techniques for this model. For example, an S/KS-design as in (3.59) with $W_P = w_P I$ (using $M = 2$ and $\omega_B = 0.05$ in the performance weight (3.60)) and $W_u = I$, yields a good nominal response (although not decoupled), but the system is very sensitive to input uncertainty, and the outputs go up to about 3.4 and settle very slowly when there is 20% input gain error.

Remark 4 Attempts to make the inverse-based controller robust using the second step of the Glover-McFarlane $\mathcal{H}_\infty$ loop-shaping procedure are also unhelpful; see Exercise 3.9. This shows that robustness with respect to coprime factor uncertainty does not necessarily imply robustness with respect to input uncertainty. In any case, the solution is to avoid inverse-based controllers for a plant with large RGA-elements.

Exercise 3.7 *Design a SVD-controller $K = W_1 K_s W_2$ for the distillation process in (3.82), i.e. select $W_1 = V$ and $W_2 = U^T$ where U and V are given in (3.46). Select K_s in the form*

$$K_s = \begin{bmatrix} c_1 \frac{75s+1}{s} & 0 \\ 0 & c_2 \frac{75s+1}{s} \end{bmatrix}$$

and try the following values:

(a) $c_1 = c_2 = 0.005$;
(b) $c_1 = 0.005$, $c_2 = 0.05$;
(c) $c_1 = 0.7/197 = 0.0036$, $c_2 = 0.7/1.39 = 0.504$.

Simulate the closed-loop reference response with and without uncertainty. Designs (a) and (b) should be robust. Which has the best performance? Design (c) should give the response in Figure 3.12. In the simulations, include high-order plant dynamics by replacing $G(s)$ by $\frac{1}{(0.02s+1)^5} G(s)$. What is the condition number of the controller in the three cases? Discuss the results. (See also the conclusion on page 244).

Exercise 3.8 *Consider again the distillation process (3.82) with the decoupling controller, but also include output gain uncertainty $\widehat{\epsilon}_i$. That is, let the perturbed loop transfer function be*

$$L'(s) = G' K_{\text{inv}} = \frac{0.7}{s} \underbrace{\begin{bmatrix} 1+\widehat{\epsilon}_1 & 0 \\ 0 & 1+\widehat{\epsilon}_2 \end{bmatrix} G \begin{bmatrix} 1+\epsilon_1 & 0 \\ 0 & 1+\epsilon_2 \end{bmatrix} G^{-1}}_{L_0} \quad (3.88)$$

where L_0 is a constant matrix for the distillation model (3.82), since all elements in G share the same dynamics, $G(s) = g(s)G_0$. The closed-loop poles of the perturbed system are solutions to $\det(I + L'(s)) = \det(I + (k_1/s)L_0) = 0$, or equivalently

$$\det(\frac{s}{k_1}I + L_0) = (s/k_1)^2 + \text{tr}(L_0)(s/k_1) + \det(L_0) = 0 \quad (3.89)$$

For $k_1 > 0$ we have from the Routh-Hurwitz stability condition indexRouth-Hurwitz stability test that instability occurs if and only if the trace and/or the determinant of L_0 are negative.

Since $\det(L_0) > 0$ *for any gain error less than 100%, instability can only occur if* $\text{tr}(L_0) < 0$. *Evaluate* $\text{tr}(L_0)$ *and show that with gain errors of equal magnitude the combination of errors which most easily yields instability is with* $\widehat{\epsilon}_1 = -\widehat{\epsilon}_2 = -\epsilon_1 = \epsilon_2 = \epsilon$. *Use this to show that the perturbed system is unstable if*

$$|\epsilon| > \sqrt{\frac{1}{2\lambda_{11} - 1}} \tag{3.90}$$

where $\lambda_{11} = g_{11}g_{22} / \det G$ *is the* $1, 1$-*element of the RGA of* G. *In our case* $\lambda_{11} = 35.1$ *and we get instability for* $|\epsilon| > 0.120$. *Check this numerically, e.g. using MATLAB.*

Remark. The instability condition in (3.90) for simultaneous input and output gain uncertainty, applies to the very special case of a 2×2 plant, in which all elements share the same dynamics, $G(s) = g(s)G_0$, and an inverse-based controller, $K(s) = (k_1/s)G^{-1}(s)$.

Exercise 3.9 *Consider again the distillation process* $G(s)$ *in (3.82). The response using the inverse-based controller* K_{inv} *in (3.84) was found to be sensitive to input gain errors. We want to see if the controller can be modified to yield a more robust system by using the Glover-McFarlane* $\mathcal{H}_\infty$ *loop-shaping procedure. To this effect, let the shaped plant be* $G_s = GK_{\text{inv}}$, *i.e.* $W_1 = K_{\text{inv}}$, *and design an* $\mathcal{H}_\infty$ *controller* K_s *for the shaped plant (see page 382 and Chapter 9), such that the overall controller becomes* $K = K_{\text{inv}}K_s$. *(You will find that* $\gamma_{min} = 1.414$ *which indicates good robustness with respect to coprime factor uncertainty, but the loop shape is almost unchanged and the system remains sensitive to input uncertainty.)*

3.7.3   Robustness conclusions

From the two motivating examples above we found that multivariable plants can display a sensitivity to uncertainty (in this case input uncertainty) which is fundamentally different from what is possible in SISO systems.

In the first example (spinning satellite), we had excellent stability margins (PM and GM) when considering one loop at a time, but small simultaneous input gain errors gave instability. This might have been expected from the peak values ($\mathcal{H}_\infty$ norms) of S and T, defined as

$$\|T\|_\infty = \max_\omega \bar{\sigma}(T(j\omega)), \quad \|S\|_\infty = \max_\omega \bar{\sigma}(S(j\omega)) \tag{3.91}$$

which were both large (about 10) for this example.

In the second example (distillation process), we again had excellent stability margins (PM and GM), and the system was also robustly stable to errors (even simultaneous) of up to 100% in the input gains. However, in this case small input gain errors gave very poor output performance, so robust performance was not satisfied, and adding simultaneous output gain uncertainty resulted in instability (see Exercise 3.8). These problems with the decoupling controller might have been expected because the plant has large RGA-elements. For this second example the $\mathcal{H}_\infty$ norms of S and T were both about 1, so the absence of peaks in S and T does not guarantee robustness.

Although sensitivity peaks, RGA-elements, etc. are useful indicators of robustness problems, they provide no exact answer to whether a given source of uncertainty will yield instability or poor performance. This motivates the need for better tools for analyzing the effects of model uncertainty. We want to avoid a trial-and-error procedure based on checking stability and performance for a large number of candidate plants. This is very time consuming, and in the end one does not know whether those plants are the limiting ones. What is desired, is a simple tool which is able to identify the worst-case plant. This will be the focus of Chapters 7 and 8 where we show how to represent model uncertainty in the $\mathcal{H}_\infty$ framework, and introduce the structured singular value μ as our tool. The two motivating examples are studied in more detail in Example 8.10 and Section 8.11.3 where a μ-analysis predicts the robustness problems found above.

3.8   General control problem formulation

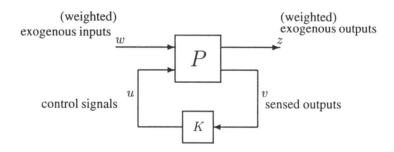

Figure 3.13: General control configuration for the case with no model uncertainty

In this section we consider a general method of formulating control problems introduced by Doyle (1983; 1984). The formulation makes use of the general control configuration in Figure 3.13, where P is the generalized plant and K is the generalized controller as explained in Table 1.1 on page 13. Note that positive feedback is used.

The overall control objective is to minimize some norm of the transfer function from w to z, for example, the $\mathcal{H}_\infty$ norm. The controller design problem is then:

- Find a controller K which based on the information in v, generates a control signal u which counteracts the influence of w on z, thereby minimizing the closed-loop norm from w to z.

The most important point of this section is to appreciate that almost any linear control problem can be formulated using the block diagram in Figure 3.13 (for the nominal case) or in Figure 3.21 (with model uncertainty).

Remark 1 The configuration in Figure 3.13 may at first glance seem restrictive. However, this is not the case, and we will demonstrate the generality of the setup with a few examples, including the design of observers (the estimation problem) and feedforward controllers.

Remark 2 We may generalize the control configuration still further by including diagnostics as additional outputs from the controller giving the *4-parameter controller* introduced by Nett (1986), but this is not considered in this book.

3.8.1   Obtaining the generalized plant P

The routines in MATLAB for synthesizing $\mathcal{H}_\infty$ and $\mathcal{H}_2$ optimal controllers assume that the problem is in the general form of Figure 3.13, that is, they assume that P is given. To derive P (and K) for a specific case we must first find a block diagram representation and identify the signals w, z, u and v. To construct P one should note that it is an *open-loop* system and remember to break all "loops" entering and exiting the controller K. Some examples are given below and further examples are given in Section 9.3 (Figures 9.9, 9.10, 9.11 and 9.12).

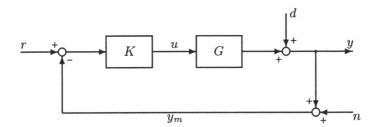

Figure 3.14: One degree-of-freedom control configuration

Example 3.13 One degree-of-freedom feedback control configuration. *We want to find P for the conventional one degree-of-freedom control configuration in Figure 3.14. The first step is to identify the signals for the generalized plant:*

$$w = \begin{bmatrix} w_1 \\ w_2 \\ w_3 \end{bmatrix} = \begin{bmatrix} d \\ r \\ n \end{bmatrix}; \quad z = e = y - r; \quad v = r - y_m = r - y - n \qquad (3.92)$$

With this choice of v, the controller only has information about the deviation $r - y_m$. Also note that $z = y - r$, which means that performance is specified in terms of the actual output y and not in terms of the measured output y_m. The block diagram in Figure 3.14 then yields

$$
\begin{aligned}
z &= y - r = Gu + d - r = Iw_1 - Iw_2 + 0w_3 + Gu \\
v &= r - y_m = r - Gu - d - n = -Iw_1 + Iw_2 - Iw_3 - Gu
\end{aligned}
$$

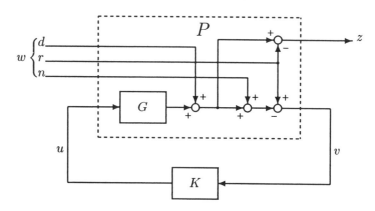

Figure 3.15: Equivalent representation of Figure 3.14 where the error signal to be minimized is $z = y - r$ and the input to the controller is $v = r - y_m$

and P which represents the transfer function matrix from $\begin{bmatrix} w & u \end{bmatrix}^T$ to $\begin{bmatrix} z & v \end{bmatrix}^T$ is

$$P = \begin{bmatrix} I & -I & 0 & G \\ -I & I & -I & -G \end{bmatrix} \tag{3.93}$$

Note that P does not *contain the controller. Alternatively, P can be obtained by inspection from the representation in Figure 3.15.*

Remark. Obtaining the generalized plant P may seem tedious. However, when performing numerical calculations P can be generated using software. For example, in MATLAB we may use the simulink program, or we may use the sysic program in the μ-toolbox. The code in Table 3.1 generates the generalized plant P in (3.93) for Figure 3.14.

Table 3.1: MATLAB program to generate P in (3.93)

```
% Uses the Mu-toolbox
systemnames = 'G';                              % G is the SISO plant.
inputvar = '[d(1);r(1);n(1);u(1)]';             % Consists of vectors w and u.
input_to_G = '[u]';
outputvar = '[G+d-r; r-G-d-n]';                 % Consists of vectors z and v.
sysoutname = 'P';
sysic;
```

3.8.2   Controller design: Including weights in P

To get a meaningful controller synthesis problem, for example, in terms of the $\mathcal{H}_\infty$ or $\mathcal{H}_2$ norms, we generally have to include weights W_z and W_w in the generalized plant P, see Figure 3.16. That is, we consider the weighted or normalized exogenous inputs w (where $\widetilde{w} = W_w w$ consists of the "physical" signals entering the system;

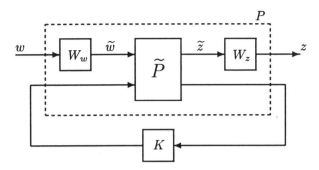

Figure 3.16: General control configuration for the case with no model uncertainty

disturbances, references and noise), and the weighted or normalized controlled outputs $z = W_z \tilde{z}$ (where $\tilde{z}$ often consists of the control error $y - r$ and the manipulated input u). The weighting matrices are usually frequency dependent and typically selected such that weighted signals w and z are of magnitude 1, that is, the norm from w to z should be less than 1. Thus, in most cases only the magnitude of the weights matter, and we may without loss of generality assume that $W_w(s)$ and $W_z(s)$ are stable and minimum phase (they need not even be rational transfer functions but if not they will be unsuitable for controller synthesis using current software).

Example 3.14 **Stacked** $S/T/KS$ **problem.** *Consider an $\mathcal{H}_\infty$ problem where we want to bound $\bar{\sigma}(S)$ (for performance), $\bar{\sigma}(T)$ (for robustness and to avoid sensitivity to noise) and $\bar{\sigma}(KS)$ (to penalize large inputs). These requirements may be combined into a stacked $\mathcal{H}_\infty$ problem*

$$\min_K \|N(K)\|_\infty, \quad N = \begin{bmatrix} W_u KS \\ W_T T \\ W_P S \end{bmatrix} \tag{3.94}$$

where K is a stabilizing controller. In other words, we have $z = Nw$ and the objective is to minimize the $\mathcal{H}_\infty$ norm from w to z. Except for some negative signs which have no effect when evaluating $\|N\|_\infty$, the N in (3.94) may be represented by the block diagram in Figure 3.17 (convince yourself that this is true). Here w represents a reference command ($w = -r$, where the negative sign does not really matter) or a disturbance entering at the output ($w = d_y$), and z consists of the weighted input $z_1 = W_u u$, the weighted output $z_2 = W_T y$, and the weighted control error $z_3 = W_P(y - r)$. We get from Figure 3.17 the following set of equations

$$\begin{aligned} z_1 &= W_u u \\ z_2 &= W_T G u \\ z_3 &= W_P w + W_P G u \\ v &= -w - G u \end{aligned}$$

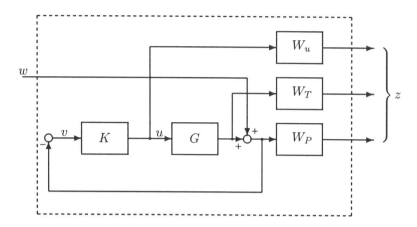

Figure 3.17: Block diagram corresponding to $z = Nw$ in (3.94)

so the generalized plant P from $\begin{bmatrix} w & u \end{bmatrix}^T$ to $\begin{bmatrix} z & v \end{bmatrix}^T$ is

$$P = \begin{bmatrix} 0 & W_u I \\ 0 & W_T G \\ W_P I & W_P G \\ -I & -G \end{bmatrix} \tag{3.95}$$

3.8.3   Partitioning the generalized plant P

We often partition P as

$$P = \begin{bmatrix} P_{11} & P_{12} \\ P_{21} & P_{22} \end{bmatrix} \tag{3.96}$$

such that its parts are compatible with the signals w, z, u and v in the generalized control configuration,

$$z = P_{11}w + P_{12}u \tag{3.97}$$
$$v = P_{21}w + P_{22}u \tag{3.98}$$

The reader should become familiar with this notation. In Example 3.14 we get

$$P_{11} = \begin{bmatrix} 0 \\ 0 \\ W_P I \end{bmatrix}, \quad P_{12} = \begin{bmatrix} W_u I \\ W_T G \\ W_P G \end{bmatrix} \tag{3.99}$$

$$P_{21} = -I, \quad P_{22} = -G \tag{3.100}$$

Note that P_{22} has dimensions compatible with the controller, i.e. if K is an $n_u \times n_v$ matrix, then P_{22} is an $n_v \times n_u$ matrix. For cases with one degree-of-freedom negative feedback control we have $P_{22} = -G$.

3.8.4 Analysis: Closing the loop to get N

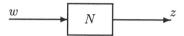

Figure 3.18: General block diagram for analysis with no uncertainty

The general feedback configurations in Figures 3.13 and 3.16 have the controller K as a separate block. This is useful when synthesizing the controller. However, for *analysis* of closed-loop performance the controller is given, and we may absorb K into the interconnection structure and obtain the system N as shown in Figure 3.18 where

$$z = Nw \tag{3.101}$$

where N is a function of K. To find N, first partition the generalized plant P as given in (3.96)-(3.98), combine this with the controller equation

$$u = Kv \tag{3.102}$$

and eliminate u and v from equations (3.97), (3.98) and (3.102) to yield $z = Nw$ where N is given by

$$N = P_{11} + P_{12}K(I - P_{22}K)^{-1}P_{21} \triangleq F_l(P, K) \tag{3.103}$$

Here $F_l(P, K)$ denotes a lower *linear fractional transformation (LFT)* of P with K as the parameter. Some properties of LFTs are given in Appendix A.7. In words, N is obtained from Figure 3.13 by using K to close a lower feedback loop around P. Since positive feedback is used in the general configuration in Figure 3.13 the term $(I - P_{22}K)^{-1}$ has a negative sign.

Remark. To assist in remembering the sequence of P_{12} and P_{21} in (3.103), notice that the first (last) index in P_{11} is the same as the first (last) index in $P_{12}K(I - P_{22}K)^{-1}P_{21}$. The lower LFT in (3.103) is also represented by the block diagram in Figure 3.2.

The reader is advised to become comfortable with the above manipulations before progressing much further.

Example 3.15 *We want to derive N for the partitioned P in (3.99) and (3.100) using the LFT-formula in (3.103). We get*

$$N = \begin{bmatrix} 0 \\ 0 \\ W_P I \end{bmatrix} + \begin{bmatrix} W_u I \\ W_T G \\ W_P G \end{bmatrix} K(I + GK)^{-1}(-I) = \begin{bmatrix} -W_u KS \\ -W_T T \\ W_P S \end{bmatrix}$$

where we have made use of the identities $S = (I + GK)^{-1}$, $T = GKS$ and $I - T = S$. With the exception of the two negative signs, this is identical to N given in (3.94). Of course, the negative signs have no effect on the norm of N.

Again, it should be noted that deriving N from P is much simpler using available software. For example in the MATLAB μ-Toolbox we can evaluate $N = F_l(P, K)$ using the command N=starp(P,K). Here starp denotes the matrix star product which generalizes the use of LFTs (see Appendix A.7.5).

Exercise 3.10 *Consider the two degrees-of-freedom feedback configuration in Figure 1.3(b). (i) Find P when*

$$w = \begin{bmatrix} d \\ r \\ n \end{bmatrix}; \quad z = \begin{bmatrix} y - r \\ u \end{bmatrix}; \quad v = \begin{bmatrix} r \\ y_m \end{bmatrix} \tag{3.104}$$

(ii) Let $z = Nw$ and derive N in two different ways; directly from the block diagram and using $N = F_l(P, K)$.

3.8.5   Generalized plant P: Further examples

To illustrate the generality of the configuration in Figure 3.13, we now present two further examples: one in which we derive P for a problem involving feedforward control, and one for a problem involving estimation.

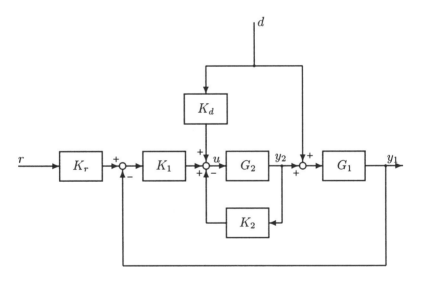

Figure 3.19: System with feedforward, local feedback and two degrees-of-freedom control

Example 3.16 *Consider the control system in Figure 3.19, where y_1 is the output we want to control, y_2 is a secondary output (extra measurement), and we also measure the disturbance d. By secondary we mean that y_2 is of secondary importance for control, that is, there is no control objective associated with it. The control configuration includes a two degrees-of-freedom controller, a feedforward controller and a local feedback controller based on the extra* *measurement y_2. To recast this into our standard configuration of Figure 3.13 we define*

$$w = \begin{bmatrix} d \\ r \end{bmatrix}; \quad z = y_1 - r; \quad v = \begin{bmatrix} r \\ y_1 \\ y_2 \\ d \end{bmatrix} \tag{3.105}$$

Note that d and r are both inputs and outputs to P and we have assumed a perfect measurement of the disturbance d. Since the controller has explicit information about r we have a two degrees-of-freedom controller. The generalized controller K may be written in terms of the individual controller blocks in Figure 3.19 as follows:

$$K = [\, K_1 K_r \quad -K_1 \quad -K_2 \quad K_d \,] \tag{3.106}$$

By writing down the equations or by inspection from Figure 3.19 we get

$$P = \begin{bmatrix} G_1 & -I & G_1 G_2 \\ 0 & I & 0 \\ G_1 & 0 & G_1 G_2 \\ 0 & 0 & G_2 \\ I & 0 & 0 \end{bmatrix} \tag{3.107}$$

Then partitioning P as in (3.97) and (3.98) yields $P_{22} = [\, 0^T \quad (G_1 G_2)^T \quad G_2^T \quad 0^T \,]^T$.

Exercise 3.11 Cascade implementation. *Consider further Example 3.16. The local feedback based on y_2 is often implemented in a cascade manner; see also Figure 10.4. In this case the output from K_1 enters into K_2 and it may be viewed as a reference signal for y_2. Derive the generalized controller K and the generalized plant P in this case.*

Remark. From Example 3.16 and Exercise 3.11, we see that a cascade *implementation* does not usually limit the achievable performance since, unless the optimal K_2 or K_1 have RHP-zeros, we can obtain from the optimal overall K the subcontrollers K_2 and K_1 (although we may have to add a small D-term to K to make the controllers proper). However, if we impose restrictions on the *design* such that, for example K_2 or K_1 are designed "locally" (without considering the whole problem), then this will limit the achievable performance. For example, for a *two degrees-of-freedom controller* a common approach is to first design the feedback controller K_y for disturbance rejection (without considering reference tracking) and then design K_r for reference tracking. This will generally give some performance loss compared to a simultaneous design of K_y and K_r.

Example 3.17 Output estimator. *Consider a situation where we have no measurement of the output y which we want to control. However, we do have a measurement of another output variable y_2. Let d denote the unknown external inputs (including noise and disturbances) and u_G the known plant inputs (a subscript G is used because in this case the output u from K is not the plant input). Let the model be*

$$y = G u_G + G_d d; \quad y_2 = F u_G + F_d d$$

The objective is to design an estimator, K_{est}, such that the estimated output $\widehat{y} = K_{\text{est}} \begin{bmatrix} y_2 \\ u_G \end{bmatrix}$ is as close as possible in some sense to the true output y; see Figure 3.20. This problem may be written in the general framework of Figure 3.13 with

$$w = \begin{bmatrix} d \\ u_G \end{bmatrix}, \ u = \widehat{y}, \ z = y - \widehat{y}, \ v = \begin{bmatrix} y_2 \\ u_G \end{bmatrix}$$

Note that $u = \widehat{y}$, that is, the output u from the generalized controller is the estimate of the plant output. Furthermore, $K = K_{\text{est}}$ and

$$P = \begin{bmatrix} G_d & G & -I \\ F_d & F & 0 \\ 0 & I & 0 \end{bmatrix} \tag{3.108}$$

We see that $P_{22} = \begin{bmatrix} 0 \\ 0 \end{bmatrix}$ since the estimator problem does not involve feedback.

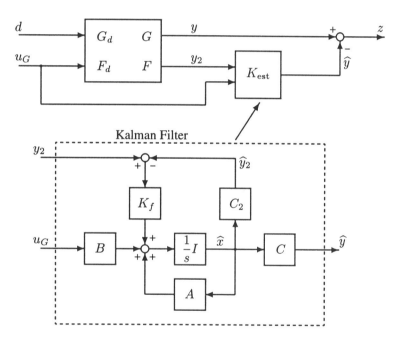

Figure 3.20: Output estimation problem. One particular estimator K_{est} is a Kalman Filter

Exercise 3.12 State estimator (observer). *In the Kalman filter problem studied in Section 9.2 the objective is to minimize $x - \widehat{x}$ (whereas in Example 3.17 the objective was to minimize $y - \widehat{y}$). Show how the Kalman filter problem can be represented in the general configuration of Figure 3.13 and find P.*

3.8.6 Deriving P from N

For cases where N is given and we wish to find a P such that

$$N = F_l(P, K) = P_{11} + P_{12}K(I - P_{22}K)^{-1}P_{21}$$

it is usually best to work from a block diagram representation. This was illustrated above for the stacked N in (3.94). Alternatively, the following procedure may be useful:

1. Set $K = 0$ in N to obtain P_{11}.
2. Define $Q = N - P_{11}$ and rewrite Q such that each term has a common factor $R = K(I - P_{22}K)^{-1}$ (this gives P_{22}).
3. Since $Q = P_{12}RP_{21}$, we can now usually obtain P_{12} and P_{21} by inspection.

Example 3.18 Weighted sensitivity. *We will use the above procedure to derive P when $N = w_P S = w_P(I + GK)^{-1}$, where w_P is a scalar weight.*

1. $P_{11} = N(K = 0) = w_P I$.
2. $Q = N - w_P I = w_P(S - I) = -w_P T = -w_P GK(I + GK)^{-1}$, and we have $R = K(I + GK)^{-1}$ so $P_{22} = -G$.
3. $Q = -w_P GR$ so we have $P_{12} = -w_P G$ and $P_{21} = I$, and we get

$$P = \begin{bmatrix} w_P I & -w_P G \\ I & -G \end{bmatrix} \tag{3.109}$$

Remark. When obtaining P from a given N, we have that P_{11} and P_{22} are unique, whereas from Step 3 in the above procedure we see that $\widetilde{P}_{12}$ and P_{21} are not unique. For instance, let α be a real scalar then we may instead choose $\widetilde{P}_{12} = \alpha P_{12}$ and $\widetilde{P}_{21} = (1/\alpha)P_{21}$. For P in (3.109) this means that we may move the negative sign of the scalar w_P from P_{12} to P_{21}.

Exercise 3.13 Mixed sensitivity. *Use the above procedure to derive the generalized plant P for the stacked N in (3.94).*

3.8.7 Problems not covered by the general formulation

The above examples have demonstrated the generality of the control configuration in Figure 3.13. Nevertheless, there are some controller design problems which are not covered. Let N be some closed-loop transfer function whose norm we want to minimize. To use the general form we must first obtain a P such that $N = F_l(P, K)$. However, this is not always possible, since there may not exist a block diagram representation for N. As a simple example, consider the stacked transfer function

$$N = \begin{bmatrix} (I + GK)^{-1} \\ (I + KG)^{-1} \end{bmatrix} \tag{3.110}$$

The transfer function $(I + GK)^{-1}$ may be represented on a block diagram with the input and output signals *after* the plant, whereas $(I + KG)^{-1}$ may be represented by another block diagram with input and output signals *before* the plant. However, in N there are no cross coupling terms between an input before the plant and an output after the plant (corresponding to $G(I + KG)^{-1}$), or between an input after the plant and an output before the plant (corresponding to $-K(I + GK)^{-1}$) so N cannot be represented in block diagram form. Equivalently, if we apply the procedure in Section 3.8.6 to N in (3.110), we are not able to find solutions to P_{12} and P_{21} in Step 3.

Another stacked transfer function which *cannot* in general be represented in block diagram form is

$$N = \begin{bmatrix} W_P S \\ S G_d \end{bmatrix} \tag{3.111}$$

Remark. The case where N cannot be written as an LFT of K, is a special case of the Hadamard weighted $\mathcal{H}_\infty$ problem studied by van Diggelen and Glover (1994a). Although the solution to this $\mathcal{H}_\infty$ problem remains intractable, van Diggelen and Glover (1994b) present a solution for a similar problem where the Frobenius norm is used instead of the singular value to "sum up the channels".

Exercise 3.14 *Show that N in (3.111) can be represented in block diagram form if $W_P = w_P I$ where w_P is a scalar.*

3.8.8   A general control configuration including model uncertainty

The general control configuration in Figure 3.13 may be extended to include model uncertainty as shown by the block diagram in Figure 3.21. Here the matrix Δ is a *block-diagonal* matrix that includes all possible perturbations (representing uncertainty) to the system. It is usually normalized in such a way that $\|\Delta\|_\infty \leq 1$.

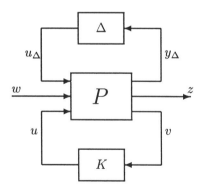

Figure 3.21: General control configuration for the case with model uncertainty

The block diagram in Figure 3.21 in terms of P (for synthesis) may be transformed into the block diagram in Figure 3.22 in terms of N (for analysis) by using K to close

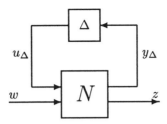

Figure 3.22: General block diagram for analysis with uncertainty included

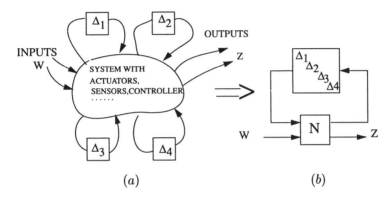

(a) (b)

Figure 3.23: Rearranging a system with multiple perturbations into the $N\Delta$-structure

a lower loop around P. If we partition P to be compatible with the controller K, then the same *lower LFT* as found in (3.103) applies, and

$$N = F_l(P, K) = P_{11} + P_{12}K(I - P_{22}K)^{-1}P_{21} \qquad (3.112)$$

To evaluate the perturbed (uncertain) transfer function from external inputs w to external outputs z, we use Δ to close the upper loop around N (see Figure 3.22), resulting in an *upper LFT* (see Appendix A.7):

$$z = F_u(N, \Delta)w; \quad F_u(N, \Delta) \triangleq N_{22} + N_{21}\Delta(I - N_{11}\Delta)^{-1}N_{12} \qquad (3.113)$$

Remark 1 Controller synthesis based on Figure 3.21 is still an unsolved problem, although good practical approaches like DK-iteration to find the "μ-optimal" controller are in use (see Section 8.12). For analysis (with a given controller), the situation is better and with the $\mathcal{H}_\infty$ norm an assessment of robust performance involves computing the structured singular value, μ. This is discussed in more detail in Chapter 8.

Remark 2 In (3.113) N has been partitioned to be compatible with Δ, that is N_{11} has dimensions compatible with Δ. Usually, Δ is square in which case N_{11} is a square matrix of the same dimension as Δ. For the nominal case with no uncertainty we have $F_u(N, \Delta) = F_u(N, 0) = N_{22}$, so N_{22} is the nominal transfer function from w to z.

Remark 3 Note that P and N here also include information about how the uncertainty affects the system, so they are *not* the same P and N as used earlier, for example in (3.103). Actually, the parts P_{22} and N_{22} of P and N in (3.112) (with uncertainty) are equal to the P and N in (3.103) (without uncertainty). Strictly speaking, we should have used another symbol for N and P in (3.112), but for notational simplicity we did not.

Remark 4 The fact that almost any control problem with uncertainty can be represented by Figure 3.21 may seem surprising, so some explanation is in order. First represent each source of uncertainty by a perturbation block, Δ_i, which is normalized such that $\|\Delta_i\| \leq 1$. These perturbations may result from parametric uncertainty, neglected dynamics, etc. as will be discussed in more detail in Chapters 7 and 8. Then "pull out" each of these blocks from the system so that an input and an output can be associated with each Δ_i as shown in Figure 3.23(a). Finally, collect these perturbation blocks into a large block-diagonal matrix having perturbation inputs and outputs as shown in Figure 3.23(b). In Chapter 8 we discuss in detail how to obtain N and Δ.

3.9   Additional exercises

Most of these exercises are based on material presented in Appendix A. The exercises illustrate material which the reader should know before reading the subsequent chapters.

Exercise 3.15 *Consider the performance specification* $\|w_P S\|_\infty < 1$. *Suggest a rational transfer function weight* $w_P(s)$ *and sketch it as a function of frequency for the following two cases:*

1. *We desire no steady-state offset, a bandwidth better than* 1 *rad/s and a resonance peak (worst amplification caused by feedback) lower than* 1.5.
2. *We desire less than* 1% *steady-state offset, less than* 10% *error up to frequency* 3 *rad/s, a bandwidth better than* 10 *rad/s, and a resonance peak lower than* 2. *Hint: See* (2.72) *and* (2.73).

Exercise 3.16 *By* $\|M\|_\infty$ *one can mean either a spatial or temporal norm. Explain the difference between the two and illustrate by computing the appropriate infinity norm for*

$$M_1 = \begin{bmatrix} 3 & 4 \\ -2 & 6 \end{bmatrix}, \quad M_2(s) = \frac{s-1}{s+1}\frac{3}{s+2}$$

Exercise 3.17 *What is the relationship between the RGA-matrix and uncertainty in the individual elements? Illustrate this for perturbations in the* 1, 1-*element of the matrix*

$$A = \begin{bmatrix} 10 & 9 \\ 9 & 8 \end{bmatrix} \tag{3.114}$$

Exercise 3.18 *Assume that A is non-singular. (i) Formulate a condition in terms of the maximum singular value of E for the matrix A + E to remain non-singular. Apply this to A in (3.114) and (ii) find an E of minimum magnitude which makes A + E singular.*

Exercise 3.19 *Compute* $\|A\|_{i1}$, $\bar{\sigma}(A) = \|A\|_{i2}$, $\|A\|_{i\infty}$, $\|A\|_F$, $\|A\|_{\max}$ *and* $\|A\|_{\mathrm{sum}}$ *for the following matrices and tabulate your results:*

$$A_1 = I; \quad A_2 = \begin{bmatrix} 1 & 0 \\ 0 & 0 \end{bmatrix}; A_3 = \begin{bmatrix} 1 & 1 \\ 1 & 1 \end{bmatrix}; A_4 = \begin{bmatrix} 1 & 1 \\ 0 & 0 \end{bmatrix}, A_5 = \begin{bmatrix} 1 & 0 \\ 1 & 0 \end{bmatrix}$$

Show using the above matrices that the following bounds are tight (i.e. we may have equality) for 2×2 matrices $(m = 2)$:

$$\bar{\sigma}(A) \leq \|A\|_F \leq \sqrt{m}\, \bar{\sigma}(A)$$

$$\|A\|_{\max} \leq \bar{\sigma}(A) \leq m\|A\|_{\max}$$

$$\|A\|_{i1}/\sqrt{m} \leq \bar{\sigma}(A) \leq \sqrt{m}\|A\|_{i1}$$

$$\|A\|_{i\infty}/\sqrt{m} \leq \bar{\sigma}(A) \leq \sqrt{m}\|A\|_{i\infty}$$

$$\|A\|_F \leq \|A\|_{\mathrm{sum}}$$

Exercise 3.20 *Find example matrices to illustrate that the above bounds are also tight when A is a square $m \times m$ matrix with $m > 2$.*

Exercise 3.21 *Do the extreme singular values bound the magnitudes of the elements of a matrix? That is, is $\bar{\sigma}(A)$ greater than the largest element (in magnitude), and is $\underline{\sigma}(A)$ smaller than the smallest element? For a non-singular matrix, how is $\underline{\sigma}(A)$ related to the largest element in A^{-1}?*

Exercise 3.22 *Consider a lower triangular $m \times m$ matrix A with $a_{ii} = -1$, $a_{ij} = 1$ for all $i > j$, and $a_{ij} = 0$ for all $i < j$.*
a) What is $\det A$?
b) What are the eigenvalues of A ?
c) Show that the smallest singular value is less than or equal to 2^{-m}.
d) What is the RGA of A?
e) Let $m = 4$ and find an E with the smallest value of $\bar{\sigma}(E)$ such that $A + E$ is singular.

Exercise 3.23 *Find two matrices A and B such that $\rho(A+B) > \rho(A)+\rho(B)$ which proves that the spectral radius does not satisfy the triangle inequality and is thus not a norm.*

Exercise 3.24 *Write $T = GK(I + GK)^{-1}$ as an LFT of K, i.e. find P such that $T = F_l(P, K)$.*

Exercise 3.25 *Write K as an LFT of $T = GK(I + GK)^{-1}$, i.e. find J such that $K = F_l(J, T)$.*

Exercise 3.26 *State-space descriptions may be represented as LFTs. To demonstrate this find H for*

$$F_l(H, 1/s) = C(sI - A)^{-1}B + D$$

Exercise 3.27 *Show that the set of all stabilizing controllers in (4.91) can be written as $K = F_l(J, Q)$ and find J.*

Exercise 3.28 *In (3.11) we stated that the sensitivity of a perturbed plant,* $S' = (I + G'K)^{-1}$, *is related to that of the nominal plant,* $S = (I + GK)^{-1}$ *by*

$$S' = S(I + E_O T)^{-1}$$

where $E_O = (G' - G)G^{-1}$. *This exercise deals with how the above result may be derived in a systematic (though cumbersome) manner using LFTs (see also (Skogestad and Morari, 1988a)).*

a) First find F *such that* $S' = (I + G'K)^{-1} = F_l(F, K)$, *and find* J *such that* $K = F_l(J, T)$ *(see Exercise 3.25).*

b) Combine these LFTs to find $S' = F_l(N, T)$. *What is* N *in terms of* G *and* G' *?. Note that since* $J_{11} = 0$ *we have from (A.156)*

$$N = \begin{bmatrix} F_{11} & F_{12}J_{12} \\ J_{21}F_{21} & J_{22} + J_{21}F_{22}J_{12} \end{bmatrix}$$

c) Evaluate $S' = F_l(N, T)$ *and show that*

$$S' = I - G'G^{-1}T(I - (I - G'G^{-1})T)^{-1}$$

d) Finally, show that this may be rewritten as $S' = S(I + E_O T)^{-1}$.

3.10   Conclusion

The main purpose of this chapter has been to give an overview of methods for analysis and design of multivariable control systems.

In terms of analysis, we have shown how to evaluate MIMO transfer functions and how to use the singular value decomposition of the frequency-dependent plant transfer function matrix to provide insight into multivariable directionality. Other useful tools for analyzing directionality and interactions are the condition number and the RGA. Closed-loop performance may be analyzed in the frequency domain by evaluating the maximum singular value of the sensitivity function as a function of frequency. Multivariable RHP-zeros impose fundamental limitations on closed-loop performance, but for MIMO systems we can often direct the undesired effect of a RHP-zero to a subset of the outputs. MIMO systems are often more sensitive to uncertainty than SISO systems, and we demonstrated in two examples the possible sensitivity to input gain uncertainty.

In terms of controller design, we discusssed some simple approaches such as decoupling and decentralized control. We also introduced a general control configuration in terms of the generalized plant P, which can be used as a basis for synthesizing multivariable controllers using a number of methods, including LQG, $\mathcal{H}_2$, $\mathcal{H}_\infty$ and μ-optimal control. These methods are discussed in much more detail in Chapters 8 and 9. In this chapter we have only discussed the $\mathcal{H}_\infty$ weighted sensitivity method.

4

ELEMENTS OF LINEAR SYSTEM THEORY

The main objective of this chapter is to summarize important results from linear system theory The treatment is thorough, but readers are encouraged to consult other books, such as Kailath (1980) or Zhou et al. (1996), for more details and background information if these results are new to them.

4.1 System descriptions

The most important property of a linear system (operator) is that it satisfies the *superposition principle*. Let $f(u)$ be a linear operator and u_1 and u_2 two independent variables, then

$$f(u_1 + u_2) = f(u_1) + f(u_2) \tag{4.1}$$

We use in this book various representations of time-invariant linear systems, all of which are equivalent for systems that can be described by linear ordinary differential equations with constant coefficients and which do not involve differentiation of the inputs (independent variables). The most important of these representations are discussed in this section.

4.1.1 State-space representation

Consider a system with m inputs (vector u) and l outputs (vector y) which has an internal description of n states (vector x). A natural way to represent many physical systems is by nonlinear state-space models of the form

$$\dot{x} = f(x, u); \quad y = g(x, u) \tag{4.2}$$

where $\dot{x} \equiv dx/dt$ and f and g are nonlinear functions. Linear state-space models may then be derived from the linearization of such models. In terms of deviation variables (where x represents a deviation from some nominal value or trajectory, etc.) we have

$$\dot{x}(t) = Ax(t) + Bu(t) \tag{4.3}$$

$$y(t) = Cx(t) + Du(t) \tag{4.4}$$

where A, B, C and D are real matrices. If (4.3) is derived by linearizing (4.2) then $A = \partial f/\partial x$ and $B = \partial f/\partial u$ (see Section 1.5 for an example of such a derivation). A is sometimes called the state matrix. These equations provide a convenient means of describing the dynamic behaviour of proper, rational, linear systems. They may be rewritten as

$$\begin{bmatrix} \dot{x} \\ y \end{bmatrix} = \begin{bmatrix} A & B \\ C & D \end{bmatrix} \begin{bmatrix} x \\ u \end{bmatrix}$$

which gives rise to the short-hand notation

$$G \overset{s}{=} \left[\begin{array}{c|c} A & B \\ \hline C & D \end{array} \right] \tag{4.5}$$

which is frequently used to describe a state-space model of a system G. Note that the representation in (4.3)–(4.4) is *not* a unique description of the input-output behaviour of a linear system. First, there exist realizations with the same input-output behaviour, but with additional unobservable and/or uncontrollable states (modes). Second, even for a minimal realization (a realization with the fewest number of states and consequently no unobservable or uncontrollable modes) there are an infinite number of possibilities. To see this, let S be an invertible constant matrix, and introduce the new states $q = Sx$, i.e. $x = S^{-1}q$. Then an equivalent state-space realization (i.e. one with the same input-output behaviour) in terms of these new states ("coordinates") is

$$A_q = SAS^{-1}, \quad B_q = SB, \quad C_q = CS^{-1}, \quad D_q = D \tag{4.6}$$

The most common realizations are given by a few canonical forms, such as the Jordan (diagonalized) canonical form, the observability canonical form, etc.

Given the linear dynamical system in (4.3) with an initial state condition $x(t_0)$ and an input $u(t)$, the dynamical system response $x(t)$ for $t \geq t_0$ can be determined from

$$x(t) = e^{A(t-t_0)}x(t_0) + \int_{t_0}^{t} e^{A(t-\tau)}Bu(\tau)d\tau \tag{4.7}$$

where the matrix exponential $e^{At} = I + \sum_{k=1}^{\infty}(At)^k/k!$. The output is then given by $y(t) = Cx(t) + Du(t)$. For a diagonalized realization (where $A_q = SAS^{-1} = \Lambda$ is a diagonal matrix) we have that $e^{A_q t} = \text{diag}\{e^{\lambda_i(A)t}\}$, where $\lambda_i(A)$ is the i'th eigenvalue of A. We will refer to the term $e^{\lambda_i(A)t}$ as the *mode* associated with the eigenvalue $\lambda_i(A)$.

For a system with disturbances d and measurement noise n the state-space-model is written as

$$\dot{x} = Ax + Bu + Ed \tag{4.8}$$

$$y = Cx + Du + Fd + n \tag{4.9}$$

Remark. The symbol n is used to represent both the noise signal and the number of states.

4.1.2 Impulse response representation

The impulse response matrix is

$$g(t) = \begin{cases} 0 & t < 0 \\ Ce^{At}B + D\delta(t) & t \geq 0 \end{cases} \tag{4.10}$$

where $\delta(t)$ is the unit impulse function which satisfies $\lim_{\epsilon \to 0} \int_0^\epsilon \delta(t)dt = 1$. The ij'th element of the impulse response matrix, $g_{ij}(t)$, represents the response $y_i(t)$ to an impulse $u_j(t) = \delta(t)$ for a system with a zero initial state.

With initial state $x(0) = 0$, the dynamic response to an arbitrary input $u(t)$ (which is zero for $t < 0$) may from (4.7) be written as

$$y(t) = g(t) * u(t) = \int_0^t g(t - \tau)u(\tau)d\tau \tag{4.11}$$

where $*$ denotes the convolution operator.

4.1.3 Transfer function representation - Laplace transforms

The transfer function representation is unique and is very useful for directly obtaining insight into the properties of a system. It is defined as the Laplace transform of the impulse response

$$G(s) = \int_0^\infty g(t)e^{-st}dt \tag{4.12}$$

Alternatively, we may start from the state-space description. With the assumption of a zero initial state, $x(t = 0) = 0$, the Laplace transforms of (4.3) and (4.4) become[1]

$$sx(s) = Ax(s) + Bu(s) \quad \Rightarrow \quad x(s) = (sI - A)^{-1}Bu(s) \tag{4.13}$$

$$y(s) = Cx(s) + Du(s) \quad \Rightarrow \quad y(s) = \underbrace{\left(C(sI - A)^{-1}B + D\right)}_{G(s)} u(s) \tag{4.14}$$

where $G(s)$ is the transfer function matrix. Equivalently,

$$G(s) = \frac{1}{\det(sI - A)}[C\text{adj}(sI - A)B + D\det(sI - A)] \tag{4.15}$$

---

[1] We make the usual abuse of notation and let $f(s)$ denote the Laplace transform of $f(t)$.

where $\mathrm{adj}(M)$ denotes the adjugate (or classical adjoint) of M which is the transpose of the matrix of cofactors of M. From Appendix A.2.1,

$$\det(sI - A) = \prod_{i=1}^{n} \lambda_i(sI - A) = \prod_{i=1}^{n}(s - \lambda_i(A)) \qquad (4.16)$$

When disturbances are treated separately, see (4.8) and (4.9), the corresponding disturbance transfer function is

$$G_d(s) = C(sI - A)^{-1}E + F \qquad (4.17)$$

Note that any system written in the state-space form of (4.3) and (4.4) has a transfer function, but the opposite is not true. For example, time delays and improper systems can be represented by Laplace transforms, but do not have a state-space representation. On the other hand, the state-space representation yields an internal description of the system which may be useful if the model is derived from physical principles. It is also more suitable for numerical calculations.

4.1.4 Frequency response

An important advantage of transfer functions is that the frequency response (Fourier transform) is directly obtained from the Laplace transform by setting $s = j\omega$ in $G(s)$. For more details on the frequency response, the reader is referred to Sections 2.1 and 3.3.

4.1.5 Coprime factorization

Another useful way of representing systems is the coprime factorization which may be used both in state-space and transfer function form. In the latter case a *right coprime factorization* of G is

$$G(s) = N_r(s)M_r^{-1}(s) \qquad (4.18)$$

where $N_r(s)$ and $M_r(s)$ are stable coprime transfer functions. The stability implies that $N_r(s)$ should contain all the RHP-zeros of $G(s)$, and $M_r(s)$ should contain as RHP-zeros all the RHP-poles of $G(s)$. The coprimeness implies that there should be no common RHP-zeros in N_r and M_r which result in pole-zero cancellations when forming $N_r M_r^{-1}$. Mathematically, coprimeness means that there exist stable $U_r(s)$ and $V_r(s)$ such that the following Bezout identity is satisfied

$$U_r N_r + V_r M_r = I \qquad (4.19)$$

Similarly, a *left coprime factorization* of G is

$$G(s) = M_l^{-1}(s)N_l(s) \qquad (4.20)$$

Here N_l and M_l are stable and coprime, that is, there exist stable $U_l(s)$ and $V_l(s)$ such that the following Bezout identity is satisfied

$$N_l U_l + M_l V_l = I \qquad (4.21)$$

For a scalar system, the left and right coprime factorizations are identical, $G = NM^{-1} = M^{-1}N$.

Remark. Two stable scalar transfer functions, $N(s)$ and $M(s)$, are coprime if and only if they have no common RHP-zeros. In this case we can always find stable U and V such that $NU + MV = 1$.

Example 4.1 *Consider the scalar system*

$$G(s) = \frac{(s-1)(s+2)}{(s-3)(s+4)} \qquad (4.22)$$

To obtain a coprime factorization, we first make all the RHP-poles of G zeros of M, and all the RHP-zeros of G zeros of N. We then allocate the poles of N and M so that N and M are both proper and the identity $G = NM^{-1}$ holds. Thus

$$N(s) = \frac{s-1}{s+4}, \quad M(s) = \frac{s-3}{s+2}$$

is a coprime factorization. Usually, we select N and M to have the same poles as each other and the same order as $G(s)$. This gives the most degrees of freedom subject to having a realization of $[\,M(s) \quad N(s)\,]^T$ with the lowest order. We then have that

$$N(s) = k\frac{(s-1)(s+2)}{s^2 + k_1 s + k_2}, \quad M(s) = k\frac{(s-3)(s+4)}{s^2 + k_1 s + k_2} \qquad (4.23)$$

is a coprime factorization of (4.22) for any k and for any $k_1, k_2 > 0$.

From the above example, we see that the coprime factorization is not unique. Now introduce the operator M^* defined as $M^*(s) = M^T(-s)$ (which for $s = j\omega$ is the same as the complex conjugate transpose $M^H = \bar{M}^T$). Then $G(s) = N_r(s)M_r^{-1}(s)$ is called a *normalized* right coprime factorization if

$$M_r^* M_r + N_r^* N_r = I \qquad (4.24)$$

In this case $X_r(s) = [\,M_r \quad N_r\,]^T$ satisfies $X_r^* X_r = I$ and is called an *inner* transfer function. The normalized left coprime factorization $G(s) = M_l^{-1}(s)N_l(s)$ is defined similarly, requiring that

$$M_l M_l^* + N_l N_l^* = I \qquad (4.25)$$

In this case $X_l(s) = [\,M_l \quad N_l\,]^T$ is *co-inner* which means $X_l X_l^* = I$. The normalized coprime factorizations are unique to within a right (left) multiplication by a unitary matrix.

Exercise 4.1 *We want to find the normalized coprime factorization for the scalar system in (4.22). Let N and M be as given in (4.23), and substitute them into (4.24). Show that after some algebra and comparing of terms one obtains:* $k = \pm 0.71$, $k_1 = 5.67$ *and* $k_2 = 8.6$.

To derive normalized coprime factorizations by hand, as in the above exercise, is in general difficult. Numerically, however, one can easily find a state-space realization. If G has a minimal state-space realization

$$G \stackrel{s}{=} \left[\begin{array}{c|c} A & B \\ \hline C & D \end{array} \right]$$

then a minimal state-space realization of a normalized left coprime factorization is given (Vidyasagar, 1985) by

$$\left[\begin{array}{cc} N_l(s) & M_l(s) \end{array} \right] \stackrel{s}{=} \left[\begin{array}{c|cc} A + HC & B + HD & H \\ \hline R^{-1/2}C & R^{-1/2}D & R^{-1/2} \end{array} \right] \qquad (4.26)$$

where

$$H \triangleq -(BD^T + ZC^T)R^{-1}, \quad R \triangleq I + DD^T$$

and the matrix Z is the unique positive definite solution to the algebraic Riccati equation

$$(A - BS^{-1}D^TC)Z + Z(A - BS^{-1}D^TC)^T - ZC^TR^{-1}CZ + BS^{-1}B^T = 0$$

where

$$S \triangleq I + D^TD.$$

Notice that the formulas simplify considerably for a strictly proper plant, i.e. when $D = 0$. The MATLAB commands in Table 4.1 can be used to find the normalized coprime factorization for $G(s)$ using (4.26).

Exercise 4.2 *Verify numerically (e.g. using the MATLAB file in Table 4.1 or the μ-toolbox command* sncfbal) *that the normalized coprime factors of $G(s)$ in (4.22) are as given in Exercise 4.1.*

4.1.6 More on state-space realizations

Inverse system. In some cases we may want to find a state-space description of the inverse of a system. For a square $G(s)$ we have

$$G^{-1} \stackrel{s}{=} \left[\begin{array}{c|c} A - BD^{-1}C & BD^{-1} \\ \hline -D^{-1}C & D^{-1} \end{array} \right] \qquad (4.27)$$

where D is assumed to be non-singular. For a non-square $G(s)$ in which D has full row (or column) rank, a right (or left) inverse of $G(s)$ can be found by replacing D^{-1} by $D^\dagger$, the pseudo-inverse of D.

Table 4.1: MATLAB commands to generate a normalized coprime factorization

```
% Uses the Mu toolbox
%
% Find Normalized Coprime factors of system [a,b,c,d] using (4.26)
%
S=eye(size(d'*d))+d'*d;
R=eye(size(d*d'))+d*d';
A1 = a-b*inv(S)*d'*c;
R1 = c'*inv(R)*c;
Q1 = b*inv(S)*b';
[z1,z2,fail,reig_min] = ric_schr([A1' -R1; -Q1 -A1]); Z = z2/z1;
% Alternative: aresolv in Robust control toolbox:
% [z1,z2,eig,zerr,zwellposed,Z] = aresolv(A1',Q1,R1);
H = -(b*d' + Z*c')*inv(R);
A = a + H*c;
Bn = b + H*d; Bm = H;
C = inv(sqrt(R))*c;
Dn = inv(sqrt(R))*d; Dm = inv(sqrt(R));
N = pck(A,Bn,C,Dn);
M = pck(A,Bm,C,Dm);
```

For a strictly proper system with $D = 0$, one may obtain an approximate inverse by including a small additional feed-through term D, preferably chosen on physical grounds. One should be careful, however, to select the signs of the terms in D such that one does not introduce RHP-zeros in $G(s)$ because this will make $G(s)^{-1}$ unstable.

Improper systems. Improper transfer functions, where the order of the s-polynomial in the numerator exceeds that of the denominator, cannot be represented in standard state-space form. To approximate improper systems by state-space models, we can include some high-frequency dynamics which we know from physical considerations will have little significance.

Realization of SISO transfer functions. Transfer functions are a good way of representing systems because they give more immediate insight into a systems behaviour. However, for numerical calculations a state-space realization is usually desired. One way of obtaining a state-space realization from a SISO transfer function is given next. Consider a strictly proper transfer function ($D = 0$) of the form

$$G(s) = \frac{\beta_{n-1}s^{n-1} + \cdots + \beta_1 s + \beta_0}{s^n + a_{n-1}s^{n-1} + \cdots + a_1 s + a_0} \tag{4.28}$$

Then, since multiplication by s corresponds to differentiation in the time domain, (4.28) and the relationship $y(s) = G(s)u(s)$ corresponds to the following differential equation

$$y^n(t) + a_{n-1}y^{n-1}(t) + \cdots + a_1 y'(t) + a_0 y(t) = \beta_{n-1}u^{n-1}(t) + \cdots + \beta_1 u'(t) + \beta_0 u(t)$$

where $y^{n-1}(t)$ and $u^{n-1}(t)$ represent $n-1$'th order derivatives, etc. We can further write this as

$$y^n = (-a_{n-1}y^{n-1} + \beta_{n-1}u^{n-1}) + \cdots + (-a_1 y' + \beta_1 u') + \underbrace{\underbrace{\underbrace{(-a_0 y + \beta_0 u)}_{x_n'}}_{x_{n-1}^2}}_{x_1^n}$$

where we have introduced new variables $x_1, x_2, \ldots x_n$ and we have $y = x_1$. Note that x_1^n is the n'th derivative of $x_1(t)$. With the notation $\dot{x} \equiv x'(t) = dx/dt$, we have the following state-space equations

$$
\begin{aligned}
\dot{x}_n &= -a_0 x_1 + \beta_0 u \\
\dot{x}_{n-1} &= -a_1 x_1 + x_n + \beta_1 u \\
&\ \vdots \\
\dot{x}_1 &= -a_{n-1} x_1 + x_2 + \beta_{n-1} u
\end{aligned}
$$

corresponding to the realization

$$
A = \begin{bmatrix}
-a_{n-1} & 1 & 0 & \cdots & 0 & 0 \\
-a_{n-2} & 0 & 1 & & 0 & 0 \\
\vdots & \vdots & & \ddots & & \vdots \\
-a_2 & 0 & 0 & & 1 & 0 \\
-a_1 & 0 & 0 & \cdots & 0 & 1 \\
-a_0 & 0 & 0 & \cdots & 0 & 0
\end{bmatrix}, \quad
B = \begin{bmatrix}
\beta_{n-1} \\
\beta_{n-2} \\
\vdots \\
\beta_2 \\
\beta_1 \\
\beta_0
\end{bmatrix}
\qquad (4.29)
$$

$$C = \begin{bmatrix} 1 & 0 & 0 & \cdots & 0 & 0 \end{bmatrix}$$

This is called the *observer canonical form*. Two advantages of this realization are that one can obtain the elements of the matrices directly from the transfer function, and that the output y is simply equal to the first state. Notice that if the transfer function is not strictly proper, then we must first bring out the constant term, i.e. write $G(s) = G_1(s) + D$, and then find the realization of $G_1(s)$ using (4.29).

Example 4.2 *To obtain the state-space realization, in observer canonical form, of the SISO transfer function $G(s) = \frac{s-a}{s+a}$, we first bring out a constant term by division to get*

$$G(s) = \frac{s-a}{s+a} = \frac{-2a}{s+a} + 1$$

Thus $D = 1$. For the term $\frac{-2a}{s+a}$ we get from (4.28) that $\beta_0 = -2a$ and $a_0 = a$, and therefore (4.29) yields $A = -a$, $B = -2a$ and $C = 1$.

Example 4.3 *Consider an ideal PID-controller*

$$K(s) = K_c(1 + \frac{1}{\tau_I s} + \tau_D s) = K_c \frac{\tau_I \tau_D s^2 + \tau_I s + 1}{\tau_I s} \qquad (4.30)$$

Since this involves differentiation of the input, it is an improper transfer function and cannot be written in state-space form. A proper PID controller may be obtained by letting the derivative action be effective over a limited frequency range. For example

$$K(s) = K_c(1 + \frac{1}{\tau_I s} + \frac{\tau_D s}{1 + \epsilon \tau_D s}) \tag{4.31}$$

where ϵ is typically 0.1 or less. This can now be realized in state-space form in an infinite number of ways. Four common forms are given below. In all cases, the D-matrix, which represents the controller gain at high frequencies ($s \to \infty$), is a scalar given by

$$D = K_c \frac{1 + \epsilon}{\epsilon} \tag{4.32}$$

1. Diagonalized form (Jordan canonical form)

$$A = \begin{bmatrix} 0 & 0 \\ 0 & -\frac{1}{\epsilon \tau_D} \end{bmatrix}, \quad B = \begin{bmatrix} K_c/\tau_I \\ K_c/(\epsilon^2 \tau_D) \end{bmatrix}, \quad C = \begin{bmatrix} 1 & -1 \end{bmatrix} \tag{4.33}$$

2. Observability canonical form

$$A = \begin{bmatrix} 0 & 1 \\ 0 & -\frac{1}{\epsilon \tau_D} \end{bmatrix}, \quad B = \begin{bmatrix} \gamma_1 \\ \gamma_2 \end{bmatrix}, \quad C = \begin{bmatrix} 1 & 0 \end{bmatrix} \tag{4.34}$$

$$\text{where} \quad \gamma_1 = K_c(\frac{1}{\tau_I} - \frac{1}{\epsilon^2 \tau_D}), \quad \gamma_2 = \frac{K_c}{\epsilon^3 \tau_D^2} \tag{4.35}$$

3. Controllability canonical form

$$A = \begin{bmatrix} 0 & 0 \\ 1 & -\frac{1}{\epsilon \tau_D} \end{bmatrix}, \quad B = \begin{bmatrix} 1 \\ 0 \end{bmatrix}, \quad C = \begin{bmatrix} \gamma_1 & \gamma_2 \end{bmatrix} \tag{4.36}$$

where γ_1 and γ_2 are as given above.
4. Observer canonical form in (4.29)

$$A = \begin{bmatrix} -\frac{1}{\epsilon \tau_D} & 1 \\ 0 & 0 \end{bmatrix}, \quad B = \begin{bmatrix} \beta_1 \\ \beta_0 \end{bmatrix}, \quad C = \begin{bmatrix} 1 & 0 \end{bmatrix} \tag{4.37}$$

$$\text{where} \quad \beta_0 = \frac{K_c}{\epsilon \tau_I \tau_D}, \quad \beta_1 = K_c \frac{\epsilon^2 \tau_D - \tau_I}{\epsilon^2 \tau_I \tau_D} \tag{4.38}$$

On comparing these four realizations with the transfer function model in (4.31), it is clear that the transfer function offers more immediate insight. One can at least see that it is a PID controller.

Time delay. A time delay (or dead time) is an infinite-dimensional system and not representable as a rational transfer function. For a state-space realization it must therefore be approximated. An n'th order approximation of a time delay θ may be obtained by putting n first-order Padé approximations in series

$$e^{-\theta s} \approx \frac{(1 - \frac{\theta}{2n} s)^n}{(1 + \frac{\theta}{2n} s)^n} \tag{4.39}$$

Alternative (and possibly better) approximations are in use, but the above approximation is often preferred because of its simplicity.

4.2   State controllability and state observability

Definition 4.1 State controllability. *The dynamical system* $\dot{x} = Ax + Bu$, *or equivalently the pair* (A, B), *is said to be state controllable if, for any initial state* $x(0) = x_0$, *any time* $t_1 > 0$ *and any final state* x_1, *there exists an input* $u(t)$ *such that* $x(t_1) = x_1$. *Otherwise the system is said to be state uncontrollable.*

There are many ways to check whether a system is state controllable.

1. We have that (A, B) is state controllable if and only if the controllability matrix

$$\mathcal{C} \triangleq [\,B \quad AB \quad A^2B \quad \cdots \quad A^{n-1}B\,] \tag{4.40}$$

   has rank n (full row rank). Here n is the number of states.
2. From (4.7) one can verify that a particular input which achieves $x(t_1) = x_1$ is

$$u(t) = -B^T e^{A^T(t_1 - t)} W_c(t_1)^{-1}(e^{At_1}x_0 - x_1) \tag{4.41}$$

   where $W_c(t)$ is the Gramian matrix at time t,

$$W_c(t) \triangleq \int_0^t e^{A\tau} BB^T e^{A^T\tau} d\tau \tag{4.42}$$

   Therefore, the system (A, B) is state controllable if and only if the Gramian matrix $W_c(t)$ has full rank (and thus is positive definite) for any $t > 0$. For a stable system (A is stable) we only need to consider $P \triangleq W_c(\infty)$, that is, the pair (A, B) is state controllable if and only if the controllability Gramian

$$P \triangleq \int_0^\infty e^{A\tau} BB^T e^{A^T\tau} d\tau \tag{4.43}$$

   is positive definite ($P > 0$) and thus has full rank n. P may also be obtained as the solution to the Lyapunov equation

$$AP + PA^T = -BB^T \tag{4.44}$$

3. Let p_i be the i'th eigenvalue of A and q_i the corresponding left eigenvector, $q_i^H A = p_i q_i^H$. Then the system is state controllable if and only if $B^H q_i \neq 0, \forall i$.

Remark. The last result has a simple interpretation. Let $u_{p_i} \triangleq B^H q_i$, then the mode $e^{p_i t}$ associated with p_i is (state) uncontrollable if and only if $u_{p_i} = 0$. This is useful for identifying the uncontrollable mode(s).

Example 4.4 *Consider a scalar system with <u>two</u> states and the following state-space realization*

$$A = \begin{bmatrix} -2 & -2 \\ 0 & -4 \end{bmatrix}, \ B = \begin{bmatrix} 1 \\ 1 \end{bmatrix}, \ C = [1 \quad 0], \ D = 0$$

The transfer function is

$$G(s) = C(sI - A)^{-1}B = \frac{1}{s+4}$$

which has only __one__ state. In fact, the first state is not controllable. This is verified by considering state controllability.

1. *The controllability matrix has two linearly dependent rows:*

$$C = [B \quad AB] = \begin{bmatrix} 1 & -4 \\ 1 & -4 \end{bmatrix}.$$

2. *The controllability Gramian is also singular*

$$P = \begin{bmatrix} 0.125 & 0.125 \\ 0.125 & 0.125 \end{bmatrix}$$

3. *The eigenvalues of A are $p_1 = -2$ and $p_2 = -4$, and the corresponding left eigenvectors are $q_1 = [0.707 \quad -0.707]^T$ and $q_2 = [0 \quad 1]^T$. We get*

$$B^H q_1 = 0, \quad B^H q_2 = 1$$

and we find that the first mode (eigenvalue) is not state controllable.

In words, if a system is state controllable we can by use of its inputs u bring it from any initial state to any final state within any given finite time. State controllability would therefore seem to be an important property for control, but it rarely is for the following four reasons:

1. It says nothing about how the states behave at earlier and later times, e.g. it does not imply that one can hold (as $t \to \infty$) the states at a given value.
2. The required inputs may be very large with sudden changes.
3. Some of the states may be of no practical importance.
4. The definition is an existence result which provides no degree of controllability (see Hankel singular values for this).

The first two objections are illustrated in the following example.

Example 4.5 Controllability of tanks in series.
Consider a system with one input and four states arising from four first-order systems in series,

$$G(s) = 1/(\tau s + 1)^4$$

A physical example could be four identical tanks (e.g. bath tubs) in series where water flows from one tank to the next. Energy balances, assuming no heat loss, yield $T_4 = \frac{1}{\tau s+1}T_3$, $T_3 = \frac{1}{\tau s+1}T_2$, $T_2 = \frac{1}{\tau s+1}T_1$, $T_1 = \frac{1}{\tau s+1}T_0$ where the states $x = [T_1 \quad T_2 \quad T_3 \quad T_4]^T$ are the four tank temperatures, the input $u = T_0$ is the inlet temperature, and $\tau = 100$ s is the residence time in each tank. A state-space realization is

$$A = \begin{bmatrix} -0.01 & 0 & 0 & 0 \\ 0.01 & -0.01 & 0 & 0 \\ 0 & 0.01 & -0.01 & 0 \\ 0 & 0 & 0.01 & -0.01 \end{bmatrix} \quad B = \begin{bmatrix} 0.01 \\ 0 \\ 0 \\ 0 \end{bmatrix} \qquad (4.45)$$

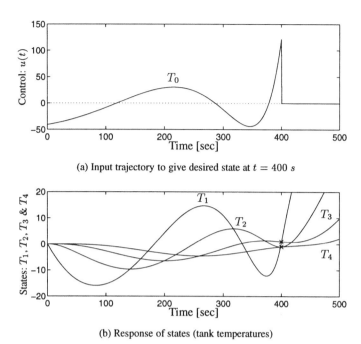

(a) Input trajectory to give desired state at $t = 400$ s

(b) Response of states (tank temperatures)

Figure 4.1: State controllability of four first-order systems in series

In practice, we know that it is very difficult to control the four temperatures independently, since at steady-state all temperatures must be equal. However, the controllability matrix C in (4.40) has full rank, so the system is state controllable and it must be possible to achieve at any given time any desired temperature in each of the four tanks simply by adjusting the inlet temperature. This sounds almost too good to be true, so let us consider a specific case. Assume that the system is initially at steady-state (all temperatures are zero), and that we want to achieve at $t = 400$ s the following temperatures: $T_1(400) = 1$, $T_2(400) = -1$, $T_3(400) = 1$ and $T_4(400) = -1$. The change in inlet temperature, $T_0(t)$, to achieve this was computed from (4.41) and is shown as a function of time in Figure 4.1(a). The corresponding tank temperatures are shown in Figure 4.1(b). Two things are worth noting:

1. *The required change in inlet temperature is more than 100 times larger than the desired temperature changes in the tanks and it also varies widely with time.*
2. *Although the tank temperatures are indeed at their desired values of ± 1 at $t = 400$ s they quickly diverge from these values for $t > 400$ s. (Since T_0 is reset to 0 at $t = 400$ s, all temperatures will eventually approach 0 as $t \to \infty$).*

It is quite easy to explain the shape of the input $T_0(t)$: The fourth tank is furthest away and we want its temperature to decrease ($T_4(400) = -1$) and therefore the inlet temperature T_0 is initially decreased to about -40. Then, since $T_3(400) = 1$ is positive, T_0 is increased to about 30 at $t = 220$ s; it is subsequently decreased to about -40, since $T_2(400) = -1$, and *finally increased to more than* 100 *to achieve* $T_1(400) = 1$.

From the above example, we see clearly that the property of state controllability may not imply that the system is "controllable" in a practical sense [2]. This is because state controllability is concerned only with the value of the states at *discrete* values of time (target hitting), while in most cases we want the outputs to remain close to some desired value (or trajectory) for all values of time, and without using inappropriate control signals.

So now we know that state controllability does not imply that the system is controllable from a practical point of view. But what about the reverse: If we do *not* have state controllability, is this an indication that the system is not controllable in a practical sense? In other words, should we be concerned if a system is not state controllable? In many cases the answer is "no", since we may not be concerned with the behaviour of the uncontrollable states which may be outside our system boundary or of no practical importance. If we are indeed concerned about these states then they should be included in the output set y. State uncontrollability will then appear as a rank deficiency in the transfer function matrix $G(s)$ (see functional controllability).

So is the issue of state controllability of any value at all? Yes, because it tells us whether we have included some states in our model which we have no means of affecting. It also tells us when we can save on computer time by deleting uncontrollable states which have no effect on the output for a zero initial state.

In summary, state controllability is a system theoretical concept which is important when it comes to computations and realizations. However, its name is somewhat misleading, and most of the above discussion might have been avoided if only Kalman, who originally defined state controllability, had used a different terminology. For example, better terms might have been "point-wise controllability" or "state affect-ability" from which it would have been understood that although all the states could be individually affected, we might not be able to control them independently over a period of time.

Definition 4.2 State observability. *The dynamical system* $\dot{x} = Ax + Bu$, $y = Cx + Du$ *(or the pair* (A, C)*) is said to be state observable if, for any time* $t_1 > 0$, *the initial state* $x(0) = x_0$ *can be determined from the time history of the input* $u(t)$ *and the output* $y(t)$ *in the interval* $[0, t_1]$. *Otherwise the system, or* (A, C), *is said to be state unobservable.*

1. The system (A, C) is state observable if and only if the observability matrix

$$\mathcal{O} \triangleq \begin{bmatrix} C \\ CA \\ \vdots \\ CA^{n-1} \end{bmatrix} \tag{4.46}$$

---

[2] In Chapter 5, we introduce a more practical concept of controllability which we call "input-output controllability".

has rank n (full column rank).

2. For a stable system we may consider the observability Gramian

$$Q \triangleq \int_0^\infty e^{A^T \tau} C^T C e^{A\tau} d\tau \tag{4.47}$$

which must have full rank n (and thus be positive definite) for the system to be state observable. Q can also be found as the solution to the following Lyapunov equation

$$A^T Q + QA = -C^T C \tag{4.48}$$

3. Let p_i be the i'th eigenvalue of A and t_i the corresponding eigenvector, $At_i = p_i t_i$. Then the system is state observable if and only if $Ct_i \neq 0, \forall i$.

Remark. The last result has a simple interpretation. Let $y_{p_i} \triangleq Ct_i$, then the mode $e^{p_i t}$ associated with p_i is (state) unobservable if and only if $y_{p_i} = 0$.

A system is state observable if we can obtain the value of all individual states by measuring the output $y(t)$ over some time period. However, even if a system is state observable it may not be observable in a practical sense. For example, obtaining $x(0)$ may require taking high-order derivatives of $y(t)$ which may be numerically poor and sensitive to noise. This is illustrated in the following example.

Example 4.5 (tanks in series) **continued.** *If we define* $y = T_4$ *(the temperature of the last tank), then* $C = \begin{bmatrix} 0 & 0 & 0 & 1 \end{bmatrix}$ *and we find that the observability matrix* $\mathcal{O}$ *has full column rank so all states are observable from* y. *However, consider a case where the initial temperatures in the tanks,* $T_i(0), i = 1, \ldots, 4$, *are non-zero (and unknown), and the inlet temperature* $T_0(t) = u(t)$ *is zero for* $t \geq 0$. *Then, from a practical point of view, it is clear that it is numerically very difficult to back-calculate, for example* $T_1(0)$ *based on measurements of* $y(t) = T_4(t)$ *over some interval* $[0, t_1]$, *although in theory all states are observable from the output.*

Definition 4.3 Minimal realization, McMillan degree and hidden mode. *A state-space realization* (A, B, C, D) *of* $G(s)$ *is said to be a minimal realization of* $G(s)$ *if A has the smallest possible dimension (i.e. the fewest number of states). The smallest dimension is called the* **McMillan** *degree of* $G(s)$. *A mode is hidden if it is not state controllable or observable and thus does not appear in the minimal realization.*

Since only controllable and observable states contribute to the input-output behaviour from u to y, it follows that a state-space realization is minimal if and only if (A, B) is state controllable and (A, C) is state observable.

Remark 1 Note that, uncontrollable states will contribute to the output response $y(t)$ if $x(t = 0) \neq 0$, but their effect will die out if the uncontrollable states are stable.

Remark 2 Unobservable states have no effect on the outputs whatsoever, and may be viewed as outside the system boundary, and thus of no direct interest from a control point of view (unless the unobservable state is unstable, because we want to avoid the system "blowing up"). However, observability is important for measurement selection and when designing state estimators (observers).

4.3 Stability

There are a number of ways in which stability may be defined, e.g. see Willems (1970). Fortunately, for linear time-invariant systems these differences have no practical significance, and we use the following definition:

Definition 4.4 *A system is* (**internally**) **stable** *if none of its components contain hidden unstable modes and the injection of bounded external signals at any place in the system result in bounded output signals measured anywhere in the system.*

We here define a signal $u(t)$ to be "bounded" if there exists a constant c such that $|u(t)| < c$ for all t. The word *internally* is included in the definition to stress that we do not only require the response from one particular input to another particular output to be stable, but require stability for signals injected or measured at any point of the system. This is discussed in more detail for feedback systems in Section 4.7. Similarly, the components must contain no hidden unstable modes, that is, any instability in the components must be contained in their input-output behaviour.

Definition 4.5 State stabilizable, state detectable and hidden unstable modes.
A system is state stabilizable if all unstable modes are state controllable. A system is state detectable if all unstable modes are state observable. A system with unstabilizable or undetectable modes is said to contain hidden unstable modes.

A linear system with a pair (A, B) is *state stabilizable* if and only if there exists a matrix F such that $A + BF$ is stable (Hurwitz, see Theorem 4.1), that is, there exists a state feedback $u = Fx$ such that the system is stable. If a system is not stabilizable then no such F exists. Similarly, a pair (A, C) is *state detectable* if and only if there exists a matrix L such that $A + LC$ is stable (Hurwitz). If a system is not state detectable, then there is a state within the system which will eventually grow out of bounds, but we have no way of observing this from the outputs $y(t)$.

Remark 1 Any unstable linear system can be stabilized by feedback control (at least in theory) provided the system contains no hidden unstable mode(s). However, this may require an unstable controller, see also the remark on page 185.

Remark 2 Systems with hidden unstable modes must be avoided both in practice and in computations (since variables will eventually blow up on our computer if not on the factory floor). In the book we always assume, unless otherwise stated, that our systems contain no hidden unstable modes.

4.4 Poles

For simplicity, we here define the poles of a system in terms of the eigenvalues of the state-space A-matrix. More generally, the poles of $G(s)$ may be somewhat loosely defined as the finite values $s = p$ where $G(p)$ has a singularity ("is infinite"), see also Theorem 4.2 below.

Definition 4.6 Poles. *The poles p_i of a system with state-space description (4.3)–(4.4) are the eigenvalues $\lambda_i(A), i = 1, \ldots, n$ of the matrix A. The pole or characteristic polynomial $\phi(s)$ is defined as $\phi(s) \triangleq \det(sI - A) = \prod_{i=1}^{n}(s - p_i)$. Thus the poles are the roots of the characteristic equation*

$$\phi(s) \triangleq \det(sI - A) = 0 \tag{4.49}$$

To see that this definition is reasonable, recall (4.16). Note that if A does not correspond to a minimal realization then the poles by this definition will include the poles (eigenvalues) corresponding to uncontrollable and/or unobservable states.

4.4.1   Poles and stability

For linear systems, the poles determine stability:

Theorem 4.1 *A linear dynamic system $\dot{x} = Ax + Bu$ is stable if and only if all the poles are in the open left-half plane (LHP), that is, $\mathrm{Re}\{\lambda_i(A)\} < 0, \forall i$. A matrix A with such a property is said to be "stable" or Hurwitz.*

Proof: Consider a diagonalized realization (Jordan form), from which we see that the time response in (4.7) can be written as a sum of terms each containing a *mode* $e^{\lambda_i(A)t}$, see also Appendix A.2.2. Eigenvalues in the RHP with $\mathrm{Re}\{\lambda_i(A)\} > 0$ give rise to *unstable modes* since in this case $e^{\lambda_i(A)t}$ is unbounded as $t \to \infty$. Eigenvalues in the open LHP give rise to stable modes where $e^{\lambda_i(A)t} \to 0$ as $t \to \infty$. Systems with poles on the $j\omega$-axis, including integrators, are unstable from our Definition 4.4 of stability. For example, consider $y = Gu$ and assume $G(s)$ has imaginary poles $s = \pm j\omega_o$. Then with a bounded sinusoidal input, $u(t) = \sin \omega_o t$, the output $y(t)$ grows unbounded as $t \to \infty$.   □

4.4.2   Poles from state-space realizations

Poles are usually obtained numerically by computing the eigenvalues of the A-matrix. To get the fewest number of poles we should use a minimal realization of the system.

4.4.3   Poles from transfer functions

The following theorem from MacFarlane and Karcanias (1976) allows us to obtain the poles directly from the transfer function matrix $G(s)$ and is also useful for hand calculations. It also has the advantage of yielding only the poles corresponding to a minimal realization of the system.

Theorem 4.2 *The pole polynomial $\phi(s)$ corresponding to a minimal realization of a system with transfer function $G(s)$, is the least common denominator of all non-identically-zero minors of all orders of $G(s)$.*

A *minor* of a matrix is the determinant of the matrix obtained by deleting certain rows and/or columns of the matrix. We will use the notation M_c^r to denote the minor corresponding to the deletion of rows r and columns c in $G(s)$. In the procedure defined by the theorem we cancel common factors in the numerator and denominator of each minor. It then follows that only observable and controllable poles will appear in the pole polynomial.

Example 4.6 *Consider the plant:* $G(s) = \frac{(3s+1)^2}{(s+1)} e^{-\theta s}$ *which has no state-space realization as it contains a delay and is also improper. Thus we can not compute the poles from (4.49). However from Theorem 4.2 we have that the denominator is $(s+1)$ and as expected $G(s)$ has a pole at $s = -1$.*

Example 4.7 *Consider the square transfer function matrix*

$$G(s) = \frac{1}{1.25(s+1)(s+2)} \begin{bmatrix} s-1 & s \\ -6 & s-2 \end{bmatrix} \tag{4.50}$$

The minors of order 1 are the four elements all have $(s+1)(s+2)$ in the denominator. The minor of order 2 is the determinant

$$\det G(s) = \frac{(s-1)(s-2) + 6s}{1.25^2(s+1)^2(s+2)^2} = \frac{1}{1.25^2(s+1)(s+2)} \tag{4.51}$$

Note the pole-zero cancellation when evaluating the determinant. The least common denominator of all the minors is then

$$\phi(s) = (s+1)(s+2) \tag{4.52}$$

so a minimal realization of the system has two poles: one at $s = -1$ and one at $s = -2$.

Example 4.8 *Consider the 2×3 system, with 3 inputs and 2 outputs,*

$$G(s) = \frac{1}{(s+1)(s+2)(s-1)} \begin{bmatrix} (s-1)(s+2) & 0 & (s-1)^2 \\ -(s+1)(s+2) & (s-1)(s+1) & (s-1)(s+1) \end{bmatrix} \tag{4.53}$$

The minors of order 1 are the five non-zero elements (e.g. $M_{2,3}^2 = g_{11}(s)$):

$$\frac{1}{s+1}, \quad \frac{s-1}{(s+1)(s+2)}, \quad \frac{-1}{s-1}, \quad \frac{1}{s+2}, \quad \frac{1}{s+2} \tag{4.54}$$

The minor of order 2 corresponding to the deletion of column 2 is

$$M_2 = \frac{(s-1)(s+2)(s-1)(s+1) + (s+1)(s+2)(s-1)^2}{((s+1)(s+2)(s-1))^2} = \frac{2}{(s+1)(s+2)} \tag{4.55}$$

The other two minors of order two are

$$M_1 = \frac{-(s-1)}{(s+1)(s+2)^2}, \quad M_3 = \frac{1}{(s+1)(s+2)} \tag{4.56}$$

By considering all minors we find their least common denominator to be

$$\phi(s) = (s+1)(s+2)^2(s-1) \tag{4.57}$$

The system therefore has four poles: one at $s = -1$, one at $s = 1$ and two at $s = -2$.

From the above examples we see that the MIMO-poles are essentially the poles of the elements. However, by looking at only the elements it is not possible to determine the multiplicity of the poles. For instance, let $G_0(s)$ be a square $m \times m$ transfer function matrix with no pole at $s = -a$, and consider

$$G(s) = \frac{1}{s+a} G_0(s) \tag{4.58}$$

How many poles at $s = -a$ does a minimal realization of $G(s)$ have? From (A.10),

$$\det\left(G(s)\right) = \det\left(\frac{1}{s+a} G_0(s)\right) = \frac{1}{(s+a)^m} \det\left(G_0(s)\right) \tag{4.59}$$

so if G_0 has no zeros at $s = -a$, then $G(s)$ has m poles at $s = -a$. However, G_0 may have zeros at $s = -a$. As an example, consider a 2×2 plant in the form given by (4.58). It may have two poles at $s = -a$ (as for $G(s)$ in (3.82)), one pole at $s = -a$ (as in (4.50) where $\det G_0(s)$ has a zero at $s = -a$) or no pole at $s = -a$ (if all the elements of $G_0(s)$ have a zero at $s = -a$).

As noted above, the poles are obtained numerically by computing the eigenvalues of the A-matrix. Thus, to compute the poles of a transfer function $G(s)$, we must first obtain a state-space realization of the system. Preferably this should be a minimal realization. For example, if we make individual realizations of the five non-zero elements in Example 4.8 and then simply combine them to get an overall state space realization, we will get a system with 15 states, where each of the three poles (in the common denominator) are repeated five times. A model reduction to obtain a minimal realization will subsequently yield a system with four poles as given in (4.57).

4.5   Zeros

Zeros of a system arise when competing effects, internal to the system, are such that the output is zero even when the inputs (and the states) are not themselves identically zero. For a SISO system the zeros z_i are the solutions to $G(z_i) = 0$. In general, it can be argued that zeros are values of s at which $G(s)$ loses rank (from rank 1 to rank 0 for a SISO system). This is the basis for the following definition of zeros for a multivariable system (MacFarlane and Karcanias, 1976).

Definition 4.7 Zeros. z_i *is a zero of* $G(s)$ *if the rank of* $G(z_i)$ *is less than the normal rank of* $G(s)$. *The zero polynomial is defined as* $z(s) = \prod_{i=1}^{n_z}(s - z_i)$ *where* n_z *is the number of finite zeros of* $G(s)$.

In this book we do not consider zeros at infinity; we require that z_i is finite. The normal rank of $G(s)$ is defined as the rank of $G(s)$ at all values of s except at a finite number of singularities (which are the zeros).

This definition of zeros is based on the transfer function matrix, corresponding to a minimal realization of a system. These zeros are sometimes called "transmission zeros", but we will simply call them "zeros". We may sometimes use the term "multivariable zeros" to distinguish them from the zeros of the elements of the transfer function matrix.

4.5.1   Zeros from state-space realizations

Zeros are usually computed from a state-space description of the system. First note that the state-space equations of a system may be written as

$$P(s)\begin{bmatrix} x \\ u \end{bmatrix} = \begin{bmatrix} 0 \\ y \end{bmatrix}, \quad P(s) = \begin{bmatrix} sI - A & -B \\ C & D \end{bmatrix} \tag{4.60}$$

The zeros are then the values $s = z$ for which the polynomial system matrix, $P(s)$, loses rank, resulting in zero output for some non-zero input. Numerically, the zeros are found as non-trivial solutions (with $u_z \neq 0$ and $x_z \neq 0$) to the following problem

$$(zI_g - M)\begin{bmatrix} x_z \\ u_z \end{bmatrix} = 0 \tag{4.61}$$

$$M = \begin{bmatrix} A & B \\ C & D \end{bmatrix}; \quad I_g = \begin{bmatrix} I & 0 \\ 0 & 0 \end{bmatrix} \tag{4.62}$$

This is solved as a generalized eigenvalue problem – in the conventional eigenvalue problem we have $I_g = I$. Note that we usually get additional zeros if the realization is not minimal.

4.5.2   Zeros from transfer functions

The following theorem from MacFarlane and Karcanias (1976) is useful for hand calculating the zeros of a transfer function matrix $G(s)$.

Theorem 4.3 *The zero polynomial $z(s)$, corresponding to a minimal realization of the system, is the greatest common divisor of all the numerators of all order-r minors of $G(s)$, where r is the normal rank of $G(s)$, provided that these minors have been adjusted in such a way as to have the pole polynomial $\phi(s)$ as their denominators.*

Example 4.9 *Consider the 2×2 transfer function matrix*

$$G(s) = \frac{1}{s+2}\begin{bmatrix} s-1 & 4 \\ 4.5 & 2(s-1) \end{bmatrix} \tag{4.63}$$

The normal rank of $G(s)$ is 2, and the minor of order 2 is the determinant, $\det G(s) = \frac{2(s-1)^2 - 18}{(s+2)^2} = 2\frac{s-4}{s+2}$. From Theorem 4.2, the pole polynomial is $\phi(s) = s + 2$ and therefore the zero polynomial is $z(s) = s - 4$. Thus, $G(s)$ has a single RHP-zero at $s = 4$.

This illustrates that in general multivariable zeros have no relationship with the zeros of the transfer function elements. This is also shown by the following example where the system has no zeros.

Example 4.7 continued. *Consider again the 2×2 system in (4.50) where* $\det G(s)$ *in (4.51) already has* $\phi(s)$ *as its denominator. Thus the zero polynomial is given by the numerator of (4.51), which is 1, and we find that the system has no multivariable zeros.*

The next two examples consider non-square systems.

Example 4.10 *Consider the 1×2 system*

$$G(s) = \left[\frac{s-1}{s+1} \quad \frac{s-2}{s+2} \right] \qquad (4.64)$$

The normal rank of $G(s)$ is 1, and since there is no value of s for which both elements become zero, $G(s)$ has no zeros.

In general, non-square systems are less likely to have zeros than square systems. For instance, for a square 2×2 system to have a zero, there must be a value of s for which the two columns in $G(s)$ are linearly dependent. On the other hand, for a 2×3 system to have a zero, we need all three columns in $G(s)$ to be linearly dependent.

The following is an example of a non-square system which does have a zero.

Example 4.8 continued. *Consider again the 2×3 system in (4.53), and adjust the minors of order 2 in (4.55) and (4.56) so that their denominators are* $\phi(s) = (s+1)(s+2)^2(s-1)$. *We get*

$$M_1(s) = \frac{-(s-1)^2}{\phi(s)}, \ M_2(s) = \frac{2(s-1)(s+2)}{\phi(s)}, \ M_3(s) = \frac{(s-1)(s+2)}{\phi(s)} \qquad (4.65)$$

The common factor for these minors is the zero polynomial

$$z(s) = (s-1) \qquad (4.66)$$

Thus, the system has a single RHP-zero located at $s = 1$.

We also see from the last example that a minimal realization of a MIMO system can have poles and zeros at the same value of s, provided their directions are different. This is discussed next.

4.6　More on poles and zeros

4.6.1　Directions of poles and zeros

In the following let s be a fixed complex scalar and consider $G(s)$ as a complex matrix. For example, given a state-space realization, we can evaluate $G(s) = C(sI - A)^{-1}B + D$.

Zero directions. Let $G(s)$ have a zero at $s = z$. Then $G(s)$ loses rank at $s = z$, and there will exist non-zero vectors u_z and y_z such that

$$G(z)u_z = 0, \quad y_z^H G(z) = 0 \qquad (4.67)$$

Here u_z is defined as the input zero direction, and y_z is defined as the output zero direction. We usually normalize the direction vectors to have unit length, i.e. $\|u_z\|_2 = 1$ and $\|y_z\|_2 = 1$. From a practical point of view, the output zero direction, y_z, is usually of more interest than u_z, because y_z gives information about which output (or combination of outputs) may be difficult to control.

In principle, we may obtain u_z and y_z from an SVD of $G(z) = U\Sigma V^H$, and we have that u_z is the last column in V (corresponding to the zero singular value of $G(z)$) and y_z is the last column of U. An example was given earlier in (3.64). A better approach numerically, is to obtain u_z from a state-space description using the generalized eigenvalue problem in (4.61). Similarly, y_z may be obtained from the transposed state-space description, using M^T in (4.62).

Pole directions. Let $G(s)$ have a pole at $s = p$. Then $G(p)$ is infinite, and we may somewhat crudely write

$$G(p)u_p = \infty, \quad y_p^H G(p) = \infty \qquad (4.68)$$

where u_p is defined as the input pole direction, and y_p is defined as the output pole direction. As for u_z and y_z, the vectors u_p and y_p may in principle be obtained from an SVD of $G(p) = U\Sigma V^H$. Then u_p is the first column in V (corresponding to the infinite singular value), and y_p the first column in U.

If we have a state-space realization of $G(s)$, then it is better to determine the pole directions from the right and left eigenvectors of A (Havre, 1995). Specifically, if p is a pole of $G(s)$, then p is an eigenvalue of A. Let t and q be the corresponding right and left eigenvectors, i.e.

$$At = pt, \quad q^H A = pq^H$$

then the pole directions are

$$y_p = Ct, \quad u_p = B^H q \qquad (4.69)$$

Remark 1 As one may expect, if $u_p = B^H q = 0$ then the corresponding pole is not state controllable, and if $y_p = Ct = 0$ the corresponding pole is not state observable (see also Zhou et al. (1996, p.52)).

Remark 2 If the inverse of $G(p)$ exists then it follows from the SVD that

$$G^{-1}(p)y_p = 0, \quad u_p^H G^{-1}(p) = 0 \qquad (4.70)$$

Example 4.11 *Consider the 2×2 plant in (4.63), which has a RHP-zero at $z = 4$ and a LHP-pole at $p = -2$. We will use an SVD of $G(z)$ and $G(p)$ to determine the zero and pole directions (but we stress that this is not a reliable method numerically). To find the zero direction consider*

$$G(z) = G(4) = \frac{1}{6} \begin{bmatrix} 3 & 4 \\ 4.5 & 6 \end{bmatrix} = \frac{1}{6} \begin{bmatrix} 0.55 & -0.83 \\ 0.83 & 0.55 \end{bmatrix} \begin{bmatrix} 9.01 & 0 \\ 0 & 0 \end{bmatrix} \begin{bmatrix} 0.6 & -0.8 \\ 0.8 & 0.6 \end{bmatrix}^H$$

The zero input and output directions are associated with the zero singular value of $G(z)$ and we get $u_z = \begin{bmatrix} -0.80 \\ 0.60 \end{bmatrix}$ and $y_z = \begin{bmatrix} -0.83 \\ 0.55 \end{bmatrix}$. We see from y_z that the zero has a slightly larger component in the first output. Next, to determine the pole directions consider

$$G(p + \epsilon) = G(-2 + \epsilon) = \frac{1}{\epsilon^2} \begin{bmatrix} -3 + \epsilon & 4 \\ 4.5 & 2(-3 + \epsilon) \end{bmatrix} \tag{4.71}$$

The SVD as $\epsilon \to 0$ yields

$$G(-2 + \epsilon) = \frac{1}{\epsilon^2} \begin{bmatrix} -0.55 & -0.83 \\ 0.83 & -0.55 \end{bmatrix} \begin{bmatrix} 9.01 & 0 \\ 0 & 0 \end{bmatrix} \begin{bmatrix} 0.6 & -0.8 \\ -0.8 & -0.6 \end{bmatrix}^H$$

The pole input and output directions are associated with the largest singular value, $\sigma_1 = 9.01/\epsilon^2$, and we get $u_p = \begin{bmatrix} 0.60 \\ -0.80 \end{bmatrix}$ and $y_p = \begin{bmatrix} -0.55 \\ 0.83 \end{bmatrix}$. We note from y_p that the pole has a slightly larger component in the second output.

Remark. It is important to note that although the locations of the poles and zeros are independent of input and output scalings, their directions are *not*. Thus, the inputs and outputs need to be scaled properly before making any interpretations based on pole and zero directions.

4.6.2   Remarks on poles and zeros

1. The zeros resulting from a minimal realization are sometimes called the *transmission zeros*. If one does *not* have a minimal realization, then numerical computations (e.g. using MATLAB) may yield additional *invariant zeros*. These invariant zeros plus the transmission zeros are sometimes called the *system zeros*. The invariant zeros can be further subdivided into *input and output decoupling zeros*. These cancel poles associated with uncontrollable or unobservable states and hence have limited practical significance. We recommend that a minimal realization is found before computing the zeros. Kailath). and

2. Rosenbrock (1966; 1970) first defined multivariable zeros using something similar to the Smith-McMillan form. Poles and zeros are defined in terms of the McMillan form in Zhou et al. (1996).

3. The presence of zeros implies blocking of certain input signals (MacFarlane and Karcanias, 1976). If z is a zero of $G(s)$, then there exists an input signal of the form $u_z e^{zt} 1_+(t)$, where u_z is a (complex) vector and $1_+(t)$ is a unit step, and a set of initial conditions (states) x_z, such that $y(t) = 0$ for $t > 0$.

4. For square systems we essentially have that the poles and zeros of $G(s)$ are the poles and zeros of det $G(s)$. However, this crude definition may fail in a few cases. For instance, when there is a zero and pole in different parts of the system which happen to cancel when forming

$\det G(s)$. For example, the system

$$G(s) = \begin{bmatrix} (s+2)/(s+1) & 0 \\ 0 & (s+1)/(s+2) \end{bmatrix} \tag{4.72}$$

has $\det G(s) = 1$, although the system obviously has poles at -1 and -2 and (multivariable) zeros at -1 and -2.

5. $G(s)$ in (4.72) provides a good example for illustrating the importance of *directions* when discussing poles and zeros of multivariable systems. We note that although the system has poles and zeros at the same locations (at -1 and -2), their directions are different and so they do not cancel or otherwise interact with each other. In (4.72) the pole at -1 has directions $u_p = y_p = \begin{bmatrix} 1 & 0 \end{bmatrix}^T$, whereas the zero at -1 has directions $u_z = y_z = \begin{bmatrix} 0 & 1 \end{bmatrix}^T$.

6. For square systems with a non-singular D-matrix, the number of poles is the same as the number of zeros, and the zeros of $G(s)$ are equal to the poles $G^{-1}(s)$, and *vice versa*.

7. There are no zeros if the outputs contain direct information about all the states; that is, if from y we can directly obtain x (e.g. $C = I$ and $D = 0$); see Example 4.13. This probably explains why zeros were given very little attention in the optimal control theory of the 1960's which was based on state feedback.

8. Zeros usually appear when there are fewer inputs or outputs than states, or when $D \neq 0$. Consider a square $m \times m$ plant $G(s) = C(sI - A)^{-1}B + D$ with n states. We then have for the number of (finite) zeros of $G(s)$ (Maciejowski, 1989, p.55)

$$\begin{array}{lll} D \neq 0: & \text{At most } n - m + \text{rank}(D) \text{ zeros} & \\ D = 0: & \text{At most } n - 2m + \text{rank}(CB) \text{ zeros} & (4.73) \\ D = 0 \text{ and rank}(CB) = m: & \text{Exactly } n - m \text{ zeros} & \end{array}$$

9. **Moving poles.** How are the poles affected by (a) feedback ($G(I + KG)^{-1}$), (b) series compensation (GK, feedforward control) and (c) parallel compensation ($G + K$)? The answer is that (a) feedback control moves the poles, (b) series compensation can cancel poles in G by placing zeros in K (but not move them), and (c) parallel compensation cannot affect the poles in G.

10. For a strictly proper plant $G(s) = C(sI - A)^{-1}B$, the open-loop poles are determined by the characteristic polynomial $\phi_{ol}(s) = \det(sI - A)$. If we apply constant gain negative feedback $u = -K_0 y$, the poles are determined by the corresponding closed-loop characteristic polynomial $\phi_{cl}(s) = \det(sI - A + BK_0C)$. Thus, unstable plants may be stabilized by use of feedback control. See also Example 4.12.

11. **Moving zeros.** Consider next the effect of feedback, series and parallel compensation on the zeros. (a) With feedback, the zeros of $G(I + KG)^{-1}$ are the zeros of G plus the poles of K. This means that the zeros in G, including their output directions y_z, are unaffected by feedback. However, even though y_z is fixed it is still possible with feedback control to move the deteriorating effect of a RHP-zero to a given output channel, provided y_z has a non-zero element for this output. This was illustrated by the example in Section 3.8, and is discussed in more detail in Section 6.5.1.

(b) Series compensation can counter the effect of zeros in G by placing poles in K to cancel them, but cancellations are not possible for RHP-zeros due to internal stability (see Section 4.7).

(c) The only way to move zeros is by parallel compensation, $y = (G + K)u$, which, if y is a physical output, can only be accomplished by adding an extra input (actuator).

12. **Pinned zeros.** A zero is pinned to a subset of the outputs if y_z has one or more elements equal to zero. In most cases, pinned zeros have a scalar origin. Pinned zeros are quite common in practice, and their effect cannot be moved freely to any output. For example, the effect of a measurement delay for output y_1 cannot be moved to output y_2. Similarly, a zero is pinned to certain inputs if u_z has one or more elements equal to zero. An example is $G(s)$ in (4.72), where the zero at -2 is pinned to input u_1 and to output y_1.

13. **Zeros of non-square systems.** The existence of zeros for non-square systems is common in practice in spite of what is sometimes claimed in the literature. In particular, they appear if we have a zero pinned to the side of the plant with the fewest number of channels. As an example consider a plant with three inputs and two outputs $G_1(s) = \begin{bmatrix} h_{11} & h_{12} & h_{13} \\ h_{21}(s-z) & h_{22}(s-z) & h_{23}(s-z) \end{bmatrix}$ which has a zero at $s = z$ which is pinned to output y_2, i.e. $y_z = \begin{bmatrix} 0 & 1 \end{bmatrix}^T$. This follows because the second row of $G_1(z)$ is equal to zero, so the rank of $G_1(z)$ is 1, which is less than the normal rank of $G_1(s)$, which is 2. On the other hand, $G_2(s) = \begin{bmatrix} h_{11}(s-z) & h_{12} & h_{13} \\ h_{21}(s-z) & h_{22} & h_{23} \end{bmatrix}$ does *not* have a zero at $s = z$ since $G_2(z)$ has rank 2 which is equal to the normal rank of $G_2(s)$ (assuming that the last two columns of $G_2(s)$ have rank 2).

14. The concept of functional controllability, see page 219, is related to zeros. Loosely speaking, one can say that a system which is functionally uncontrollable has in a certain output direction "a zero for all values of s".

The control implications of RHP-zeros and RHP-poles are discussed for SISO systems on pages 174-187 and for MIMO systems on pages 221-224.

Example 4.12 Effect of feedback on poles and zeros. *Consider a SISO negative feedback system with plant* $G(s) = z(s)/\phi(s)$ *and a constant gain controller,* $K(s) = k$. *The closed-loop response from reference* r *to output* y *is*

$$T(s) = \frac{L(s)}{1 + L(s)} = \frac{kG(s)}{1 + kG(s)} = \frac{kz(s)}{\phi(s) + kz(s)} = k\frac{z_{cl}(s)}{\phi_{cl}(s)} \qquad (4.74)$$

Note the following:

1. The zero polynomial is $z_{cl}(s) = z(s)$, *so the zero locations are unchanged by feedback.*
2. The pole locations are changed by feedback. For example,

$$k \to 0 \quad \Rightarrow \quad \phi_{cl}(s) \to \phi(s) \qquad (4.75)$$

$$k \to \infty \quad \Rightarrow \quad \phi_{cl}(s) \to kz(s) \qquad (4.76)$$

That is, as we increase the feedback gain, the closed-loop poles move from open-loop poles to the open-loop zeros. RHP-zeros therefore imply high gain instability. These results are well known from a classical root locus analysis.

Example 4.13 *We want to prove that* $G(s) = C(sI - A)^{-1}B + D$ *has no zeros if* $D = 0$ *and rank* $(C) = n$, *where* n *is the number of states. Solution: Consider the polynomial system matrix* $P(s)$ *in (4.60). The first* n *columns of* P *are independent because* C *has rank* n. *The last* m *columns are independent of* s. *Furthermore, the first* n *and last* m *columns are independent* *of each other, since $D = 0$ and C has full column rank and thus cannot have any columns equal to zero. In conclusion, $P(s)$ always has rank $n + m$ and there are no zeros. (We need $D = 0$ because if D is non-zero then the first n columns of P may depend on the last m columns for some value of s).*

Exercise 4.3 *Consider a SISO system $G(s) = C(sI - A)^{-1}B + D$ with just one state, i.e. A is a scalar. Find the zeros. Does $G(s)$ have any zeros for $D = 0$?*

Exercise 4.4 *Determine the poles and zeros of*

$$G(s) = \begin{bmatrix} \frac{11s^3 - 18s^2 - 70s - 50}{s(s+10)(s+1)(s-5)} & \frac{(s+2)}{(s+1)(s-5)} \\ \frac{5(s+2)}{(s+1)(s-5)} & \frac{5(s+2)}{(s+1)(s-5)} \end{bmatrix}$$

given that

$$\det G(s) = \frac{50(s^4 - s^3 - 15s^2 - 23s - 10)}{s(s+1)^2(s+10)(s-5)^2} = \frac{50(s+1)^2(s+2)(s-5)}{s(s+1)^2(s+10)(s-5)^2}$$

How many poles does $G(s)$ have?

Exercise 4.5 *Given $y(s) = G(s)u(s)$, with $G(s) = \frac{1-s}{1+s}$, determine a state-space realization of $G(s)$ and then find the zeros of $G(s)$ using the generalized eigenvalue problem. What is the transfer function from $u(s)$ to $x(s)$, the single state of $G(s)$, and what are the zeros of this transfer function?*

Exercise 4.6 *Find the zeros for a 2×2 plant with*

$$A = \begin{bmatrix} a_{11} & a_{12} \\ a_{21} & a_{22} \end{bmatrix}, \quad B = \begin{bmatrix} 1 & 1 \\ b_{21} & b_{22} \end{bmatrix}, \quad C = I, \quad D = 0$$

Exercise 4.7 *For what values of c_1 does the following plant have RHP-zeros?*

$$A = \begin{bmatrix} 10 & 0 \\ 0 & -1 \end{bmatrix}, \quad B = I, \quad C = \begin{bmatrix} 10 & c_1 \\ 10 & 0 \end{bmatrix}, \quad D = \begin{bmatrix} 0 & 0 \\ 0 & 1 \end{bmatrix} \quad (4.77)$$

Exercise 4.8 *Consider the plant in (4.77), but assume that both states are measured and used for feedback control, i.e. $y_m = x$ (but the controlled output is still $y = Cx + Du$). Can a RHP-zero in $G(s)$ give problems with stability in the feedback system? Can we achieve "perfect" control of y in this case? (Answers: No and no).*

4.7 Internal stability of feedback systems

To test for closed-loop stability of a feedback system, it is usually enough to check just one closed-loop transfer function, e.g. $S = (I + GK)^{-1}$. However, this assumes that there are no internal RHP pole-zero cancellations between the controller and the plant. The point is best illustrated by an example.

Example 4.14 *Consider the feedback system shown in Figure 4.2 where $G(s) = \frac{s-1}{s+1}$ and $K(s) = \frac{k}{s}\frac{s+1}{s-1}$. In forming the loop transfer function $L = GK$ we cancel the term $(s-1)$, a RHP pole-zero cancellation, to obtain*

$$L = GK = \frac{k}{s}, \text{ and } S = (I+L)^{-1} = \frac{s}{s+k} \qquad (4.78)$$

$S(s)$ is stable, that is, the transfer function from d_y to y is stable. However, the transfer function from d_y to u is unstable:

$$u = -K(I+GK)^{-1}d_y = -\frac{k(s+1)}{(s-1)(s+k)}d_y \qquad (4.79)$$

Consequently, although the system appears to be stable when considering the output signal y, it is unstable when considering the "internal" signal u, so the system is (internally) unstable.

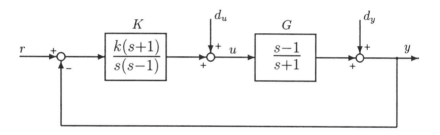

Figure 4.2: Internally unstable system

Remark. In practice, it is not possible to cancel exactly a plant zero or pole because of modelling errors. In the above example, therefore, L and S will in practice also be unstable. However, it is important to stress that even in the ideal case with a perfect RHP pole-zero cancellation, as in the above example, we would still get an internally unstable system. This is a subtle but important point. In this ideal case the state-space descriptions of L and S contain an unstable hidden mode corresponding to an unstabilizable or undetectable state.

From the above example, it is clear that to be rigorous we must consider *internal* stability of the feedback system, see Definition 4.4. To this effect consider the system in Figure 4.3 where we inject and measure signals at both locations between the two components, G and K. We get

$$u = (I+KG)^{-1}d_u - K(I+GK)^{-1}d_y \qquad (4.80)$$

$$y = G(I+KG)^{-1}d_u + (I+GK)^{-1}d_y \qquad (4.81)$$

The theorem below follows immediately:

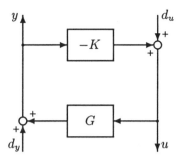

Figure 4.3: Block diagram used to check internal stability of feedback system

Theorem 4.4 *The feedback system in Figure 4.3 is* **internally stable** *if and only if all four closed-loop transfer matrices in (4.80) and (4.81) are stable.*

In addition, we assume as usual that the components G and K contain no unstable hidden modes. The signal relationship in the block diagram of Figure 4.3 may also be written as (see Exercise 4.9)

$$\begin{bmatrix} u \\ y \end{bmatrix} = M(s) \begin{bmatrix} d_u \\ d_y \end{bmatrix}; \quad M(s) = \begin{bmatrix} I & K \\ -G & I \end{bmatrix}^{-1} \tag{4.82}$$

and we get that the system is internally stable if and only if $M(s)$ is stable.

If we disallow RHP pole-zero cancellations between system components, such as G and K, then stability of *one* closed-loop transfer function implies stability of the others. This is stated in the following theorem.

Theorem 4.5 *Assume there are no RHP pole-zero cancellations between $G(s)$ and $K(s)$, that is, all RHP-poles in $G(s)$ and $K(s)$ are contained in the minimal realizations of GK and KG. Then the feedback system in Figure 4.3 is internally stable if and only if <u>one</u> of the four closed-loop transfer function matrices in (4.80) and (4.81) is stable.*

Proof: A proof is given by Zhou et al. (1996, p.125).   □

Note how we define pole-zero cancellations in the above theorem. In this way, RHP pole-zero cancellations resulting from G or K not having full normal rank are also disallowed. For example, with $G(s) = 1/(s - a)$ and $K = 0$ we get $GK = 0$ so the RHP-pole at $s = a$ has disappeared and there is effectively a RHP pole-zero cancellation. In this case, we get $S(s) = 1$ which is stable, but internal stability is clearly not possible.

Exercise 4.9 *Use (A.7) to show that $M(s)$ in (4.82) is identical to the block transfer function matrix implied by (4.80) and (4.81).*

4.7.1   Implications of the internal stability requirement

The requirement of internal stability in a feedback system leads to a number of interesting results, some of which are investigated below. Note in particular Exercise 4.12, where we discuss alternative ways of implementing a two degrees-of-freedom controller.

We first prove the following statements which apply when the overall feedback system is internally stable (Youla et al., 1974):

1. *If $G(s)$ has a RHP-zero at z, then $L = GK$, $T = GK(I + GK)^{-1}$, $SG = (I+GK)^{-1}G$, $L_I = KG$ and $T_I = KG(I+KG)^{-1}$ will each have a RHP-zero at z.*
2. *If $G(s)$ has a RHP-pole at p, then $L = GK$ and $L_I = KG$ also have a RHP-pole at p, while $S = (I+GK)^{-1}$, $KS = K(I+GK)^{-1}$ and $S_I = (I+KG)^{-1}$ have a RHP-zero at p.*

Proof of 1: To achieve internal stability, RHP pole-zero cancellations between system components, such as G and K, are not allowed. Thus $L = GK$ must have a RHP-zero when G has a RHP-zero. Now S is stable and thus has no RHP-pole which can cancel the RHP-zero in L, and so $T = LS$ must have a RHP-zero at z. Similarly, $SG = (I + GK)^{-1}G$ must have a RHP-zero, etc.   □

Proof of 2: Clearly, L has a RHP-pole at p. Since T is stable, it follows from $T = LS$ that S must have a RHP-zero which exactly cancels the RHP-pole in L, etc.   □

We notice from this that a RHP pole-zero cancellation between two transfer functions, such as between L and $S = (I+L)^{-1}$, does not necessarily imply internal instability. It is only between separate physical components (e.g. controller, plant) that RHP pole-zero cancellations are not allowed.

Exercise 4.10 Interpolation constraints. *Prove the following interpolation constraints which apply for SISO feedback systems when the plant $G(s)$ has a RHP-zero z or a RHP-pole p:*

$$G(z) = 0 \quad \Rightarrow \quad L(z) = 0 \quad \Leftrightarrow \quad T(z) = 0, S(z) = 1 \qquad (4.83)$$
$$G^{-1}(p) = 0 \quad \Rightarrow \quad L(p) = \infty \quad \Leftrightarrow \quad T(p) = 1, S(p) = 0 \qquad (4.84)$$

Exercise 4.11 *Given the complementary sensitivity functions*

$$T_1(s) = \frac{2s + 1}{s^2 + 0.8s + 1} \quad T_2(s) = \frac{-2s + 1}{s^2 + 0.8s + 1}$$

what can you say about possible RHP-poles or RHP-zeros in the corresponding loop transfer functions, $L_1(s)$ and $L_2(s)$?

Remark. A discussion of the significance of these interpolation constraints is relevant. Recall that for "perfect control" we want $S \approx 0$ and $T \approx 1$. We note from (4.83) that a RHP-zero z puts constraints on S and T which are incompatible with perfect control. On the other hand, the constraints imposed by the RHP-pole are consistent with what we would like for perfect control. Thus the presence of RHP-poles mainly impose problems when tight (high gain) control is *not* possible. We discuss this in more detail in Chapters 5 and 6.

The following exercise demonstrates another application of the internal stability requirement.

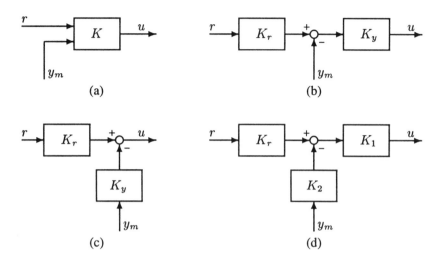

Figure 4.4: Different forms of two degrees-of-freedom controller
    (a) General form
    (b) Suitable when $K_y(s)$ has no RHP-zeros
    (c) Suitable when $K_y(s)$ is stable (no RHP-poles)
    (d) Suitable when $K_y(s) = K_1(s)K_2(s)$ where $K_1(s)$ contains no RHP-zeros and $K_2(s)$ no RHP poles

Exercise 4.12 Internal stability of two degrees-of-freedom control configurations. *A two degrees-of-freedom controller allows one to improve performance by treating disturbance rejection and command tracking separately (at least to some degree). The general form shown in Figure 4.4(a) is usually preferred both for implementation and design. However, in some cases one may want to first design the pure feedback part of the controller, here denoted $K_y(s)$, for disturbance rejection, and then to add a simple precompensator, $K_r(s)$, for command tracking. This approach is in general not optimal, and may also yield problems when it comes to implementation, in particular, if the feedback controller $K_y(s)$ contains RHP poles or zeros, which can happen. This implementation issue is dealt with in this exercise by considering the three possible schemes in Figure 4.4(b)–4.4(d). In all these schemes K_r must clearly be stable.*

1) Explain why the configuration in Figure 4.4(b) should not be used if K_y contains RHP-zeros (Hint: Avoid a RHP-zero between r and y).

2) Explain why the configuration in Figure 4.4(c) should not be used if K_y contains RHP-poles. This implies that this configuration should not be used if we want integral action in K_y (Hint: Avoid a RHP-zero between r and y).

3) Show that for a feedback controller K_y the configuration in Figure 4.4(d) may be used, provided the RHP-poles (including integrators) of K_y are contained in K_1 and the RHP-zeros in K_2. Discuss why one may often set $K_r = I$ in this case (to give a fourth possibility).

The requirement of internal stability also dictates that we must exercise care when we use a separate unstable disturbance model $G_d(s)$. To avoid this problem one should for state-space computations use a combined model for inputs and disturbances, i.e. write the model $y = Gu + G_d d$ in the form

$$y = [\, G \quad G_d \,] \begin{bmatrix} u \\ d \end{bmatrix}$$

where G and G_d share the same states, see (4.14) and (4.17).

4.8   Stabilizing controllers

In this section, we introduce a parameterization, known as the Q-parameterization or Youla-parameterization (Youla et al., 1976) of all stabilizing controllers for a plant. By all stabilizing controllers we mean all controllers that yield internal stability of the closed-loop system. We first consider stable plants, for which the parameterization is easily derived, and then unstable plants where we make use of the coprime factorization.

4.8.1   Stable plants

The following lemma forms the basis.

Lemma 4.6 *For a stable plant $G(s)$ the negative feedback system in Figure 4.3 is internally stable if and only if $Q = K(I + GK)^{-1}$ is stable.*

Proof: The four transfer functions in (4.80) and (4.81) are easily shown to be

$$K(I + GK)^{-1} = Q \tag{4.85}$$

$$(I + GK)^{-1} = I - GQ \tag{4.86}$$

$$(I + KG)^{-1} = I - QG \tag{4.87}$$

$$G(I + KG)^{-1} = G(I - QG) \tag{4.88}$$

which are clearly all stable if G and Q are stable. Thus, with G stable the system is internally stable if and only if Q is stable.   □

As proposed by Zames (1981), by solving (4.85) with respect to the controller K, we find that a parameterization of *all stabilizing negative feedback controllers for the stable plant $G(s)$* is given by

$$K = (I - QG)^{-1}Q = Q(I - GQ)^{-1} \tag{4.89}$$

where the "parameter" Q is *any stable transfer function matrix*.

Remark 1 If only proper controllers are allowed then Q must be proper since the term $(I - QG)^{-1}$ is semi-proper.

Remark 2 We have shown that by varying Q freely (but stably) we will always have internal stability, and thus avoid internal RHP pole-zero cancellations between K and G. This means that although Q may generate unstable controllers K, there is no danger of getting a RHP-pole in K that cancels a RHP-zero in G.

The parameterization in (4.89) is identical to the internal model control (IMC) parameterization (Morari and Zafiriou, 1989) of stabilizing controllers. It may be derived directly from the IMC structure given in Figure 4.5. The idea behind the IMC-structure is that the "controller" Q can be designed in an open-loop fashion since the feedback signal only contains information about the difference between the actual output and the output predicted from the model.

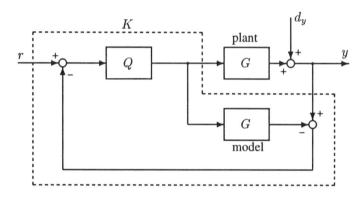

Figure 4.5: The internal model control (IMC) structure

Exercise 4.13 *Show that the IMC-structure in Figure 4.5 is internally unstable if either Q or G is unstable.*

Exercise 4.14 *Show that testing internal stability of the IMC-structure is equivalent to testing for stability of the four closed-loop transfer functions in (4.85)-(4.88).*

Exercise 4.15 *Given a stable controller K. What set of plants can be stabilized by this controller? (Hint: interchange the roles of plant and controller.)*

4.8.2 Unstable plants

For an unstable plant $G(s)$, consider its left coprime factorization

$$G(s) = M_l^{-1} N_l \tag{4.90}$$

A parameterization of *all stabilizing negative feedback controllers for the plant* $G(s)$ is then (Vidyasagar, 1985)

$$K(s) = (V_r - QN_l)^{-1}(U_r + QM_l) \qquad (4.91)$$

where V_r and U_r satisfy the Bezout identity (4.19) for the right coprime factorization, and $Q(s)$ is *any stable transfer function* satisfying the technical condition $\det(V_r(\infty) - Q(\infty)N_l(\infty)) \neq 0$.

Remark 1 With $Q = 0$ we have $K_0 = V_r^{-1}U_r$, so V_r and U_r can alternatively be obtained from a left coprime factorization of some initial stabilizing controller K_0.

Remark 2 For a stable plant, we may write $G(s) = N_l(s)$ corresponding to $M_l = I$. In this case $K_0 = 0$ is a stabilizing controller, so we may select $U_r = 0$ and $V_r = I$, and (4.91) yields $K = (I - QG)^{-1}Q$ as found before in (4.89).

Remark 3 All *closed-loop* transfer functions (S, T, etc.) will be in the form $H_1 + H_2QH_3$, so they are affine[3] in Q. This can be useful when Q is varied to minimize the norm of some closed-loop transfer function.

Remark 4 We can also formulate the parameterization of all stabilizing controllers in state-space form, e.g. see page 312 in Zhou et al. (1996) for details.

4.9   Stability analysis in the frequency domain

As noted above the stability of a linear system is equivalent to the system having no poles in the closed right-half plane (RHP). This test may be used for any system, be it open-loop or closed-loop. In this section we will study the use of frequency-domain techniques to derive information about *closed-loop* stability from the *open-loop* transfer matrix $L(j\omega)$. This provides a direct generalization of Nyquist's stability test for SISO systems.

Note that when we talk about eigenvalues in this section, we refer to the eigenvalues of a complex matrix, usually of $L(j\omega) = GK(j\omega)$, and not those of the state matrix A.

4.9.1   Open and closed-loop characteristic polynomials

We first derive some preliminary results involving the determinant of the return difference operator $I + L$. Consider the feedback system shown in Figure 4.6, where $L(s)$ is the loop transfer function matrix. Stability of the open-loop system is

---

[3] A function $f(x)$ is affine in x if $f(x) = ax + b$, and is linear in x if $f(x) = ax$.

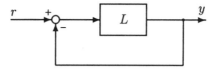

Figure 4.6: Negative feedback system

determined by the poles of $L(s)$. If $L(s)$ has a state-space realization $\left[\begin{array}{c|c} A_{ol} & B_{ol} \\ \hline C_{ol} & D_{ol} \end{array}\right]$, that is

$$L(s) = C_{ol}(sI - A_{ol})^{-1}B_{ol} + D_{ol} \qquad (4.92)$$

then the poles of $L(s)$ are the roots of the *open-loop* characteristic polynomial

$$\phi_{ol}(s) = \det(sI - A_{ol}) \qquad (4.93)$$

Assume there are no RHP pole-zero cancellations between $G(s)$ and $K(s)$. Then from Theorem 4.5 internal stability of the *closed-loop* system is equivalent to the stability of $S(s) = (I + L(s))^{-1}$. The state matrix of $S(s)$ is given (assuming $L(s)$ is well-posed, i.e. $D_{ol} \neq -I$) by

$$A_{cl} = A_{ol} - B_{ol}(I + D_{ol})^{-1}C_{ol} \qquad (4.94)$$

This equation may be derived by writing down the state-space equations for the transfer function from r to y in Figure 4.6

$$\dot{x} = A_{ol}x + B_{ol}(r - y) \qquad (4.95)$$

$$y = C_{ol}x + D_{ol}(r - y) \qquad (4.96)$$

and using (4.96) to eliminate y from (4.95). The closed-loop characteristic polynomial is thus given by

$$\phi_{cl}(s) \triangleq \det(sI - A_{cl}) = \det(sI - A_{ol} + B_{ol}(I + D_{ol})^{-1}C_{ol}) \qquad (4.97)$$

Relationship between characteristic polynomials

The above identities may be used to express the determinant of the return difference operator, $I + L$, in terms of $\phi_{cl}(s)$ and $\phi_{ol}(s)$. From (4.92) we get

$$\det(I + L(s)) = \det(I + C_{ol}(sI - A_{ol})^{-1}B_{ol} + D_{ol}) \qquad (4.98)$$

Schur's formula (A.14) then yields (with $A_{11} = I + D_{ol}, A_{12} = -C_{ol}, A_{22} = sI - A_{ol}, A_{21} = B_{ol}$)

$$\det(I + L(s)) = \frac{\phi_{cl}(s)}{\phi_{ol}(s)} \cdot c \qquad (4.99)$$

where $c = \det(I + D_{ol})$ is a constant which is of no significance when evaluating the poles. Note that $\phi_{cl}(s)$ and $\phi_{ol}(s)$ are polynomials in s which have zeros only, whereas $\det(I + L(s))$ is a transfer function with both poles and zeros.

Example 4.15 *We will rederive expression (4.99) for SISO systems. Let $L(s) = k\frac{z(s)}{\phi_{ol}(s)}$ The sensitivity function is given by*

$$S(s) = \frac{1}{1 + L(s)} = \frac{\phi_{ol}(s)}{kz(s) + \phi_{ol}(s)} \tag{4.100}$$

and the denominator is

$$d(s) = kz(s) + \phi_{ol}(s) = \phi_{ol}(s)(1 + \frac{kz(s)}{\phi_{ol}(s)}) = \phi_{ol}(s)(1 + L(s)) \tag{4.101}$$

which is the same as $\phi_{cl}(s)$ in (4.99) (except for the constant c which is necessary to make the leading coefficient of $\phi_{cl}(s)$ equal to 1, as required by its definition).

Remark 1 One may be surprised to see from (4.100) that the zero polynomial of $S(s)$ is equal to the open-loop pole polynomial, $\phi_{ol}(s)$, but this is indeed correct. On the other hand, note from (4.74) that the zero polynomial of $T(s) = L(s)/(1 + L(s))$ is equal to $z(s)$, the open-loop zero polynomial.

Remark 2 From (4.99), for the case when there are no cancellations between $\phi_{ol}(s)$ and $\phi_{cl}(s)$, we have that the closed-loop poles are solutions to

$$\det(I + L(s)) = 0 \tag{4.102}$$

4.9.2   MIMO Nyquist stability criteria

We will consider the negative feedback system of Figure 4.6, and assume there are no internal RHP pole-zero cancellations in the loop transfer function matrix $L(s)$, i.e. $L(s)$ contains no unstable hidden modes. Expression (4.99) for $\det(I + L(s))$ then enables a straightforward generalization of Nyquist's stability condition to multivariable systems.

Theorem 4.7 Generalized (MIMO) Nyquist theorem. *Let P_{ol} denote the number of open-loop unstable poles in $L(s)$. The closed-loop system with loop transfer function $L(s)$ and negative feedback is stable if and only if the Nyquist plot of $\det(I + L(s))$*

*   *i) makes P_{ol} anti-clockwise encirclements of the origin, and*
*   *ii) does not pass through the origin.*

The theorem is proved below, but let us first make some important remarks.

Remark 1 By "Nyquist plot of $\det(I + L(s))$" we mean "the image of $\det(I + L(s))$ as s goes clockwise around the Nyquist D-contour". The Nyquist D-contour includes the entire $j\omega$-axis ($s = j\omega$) and an infinite semi-circle into the right-half plane as illustrated in Figure 4.7. The D-contour must also avoid locations where $L(s)$ has $j\omega$-axis poles by making small indentations (semi-circles) around these points.

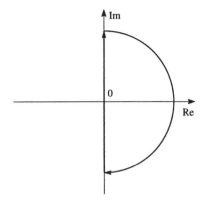

Figure 4.7: Nyquist D-contour for system with no open-loop $j\omega$-axis poles

Remark 2 In the following we define for practical reasons *unstable poles* or *RHP-poles* as poles in the *open* RHP, excluding the $j\omega$-axis. In this case the Nyquist D-contour should make a small semicircular indentation into the RHP at locations where $L(s)$ has $j\omega$-axis poles, thereby avoiding the extra count of encirclements due to $j\omega$-axis poles.

Remark 3 Another practical way of avoiding the indentation is to shift all $j\omega$-axis poles into the LHP, for example, by replacing the integrator $1/s$ by $1/(s+\epsilon)$ where ϵ is a small positive number.

Remark 4 We see that for stability $\det(I + L(j\omega))$ should make no encirclements of the origin if $L(s)$ is open-loop stable, and should make P_{ol} anti-clockwise encirclements if $L(s)$ is unstable. If this condition is not satisfied then the number of closed-loop unstable poles of $(I + L(s))^{-1}$ is $P_{cl} = \mathcal{N} + P_{ol}$, where $\mathcal{N}$ is the number of clockwise encirclements of the origin by the Nyquist plot of $\det(I + L(j\omega))$.

Remark 5 For any real system, $L(s)$ is proper and so to plot $\det(I + L(s))$ as s traverses the D-contour we need only consider $s = j\omega$ along the imaginary axis. This follows since $\lim_{s\to\infty} L(s) = D_{ol}$ is finite, and therefore for $s = \infty$ the Nyquist plot of $\det(I + L(s))$ converges to $\det(I + D_{ol})$ which is on the real axis.

Remark 6 In many cases $L(s)$ contains integrators so for $\omega = 0$ the plot of $\det(I + L(j\omega))$ may "start" from $\pm j\infty$. A typical plot for positive frequencies is shown in Figure 4.8 for the system

$$L = GK, \; G = \frac{3(-2s+1)}{(5s+1)(10s+1)}, \; K = 1.14\frac{12.7s+1}{12.7s} \tag{4.103}$$

Note that the solid and dashed curves (positive and negative frequencies) need to be connected as ω approaches 0, so there is also a large (infinite) semi-circle (not shown) corresponding to the indentation of the D-contour into the RHP at $s = 0$ (the indentation is to avoid the integrator in $L(s)$). To find which way the large semi-circle goes, one can use the rule (based on conformal mapping arguments) that a right-angled turn in the D-contour will result in a right-angled turn

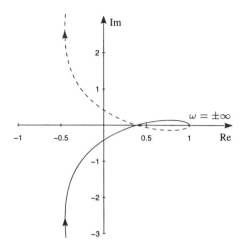

Figure 4.8: Typical Nyquist plot of $1 + \det L(j\omega)$

in the Nyquist plot. It then follows for the example in (4.103) that there will be an infinite semi-circle into the RHP. There are therefore no encirclements of the origin. Since there are no open-loop unstable poles ($j\omega$-axis poles are excluded in the counting), $P_{ol} = 0$, and we conclude that the closed-loop system is stable.

Proof of Theorem 4.7: The proof makes use of the following result from complex variable theory (Churchill et al., 1974):

Lemma 4.8 Argument Principle. *Consider a (transfer) function $f(s)$ and let C denote a closed contour in the complex plane. Assume that:*

1. *$f(s)$ is analytic along C, that is, $f(s)$ has no poles on C.*
2. *$f(s)$ has Z zeros inside C.*
3. *$f(s)$ has P poles inside C.*

Then the image $f(s)$ as the complex argument s traverses the contour C once in a clockwise direction will make $Z - P$ clockwise encirclements of the origin.

Let $\mathcal{N}(A, f(s), C)$ denote the number of clockwise encirclements of the point A by the image $f(s)$ as s traverses the contour C clockwise. Then a restatement of Lemma 4.8 is

$$\mathcal{N}(0, f(s), C) = Z - P \qquad (4.104)$$

We now recall (4.99) and apply Lemma 4.8 to the function $f(s) = \det(I + L(s)) = \frac{\phi_{cl}(s)}{\phi_{ol}(s)}c$ selecting C to be the Nyquist D-contour. We assume $c = \det(I + D_{ol}) \neq 0$ since otherwise the feedback system would be ill-posed. The contour D goes along the $j\omega$-axis and around the entire RHP, but avoids open-loop poles of $L(s)$ on the $j\omega$-axis (where $\phi_{ol}(j\omega) = 0$) by making small semi-circles into the RHP. This is needed to make $f(s)$ analytic along D. We then have that $f(s)$ has $P = P_{ol}$ poles and $Z = P_{cl}$ zeros inside D. Here P_{cl} denotes the number of unstable closed-loop poles (in the open RHP). (4.104) then gives

$$\mathcal{N}(0, \det(I + L(s)), D) = P_{cl} - P_{ol} \qquad (4.105)$$

Since the system is stable if and only if $P_{cl} = 0$, condition i) of Theorem 4.7 follows. However, we have not yet considered the possibility that $f(s) = \det(I + L(s))$, and hence $\phi_{cl}(s)$ has zeros on the D-contour itself, which will also correspond to a closed-loop unstable pole. To avoid this, $\det(I + L(j\omega))$ must not be zero for any value of ω and condition ii) in Theorem 4.7 follows.   □

Example 4.16 SISO stability conditions. *Consider an open-loop stable SISO system. In this case, the Nyquist stability condition states that for closed-loop stability the Nyquist plot of $1 + L(s)$ should not encircle the origin. This is equivalent to the Nyquist plot of $L(j\omega)$ not encircling the point -1 in the complex plane*

4.9.3 Eigenvalue loci

The eigenvalue loci (sometimes called characteristic loci) are defined as the eigenvalues of the frequency response of the open-loop transfer function, $\lambda_i(L(j\omega))$. They partly provide a generalization of the Nyquist plot of $L(j\omega)$ from SISO to MIMO systems, and with them gain and phase margins can be defined as in the classical sense. However, these margins are not too useful as they only indicate stability with respect to a *simultaneous parameter change* in all of the loops. Therefore, although characteristic loci were well researched in the 70's and greatly influenced the British developments in multivariable control, e.g. see Postlethwaite and MacFarlane (1979), they will not be considered further in this book.

4.9.4 Small gain theorem

The Small Gain Theorem is a very general result which we will find useful in the book. We present first a generalized version of it in terms of the spectral radius, $\rho(L(j\omega))$, which at each frequency is defined as the maximum eigenvalue magnitude

$$\rho(L(j\omega)) \triangleq \max_i |\lambda_i(L(j\omega))| \qquad (4.106)$$

Theorem 4.9 Spectral radius stability condition. *Consider a system with a stable loop transfer function $L(s)$. Then the closed-loop system is stable if*

$$\rho(L(j\omega)) < 1 \quad \forall \omega \qquad (4.107)$$

Proof: The generalized Nyquist theorem (Theorem 4.7) says that if $L(s)$ is stable, then the closed-loop system is stable if and only if the Nyquist plot of $\det(I + L(s))$ does not encircle the origin. To prove condition (4.107) we will prove the "reverse", that is, if the system is unstable and therefore $\det(I + L(s))$ does encircle the origin, then there is an eigenvalue, $\lambda_i(L(j\omega))$ which is larger than 1 at some frequency. If $\det(I + L(s))$ does encircle the origin, then there must exists a gain $\epsilon \in (0, 1]$ and a frequency ω' such that

$$\det(I + \epsilon L(j\omega')) = 0 \qquad (4.108)$$

This is easily seen by geometric arguments since $\det(I + \epsilon L(j\omega')) = 1$ for $\epsilon = 0$. (4.108) is equivalent to (see eigenvalue properties in Appendix A.2.1)

$$\prod_i \lambda_i(I + \epsilon L(j\omega')) = 0 \qquad (4.109)$$

$$\Leftrightarrow \quad 1 + \epsilon\lambda_i(L(j\omega')) = 0 \quad \text{for some } i \qquad (4.110)$$

$$\Leftrightarrow \quad \lambda_i(L(j\omega')) = -\frac{1}{\epsilon} \quad \text{for some } i \qquad (4.111)$$

$$\Rightarrow \quad |\lambda_i(L(j\omega'))| \geq 1 \quad \text{for some } i \qquad (4.112)$$

$$\Leftrightarrow \quad \rho(L(j\omega')) \geq 1 \qquad (4.113)$$

$$\square$$

Theorem 4.9 is quite intuitive, as it simply says that if the system gain is less than 1 in all directions (all eigenvalues) and for all frequencies ($\forall\omega$), then all signal deviations will eventually die out, and the system is stable.

In general, the spectral radius theorem is conservative because phase information is not considered. For SISO systems $\rho(L(j\omega)) = |L(j\omega)|$, and consequently the above stability condition requires that $|L(j\omega)| < 1$ for all frequencies. This is clearly conservative, since from the Nyquist stability condition for a stable $L(s)$, we need only require $|L(j\omega)| < 1$ at frequencies where the phase of $L(j\omega)$ is $-180° \pm n \cdot 360°$. As an example, let $L = k/(s + \epsilon)$. Since the phase never reaches $-180°$ the system is closed-loop stable for any value of $k > 0$. However, to satisfy (4.107) we need $k \leq \epsilon$, which for a small value of ϵ is very conservative indeed.

Remark. Later we will consider cases where the phase of L is allowed to vary freely, and in which case Theorem 4.9 is not conservative. Actually, a clever use of the above theorem is the main idea behind most of the conditions for robust stability and robust performance presented later in this book.

The small gain theorem below follows directly from Theorem 4.9 if we consider a matrix norm satisfying $\|AB\| \leq \|A\| \cdot \|B\|$, since at any frequency we then have $\rho(L) \leq \|L\|$ (see (A.116)).

Theorem 4.10 Small Gain Theorem. *Consider a system with a stable loop transfer function $L(s)$. Then the closed-loop system is stable if*

$$\|L(j\omega)\| < 1 \quad \forall\omega \qquad (4.114)$$

where $\|L\|$ denotes any matrix norm satisfying $\|AB\| \leq \|A\| \cdot \|B\|$.

Remark 1 This result is only a special case of a more general small gain theorem which also applies to many nonlinear systems (Desoer and Vidyasagar, 1975).

Remark 2 The small gain theorem does not consider phase information, and is therefore independent of the sign of the feedback.

Remark 3 Any induced norm can be used, for example, the singular value, $\bar{\sigma}(L)$.

Remark 4 The small gain theorem can be extended to include more than one block in the loop, e.g. $L = L_1 L_2$. In this case we get from (A.97) that the system is stable if $\|L_1\| \cdot \|L_2\| < 1$, $\forall \omega$.

Remark 5 The small gain theorem is generally more conservative than the spectral radius condition in Theorem 4.9. Therefore, the arguments on conservatism made Theorem 4.9 also apply to Theorem 4.10.

4.10 System norms

Figure 4.9: System G

Consider the system in Figure 4.9, with a stable transfer function matrix $G(s)$ and impulse response matrix $g(t)$. To evaluate the performance we ask the question: given information about the allowed input signals $w(t)$, how large can the outputs $z(t)$ become? To answer this, we must evaluate the relevant system norm.

We will here evaluate the output signal in terms of the usual 2-norm,

$$\|z(t)\|_2 = \sqrt{\sum_i \int_{-\infty}^{\infty} |z_i(\tau)|^2 d\tau} \qquad (4.115)$$

and consider three different choices for the inputs:

1. $w(t)$ is a series of unit impulses.
2. $w(t)$ is any signal satisfying $\|w(t)\|_2 = 1$.
3. $w(t)$ is any signal satisfying $\|w(t)\|_2 = 1$, but $w(t) = 0$ for $t \geq 0$, and we only measure $z(t)$ for $t \geq 0$.

The relevant system norms in the three cases are the $\mathcal{H}_2$, $\mathcal{H}_\infty$, and Hankel norms, respectively. The $\mathcal{H}_2$ and $\mathcal{H}_\infty$ norms also have other interpretations as are discussed below. We introduced the $\mathcal{H}_2$ and $\mathcal{H}_\infty$ norms in Section 2.7, where we also discussed the terminology. In Appendix A.5.7 we present a more detailed interpretation and comparison of these and other norms.

4.10.1 $\mathcal{H}_2$ norm

Consider a strictly proper system $G(s)$, i.e. $D = 0$ in a state-space realization. For the $\mathcal{H}_2$ norm we use the Frobenius norm spatially (for the matrix) and integrate over frequency, i.e.

$$\|G(s)\|_2 \triangleq \sqrt{\frac{1}{2\pi} \int_{-\infty}^{\infty} \underbrace{\text{tr}(G(j\omega)^H G(j\omega))}_{\|G(j\omega)\|_F^2 = \sum_{ij} |G_{ij}(j\omega)|^2} \, d\omega} \qquad (4.116)$$

We see that $G(s)$ must be strictly proper, otherwise the $\mathcal{H}_2$ norm is infinite. The $\mathcal{H}_2$ norm can also be given another interpretation. By Parseval's theorem, (4.116) is equal to the $\mathcal{H}_2$ norm of the impulse response

$$\|G(s)\|_2 = \|g(t)\|_2 \triangleq \sqrt{\int_0^{\infty} \underbrace{\text{tr}(g^T(\tau)g(\tau))}_{\|g(\tau)\|_F^2 = \sum_{ij} |g_{ij}(\tau)|^2} \, d\tau} \qquad (4.117)$$

Remark 1 Note that $G(s)$ and $g(t)$ are dynamic *systems* while $G(j\omega)$ and $g(\tau)$ are constant *matrices* (for a given value of ω or τ).

Remark 2 We can change the order of integration and summation in (4.117) to get

$$\|G(s)\|_2 = \|g(t)\|_2 = \sqrt{\sum_{ij} \int_0^{\infty} |g_{ij}(\tau)|^2 d\tau} \qquad (4.118)$$

where $g_{ij}(t)$ is the ij'th element of the impulse response matrix, $g(t)$. From this we see that the $\mathcal{H}_2$ norm can be interpreted as the 2-norm output resulting from applying unit impulses $\delta_j(t)$ to each input, one after another (allowing the output to settle to zero before applying an impulse to the next input). This is more clearly seen by writing $\|G(s)\|^2 = \sqrt{\sum_{i=1}^m \|z_i(t)\|_2^2}$ where $z_i(t)$ is the output vector resulting from applying a unit impulse $\delta_i(t)$ to the i'th input.

In summary, we have the following deterministic performance interpretation of the $\mathcal{H}_2$ norm:

$$\|G(s)\|_2 = \max_{w(t)=\text{ unit impulses}} \|z(t)\|_2 \qquad (4.119)$$

The $\mathcal{H}_2$ norm can also be given a stochastic interpretation (see page 365) in terms of the quadratic criterion in optimal control (LQG) where we measure the expected root mean square (rms) value of the output in response to white noise excitation.

For numerical computations of the $\mathcal{H}_2$ norm, consider the state-space realization $G(s) = C(sI - A)^{-1}B$. By substituting (4.10) into (4.117) we find

$$\|G(s)\|_2 = \sqrt{\text{tr}(B^T Q B)} \quad \text{or} \quad \|G(s)\|_2 = \sqrt{\text{tr}(CPC^T)} \qquad (4.120)$$

where Q and P are the observability and controllability Gramians, respectively, obtained as solutions to the Lyapunov equations (4.48) and (4.44).

4.10.2 $\mathcal{H}_\infty$ norm

Consider a proper linear stable system $G(s)$ (i.e. $D \neq 0$ is allowed). For the $\mathcal{H}_\infty$ norm we use the singular value (induced 2-norm) spatially (for the matrix) and pick out the peak value as a function of frequency

$$\|G(s)\|_\infty \triangleq \max_\omega \bar{\sigma}(G(j\omega)) \qquad (4.121)$$

In terms of *performance* we see from (4.121) that the $\mathcal{H}_\infty$ norm is the peak of the transfer function "magnitude", and by introducing weights, the $\mathcal{H}_\infty$ norm can be interpreted as the magnitude of some closed-loop transfer function relative to a specified upper bound. This leads to specifying performance in terms of weighted sensitivity, mixed sensitivity, and so on.

However, the $\mathcal{H}_\infty$ norm also has several time domain performance interpretations. First, as discussed in Section 3.3.5, it is the worst-case steady-state gain for sinusoidal inputs at any frequency. Furthermore, from Tables A.1 and A.2 in the Appendix we see that the $\mathcal{H}_\infty$ norm is equal to the induced (worst-case) 2-norm in the time domain:

$$\|G(s)\|_\infty = \max_{w(t)\neq 0} \frac{\|z(t)\|_2}{\|w(t)\|_2} = \max_{\|w(t)\|_2=1} \|z(t)\|_2 \qquad (4.122)$$

This is a fortunate fact from functional analysis which is proved, for example, in Desoer and Vidyasagar (1975). In essence, (4.122) arises because the worst input signal $w(t)$ is a sinusoid with frequency ω^* and a direction which gives $\bar{\sigma}(G(j\omega^*))$ as the maximum gain.

The $\mathcal{H}_\infty$ norm is also equal to the induced power (rms) norm, and also has an interpretation as an induced norm in terms of the expected values of stochastic signals. All these various interpretations make the $\mathcal{H}_\infty$ norm useful in engineering applications.

The $\mathcal{H}_\infty$ norm is usually computed numerically from a state-space realization as the smallest value of γ such that the Hamiltonian matrix H has no eigenvalues on the imaginary axis, where

$$H = \begin{bmatrix} A + BR^{-1}D^T C & BR^{-1}B^T \\ -C^T(I + DR^{-1}D^T)C & -(A + BR^{-1}D^T C)^T \end{bmatrix} \qquad (4.123)$$

and $R = \gamma^2 I - D^T D$, see Zhou et al. (1996, p.115). This is an iterative procedure, where one may start with a large value of γ and reduce it until imaginary eigenvalues for H appear.

4.10.3 Difference between the $\mathcal{H}_2$ and $\mathcal{H}_\infty$ norms

To understand the difference between the $\mathcal{H}_2$ and $\mathcal{H}_\infty$ norms, note that from (A.126) we can write the Frobenius norm in terms of singular values. We then have

$$\|G(s)\|_2 = \sqrt{\frac{1}{2\pi} \int_{-\infty}^{\infty} \sum_i \sigma_i^2(G(j\omega))d\omega} \qquad (4.124)$$

From this we see that minimizing the $\mathcal{H}_\infty$ norm corresponds to minimizing the peak of the largest singular value ("worst direction, worst frequency"), whereas minimizing the $\mathcal{H}_2$ norm corresponds to minimizing the sum of the square of all the singular values over all frequencies ("average direction, average frequency"). In summary, we have

- $\mathcal{H}_\infty$: "push down peak of largest singular value".
- $\mathcal{H}_2$: "push down whole thing" (all singular values over all frequencies).

Example 4.17 *We will compute the $\mathcal{H}_\infty$ and $\mathcal{H}_2$ norms for the following SISO plant*

$$G(s) = \frac{1}{s+a} \tag{4.125}$$

The $\mathcal{H}_2$ norm is

$$\|G(s)\|_2 = \left(\frac{1}{2\pi}\int_{-\infty}^{\infty}\underbrace{|G(j\omega)|^2}_{\frac{1}{\omega^2+a^2}}\,d\omega\right)^{\frac{1}{2}} = \left(\frac{1}{2\pi a}\left[\tan^{-1}(\frac{\omega}{a})\right]_{-\infty}^{\infty}\right)^{\frac{1}{2}} = \sqrt{\frac{1}{2a}} \tag{4.126}$$

To check Parseval's theorem we consider the impulse response

$$g(t) = \mathcal{L}^{-1}\left(\frac{1}{s+a}\right) = e^{-at}, t \geq 0 \tag{4.127}$$

and we get

$$\|g(t)\|_2 = \sqrt{\int_0^\infty (e^{-at})^2 dt} = \sqrt{\frac{1}{2a}} \tag{4.128}$$

as expected. The $\mathcal{H}_\infty$ norm is

$$\|G(s)\|_\infty = \max_\omega |G(j\omega)| = \max_\omega \frac{1}{(\omega^2+a^2)^{\frac{1}{2}}} = \frac{1}{a} \tag{4.129}$$

For interest, we also compute the 1-norm of the impulse response (which is equal to the induced ∞-norm in the time domain):

$$\|g(t)\|_1 = \int_0^\infty |\underbrace{g(t)}_{e^{-at}}|dt = \frac{1}{a} \tag{4.130}$$

In general, it can be shown that $\|G(s)\|_\infty \leq \|g(t)\|_1$, and this example illustrates that we may have equality.

Example 4.18 *There exists no general relationship between the $\mathcal{H}_2$ and $\mathcal{H}_\infty$ norms. As an example consider the two systems*

$$f_1(s) = \frac{1}{\epsilon s + 1}, \quad f_2(s) = \frac{\epsilon s}{s^2 + \epsilon s + 1} \tag{4.131}$$

and let $\epsilon \to 0$. Then we have for f_1 that the $\mathcal{H}_\infty$ norm is 1 and the $\mathcal{H}_2$ norm is infinite. For f_2 the $\mathcal{H}_\infty$ norm is again 1, but now the $\mathcal{H}_2$ norm is zero.

Why is the $\mathcal{H}_\infty$ norm so popular? In robust control we use the $\mathcal{H}_\infty$ norm mainly because it is convenient for representing unstructured model uncertainty, and because it satisfies the multiplicative property (A.97):

$$\|A(s)B(s)\|_\infty \leq \|A(s)\|_\infty \cdot \|B(s)\|_\infty \tag{4.132}$$

This follows from (4.122) which shows that the $\mathcal{H}_\infty$ norm is an induced norm.

What is wrong with the $\mathcal{H}_2$ norm? The $\mathcal{H}_2$ norm has a number of good mathematical and numerical properties, and its minimization has important engineering implications. However, the $\mathcal{H}_2$ norm is *not* an induced norm and does *not* satisfy the multiplicative property.

Example 4.19 *Consider again $G(s) = 1/(s+a)$ in (4.125), for which we found $\|G(s)\|_2 = \sqrt{1/2a}$. Now consider the $\mathcal{H}_2$ norm of $G(s)G(s)$:*

$$\|G(s)G(s)\|_2 = \sqrt{\int_0^\infty |\mathcal{L}^{-1}[(\underbrace{\frac{1}{s+a}}_{te^{-at}})^2]|^2} = \sqrt{\frac{1}{a}\frac{1}{2a}} = \sqrt{\frac{1}{a}}\|G(s)\|_2^2$$

and we find, for $a < 1$, that

$$\|G(s)G(s)\|_2 > \|G(s)\|_2 \cdot \|G(s)\|_2 \tag{4.133}$$

which does not satisfy the multiplicative property (A.97). On the other hand, the $\mathcal{H}_\infty$ norm does satisfy the multiplicative property, and for the specific example we have equality with $\|G(s)G(s)\|_\infty = \frac{1}{a^2} = \|G(s)\|_\infty \cdot \|G(s)\|_\infty$.

4.10.4   Hankel norm

In the following discussion, we aim at developing an understanding of the Hankel norm. The Hankel norm of a stable system $G(s)$ is obtained when one applies an input $w(t)$ up to $t = 0$ and measures the output $z(t)$ for $t > 0$, and selects $w(t)$ to maximize the ratio of the 2-norms of these two signals:

$$\|G(s)\|_H \triangleq \max_{w(t)} \frac{\sqrt{\int_0^\infty \|z(\tau)\|_2^2 d\tau}}{\sqrt{\int_{-\infty}^0 \|w(\tau)\|_2^2 d\tau}} \tag{4.134}$$

The Hankel norm is a kind of induced norm from past inputs to future outputs. Its definition is analogous to trying to pump a swing with limited input energy such that the subsequent length of jump is maximized as illustrated in Figure 4.10.

It may be shown that the *Hankel norm* is equal to

$$\|G(s)\|_H = \sqrt{\rho(PQ)} \tag{4.135}$$

where ρ is the spectral radius (maximum eigenvalue), P is the controllability Gramian defined in (4.43) and Q the observability Gramian defined in (4.47). The name

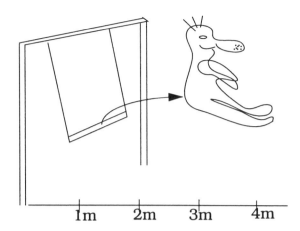

Figure 4.10: Pumping a swing: illustration of Hankel norm. The input is applied for $t \leq 0$ and the jump starts at $t = 0$.

"Hankel" is used because the matrix PQ has the special structure of a Hankel matrix (which has identical elements along the "wrong-way" diagonals). The corresponding *Hankel singular values* are the positive square roots of the eigenvalues of PQ,

$$\sigma_i = \sqrt{\lambda_i(PQ)} \qquad (4.136)$$

The Hankel and $\mathcal{H}_\infty$ norms are closely related and we have (Zhou et al., 1996, p.111)

$$\|G(s)\|_H \equiv \sigma_1 \leq \|G(s)\|_\infty \leq 2 \sum_{i=1}^n \sigma_i \qquad (4.137)$$

Thus, the Hankel norm is always smaller than (or equal to) the $\mathcal{H}_\infty$ norm, which is also reasonable by comparing the definitions in (4.122) and (4.134).

Model reduction. Consider the following problem: given a state-space description $G(s)$ of a system, find a model $G_a(s)$ with fewer states such that the input-output behaviour (from w to z) is changed as little as possible. Based on the discussion above it seems reasonable to make use of the Hankel norm, since the inputs only affect the outputs through the states at $t = 0$. For model reduction, we usually start with a realization of G which is internally balanced, that is, such that $Q = P = \Sigma$, where Σ is the matrix of Hankel singular values. We may then discard states (or rather combinations of states corresponding to certain subspaces) corresponding to the smallest Hankel singular values. The change in $\mathcal{H}_\infty$ norm caused by deleting states in $G(s)$ is less than twice the sum of the discarded Hankel singular values, i.e.

$$\|G(s) - G_a(s)\|_\infty \leq 2(\sigma_{k+1} + \sigma_{k+2} + \cdots) \qquad (4.138)$$

where $G_a(s)$ denotes a truncated or residualized balanced realization with k states; see Chapter 11. The method of Hankel norm minimization gives a somewhat improved error bound, where we are guaranteed that $\|G(s) - G_a(s)\|_\infty$ is less than the sum of the discarded Hankel singular values. This and other methods for model reduction are discussed in detail in Chapter 11 where a number of examples can be found.

Example 4.20 *We want to compute analytically the various system norms for $G(s) = 1/(s+a)$ using state-space methods. A state-space realization is $A = -a$, $B = 1$, $C = 1$ and $D = 0$. The controllability Gramian P is obtained from the Lyapunov equation $AP + PA^T = -BB^T \Leftrightarrow -aP - aP = -1$, so $P = 1/2a$. Similarly, the observability Gramian is $Q = 1/2a$. From (4.120) the $\mathcal{H}_2$ norm is then*

$$\|G(s)\|_2 = \sqrt{\text{tr}(B^T Q B)} = \sqrt{1/2a}$$

The eigenvalues of the Hamiltonian matrix H in (4.123) are

$$\lambda(H) = \lambda \begin{bmatrix} -a & 1/\gamma^2 \\ -1 & a \end{bmatrix} = \pm\sqrt{a^2 - 1/\gamma^2}$$

We find that H has no imaginary eigenvalues for $\gamma > 1/a$, so

$$\|G(s)\|_\infty = 1/a$$

The Hankel matrix is $PQ = 1/4a^2$ and from (4.135) the Hankel norm is

$$\|G(s)\|_H = \sqrt{\rho(PQ)} = 1/2a$$

These results agree with the frequency-domain calculations in Example 4.17.

Exercise 4.16 *Let $a = 0.5$ and $\epsilon = 0.0001$ and check numerically the results in Examples 4.17, 4.18, 4.19 and 4.20 using, for example, the MATLAB μ-toolbox commands* h2norm, hinfnorm, *and for the Hankel norm,* [sysb,hsig]=sysbal(sys); max(hsig).

4.11   Conclusion

This chapter has covered the following important elements of linear system theory: system descriptions, state controllability and observability, poles and zeros, stability and stabilization, and system norms. The topics are standard and the treatment is complete for the purposes of this book.

5

LIMITATIONS ON PERFORMANCE IN SISO SYSTEMS

In this chapter, we discuss the fundamental limitations on performance in SISO systems. We summarize these limitations in the form of a procedure for input-output controllability analysis, which is then applied to a series of examples. Input-output controllability of a plant is the ability to achieve acceptable control performance. Proper scaling of the input, output and disturbance variables prior to this analysis is critical.

5.1 Input-Output Controllability

In university courses on control, methods for controller design and stability analysis are usually emphasized. However, in practice the following three questions are often more important:

I. How well can the plant be controlled? Before starting any controller design one should first determine how easy the plant actually is to control. Is it a difficult control problem? Indeed, does there even exist a controller which meets the required performance objectives?

II. What control structure should be used? By this we mean what variables should we measure, which variables should we manipulate, and how are these variables best paired together? In other textbooks one can find qualitative rules for these problems. For example, in Seborg et al. (1989) in a chapter called "The art of process control", the following rules are given:

1. Control the outputs that are not self-regulating.
2. Control the outputs that have favourable dynamic and static characteristics, i.e. for each output, there should exist an input which has a significant, direct and rapid effect on it.
3. Select the inputs that have large effects on the outputs.
4. Select the inputs that rapidly affect the controlled variables

These rules are reasonable, but what is "self-regulating", "large", "rapid" and "direct". A major objective of this chapter is to quantify these terms.

III. How might the process be changed to improve control? For example, to reduce the effects of a disturbance one may in process control consider changing the size of a buffer tank, or in automotive control one might decide to change the properties of a spring. In other situations, the speed of response of a measurement device might be an important factor in achieving acceptable control.

The above three questions are each related to the inherent control characteristics of the process itself. We will introduce the term *input-output controllability* to capture these characteristics as described in the following definition.

Definition 5.1 (Input-output) controllability *is the ability to achieve acceptable control performance; that is, to keep the outputs (y) within specified bounds or displacements from their references (r), in spite of unknown but bounded variations, such as disturbances (d) and plant changes, using available inputs (u) and available measurements (y_m or d_m).*

In summary, a plant is controllable if there *exists* a controller (connecting plant measurements and plant inputs) that yields acceptable performance for all expected plant variations. Thus, *controllability is independent of the controller, and is a property of the plant (or process) alone.* It can only be affected by changing the plant itself, that is, by (plant) design changes. These may include:

- changing the apparatus itself, e.g. type, size, etc.
- relocating sensors and actuators
- adding new equipment to dampen disturbances
- adding extra sensors
- adding extra actuators
- changing the control objectives
- changing the configuration of the lower layers of control already in place

Whether or not the last two actions are design modifications is arguable, but at least they address important issues which are relevant before the controller is designed.

Early work on input-output controllability analysis includes that of Ziegler and Nichols (1943), Rosenbrock (1970) and Morari (1983) who made use of the concept of "perfect control". Important ideas on performance limitations are also found in Bode (1945), Horowitz (1963), Frank (1968a; 1968b), Horowitz and Shaked (1975), Zames (1981), Doyle and Stein (1981), Francis and Zames (1984), Boyd and Desoer (1985), Kwakernaak (1985), Freudenberg (1985; 1988), Engell (1988), Morari and Zafiriou (1989), Boyd and Barratt (1991), and Chen (1995). We also refer the reader to two IFAC workshops on *Interactions between process design and process control* (Perkins, 1992; Zafiriou, 1994).

5.1.1 Input-output controllability analysis

Input-output controllability analysis is applied to a plant to find out what control performance can be expected. Another term for input-output controllability analysis is *performance targeting*.

Surprisingly, given the plethora of mathematical methods available for control system design, the methods available for controllability analysis are largely qualitative. In most cases the "simulation approach" is used i.e. performance is assessed by exhaustive simulations. However, this requires a specific controller design and specific values of disturbances and setpoint changes. Consequently, with this approach, one can never know if the result is a fundamental property of the plant, or if it depends on the specific controller designed, the disturbances or the setpoints.

A rigorous approach to controllability analysis would be to formulate mathematically the control objectives, the class of disturbances, the model uncertainty, etc., and then to synthesize controllers to see whether the objectives can be met. With model uncertainty this involves designing a μ-optimal controller (see Chapter 8). However, in practice such an approach is difficult and time consuming, especially if there are a large number of candidate measurements or actuators; see Chapter 10. More desirable, is to have a few simple tools which can be used to get a rough idea of how easy the plant is to control, i.e. to determine whether or not a plant is controllable, without performing a detailed controller design. The main objective of this chapter is to derive such controllability tools based on appropriately scaled models of $G(s)$ and $G_d(s)$.

An apparent shortcoming of the controllability analysis presented in this book is that all the tools are linear. This may seem restrictive, but usually it is not. In fact, one of the most important nonlinearities, namely that associated with input constraints, can be handled quite well with a linear analysis. Also, to deal with slowly varying changes one may perform a controllability analysis at several selected operating points. Nonlinear simulations to validate the linear controllability analysis are of course still recommended. Experience from a large number of case studies confirms that the linear measures are often very good.

5.1.2 Scaling and performance

The above definition of controllability does not specify the allowed bounds for the displacements or the expected variations in the disturbance; that is, no definition of the desired performance is included. Throughout this chapter and the next, when we discuss controllability, we will assume that the variables and models have been scaled as outlined in Section 1.4, so that the requirement for acceptable performance is:

- To keep the output $y(t)$ within the range $r-1$ to $r+1$ (at least most of the time) for any disturbance $d(t)$ between -1 and 1 and any reference $r(t)$ between $-R$ and R, using an input $u(t)$ within the range -1 to 1.

We will interpret this definition from a frequency-by-frequency sinusoidal point of view, i.e. $d(t) = \sin \omega t$, and so on. We then have:

- At each frequency the performance requirement is to keep the control error $|e(\omega)| \leq 1$, for any disturbance $|d(\omega)| \leq 1$ and any reference $|r(\omega)| \leq R(\omega)$, using an input $|u(\omega)| \leq 1$.

It is impossible to track very fast reference changes, so we will assume that $R(\omega)$ is frequency-dependent; for simplicity we assume that $R(\omega)$ is R (a constant) up to the frequency ω_r and is 0 above that frequency.

It could also be argued that the magnitude of the sinusoidal disturbances should approach zero at high frequencies. While this may be true, we really only care about frequencies within the bandwidth of the system, and in most cases it is reasonable to assume that the plant experiences sinusoidal disturbances of constant magnitude up to this frequency. Similarly, it might also be argued that the allowed control error should be frequency dependent. For example, we may require no steady-state offset, i.e. e should be zero at low frequencies. However, including frequency variations is not recommended when doing a preliminary analysis (however, one may take such considerations into account when interpreting the results).

Recall that with $r = R\tilde{r}$ (see Section 1.4) the control error may be written as

$$e = y - r = Gu + G_d d - R\tilde{r} \qquad (5.1)$$

where R is the magnitude of the reference and $|\tilde{r}(\omega)| \leq 1$ and $|d(\omega)| \leq 1$ are unknown signals. We will use (5.1) to unify our treatment of disturbances and references. Specifically, we will derive results for disturbances, which can then be applied directly to the references by replacing G_d by $-R$.

5.1.3   Remarks on the term controllability

The above definition of (input-output) controllability is in tune with most engineers' intuitive feeling about what the term means, and was also how the term was used historically in the control literature. For example, Ziegler and Nichols (1943) defined *controllability* as *"the ability of the process to achieve and maintain the desired equilibrium value"*. Unfortunately, in the 60's "controllability" became synonymous with the rather narrow concept of "state controllability" introduced by Kalman, and the term is still used in this restrictive manner by the system theory community. *State controllability* is the ability to bring a system from a given initial state to any final state within a finite time. However, as shown in Example 4.5 this gives no regard to the quality of the response between and after these two states and the required inputs may be excessive. The concept of *state controllability* is important for realizations and numerical calculations, but as long as we know that all the unstable modes are both controllable and observable, it usually has little practical significance. For example, Rosenbrock (1970, p. 177) notes that "most industrial plants are controlled quite satisfactorily though they are not [state] controllable". And conversely, there are many systems, like the tanks in series Example 4.5, which are state controllable, but which are not input-output controllable. To avoid any confusion between practical controllability and Kalman's state controllability, Morari (1983) introduced the term *dynamic resilience*. However, this term does not capture the fact that it is related to control, so instead we prefer the term *input-output controllability*, or simply *controllability* when it is clear we are not referring to state controllability.

Where are we heading? In this chapter we will discuss a number of results related to achievable performance. Many of the results can be formulated as upper and lower bounds on the bandwidth of the system. However, as noted in Section 2.4.5, there are several definitions of bandwidth (ω_B, ω_c and ω_{BT}) in terms of the transfer functions S, L and T, but since we are looking for approximate bounds we will not be too concerned with these differences. The main results are summarized at end of the chapter in terms of eight controllability rules.

5.2 Perfect control and plant inversion

A good way of obtaining insight into the inherent limitations on performance originating in the plant itself, is to consider the inputs needed to achieve *perfect control*. Let the plant model be

$$y = Gu + G_d d \qquad (5.2)$$

"Perfect control" (which, of course, cannot be realized in practice) is achieved when the output is identically equal to the reference, i.e. $y = r$. To find the corresponding plant input set $y = r$ and solve for u in (5.2):

$$u = G^{-1}r - G^{-1}G_d d \qquad (5.3)$$

(5.3) represents a perfect feedforward controller, assuming d is measurable. When feedback control $u = K(r - y)$ is used, we have from (2.20) that

$$u = KSr - KSG_d d$$

or since the complementary sensitivity function is $T = GKS$,

$$u = G^{-1}Tr - G^{-1}TG_d d \qquad (5.4)$$

We see that at frequencies where feedback is effective and $T \approx I$ (these arguments also apply to MIMO systems), the input generated by feedback in (5.4) is the same as the perfect control input in (5.3). That is, high gain feedback generates an inverse of G even though the controller K may be very simple.

An important lesson therefore is that perfect control requires the controller to somehow generate an inverse of G. From this we get that perfect control *cannot* be achieved if

- G contains RHP-zeros (since then G^{-1} is unstable)
- G contains time delay (since then G^{-1} contains a prediction)
- G has more poles than zeros (since then G^{-1} is unrealizable)

In addition, for feedforward control we have that perfect control *cannot* be achieved if

- G is uncertain (since then G^{-1} cannot be obtained exactly)

The last restriction may be overcome by high gain feedback, but we know that we cannot have high gain feedback at all frequencies.

The required input in (5.3) must not exceed the maximum physically allowed value. Therefore, perfect control *cannot* be achieved if

- $|G^{-1}G_d|$ is large
- $|G^{-1}R|$ is large

where "large" with our scaled models means larger than 1. There are also other situations which make control difficult such as

- G is unstable
- $|G_d|$ is large

If the plant is unstable, the outputs will "take off", and eventually hit physical constraints, unless feedback control is applied to stabilize the system. Similarly, if $|G_d|$ is large, then without control a disturbance will cause the outputs to move far away from their desired values. So in both cases control is required, and problems occur if this demand for control is somehow in conflict with the other factors mentioned above which also make control difficult. We have assumed perfect measurements in the discussion so far, but in practice, noise and uncertainty associated with the measurements of disturbances and outputs will present additional problems for feedforward and feedback control, respectively.

5.3   Constraints on S and T

In this section, we present some fundamental algebraic and analytic constraints which apply to the sensitivity S and complementary sensitivity T.

5.3.1 S plus T is one

From the definitions $S = (I + L)^{-1}$ and $T = L(I + L)^{-1}$ we derive

$$S + T = 1 \tag{5.5}$$

(or $S + T = I$ for a MIMO system). Ideally, we want S small to obtain the benefits of feedback (small control error for commands and disturbances), and T small to avoid sensitivity to noise which is one of the disadvantages of feedback. Unfortunately, these requirements are not simultaneously possible at any frequency as is clear from (5.5). Specifically, (5.5) implies that at any frequency either $|S(j\omega)|$ or $|T(j\omega)|$ must be larger than or equal to 0.5.

5.3.2 The waterbed effects (sensitivity integrals)

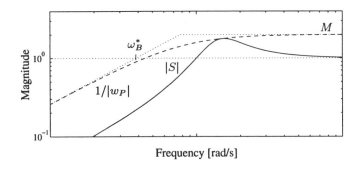

Figure 5.1: Plot of typical sensitivity, $|S|$, with upper bound $1/|w_P|$

A typical sensitivity function is shown by the solid line in Figure 5.1. We note that $|S|$ has a peak value greater than 1; we will show that this peak is unavoidable in practice. Two formulas are given, in the form of theorems, which essentially say that if we push the sensitivity down at some frequencies then it will have to increase at others. The effect is similar to sitting on a waterbed: pushing it down at one point, which reduces the water level locally will result in an increased level somewhere else on the bed. In general, a trade-off between sensitivity reduction and sensitivity increase must be performed whenever:

1. $L(s)$ has at least two more poles than zeros (first waterbed formula), or
2. $L(s)$ has a RHP-zero (second waterbed formula).

Pole excess of two: First waterbed formula

To motivate the first waterbed formula consider the open-loop transfer function $L(s) = \frac{1}{s(s+1)}$. As shown in Figure 5.2, there exists a frequency range over which the Nyquist plot of $L(j\omega)$ is inside the unit circle centred on the point -1, such that $|1 + L|$, which is the distance between L and -1, is less than one, and thus $|S| = |1 + L|^{-1}$ is greater than one. In practice, $L(s)$ will have *at least* two more poles than zeros (at least at sufficiently high frequency, e.g. due to actuator and measurement dynamics), so there will always exist a frequency range over which $|S|$ is greater than one. This behaviour may be quantified by the following theorem, of which the stable case is a classical result due to Bode.

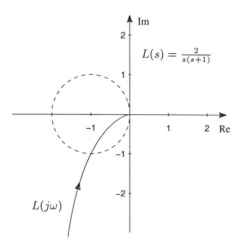

Figure 5.2: $|S| > 1$ whenever the Nyquist plot of L is inside the circle

Theorem 5.1 Bode Sensitivity Integral. *Suppose that the open-loop transfer function $L(s)$ is rational and has at least two more poles than zeros (relative degree of two or more). Suppose also that $L(s)$ has N_p RHP-poles at locations p_i. Then for closed-loop stability the sensitivity function must satisfy*

$$\int_0^\infty \ln|S(j\omega)| dw = \pi \cdot \sum_{i=1}^{N_p} Re(p_i) \tag{5.6}$$

where $Re(p_i)$ denotes the real part of p_i.

Proof: See Doyle et al. (1992, p. 100) or Zhou et al. (1996). The generalization of Bode's criterion to unstable plants is due to Freudenberg (1985; 1988).   □

For a graphical interpretation of (5.6) note that the magnitude scale is logarithmic whereas the frequency-scale is <u>linear</u>.

Stable plant. For a stable plant we must have

$$\int_0^\infty \ln|S(j\omega)| dw = 0 \tag{5.7}$$

and the area of sensitivity reduction ($\ln|S|$ negative) must *equal* the area of sensitivity increase ($\ln|S|$ positive). In this respect, the benefits and costs of feedback are balanced exactly, as in the waterbed analogy. From this we expect that an increase in the bandwidth (S smaller than 1 over a larger frequency range) must come at the expense of a larger peak in $|S|$.

Remark. Although this is true in most practical cases, the effect may not be so striking in some cases, and it is not strictly implied by (5.6) anyway. This is because the increase in area may come over an infinite frequency range; imagine a waterbed of infinite size. Consider $|S(j\omega)| = 1+\delta$ for $\omega \in [\omega_1, \omega_2]$, where δ is arbitrarily small (small peak), then we can choose ω_1 arbitrary large (high bandwidth) simply by selecting the interval $[\omega_1, \omega_2]$ to be sufficiently large. However, in practice the frequency response of L has to roll off at high frequencies so ω_2 is limited, and (5.6) and (5.7) impose real design limitations.

Unstable plant. The presence of unstable poles usually increases the peak of the sensitivity, as seen from the positive contribution $\pi \cdot \sum_{i=1}^{N_p} Re(p_i)$ in (5.6). Specifically, the area of sensitivity increase ($|S| > 1$) *exceeds* that of sensitivity reduction by an amount proportional to the sum of the distance from the unstable poles to the left-half plane. This is plausible since we might expect to have to pay a price for stabilizing the system.

RHP-zeros: Second waterbed formula

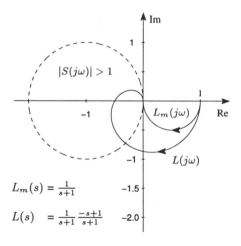

$$L_m(s) = \frac{1}{s+1}$$

$$L(s) = \frac{1}{s+1}\frac{-s+1}{s+1}$$

Figure 5.3: Additional phase lag contributed by RHP-zero causes $|S| > 1$

For plants with RHP-zeros the sensitivity function must satisfy an additional integral relationship, which has stronger implications for the peak of S. Before stating the result, let us illustrate why the presence of a RHP-zero implies that the peak of S must exceed one. First, consider the non-minimum phase loop transfer function $L(s) = \frac{1}{1+s}\frac{1-s}{1+s}$ and its minimum phase counterpart $L_m(s) = \frac{1}{1+s}$. From Figure 5.3 we see

that the additional phase lag contributed by the RHP-zero and the extra pole causes
the Nyquist plot to penetrate the unit circle and hence causes the sensitivity function
to be larger than one.

As a further example, consider Figure 5.4 which shows the magnitude of the
sensitivity function for the following loop transfer function

$$L(s) = \frac{k}{s}\frac{2-s}{2+s} \quad k = 0.1, 0.5, 1.0, 2.0 \tag{5.8}$$

The plant has a RHP-zero $z = 2$, and we see that an increase in the controller gain
k, corresponding to a higher bandwidth, results in a larger peak for S. For $k = 2$ the
closed-loop system becomes unstable with two poles on the imaginary axis, and the
peak of S is infinite.

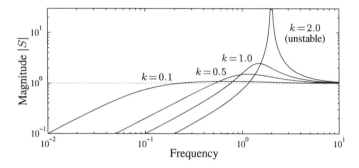

Figure 5.4: Effect of increased controller gain on $|S|$ for system with RHP-zero at $z = 2$,
$L(s) = \frac{k}{s}\frac{2-s}{2+s}$

Theorem 5.2 Weighted sensitivity integral.   *Suppose that $L(s)$ has a single real
RHP-zero z or a complex conjugate pair of zeros $z = x \pm jy$, and has N_p RHP-
poles, p_i. Let $\bar{p}_i$ denote the complex conjugate of p_i. Then for closed-loop stability
the sensitivity function must satisfy*

$$\int_0^\infty \ln |S(j\omega)| \cdot w(z,\omega)d\omega = \pi \cdot \ln \prod_{i=1}^{N_p} \left| \frac{p_i + z}{\bar{p}_i - z} \right| \tag{5.9}$$

where if the zero is real

$$w(z,\omega) = \frac{2z}{z^2 + \omega^2} = \frac{2}{z}\frac{1}{1 + (\omega/z)^2} \tag{5.10}$$

and if the zero pair is complex ($z = x \pm jy$)

$$w(z,\omega) = \frac{x}{x^2 + (y - \omega)^2} + \frac{x}{x^2 + (y + \omega)^2} \tag{5.11}$$

Proof: See Freudenberg and Looze (1985; 1988).   □

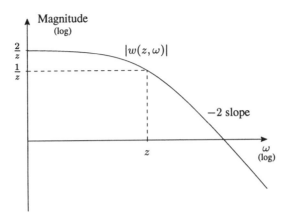

Figure 5.5: Plot of weight $w(z, \omega)$ for case with real zero at $s = z$

As shown graphically in Figure 5.5, the weight $w(z, \omega)$ effectively "cuts off" the contribution of $\ln|S|$ to the sensitivity integral at frequencies $\omega > z$. Thus, for a stable plant where $|S|$ is reasonably close to 1 at high frequencies we have the approximate relationship

$$\int_0^z \ln|S(j\omega)|d\omega \approx 0 \quad (\text{ for } |S| \approx 1 \text{ at } \omega > z) \tag{5.12}$$

This is similar to Bode's sensitivity integral relationship in (5.7), except that the trade-off between S less than 1 and S larger than 1, is done over a limited frequency range. Thus, in this case the waterbed is finite, and a large peak for $|S|$ is unavoidable if we try to reduce $|S|$ at low frequencies. This is illustrated by the example in Figure 5.4.

Note that when there is a RHP-pole close to the RHP-zero ($p_i \to z$) then $\frac{p_i+z}{p_i-z} \to \infty$. This is not surprising as such plants are in practice impossible to stabilize.

Exercise 5.1 Kalman inequality *The Kalman inequality for optimal state feedback, which also applies to unstable plants, says that $|S| \leq 1 \; \forall\omega$, see Example 9.1. Explain why this does not conflict with the above sensitivity integrals. (Solution: 1. Optimal control with state feedback yields a loop transfer function with a pole-zero excess of one so (5.6) does not apply. 2. There are no RHP-zeros when all states are measured so (5.9) does not apply).*

The two sensitivity integrals (waterbed formulas) presented above are interesting and provide valuable insights, but for a quantitative analysis of achievable performance they are less useful. Fortunately, however, we can derive lower bounds on the weighted sensitivity and weighted complementary sensitivity, see (5.21), which are more useful for analyzing the effects of RHP-zeros and RHP-poles. The basis for these bounds is the interpolation constraints which we discuss first.

5.3.3 Interpolation constraints

If p is a RHP-pole of the loop transfer function $L(s)$ then

$$T(p) = 1, \quad S(p) = 0 \qquad (5.13)$$

Similarly, if z is a RHP-zero of $L(s)$ then

$$T(z) = 0, \quad S(z) = 1 \qquad (5.14)$$

These *interpolation constraints* follow from the requirement of internal stability as shown in (4.83) and (4.84). The conditions clearly restrict the allowable S and T and prove very useful in the next subsection.

5.3.4 Sensitivity peaks

In Theorem 5.2, we found that a RHP-zero implies that a peak in $|S|$ is inevitable, and that the peak will increase if we reduce $|S|$ at other frequencies. Here we will derive explicit bounds on the weighted peak of S, which are more useful in applications than the integral relationship. These lower bounds on sensitivity were originally derived by Zames (1981). We will present for three cases:

1. A bound on weighted sensitivity for a plant with a RHP-zero.
2. A bound on weighted complementary sensitivity for a plant with a RHP-pole.
3. Bounds for plants with both RHP-zeros and RHP-poles.

The results are based on the interpolation constraints $S(z) = 1$ and $T(p) = 1$ given above. In addition, we make use of the maximum modulus principle for complex analytic functions (e.g. see maximum principle in Churchill et al., 1974) which for our purposes can be stated as follows:

Maximum modulus principle. *Suppose $f(s)$ is stable (i.e. $f(s)$ is analytic in the complex RHP). Then the maximum value of $|f(s)|$ for s in the right-half plane is attained on the region's boundary, i.e. somewhere along the $j\omega$-axis. Hence, we have for a stable $f(s)$*

$$\|f(j\omega)\|_\infty = \max_\omega |f(j\omega)| \geq |f(s_0)| \quad \forall s_0 \in \text{RHP} \qquad (5.15)$$

Remark. (5.15) can be understood by imagining a 3-D plot of $|f(s)|$ as a function of the complex variable s. In such a plot $|f(s)|$ has "peaks" at its poles and "valleys" at its zeros. Thus, if $f(s)$ has no poles (peaks) in the RHP, and we find that $|f(s)|$ slopes downwards from the LHP and into the RHP.

To derive the results below we first consider $f(s) = w_P(s)S(s)$ (weighted sensitivity) and then $f(s) = w_T(s)T(s)$ (weighted complementary sensitivity). The weights are included to make the results more general, but if required we may of course select $w_P(s) = 1$ and $w_T(s) = 1$.

Theorem 5.3 Weighted sensitivity peak. *Suppose that $G(s)$ has a RHP-zero z and let $w_P(s)$ be any stable weight function. Then for closed-loop stability the weighted sensitivity function must satisfy*

$$\|w_P S\|_\infty \geq |w_P(z)| \qquad (5.16)$$

Proof: Applying (5.15) to $f(s) = w_P(s)S(s)$ and using the interpolation constraint $S(z) = 1$, gives $\|w_P S\|_\infty \geq |w_P(z)S(z)| = |w_P(z)|$.   □

Note that $w_P(s) = 1$ in (5.16) yields the requirement $\|S\|_\infty \geq 1$ which we already know must hold, and which in any case is satisfied for any real system since $|S(j\omega)|$ must approach 1 at high frequencies. However, many useful relationships are derived by making other choices for the weight $w_P(s)$. This is discussed later in this chapter.

The following result is similar to Theorem 5.3, but it involves the complementary sensitivity T, rather than S. The basis for the result is that if $G(s)$ has a RHP-pole at $s = p$, then for internal stability $S(p)$ must have a RHP-zero at $s = p$, and from (5.13) we have $T(p) = 1$.

Theorem 5.4 Weighted complementary sensitivity peak. *Suppose that $G(s)$ has a RHP-pole p and let $w_T(s)$ be any stable weight function. Then for closed-loop stability the weighted complementary sensitivity function must satisfy*

$$\|w_T T\|_\infty \geq |w_T(p)| \qquad (5.17)$$

Proof: Applying the maximum modulus principle in (5.15) to $f(s) = w_T(s)T(s)$ and using the interpolation constraint $T(p) = 1$ gives $\|w_T T\|_\infty \geq |w_T(p)T(p)| = |w_T(p)|$.   □

Consider $w_T(s) = 1$ which yields the requirement $\|T\|_\infty \geq 1$, and illustrates that some control is needed to stabilize an unstable plant (since no control, $K = 0$, makes $T = 0$).

The following theorem provides a generalization of Theorems 5.3 and 5.4.

Theorem 5.5 Combined RHP-poles and RHP-zeros. *Suppose that $G(s)$ has N_z RHP-zeros z_j, and has N_p RHP-poles p_i. Then for closed-loop stability the weighted sensitivity function must satisfy for each RHP-zero z_j*

$$\|w_P S\|_\infty \geq c_{1j}|w_P(z_j)|, \quad c_{1j} = \prod_{i=1}^{N_p} \frac{|z_j + \bar{p}_i|}{|z_j - p_i|} \geq 1 \qquad (5.18)$$

and the weighted complementary sensitivity function must satisfy for each RHP-pole p_i

$$\|w_T T\|_\infty \geq c_{2i}|w_T(p_i)|, \quad c_{2i} = \prod_{j=1}^{N_z} \frac{|\bar{z}_j + p_i|}{|z_j - p_i|} \geq 1 \qquad (5.19)$$

Proof: The basis for the theorem is a "trick" where we first factor out the RHP-zeros in S or T into an all-pass part (with magnitude 1 at all points on the $j\omega$-axis). Consider S first. Since G have RHP-poles at p_i, $S(s)$ has RHP-zeros at p_i and we may write

$$S = S_a S_m, \quad S_a(s) = \prod_i \frac{s - p_i}{s + \bar{p}_i} \tag{5.20}$$

where, since $S_a(s)$ is all-pass, $|S_a(j\omega)| = 1$ at all frequencies. (Remark: There is a technical problem here with $j\omega$-axis poles; these must first be moved slightly into the RHP). Consider a RHP-zero located at z, for which we get from the maximum modulus principle $\|w_P S\|_\infty = \max_\omega |w_P S(j\omega)| = \max_\omega |w_P S_m(j\omega)| \geq |w_P(z) S_m(z)|$, where since $S(z) = 1$ we get $S_m(z) = 1/S_a(z) = c_1$. This proves (5.18).

Next consider T. G has RHP-zeros at z_j, and therefore T must have RHP-zeros at z_j, so write $T = T_a T_m$ with $T_a = \prod_j \frac{s - z_j}{s + \bar{z}_j}$. (Remark: There is a technical problem here with $j\omega$-axis zeros; these must first be moved slightly into the RHP). Consider a RHP-pole located at p. Since $|T_a(j\omega)| = 1$ at all frequencies, we get from the maximum modulus principle $\|w_T T\|_\infty = \max_\omega |w_T T(j\omega)| = \max_\omega |w_T T_m(j\omega)| \geq |w_T(p) T_m(p)|$, where since $T(p) = 1$ we get $T_m(p) = 1/T_a(p) = c_2$. This proves (5.19).   □

Notice that (5.16) is a special case of (5.18) with $c_1 = 1$, and (5.17) is a special case of (5.19) with $c_2 = 1$. Also, if we select $w_P = w_T = 1$, we derive the following useful bounds on the peaks of S and T:

$$\|S\|_\infty \geq \max_j c_{1j}, \quad \|T\|_\infty \geq \max_i c_{2i} \tag{5.21}$$

This shows that large peaks for S and T are unavoidable if we have a RHP-zero and RHP-pole located close to each other. This is illustrated by examples in Section 5.9.

These bounds may be generalized to MIMO systems if the directions of poles and zeros are taken into account, see Chapter 6.

5.4 Ideal ISE optimal control

Another good way of obtaining insight into performance limitations, is to consider an "ideal" controller which is integral square error (ISE) optimal. That is, for a given command $r(t)$ (which is zero for $t < 0$), the "ideal" controller is the one that generates the plant input $u(t)$ (zero for $t < 0$) which minimizes

$$\text{ISE} = \int_0^\infty |y(t) - r(t)|^2 dt \tag{5.22}$$

This controller is "ideal" in the sense that it may not be realizable in practice because the cost function includes no penalty on the input $u(t)$. This particular problem is considered in detail by Frank (1968a; 1968b) and Morari and Zafiriou (1989), and also Qiu and Davison (1993) who study "cheap" LQR control. Morari and Zafiriou show that for stable plants with RHP-zeros at z_j (real or complex) and a time delay θ, the "ideal" response $y = Tr$ when $r(t)$ is a *unit step* is given by

$$T(s) = \prod_i \frac{-s + z_j}{s + \bar{z}_j} e^{-\theta s} \tag{5.23}$$

where $\bar{z}_j$ is the complex conjugate of z_j.

Remark 1 The result in (5.23) is derived by considering an "open-loop" optimization problem, and applies to feedforward as well as feedback control.

Remark 2 The ideal $T(s)$ is "all-pass" with $|T(j\omega)| = 1$ at all frequencies. In the feedback case the ideal sensitivity function is $|S(j\omega)| = |L^{-1}(j\omega)T(j\omega)| = 1/|L(j\omega)|$ at all frequencies.

Remark 3 If $r(t)$ is not a step then other expressions for T rather than that in (5.23) are derived; see Morari and Zafiriou (1989) for details.

The optimal values of the cost function for three simple stable plants are:

1. with a delay θ :   ISE $= \theta$
2. with a RHP$-$zero z :   ISE $= 2/z$
3. with complex RHP$-$zeros $z = x \pm jy$:   ISE $= 4x/(x^2 + y^2)$

This quantifies nicely the limitations imposed by non-minimum phase behaviour, and the implications in terms of the achievable bandwidth are considered below.

5.5   Limitations imposed by time delays

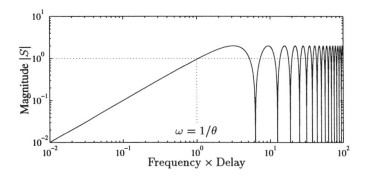

Figure 5.6: "Ideal" sensitivity function (5.24) for a plant with delay

Consider a plant $G(s)$ that contains a time delay $e^{-\theta s}$ (and no RHP-zeros). Even the "ideal" controller cannot remove this delay. For a step change in the reference $r(t)$, we have to wait a time θ until perfect control is achieved. Thus, as shown in (5.23), the "ideal" complementary sensitivity function will be $T = e^{-\theta s}$. The corresponding "ideal" sensitivity function is

$$S = 1 - T = 1 - e^{-\theta s} \qquad (5.24)$$

The magnitude $|S|$ is plotted in Figure 5.6. At low frequencies, $\omega \theta < 1$, we have $1 - e^{-\theta s} \approx \theta s$ (by a Taylor series expansion of the exponential) and the low-frequency asymptote of $|S(j\omega)|$ crosses 1 at a frequency of about $1/\theta$ (the exact frequency where $|S(j\omega)|$ crosses 1 in Figure 5.6 is $\frac{\pi}{3}\frac{1}{\theta} = 1.05/\theta$). Since in this case $|S| = 1/|L|$, we also have that $1/\theta$ is equal to the gain crossover frequency for L. In practice, the "ideal" controller cannot be realized, so we expect this value to provide an approximate upper bound on w_c, namely

$$\omega_c < 1/\theta \qquad (5.25)$$

This approximate bound is the same as derived in Section 2.6.2 by considering the limitations imposed on a loop-shaping design by a time delay θ.

5.6 Limitations imposed by RHP-zeros

We will here consider plants with a zero z in the closed right-half plane (and no pure time delay). In the following we attempt to build up insight into the performance limitations imposed by RHP-zeros using a number of different results in both the time and frequency domains.

RHP-zeros typically appear when we have competing effects of slow and fast dynamics. For example, the plant

$$G(s) = \frac{1}{s+1} - \frac{2}{s+10} = \frac{-s+8}{(s+1)(s+10)}$$

has a real RHP-zero at $z = 8$. We may also have complex zeros, and since these always occur in complex conjugate pairs we have $z = x \pm jy$ where $x \geq 0$ for RHP-zeros.

5.6.1 Inverse response

For a stable plant with n_z RHP-zeros, it may be proven (Holt and Morari, 1985b; Rosenbrock, 1970) that the output in response to a step change in the input will cross zero (its original value) n_z times, that is, we have *inverse response* behaviour. A typical response for the case with one RHP-zero is shown in Figure 2.14, page 38. We see that the output initially decreases before increasing to its positive steady-state value. With two RHP-zeros the output will initially increase, then decrease below its original value, and finally increase to its positive steady-state value.

5.6.2 High-gain instability

It is well-known from classical root-locus analysis that as the feedback gain increases towards infinity, the closed-loop poles migrate to the positions of the open-loop zeros; also see (4.76). Thus, the presence of RHP-zeros implies high-gain instability.

5.6.3 Bandwidth limitation I

For a step change in the reference we have from (5.23) that the "ideal" complementary sensitivity function T is all-pass, and for a single *real RHP-zero* the "ideal" sensitivity function is

$$S = 1 - T = \frac{2s}{s+z} \tag{5.26}$$

The Bode magnitude plot of $|S|$ $(= 1/|L|)$ is shown in Figure 5.7(a). The low-frequency asymptote of $|S(j\omega)|$ crosses 1 at the frequency $z/2$. In practice, the "ideal" ISE optimal controller cannot be realized, and we derive (for a real RHP-zero) the approximate requirement

$$\omega_B \approx \omega_c < \frac{z}{2} \tag{5.27}$$

which we also derived in Section 2.6.2 using loop-shaping arguments.

For a *complex pair of RHP-zeros*, $z = x \pm jy$, we get from (5.23) the "ideal" sensitivity function

$$S = \frac{4xs}{(s+x+jy)(s+x-jy)} \tag{5.28}$$

In Figure 5.7(b) we plot $|S|$ for y/x equal to 0.1, 1, 10 and 50. An analysis of (5.28) and the figure yields the following approximate bounds

$$\omega_B \approx \omega_c < \begin{cases} |z|/4 & \mathrm{Re}(z) \gg \mathrm{Im}(z) \\ |z|/2.8 & \mathrm{Re}(z) = \mathrm{Im}(z) \\ |z| & \mathrm{Re}(z) \ll \mathrm{Im}(z) \end{cases} \tag{5.29}$$

In summary, RHP-zeros located close to the origin (with $|z|$ small) are bad for control, and it is worse for them to be located closer to the real axis than the imaginary axis.

Remark. For a complex pair of zeros, $z = x \pm jy$, we notice from (5.28) and Figure 5.7 that the resonance peak of S at $\omega \approx y$ becomes increasingly "thin" as the zero approaches the imaginary axis ($x \to 0$). Thus, for a zero located on the imaginary axis ($x = 0$) the ideal sensitivity function is zero at all frequencies, except for a single "spike" at $\omega = y$ where it jumps up to 2. The integral under the curve for $|S(j\omega)|^2$ thus approaches zero, as does the ideal ISE-value in response to a step in the reference, ISE $= 4x/(x^2 + y^2)$; see Section 5.4. This indicates that purely imaginary zeros do not always impose limitations. This is also confirmed by the flexible structure in Example 2.10, for which the response to an input disturbance is satisfactory, even though the plant has a pair of imaginary zeros. However, the flexible structure is a rather special case where the plant also has imaginary poles which counteracts most of

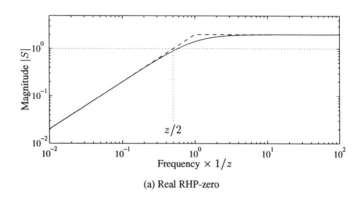

(a) Real RHP-zero

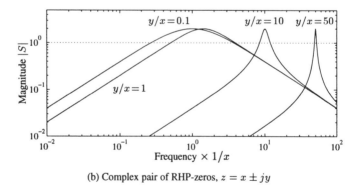

(b) Complex pair of RHP-zeros, $z = x \pm jy$

Figure 5.7: "Ideal" sensitivity functions for plants with RHP-zeros

the effect of the imaginary zeros. Therefore, in other cases, the presence of imaginary zeros may limit achievable performance, for example, in the presence of uncertainty which makes it difficult to place poles in the controller to counteract the zeros.

5.6.4 Bandwidth limitation II

Another way of deriving a bandwidth limitation is to consider the bound (5.16) on weighted sensitivity given in Theorem 5.3. The idea is to select a form for the performance weight $w_P(s)$, and then to derive a bound for the "bandwidth parameter" in the weight.

As usual, we select $1/|w_P|$ as an upper bound on the sensitivity function (see Figure 5.1 on page 165), that is, we require

$$|S(j\omega)| < 1/|w_P(j\omega)| \quad \forall \omega \quad \Leftrightarrow \quad \|w_P S\|_\infty < 1 \tag{5.30}$$

However, from (5.16) we have that $\|w_P S\|_\infty \geq |w_P(z)|$, so to be able to satisfy (5.30) we must *at least* require that the weight satisfies

$$\boxed{|w_P(z)| < 1} \tag{5.31}$$

(We say "at least" because condition (5.16) is not an equality). We will now use (5.31) to gain insight into the limitations imposed by RHP-zeros; first by considering a weight that requires good performance at low frequencies, and then by considering a weight that requires good performance at high frequencies.

Performance at low frequencies

Consider the following performance weight

$$w_P(s) = \frac{s/M + \omega_B^*}{s + \omega_B^* A} \tag{5.32}$$

From (5.30) it specifies a minimum bandwidth ω_B^* (actually, ω_B^* is the frequency where the straight-line approximation of the weight crosses 1), a maximum peak of $|S|$ less than M, a steady-state offset less than $A < 1$, and at frequencies lower than the bandwidth the sensitivity is required to improve by at least 20 dB/decade (i.e. $|S|$ has slope 1 or larger on a log-log plot). If the plant has a RHP-zero at $s = z$, then from (5.31) we must require

$$|w_P(z)| = \left| \frac{z/M + \omega_B^*}{z + \omega_B^* A} \right| < 1 \tag{5.33}$$

Real zero. Consider the case when z is real. Then all variables are real and positive and (5.33) is equivalent to

$$\omega_B^*(1 - A) < z \left(1 - \frac{1}{M}\right) \tag{5.34}$$

For example, with $A = 0$ (no steady-state offset) and $M = 2$ ($\|S\|_\infty < 2$) we must at least require $\omega_B^* < 0.5z$, which is consistent with the requirement $\omega_B < 0.5z$ in (5.27).

Imaginary zero. For a RHP-zero on the imaginary axis, $z = j|z|$, a similar derivation yields with $A = 0$:

$$\boxed{\omega_B^* < |z|\sqrt{1 - \frac{1}{M^2}}} \qquad (5.35)$$

For example, with $M = 2$ we require $\omega_B^* < 0.86|z|$, which is very similar to the requirement $\omega_B < |z|$ given in (5.29). The next two exercises show that the bound on ω_B^* does not depend much on the slope of the weight at low frequencies, or on how the weight behaves at high frequencies.

Exercise 5.2 *Consider the weight*

$$w_P(s) = \frac{s + M\omega_B^*}{s} \frac{s + fM\omega_B^*}{s + fM^2\omega_B^*} \qquad (5.36)$$

with $f > 1$. This is the same weight as (5.32) with $A = 0$ except that it approaches 1 at high frequencies, and f gives the frequency range over which we allow a peak. Plot the weight for $f = 10$ and $M = 2$. Derive an upper bound on ω_B^* for the case with $f = 10$ and $M = 2$.

Exercise 5.3 *Consider the weight $w_P(s) = \frac{1}{M} + (\frac{\omega_B^*}{s})^n$ which requires $|S|$ to have a slope of n at low frequencies and requires its low-frequency asymptote to cross 1 at a frequency ω_B^*. Note that $n = 1$ yields the weight (5.32) with $A = 0$. Derive an upper bound on ω_B^* when the plant has a RHP-zero at z. Show that the bound becomes $\omega_B^* \leq |z|$ as $n \to \infty$.*

Remark. The result for $n \to \infty$ in exercise 5.3 is a bit surprising. It says that the bound $\omega_B^* < |z|$, is independent of the required slope (n) at low frequency and is also independent of M. This is surprising since from Bode's integral relationship (5.6) we expect to pay something for having the sensitivity smaller at low frequencies, so we would expect ω_B^* to be smaller for larger n. This illustrates that $|w_P(z)| \leq 1$ in (5.31) is a necessary condition on the weight (i.e. it must at least satisfy this condition), but since it is not sufficient it can be optimistic. For the simple weight (5.32), with $n = 1$, condition (5.31) is not very optimistic (as is confirmed by other results), but apparently it is optimistic for n large.

Performance at high frequencies

The bounds (5.34) and (5.35) derived above assume tight control at low frequencies. Here, we consider a case where we want tight control at high frequencies, by use of the performance weight

$$w_P(s) = \frac{1}{M} + \frac{s}{\omega_B^*} \qquad (5.37)$$

This requires tight control ($|S(j\omega)| < 1$) at frequencies higher than ω_B^*, whereas the only requirement at low frequencies is that the peak of $|S|$ is less than M. Admittedly, the weight in (5.37) is unrealistic in that it requires $S \to 0$ at high frequencies, but this does not affect the result as is confirmed in Exercise 5.5 where a more realistic weight is studied. In any case, to satisfy $\|w_P S\|_\infty < 1$ we must at least require that the weight satisfies $|w_P(z)| < 1$, and with a *real RHP-zero* we derive for the weight in (5.37)

$$\omega_B^* > z \frac{1}{1 - 1/M} \qquad (5.38)$$

For example, with $M = 2$ the requirement is $\omega_B^* > 2z$, so we can only achieve tight control at frequencies *beyond* the frequency of the RHP-zero.

Exercise 5.4 *Draw an asymptotic magnitude Bode-plot of $w_P(s)$ in (5.37).*

Exercise 5.5 *Consider the case of a plant with a RHP-zero where we want to limit the sensitivity function over some frequency range. To this effect let*

$$w_P(s) = \frac{(1000s/\omega_B^* + \frac{1}{M})(s/(M\omega_B^*) + 1)}{(10s/\omega_B^* + 1)(100s/\omega_B^* + 1)} \qquad (5.39)$$

This weight is equal to $1/M$ at low and high frequencies, has a maximum value of about $10/M$ at intermediate frequencies, and the asymptote crosses 1 at frequencies $\omega_B^*/1000$ and ω_B^*. Thus we require "tight" control, $|S| < 1$, in the frequency range between $\omega_{BL}^* = \omega_B^*/1000$ and $\omega_{BH}^* = \omega_B^*$.

a) Make a sketch of $1/|w_P|$ (which provides an upper bound on $|S|$).

b) Show that the RHP-zero cannot be in the frequency range where we require tight control, and that we can achieve tight control either at frequencies below about $z/2$ (the usual case) or above about $2z$. To see this select $M = 2$ and evaluate $w_P(z)$ for various values of $\omega_B^* = kz$, e.g. $k = .1, .5, 1, 10, 100, 1000, 2000, 10000$. (You will find that $w_P(z) = 0.95$ (≈ 1) for $k = 0.5$ (corresponding to the requirement $\omega_{BH}^* < z/2$) and for $k = 2000$ (corresponding to the requirement $\omega_{BL}^* > 2z$))

5.6.5 Limitations at low or high frequencies

Based on (5.34) and (5.38) we see that a RHP-zero will pose control limitations *either* at low *or* high frequencies. In most cases we desire tight control at low frequencies, and with a RHP-zero this may be achieved at frequencies lower than about $z/2$. However, if we do not need tight control at low frequencies, then we may usually reverse the sign of the controller gain, and instead achieve tight control at frequencies higher than about $2z$.

Remark. The reversal of the sign in the controller is probably best understood by considering the inverse response behaviour of a plant with a RHP-zero. Normally, we want tight control at low frequencies, and the sign of the controller is based on the steady-state gain of the plant. However, if we instead want tight control at high frequencies (and have no requirements at low frequencies) then we base the controller design on the plants initial response where the gain is reversed because of the inverse response.

Example 5.1 *To illustrate this, consider in Figures 5.8 and 5.9 the use of negative and positive feedback for the plant*

$$G(s) = \frac{-s+z}{s+z}, \quad z = 1 \tag{5.40}$$

Note that $G(s) \approx 1$ at low frequencies ($\omega \ll z$), whereas $G(s) \approx -1$ at high frequencies ($\omega \gg z$). The negative plant gain in the latter case explains why we then use positive feedback in order to achieve tight control at high frequencies.

More precisely, we show in the figures the sensitivity function and the time response to a step change in the reference using

1. PI-control with negative feedback (Figure 5.8)
2. derivative control with positive feedback (Figure 5.9).

Note that the time scales for the simulations are different. For positive feedback the step change in reference only has a duration of 0.1 s. This is because we cannot track references over longer times than this since the RHP-zero then causes the output to start drifting away (as can be seen in Figure 5.9(b)).

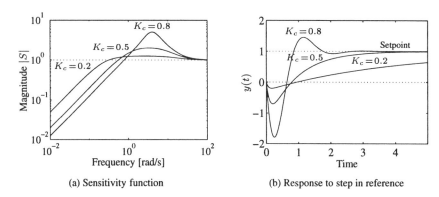

(a) Sensitivity function   (b) Response to step in reference

Figure 5.8: Control of plant with RHP-zero at $z = 1$ using negative feedback
$G(s) = \frac{-s+1}{s+1}$, $K_1(s) = K_c \frac{s+1}{s} \frac{1}{0.05s+1}$

An important case, where we can *only* achieve tight control at high frequencies, is characterized by plants with a zero at the origin, for example $G(s) = s/(5s+1)$. In this case, good transient control is possible, but the control has no effect at steady-state. The only way to achieve tight control at low frequencies is to use an additional actuator (input) as is often done in practice.

Short-term control. In this book, we generally assume that the system behaviour as $t \to \infty$ is important. However, this is not true in some cases because the system may only be under closed-loop control for a finite time t_f. In which case, the presence of a "slow" RHP-zero (with $|z|$ small), may not be significant provided $t_f \ll 1/|z|$. For example, in Figure 5.9(b) if the total control time is $t_f = 0.01$ [s], then the RHP-zero at $z = 1$ [rad/s] is insignificant.

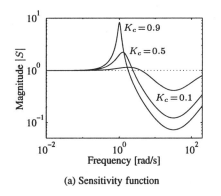

(a) Sensitivity function

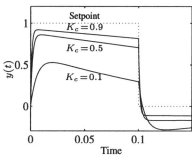

(b) Response to step in reference

Figure 5.9: Control of plant with RHP-zero at $z = 1$ using positive feedback.
$G(s) = \frac{-s+1}{s+1}$, $K_2(s) = -K_c\frac{s}{(0.05s+1)(0.02s+1)}$

Remark. As an example of short-term control, consider treating a patient with some medication. Let u be the dosage of medication and y the condition of the patient. With most medications we find that in the short-term the treatment has a positive effect, whereas in the long-term the treatment has a negative effect (due to side effects which may eventually lead to death). However, this inverse-response behaviour (characteristic of a plant with a RHP-zero) may be largely neglected during limited treatment, although one may find that the dosage has to be increased during the treatment to have the desired effect. Interestingly, the last point is illustrated by the upper left curve in Figure 5.10, which shows the input $u(t)$ using an internally unstable controller which over some finite time may eliminate the effect of the RHP-zero.

Exercise 5.6 *In the simulations in Figures 5.8 and 5.9, we use simple PI- and derivative controllers. As an alternative use the S/KS method in (3.59) to synthesize $\mathcal{H}_\infty$ controllers for both the negative and positive feedback cases. Use performance weights in the form given by (5.32) and (5.37), respectively. With $\omega_B^* = 1000$ and $M = 2$ in (5.37) and $w_u = 1$ (for the weight on KS) you will find that the time response is quite similar to that in Figure 5.9 with $K_c = 0.5$. Try to improve the response, for example, by letting the weight have a steeper slope at the crossover near the RHP-zero.*

5.6.6 Remarks on the effects of RHP-zeros

1. The derived bound $\omega_c < z/2$ for the case of tight control at low frequencies is consistent with the bound $\omega_c < 1/\theta$ found earlier for a time delay. This is seen from the following Padé approximation of a delay, $e^{-\theta s} \approx (1 - \frac{\theta}{2}s)/(1 + \frac{\theta}{2}s)$, which has a RHP-zero at $2/\theta$.
2. **LHP-zero.** Zeros in the left-half plane, usually corresponding to "overshoots" in the time response, do not present a *fundamental* limitation on control, but *in practice* a LHP-zero located close to the origin may cause problems. First, one may encounter problems with input constraints at low frequencies (because the steady-state gain is small). Second, a simple controller can probably not then be used. For example, a simple PID controller as in

(5.66) contains no adjustable poles that can be used to counteract the effect of a LHP-zero.
3. For uncertain plants, zeros can cross from the LHP into the RHP both through zero and through infinity. We discuss this in Chapter 7.

5.7 Non-causal controllers

Perfect control can be achieved for a plant with a time delay or RHP-zero if we use a non-causal controller[1], i.e. a controller which uses information about the future. This may be relevant for certain servo problems, e.g. in robotics and for batch processing. A brief discussion is given here, but non-causal controllers are not considered in the rest of the book because future information is usually not available.

Time delay. For a delay $e^{-\theta s}$ we may achieve perfect control with a non-causal feedforward controller $K_r = e^{\theta s}$ (a prediction). Such a controller may be used if we have knowledge about future changes in $r(t)$ or $d(t)$. For example, if we know that we should be at work at 8:00, and we know that it takes 30 min to get to work, then we make a prediction and leave home at 7:30.

RHP-zero. Future knowledge can also be used to give perfect control in the presence of a RHP-zero. As an example, consider a plant with a real RHP-zero given by

$$G(s) = \frac{-s + z}{s + z}; \quad z > 0 \tag{5.41}$$

and a desired reference change

$$r(t) = \begin{cases} 0 & t < 0 \\ 1 & t \geq 0 \end{cases}$$

With a feedforward controller K_r the response from r to y is $y = G(s)K_r(s)r$. In theory we may achieve perfect control (y(t)=r(t)) with the following two controllers (Eaton and Rawlings, 1992).

1. A causal *unstable controller*

$$K_r(s) = \frac{s + z}{-s + z}$$

For a step in r from 0 to 1 at $t = 0$, this controller generates the following input signal

$$u(t) = \begin{cases} 0 & t < 0 \\ 1 - 2e^{zt} & t \geq 0 \end{cases}$$

However, since the controller cancels the RHP-zero in the plant it yields an internally unstable system.

---

[1] A system is causal if its outputs depend only on past inputs, and non-causal if its outputs also depend on future inputs.

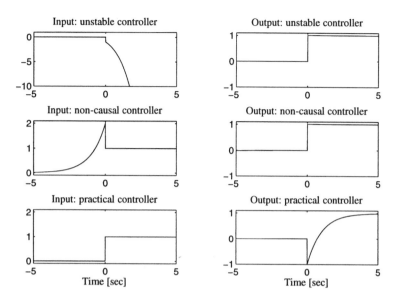

Figure 5.10: Feedforward control of plant with RHP-zero

2. A stable *non-causal controller* that assumes that the future setpoint change is known. This controller cannot be represented in the usual transfer function form, but it will generate the following input

$$u(t) = \begin{cases} 2e^{zt} & t < 0 \\ 1 & t \geq 0 \end{cases}$$

These input signals $u(t)$ and the corresponding outputs $y(t)$ are shown in Figure 5.10 for a plant with $z = 1$. Note that for perfect control the non-causal controller needs to start changing the input at $t = -\infty$, but for practical reasons we started the simulation at $t = -5$ where $u(t) = 2e^{-5} = 0.013$.

The first option, the unstable controller, is not acceptable as it yields an internally unstable system in which $u(t)$ goes to infinity as t increases (an exception may be if we want to control the system only over a limited time t_f; see page 180).

The second option, the non-causal controller, is usually not possible because future setpoint changes are unknown. However, if we have such information, it is certainly beneficial for plants with RHP-zeros.

However, in most cases we have to accept the poor performance resulting from the RHP-zero and use a stable causal controller, the third option. The ideal causal feedforward controller in terms of minimizing the ISE ($\mathcal{H}_2$ norm) of $y(t)$ for the plant in (5.41) is to use $K_r = 1$, and the corresponding plant input and output responses are shown in the lower plots in Figure 5.10.

5.8   Limitations imposed by RHP-poles

We here consider the limitations imposed when the plant has a RHP-pole (unstable pole) at $s = p$. For example, the plant $G(s) = 1/(s - 3)$ has a RHP-pole at $s = 3$.

For unstable plants we *need* feedback for stabilization. Thus, whereas the presence of RHP-zeros usually places an upper bound on the allowed bandwidth, the presence of RHP-poles generally imposes a lower bound. It also follows that it may be more difficult to stabilize an unstable plant if there are RHP-zeros or a time delay present ("the system goes unstable before we have time to react"). These qualitative statements are quantified below where we derive bounds on the bandwidth in terms of T.

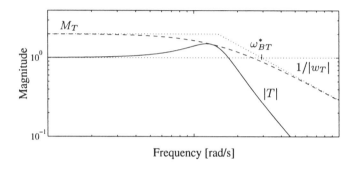

Figure 5.11: Typical complementary sensitivity, $|T|$, with upper bound $1/|w_T|$

We start by selecting a weight $w_T(s)$ such that $1/|w_T|$ is a reasonable upper bound on the complementary sensitivity function.

$$|T(j\omega)| < 1/|w_T(j\omega)| \quad \forall \omega \quad \Leftrightarrow \quad \|w_T T\|_\infty < 1 \qquad (5.42)$$

However, from (5.17) we have that $\|w_T T\|_\infty \geq |w_T(p)|$, so to be able to satisfy (5.42) we must *at least* require that the weight satisfies

$$\boxed{|w_T(p)| < 1} \qquad (5.43)$$

Now consider the following weight

$$w_T(s) = \frac{s}{\omega_{BT}^*} + \frac{1}{M_T} \qquad (5.44)$$

which requires T (like $|L|$) to have a roll-off rate of at least 1 at high frequencies (which must be satisfied for any real system), that $|T|$ is less than M_T at low frequencies, and that $|T|$ drops below 1 at frequency ω_{BT}^*. The requirements on $|T|$ are shown graphically in Figure 5.11.

Real RHP-pole at $s = p$. For the weight (5.44) condition (5.43) yields

$$\omega_{BT}^* > p\frac{M_T}{M_T - 1}$$ (5.45)

Thus, the presence of the RHP-pole puts a lower limit on the bandwidth in terms of T; that is, we cannot let the system roll-off at frequencies lower than p. For example, with $M_T = 2$ we get $\omega_{BT}^* > 2p$, which is approximately achieved if

$$\omega_c > 2p$$ (5.46)

Imaginary RHP-pole. For a purely imaginary pole located at $p = j|p|$ a similar analysis of the weight (5.44) with $M_T = 2$, shows that we must at least require $\omega_{BT}^* > 1.15|p|$, which is achieved if

$$\omega_c > 1.15|p|$$ (5.47)

In conclusion, we find that stabilization with reasonable performance requires a bandwidth which is larger than the distance $|p|$ of the RHP-pole from the origin.

Exercise 5.7 *Derive the bound in (5.47).*

5.9   Combined RHP-poles and RHP-zeros

We have above considered separately the effects of RHP-zeros and RHP-poles. In order to get acceptable performance and robustness, we derived for a real RHP-zero, the approximate bound $\omega_c < z/2$ (assuming we want tight control below this frequency); and for a real RHP-pole, we derived the approximate bound $\omega_c > 2p$. From these we get that for a system with a single real RHP-pole and a RHP-zero we must approximately require $z > 4p$ in order to get acceptable performance and robustness.

Remark. Stabilization. In theory, any linear plant may be stabilized irrespective of the location of its RHP-poles and RHP-zeros, provided the plant does not contain unstable hidden modes. However, this may require an unstable controller, and for practical purposes it is sometimes desirable that the controller is stable. If such a controller exists the plant is said to be *strongly stabilizable*. It has been proved by Youla et al. (1974) that a strictly proper SISO plant is strongly stabilizable if and only if every real RHP-zero in $G(s)$ lies to the left of an even number (including zero) of real RHP-poles in $G(s)$. Note that the presence of any complex RHP-poles or complex RHP-zeros does not affect this result. We then have:

- *A strictly proper plant with a single real RHP-zero z and a single real RHP-pole p, e.g.*
  $G(s) = \frac{s-z}{(s-p)(\tau s+1)}$, *can be stabilized by a stable controller if and only if $z > p$.*

Note the requirement that $G(s)$ is strictly proper. For example, the plant $G(s) = (s-1)/(s-2)$ with $z = 1 < p = 2$ is stabilized with a stable constant gain controller with $-2 < K < -1$, but this plant is not strictly proper.

In summary, for a plant with a single RHP-pole and RHP-zero, our approximate performance requirement $z > 4p$ is quite close to the strong stabilizability requirement $z > p$. The following example for a plant with $z = 4p$ shows that we can indeed get acceptable performance when the RHP-pole and zero are located this close.

Example 5.2 $\mathcal{H}_\infty$ **design for plant with RHP-pole and RHP-zero.** *We want to design an*

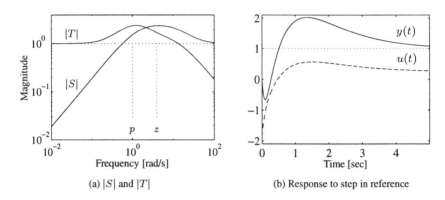

(a) $|S|$ and $|T|$ (b) Response to step in reference

Figure 5.12: $\mathcal{H}_\infty$ design for a plant with RHP-zero at $z = 4$ and RHP-pole at $p = 1$

$\mathcal{H}_\infty$ *controller for a plant with $z = 4$ and $p = 1$,*

$$G(s) = \frac{s - 4}{(s - 1)(0.1s + 1)} \tag{5.48}$$

We use the S/KS design method as in Example 2.11 with input weight $W_u = 1$ and performance weight (5.32) with $A=0$, $M = 2$, $\omega_B^* = 1$. The software gives a stable and minimum phase controller with an $\mathcal{H}_\infty$ norm of 1.89. The corresponding sensitivity and complementary sensitivity functions, and the time response to a unit step reference change are shown in Figure 5.12. The time response is good, taking into account the closeness of the RHP-pole and zero.

Sensitivity peaks. In Theorem 5.5 we derived lower bounds on the weighted sensitivity and complementary sensitivity. For example, for a plant with a single real RHP-pole p and a single real RHP-zero z, we always have

$$\boxed{\|S\|_\infty \geq c, \ \|T\|_\infty \geq c, \quad c = \frac{|z + p|}{|z - p|}} \tag{5.49}$$

Example 5.3 *Consider the plant in (5.48). With $z = 4p$, (5.49) gives $c = 5/3 = 1.67$ and it follows that for any controller we must at least have $\|S\|_\infty > 1.67$ and $\|T\|_\infty > 1.67$. The actual peak values for the above S/KS-design are 2.40 and 2.43, respectively.*

Example 5.4 Balancing a rod. *This example is taken from Doyle et al. (1992) Consider the problem of balancing a rod in the palm of one's hand. The objective is to keep the rod upright, by small hand movements, based on observing the rod either at its far end (output y_1) or the end in one's hand (output y_2). The linearized transfer functions for the two cases are*

$$G_1(s) = \frac{-g}{s^2 \left(Mls^2 - (M+m)g\right)}; \quad G_2(s) = \frac{ls^2 - g}{s^2 \left(Mls^2 - (M+m)g\right)}$$

Here l [m] is the length of the rod and m [kg] its mass. M [kg] is the mass of your hand and g [≈ 10 m/s^2] is the acceleration due to gravity. In both cases, the plant has three unstable poles: two at the origin and one at $p = \sqrt{\frac{(M+m)g}{Ml}}$. A short rod with a large mass gives a large value of p, and this in turn means that the system is more difficult to stabilize. For example, with $M = m$ and $l = 1$ [m] we get $p = 4.5$ [rad/s] and from (5.46) we desire a bandwidth of about 9 [rad/s] (corresponding to a response time of about 0.1 [s]).

If one is measuring y_1 (looking at the far end of the rod) then achieving this bandwidth is the main requirement. However, if one tries to balance the rod by looking at one's hand (y_2) there is also a RHP-zero at $z = \sqrt{\frac{g}{l}}$. If the mass of the rod is small (m/M is small), then p is close to z and stabilization is in practice impossible with any controller. However, even with a large mass, stabilization is very difficult because $p > z$ whereas we would normally prefer to have the RHP-zero far from the origin and the RHP-pole close to the origin ($z > p$). So although in theory the rod may be stabilized by looking at one's hand (G_2), it seems doubtful that this is possible for a human. To quantify these problems use (5.49). We get

$$c = \frac{|z+p|}{|z-p|} = \frac{|1+\gamma|}{|1-\gamma|}, \quad \gamma = \sqrt{\frac{M+m}{M}}$$

Consider a light weight rod with $m/M = 0.1$, for which we expect stabilization to be difficult. We obtain $c = 42$, and we must have $\|S\|_\infty \geq 42$ and $\|T\|_\infty \geq 42$, so poor control performance is inevitable if we try to balance the rod by looking at our hand (y_2).

The difference between the two cases, measuring y_1 and measuring y_2, highlights the importance of sensor location on the achievable performance of control.

5.10 Performance requirements imposed by disturbances and commands

The question we here want to answer is: how fast must the control system be in order to reject disturbances and track commands of a given magnitude? We find that some plants have better "built-in" disturbance rejection capabilities than others. This may be analyzed directly by considering the appropriately scaled disturbance model, $G_d(s)$. Similarly, for tracking we may consider the magnitude R of the reference change.

Disturbance rejection. Consider a single disturbance d and assume that the reference is constant, i.e. $r = 0$. Without control the steady-state sinusoidal response is $e(\omega) = G_d(j\omega)d(\omega)$; recall (2.9). If the variables have been scaled as outlined in Section 1.4 then the worst-case disturbance at any frequency is $d(t) = \sin \omega t$, i.e. $|d(\omega)| = 1$, and the control objective is that at each frequency $|e(t)| < 1$, i.e. $|e(\omega)| < 1$. From this we can immediately conclude that

- *no control is needed if* $|G_d(j\omega)| < 1$ *at all frequencies (in which case the plant is said to be "self-regulated").*

If $|G_d(j\omega)| > 1$ at some frequency, then we need control (feedforward or feedback). In the following, we consider feedback control, in which case we have

$$e(s) = S(s)G_d(s)d(s) \tag{5.50}$$

The performance requirement $|e(\omega)| < 1$ for any $|d(\omega)| \leq 1$ at any frequency, is satisfied if and only if

$$|SG_d(j\omega)| < 1 \quad \forall\omega \quad \Leftrightarrow \quad \|SG_d\|_\infty < 1 \tag{5.51}$$

$$\Leftrightarrow \quad |S(j\omega)| < 1/|G_d(j\omega)| \quad \forall\omega \tag{5.52}$$

A typical plot of $1/|G_d(j\omega)|$ is shown in Figure 5.13 (dotted line). If the plant has a RHP-zero at $s = z$, which fixes $S(z) = 1$, then using (5.16) we have the following necessary condition for satisfying $\|SG_d\|_\infty < 1$:

$$\boxed{|G_d(z)| < 1} \tag{5.53}$$

From (5.52) we also get that the frequency ω_d where $|G_d|$ crosses 1 from above yields a lower bound on the bandwidth:

$$\boxed{\omega_B > \omega_d} \quad \text{where } \omega_d \text{ is defined by } |G_d(j\omega_d)| = 1 \tag{5.54}$$

A plant with a small $|G_d|$ or a small ω_d is preferable since the need for feedback control is then less, or alternatively, given a feedback controller (which fixes S) the effect of disturbances on the output is less.

Example 5.5 *Assume that the disturbance model is $G_d(s) = k_d/(1 + \tau_d s)$ where $k_d = 10$ and $\tau_d = 100$ [seconds]. Scaling has been applied to G_d so this means that without feedback, the effect of disturbances on the outputs at low frequencies is $k_d = 10$ times larger than we desire. Thus feedback is required, and since $|G_d|$ crosses 1 at a frequency $\omega_d \approx k_d/\tau_d = 0.1$ rad/s, the minimum bandwidth requirement for disturbance rejection is $\omega_B > 0.1$ [rad/s].*

Remark. G_d **is of high order.** The actual bandwidth requirement imposed by disturbances may be higher than ω_d if $|G_d(j\omega)|$ drops with a slope steeper than -1 (on a log-log plot) just before the frequency ω_d. The reason for this is that we must, in addition to satisfying (5.52), also ensure stability with reasonable margins; so as discussed in Section 2.6.2 we cannot let the slope of $|L(j\omega)|$ around crossover be much larger than -1.

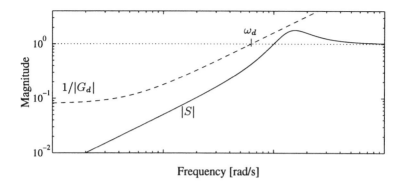

Frequency [rad/s]

Figure 5.13: Typical performance requirement on S imposed by disturbance rejection

An example, in which $G_d(s)$ is of high order, is given later in Section 5.16.3 for a neutralization process. There we actually overcome the limitation on the slope of $|L(j\omega)|$ around crossover by using local feedback loops in series. We find that, although each loop has a slope -1 around crossover, the overall loop transfer function $L(s) = L_1(s)L_2(s)\cdots L_n(s)$ has a slope of about $-n$; see the example for more details. This is a case where stability is determined by each $I + L_i$ separately, but the benefits of feedback are determined by $1 + \prod_i L_i$ (also see Horowitz (1991, p. 284) who refers to lectures by Bode).

Command tracking. Assume there are no disturbances, i.e. $d = 0$, and consider a reference change $r(t) = R\tilde{r}(t) = R\sin(\omega t)$. Since $e = Gu + G_d d - R\tilde{r}$, the same performance requirement as found for disturbances, see (5.51), applies to command tracking with G_d replaced by $-R$. Thus for acceptable control ($|e(\omega)| < 1$) we must have

$$|S(j\omega)R| < 1 \quad \forall \omega \leq \omega_r \qquad (5.55)$$

where ω_r is the frequency up to which performance tracking is required.

Remark. The bandwidth requirement imposed by (5.55) depends on on how sharply $|S(j\omega)|$ increases in the frequency range from ω_r (where $|S| < 1/R$) to ω_B (where $|S| \approx 1$). If $|S|$ increases with a slope of 1 then the approximate bandwidth requirement becomes $\omega_B > R\omega_r$, and if $|S|$ increases with a slope of 2 it becomes $\omega_B > \sqrt{R}\omega_r$.

5.11 Limitations imposed by input constraints

In all physical systems there are limits to the changes that can be made to the manipulated variables. In this section, we assume that the model has been scaled as outlined in Section 1.4, so that at any time we must have $|u(t)| \leq 1$. The question we want to answer is: can the expected disturbances be rejected and can we track the reference changes while maintaining $|u(t)| \leq 1$? We will consider separately the two cases of perfect control ($e = 0$) and acceptable control ($|e| < 1$). These results apply to both feedback and feedforward control.

At the end of the section we consider the additional problems encountered for unstable plants (where feedback control is required).

Remark 1 We use a frequency-by-frequency analysis and assume that at each frequency $|d(\omega)| \leq 1$ (or $|\tilde{r}(\omega)| \leq 1$). The worst-case disturbance at each frequency is $|d(\omega)| = 1$ and the worst-case reference is $r = R\tilde{r}$ with $|\tilde{r}(\omega)| = 1$.

Remark 2 Note that rate limitations, $|du/dt| \leq 1$, may also be handled by our analysis. This is done by considering du/dt as the plant input by including a term $1/s$ in the plant model $G(s)$.

Remark 3 Below we require $|u| < 1$ rather than $|u| \leq 1$. This has *no* practical effect, and is used to simplify the presentation.

5.11.1   Inputs for perfect control

From (5.3) the input required to achieve perfect control ($e = 0$) is

$$u = G^{-1}r - G^{-1}G_d d \tag{5.56}$$

Disturbance rejection. With $r = 0$ and $|d(\omega)| = 1$ the requirement $|u(\omega)| < 1$ is equivalent to

$$|G^{-1}(j\omega)G_d(j\omega)| < 1 \quad \forall\omega \tag{5.57}$$

In other words, to achieve perfect control and avoid input saturation we need $|G| > |G_d|$ at all frequencies. (However, as is discussed below, we do not really need control at frequencies where $|G_d| < 1$.)

Command tracking. Next let $d = 0$ and consider the worst-case reference command which is $|r(\omega)| = R$ at all frequencies up to ω_r. To keep the inputs within their constraints we must then require from (5.56) that

$$|G^{-1}(j\omega)R| < 1 \quad \forall\omega \leq \omega_r \tag{5.58}$$

In other words, to avoid input saturation we need $|G| > R$ at all frequencies where perfect command tracking is required.

Example 5.6 *Consider a process with*

$$G(s) = \frac{40}{(5s+1)(2.5s+1)}, \quad G_d(s) = 3\frac{50s+1}{(10+1)(s+1)}$$

From Figure 5.14 we see that $|G| < |G_d|$ for $\omega > \omega_1$, and $|G_d| < 1$ for $\omega > \omega_d$. Thus, condition (5.57) is not satisfied for $\omega > \omega_1$. However, for frequencies $\omega > \omega_d$ we do not really need control. Thus, in practice, we expect that disturbances in the frequency range between ω_1 and ω_d may cause input saturation.

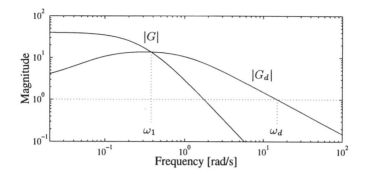

Figure 5.14: Input saturation is expected for disturbances at intermediate frequencies from ω_1 to ω_d

5.11.2 Inputs for acceptable control

For simplicity above, we assumed perfect control. However, perfect control is never really required, especially not at high frequencies, and the input magnitude required for *acceptable control* (namely $|e(j\omega)| < 1$) is somewhat smaller. For *disturbance rejection* we must then require

$$\boxed{|G| > |G_d| - 1} \quad \text{at frequencies where } |G_d| > 1 \qquad (5.59)$$

Proof: Consider a "worst-case" disturbance with $|d(\omega)| = 1$. The control error is $e = y = Gu + G_d d$. Thus at frequencies where $|G_d(j\omega)| > 1$ the smallest input needed to reduce the error to $|e(\omega)| = 1$ is found when $u(\omega)$ is chosen such that the complex vectors Gu and $G_d d$ have opposite directions. That is, $|e| = 1 = |G_d d| - |Gu|$, and with $|d| = 1$ we get $|u| = |G^{-1}|(|G_d| - 1)$, and the result follows by requiring $|u| < 1$. □

Similarly, to achieve acceptable control for *command tracking* we must require

$$\boxed{|G| > |R| - 1 < 1} \quad \forall\omega \leq \omega_r \qquad (5.60)$$

In summary, if we want "acceptable control" ($|e| < 1$) rather than "perfect control" ($e = 0$), then $|G_d|$ in (5.57) should be replaced by $|G_d| - 1$, and similarly, R in (5.58) should be replaced by $R - 1$. The differences are clearly small at frequencies where $|G_d|$ and $|R|$ are much larger than 1.

The requirements given by (5.59) and (5.60) are restrictions imposed on the *plant design* in order to avoid input constraints and they apply to any controller (feedback or feedforward control). If these bounds are violated at some frequency then performance will not be satisfactory (i.e, $|e(\omega)| > 1$) for a worst-case disturbance or reference occurring at this frequency.

5.11.3   Unstable plant and input constraints

Feedback control is required to stabilize an unstable plant. However, input constraints combined with large disturbances may make stabilization difficult. For example, *for an unstable plant with a real RHP-pole at $s = p$ we approximately need*

$$\boxed{|G| > |G_d| \quad \forall \omega < p} \tag{5.61}$$

Otherwise, the input will saturate when there is a sinusoidal disturbance $d(t) = \sin \omega t$, and we may not be able to stabilize the plant. Note that the frequency p may be *larger* than the frequency ω_d at which $|G_d(j\omega_d)| = 1$.

Proof of (5.61):. With feedback control the input signal is $u = -KSG_d d = -G^{-1}TG_d d$. We showed in (5.45) that we need $|T(j\omega)| \geq 1$ up to the frequency p. Thus we need $|u| \geq |G^{-1}G_d| \cdot |d|$ up to the frequency p, and to have $|u| \leq 1$ for $|d| = 1$ (the worst-case disturbance) we must require $|G^{-1}G_d| \leq 1$.   □

Example 5.7 *Consider*

$$G(s) = \frac{5}{(10s + 1)(s - 1)}, \quad G_d(s) = \frac{k_d}{(s + 1)(0.2s + 1)}, \quad k_d < 1 \tag{5.62}$$

Since $k_d < 1$ and the performance objective is $|e| < 1$, we do not really need control for disturbance rejection, but feedback control is required for stabilization, since the plant has a RHP-pole at $p = 1$. We have $|G| > |G_d|$ (i.e. $|G^{-1}G_d| < 1$) for frequencies lower than $0.5/k_d$, so from (5.61) we do not expect problems with input constraints at low frequencies. However, at frequencies higher than $0.5/k_d$ we have $|G| < |G_d|$, as is illustrated in Figure 5.15(a). Thus from (5.61), we may expect to have problems with instability if $0.5/k_d < p$, i.e. if $k_d > 0.5$.

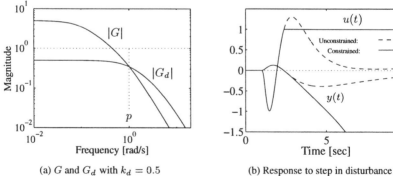

(a) G and G_d with $k_d = 0.5$ (b) Response to step in disturbance

Figure 5.15: Instability caused by input saturation for unstable plant

To check this for a particular case we select $k_d = 0.5$ (the limiting value) and use the controller

$$K(s) = \frac{0.04}{s} \frac{(10s + 1)^2}{(0.1s + 1)^2} \tag{5.63}$$

which without constraints yields a stable closed-loop system with a gain crossover frequency,
ω_c, *of about 1.7. The closed-loop response to a unit step disturbance occurring after 1 second*
is shown in Figure 5.15(b) both for the linear case when there are no input constraints (dashed
line), and where u is constrained to be within the interval $[-1, 1]$ *(solid lines). We see from*
the latter response that the system is indeed unstable when there is a disturbance that drives
the system into saturation.

Remark. *Interestingly, for this example, a small reduction in the disturbance magnitude*
from $k_d = 0.5$ *to* $k_d = 0.48$ *results in a stable closed-loop response in the presence of*
input constraints (not shown). Since $k_d = 0.5$ *is the limiting value obtained from (5.61), this*
seems to indicate that (5.61) is a very tight condition, but one should be careful about making
such a conclusion. First, (5.61) was derived by considering sinusoids and the responses in this
example are for a step disturbance. Furthermore, for other controllers the values of k_d *for*
which instability occurs will be different.

For unstable plants, reference changes can also drive the system into input saturation
and instability. But in contrast to disturbance changes, one then has the option to use
a two degrees-of-freedom controller to filter the reference signal and thus reduce the
magnitude of the manipulated input.

5.12   Limitations imposed by phase lag

We already know that phase lag from RHP-zeros and time delays is a fundamental
problem, but are there any limitations imposed by the phase lag resulting from
minimum-phase elements? The answer is both no and yes: *No*, there are no
fundamental limitations, but *Yes*, there are often limitations on practical designs.

As an example, consider a minimum-phase plant of the form

$$G(s) = \frac{k}{(1 + \tau_1 s)(1 + \tau_2 s)(1 + \tau_3 s)\cdots} = \frac{k}{\prod_{i=1}^{n}(1 + \tau_i s)} \qquad (5.64)$$

where n is three or larger. At high frequencies the gain drops sharply with frequency,
$|G(j\omega)| \approx (k/\prod \tau_i)\omega^{-n}$. From condition (5.57), it is therefore likely (at least if k is
small) that we encounter problems with *input saturation*. Otherwise, the presence of
high-order lags *does not present any fundamental limitations*.

However, *in practice* a large phase lag at high frequencies, e.g. $\angle G(j\omega) \to -n{\cdot}90°$
for the plant in (5.64), poses a problem (independent of K) even when input saturation
is not an issue. This is because for stability we need a positive phase margin, i.e. the
phase of $L = GK$ must be larger than $-180°$ at the gain crossover frequency ω_c.
That is, for stability we need $\omega_c < \omega_{180}$; see (2.27).

In principle, ω_{180} (the frequency at which the phase lag around the loop is $-180°$)
is not directly related to phase lag in the plant, but in most practical cases there is a
close relationship. Define ω_u as the frequency where the phase lag in the plant G is
$-180°$, i.e.

$$\angle G(j\omega_u) = -180°$$

Note that ω_u depends only on the plant model. Then, with a proportional controller we have that $\omega_{180} = \omega_u$, and with a PI-controller $\omega_{180} < \omega_u$. Thus with these simple controllers a phase lag in the plant *does* pose a fundamental limitation:

$$\text{Stability bound for P- or PI-control:} \quad \boxed{\omega_c < \omega_u} \tag{5.65}$$

Note that this is a strict bound to get stability, and for performance (phase and gain margin) we typically need ω_c less than bout $0.5\omega_u$.

If we want to extend the gain crossover frequency ω_c beyond ω_u, we must place zeros in the controller (e.g. "derivative action") to provide phase lead which counteracts the negative phase in the plant. A commonly used controller is the PID controller which has a maximum phase lead of 90° at high frequencies. In practice, the maximum phase lead is smaller than 90°. For example, an industrial *cascade PID controller* typically has derivative action over only one decade,

$$K(s) = K_c \frac{\tau_I s + 1}{\tau_I s} \frac{\tau_D s + 1}{0.1\tau_D s + 1} \tag{5.66}$$

and the maximum phase lead is 55° (which is the maximum phase lead of the term $\frac{\tau_D s + 1}{0.1\tau_D s + 1}$). This is also a reasonable value for the phase margin, so for performance we approximately require

$$\text{Practical performance bound (PID control):} \quad \omega_c < \omega_u \tag{5.67}$$

We stress again that plant phase lag does *not* pose a *fundamental* limitation if a more complex controller is used. Specifically, if the model is known exactly and there are no RHP-zeros or time delays, then one may in theory extend ω_c to infinite frequency. For example, one may simply invert the plant model by placing zeros in the controller at the plant poles, and then let the controller roll off at high frequencies beyond the dynamics of the plant. However, in many practical cases the bound in (5.67) applies because we may want to use a simple controller, and also because uncertainty about the plant model often makes it difficult to place controller zeros which counteract the plant poles at high frequencies.

Remark. The *relative order* (relative degree) of the plant is sometimes used as an input-output controllability measure (e.g. Daoutidis and Kravaris, 1992). The relative order may be defined also for nonlinear plants, and it corresponds for linear plants to the pole excess of $G(s)$. For a minimum-phase plant the phase lag at infinite frequency is the relative order times $-90°$. Of course, we want the inputs to directly affect the outputs, so we want the relative order to be small. However, the practical usefulness of the relative order is rather limited since it only gives information at infinite frequency. The phase lag of $G(s)$ as a function of frequency, including the value of ω_u, provides much more information.

5.13 Limitations imposed by uncertainty

The presence of uncertainty requires us to use feedback control rather than just feedforward control. The main objective of this section is to gain more insight into this statement. A further discussion is given in Section 6.10, where we consider MIMO systems.

5.13.1 Feedforward control

Consider a plant with the nominal model $y = Gu + G_d d$. Assume that $G(s)$ is minimum phase and stable and assume there are no problems with input saturation. Then *perfect control*, $e = y - r = 0$, is obtained using a perfect feedforward controller which generates the following control inputs

$$u = G^{-1} r - G^{-1} G_d d \tag{5.68}$$

Now consider applying this perfect controller to the actual plant with model

$$y' = G' u + G'_d d \tag{5.69}$$

After substituting (5.68) into (5.69), we find that the actual control error with the "perfect" feedforward controller is

$$e' = y' - r = \underbrace{\left(\frac{G'}{G} - 1 \right)}_{\text{rel. error in } G} r - \underbrace{\left(\frac{G'/G'_d}{G/G_d} - 1 \right)}_{\text{rel. error in } G/G_d} G'_d d \tag{5.70}$$

Thus, we find for feedforward control that the model error propagates directly to the control error. From (5.70) we see that to achieve $|e'| < 1$ for $|d| = 1$ we must require that the relative model error in G/G_d is less than $1/|G'_d|$. This requirement is clearly very difficult to satisfy at frequencies where $|G'_d|$ is much larger than 1, and this motivates the need for feedback control.

5.13.2 Feedback control

With feedback control the closed-loop response with no model error is $y - r = S(G_d - r)$ where $S = (I + GK)^{-1}$ is the sensitivity function. With model error we get

$$y' - r = S'(G'_d d - r) \tag{5.71}$$

where $S' = (I + G'K)^{-1}$ can be written (see (A.139)) as

$$S' = S \frac{1}{1 + ET} \tag{5.72}$$

Here $E = (G' - G)/G$ is the relative error for G, and T is the complementary sensitivity function.

From (5.71) we see that the control error is only weakly affected by model error at frequencies where feedback is effective (where $|S| << 1$ and $T \approx 1$). For example, if we have integral action in the feedback loop and if the feedback system with model error is stable, then $S(0) = S'(0) = 0$ and the steady-state control error is zero even with model error.

Uncertainty at crossover. Although feedback control counteracts the effect of uncertainty at frequencies where the loop gain is large, uncertainty in the crossover frequency region can result in poor performance and even instability. This may be analyzed by considering the effect of the uncertainty on the gain margin, GM $= 1/|L(j\omega_{180})|$, where ω_{180} is the frequency where $\angle L$ is $-180°$; see (2.33). Most practical controllers behave as a constant gain K_o in the crossover region, so $|L(j\omega_{180}| \approx K_o|G(j\omega_{180}|$ where $\omega_{180} \approx \omega_u$ (since the phase lag of the controller is approximately zero at this frequency; see also Section 5.12). Here ω_u is the frequency where $\angle G(j\omega_u) = -180°$. This observation yields the following approximate rule:

- *Uncertainty which keeps $|G(j\omega_u)|$ approximately constant will not change the gain margin. Uncertainty which increases $|G(j\omega_u)|$ may yield instability.*

This rule is useful, for example, when evaluating the effect of parametric uncertainty. This is illustrated in the following example.

Example 5.8 *Consider a stable first-order delay process, $G(s) = ke^{-\theta s}/(1 + \tau s)$, where the parameters k, τ and θ are uncertain in the sense that they may vary with operating conditions. If we assume $\tau > \theta$ then $\omega_u \approx (\pi/2)/\tau$ and we derive*

$$|G(j\omega_u)| \approx \frac{2}{\pi} k \frac{\theta}{\tau} \qquad (5.73)$$

We see that to keep $|G(j\omega_u)|$ constant we want $k\frac{\theta}{\tau}$ constant. From (5.73) we see, for example, that an increase in θ increases $|G(j\omega_u)|$, and may yield instability. However, the uncertainty in the parameters is often coupled. For example, the ratio τ/θ may be approximately constant, in which case an increase in θ may not affect stability. In another case the steady-state gain k may change with operating point, but this may not affect stability if the ratio k/τ, which determines the high-frequency gain, is unchanged.

The above example illustrates the importance of taking into account the *structure of the uncertainty*, for example, the coupling between the uncertain parameters. A robustness analysis which assumes the uncertain parameters to be uncorrelated is generally conservative. This is further discussed in Chapters 7 and 8.

5.14   Controllability analysis with feedback control

We will now summarize the results of this chapter by a set of "controllability rules". We use the term "(input-output) controllability" since the bounds depend on the

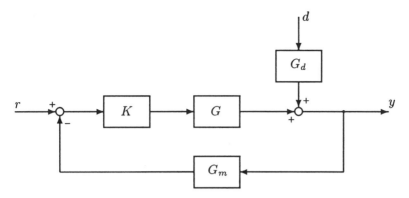

Figure 5.16: Feedback control system

plant only, that is, are independent of the specific controller. Except for Rule 7, all requirements are fundamental, although some of the expressions, as seen from the derivations, are approximate (i.e, they may be off by a factor of 2 or thereabout). However, for practical designs the bounds will need to be satisfied to get acceptable performance.

Consider the control system in Figure 5.16, for the case when all blocks are scalar. The model is

$$y = G(s)u + G_d(s)d; \quad y_m = G_m(s)y \tag{5.74}$$

Here $G_m(s)$ denotes the measurement transfer function and we assume $G_m(0) = 1$ (perfect steady-state measurement). The variables d, u, y and r are assumed to have been scaled as outlined in Section 1.4, and therefore $G(s)$ and $G_d(s)$ are the scaled transfer functions. Let ω_c denote the gain crossover frequency; defined as the frequency where $|L(j\omega)|$ crosses 1 from above. Let ω_d denote the frequency at which $|G_d(j\omega_d)|$ first crosses 1 from above. The following rules apply:

Rule 1. Speed of response to reject disturbances. *We approximately require $\omega_c > \omega_d$. More specifically, with feedback control we require $|S(j\omega)| \leq |1/G_d(j\omega)|\ \forall\omega$. (See (5.51) and (5.54)).*

Rule 2. Speed of response to track reference changes. *We require $|S(j\omega)| \leq 1/R$ up to the frequency ω_r where tracking is required. (See (5.55)).*

Rule 3. Input constraints arising from disturbances. *For acceptable control ($|e| < 1$) we require $|G(j\omega)| > |G_d(j\omega)| - 1$ at frequencies where $|G_d(j\omega)| > 1$. For perfect control ($e = 0$) the requirement is $|G(j\omega)| > |G_d(j\omega)|$. (See (5.57) and (5.59)).*

Rule 4. Input constraints arising from setpoints. *We require $|G(j\omega)| > R - 1$ up to the frequency ω_r where tracking is required. (See (5.60)).*

Rule 5. Time delay θ **in** $G(s)G_m(s)$. *We approximately require* $\omega_c < 1/\theta$. (See (5.25)).

Rule 6. Tight control at low frequencies with a RHP-zero z **in** $G(s)G_m(s)$. *For a real RHP-zero we require* $\omega_c < z/2$ *and for an imaginary RHP-zero we approximately require* $\omega_c < |z|$. (See (5.27) and (5.29)).

> **Remark.** Strictly speaking, a RHP-zero only makes it impossible to have tight control in the frequency range close to the location of the RHP-zero. If we do not need tight control at low frequencies, then we may reverse the sign of the controller gain, and instead achieve tight control at higher frequencies. In this case we must for a RHP-zero z approximately require $\omega_c > 2z$. A special case is for plants with a zero at the origin; here we can achieve good transient control even though the control has no effect at steady-state.

Rule 7. Phase lag constraint. *We require in most practical cases (e.g. with PID control):* $\omega_c < \omega_u$. *Here the* ultimate frequency ω_u *is where* $\angle GG_m(j\omega_u) = -180°$. (See (5.67)).

Since time delays (Rule 5) and RHP-zeros (Rule 6) also contribute to the phase lag, one may in in most practical cases combine Rules 5, 6 and 7 into the single rule: $\omega_c < \omega_u$ (Rule 7).

Rule 8. Real open-loop unstable pole in $G(s)$ **at** $s = p$. *We need high feedback gains to stabilize the system and we approximately require* $\omega_c > 2p$. (See (5.46)).

In addition, for unstable plants we need $|G| > |G_d|$ *up to the frequency* p *(which may be larger than* ω_d *where* $|G_d| = 1|$). Otherwise, the input may saturate when there are disturbances, and the plant cannot be stabilized. (See (5.61)).

Most of the rules are illustrated graphically in Figure 5.17.

We have not formulated a rule to guard against model uncertainty. This is because, as shown in (5.71) and (5.72), uncertainty has only a minor effect on feedback performance for SISO systems, except at frequencies where the relative uncertainty E approaches 100%, and we obviously have to detune the system. Also, since 100% uncertainty at a given frequency allows for the presence of a RHP-zero on the imaginary axis at this frequency ($G(j\omega) = 0$), it is already covered by Rule 6.

The rules are necessary conditions ("minimum requirements") to achieve acceptable control performance. They are not sufficient since among other things we have only considered one effect at a time.

The rules quantify the qualitative rules given in the introduction. For example, the rule "Control outputs that are not self-regulating" may be quantified as: "Control outputs y for which $|G_d(j\omega)| > 1$ at some frequency" (Rule 1). The rule "Select inputs that have a large effect on the outputs" may be quantified as: "In terms of scaled

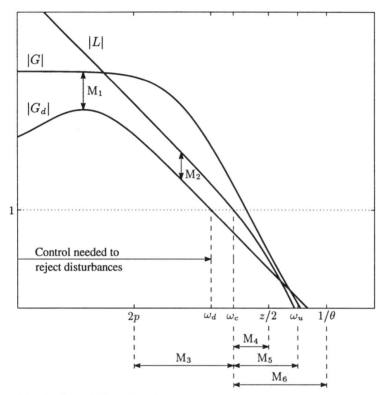

Margins for stability and performance:
  M_1 : Margin to stay within constraints, $|u| < 1$.
  M_2 : Margin for performance, $|e| < 1$.
  M_3 : Margin because of RHP-pole, p.
  M_4 : Margin because of RHP-zero, z.
  M_5 : Margin because of phase lag, $\angle G(j\omega_u) = -180°$.
  M_6 : Margin because of delay, θ.

Figure 5.17: Illustration of controllability requirements

variables we must have $|G| > |G_d| - 1$ at frequencies where $|G_d| > 1$ (Rule 3), and we must have $|G| > R - 1$ at frequencies where setpoint tracking is desired (Rule 4)". Another important insight from the above rules is that a larger disturbance or a smaller specification on the control error requires faster response (higher bandwidth).

In summary, Rules 1, 2 and 8 tell us that we need high feedback gain ("fast control") in order to reject disturbances, to track setpoints and to stabilize the plant. On the other hand, Rules 5, 6 and 7 tell us that we must use low feedback gains in the frequency range where there are RHP-zeros or delays or where the plant has a lot of phase lag. We have formulated these requirements for high and low gain as bandwidth requirements. If they somehow are in conflict then the plant is not controllable and the only remedy is to introduce design modifications to the plant.

Sometimes the problem is that the disturbances are so large that we hit input saturation, or the required bandwidth is not achievable. To avoid the latter problem, we must at least require that the effect of the disturbance is less than 1 (in terms of scaled variables) at frequencies beyond the bandwidth, (Rule 1)

$$|G_d(j\omega)| < 1 \quad \forall \omega \geq \omega_c \tag{5.75}$$

where as found above we approximately require $\omega_c < 1/\theta$ (Rule 5), $\omega_c < z/2$ (Rule 6) and $\omega_c < \omega_u$ (Rule 7). Condition (5.75) may be used, as in the example of Section 5.16.3 below, to determine the size of equipment.

5.15   Controllability analysis with feedforward control

The above controllability rules apply to feedback control, but we find that essentially the same conclusions apply to feedforward control when relevant. That is, if a plant is not controllable using feedback control, it is usually not controllable with feedforward control. A major difference, as shown below, is that a delay in $G_d(s)$ is an advantage for feedforward control ("it gives the feedforward controller more time to make the right action"). Also, a RHP-zero in $G_d(s)$ is also an advantage for feedforward control if $G(s)$ has a RHP-zero at the same location. Rules 3 and 4 on input constraints apply directly to feedforward control, but Rule 8 does not apply since unstable plants can only be stabilized by feedback control. The remaining rules in terms of performance and "bandwidth" do not apply directly to feedforward control.

Controllability can be analyzed by considering the feasibility of achieving perfect control. The feedforward controller is

$$u = K_d(s)d_m$$

where $d_m = G_{md}(s)d$ is the measured disturbance. The disturbance response with $r = 0$ becomes

$$e = Gu + G_d d = (GK_d G_{md} + G_d)d \tag{5.76}$$

(Reference tracking can be analyzed similarly by setting $G_{md} = 1$ and $G_d = -R$.)

Perfect control. From (5.76), $e = 0$ is achieved with the controller

$$K_d^{\text{perfect}} = -G^{-1}G_dG_{md}^{-1} \tag{5.77}$$

This assumes that K_d^{perfect} is stable and causal (no prediction), and so $GG_d^{-1}G_{md}$ should have no RHP-zeros and no (positive) delay. From this we find that a delay (or RHP-zero) in $G_d(s)$ is an advantage if it cancels a delay (or RHP-zero) in GG_{md}.

Ideal control. If perfect control is not possible, then one may analyze controllability by considering an "ideal" feedforward controller, K_d^{ideal}, which is (5.77) modified to be stable and causal (no prediction). The controller is ideal in that it assumes we have a perfect model. Controllability is then analyzed by using K_d^{ideal} in (5.76). An example is given below in (5.86) and (5.87) for a first-order delay process.

Model uncertainty. As discussed in Section 5.13, model uncertainty is a more serious problem for feedforward than for feedback control because there is no correction from the output measurement. For disturbance rejection, we have from (5.70) that the plant is *not* controllable with feedforward control if the relative model error for G/G_d at any frequency exceeds $1/|G_d|$. Here G_d is the scaled disturbance model. For example, if $|G_d(j\omega)| = 10$ then the error in G/G_d must not exceed 10% at this frequency. In practice, this means that feedforward control has to be combined with feedback control if the output is sensitive to the disturbance (i.e. if $|G_d|$ is much larger than 1 at some frequency).

Combined feedback and feedforward control. To analyze controllability in this case we may assume that the feedforward controller K_d has already been designed. Then from (5.76) the controllability of the remaining feedback problem can be analyzed using the rules in Section 5.14 if $G_d(s)$ is replaced by

$$\widehat{G}_d(s) = GK_dG_{md} + G_d \tag{5.78}$$

However, one must beware that the feedforward control may be very sensitive to model error, so the benefits of feedforward may be less in practice.

Conclusion. From (5.78) we see that the primary potential benefit of feedforward control is to reduce the effect of the disturbance and make $\widehat{G}_d$ less than 1 at frequencies where feedback control is not effective due to, for example, a delay or a large phase lag in $GG_m(s)$.

5.16   Applications of controllability analysis

5.16.1   First-order delay process

Problem statement. Consider disturbance rejection for the following process

$$G(s) = k\frac{e^{-\theta s}}{1 + \tau s}; \quad G_d(s) = k_d\frac{e^{-\theta_d s}}{1 + \tau_d s} \tag{5.79}$$

In addition there are measurement delays θ_m for the output and θ_{md} for the disturbance. All parameters have been appropriately scaled such that at each frequency
$|u| < 1, |d| < 1$ and we want $|e| < 1$. Assume $|k_d| > 1$. Treat separately the two
cases of i) feedback control only, and ii) feedforward control only, and carry out the
following:

a) For each of the eight parameters in this model explain qualitatively what value
you would choose from a controllability point of view (with descriptions such as
large, small, value has no effect).

b) Give quantitative relationships between the parameters which should be satisfied
to achieve controllability. Assume that appropriate scaling has been applied in such a
way that the disturbance is less than 1 in magnitude, and that the input and the output
are required to be less than 1 in magnitude.

Solution. (a) *Qualitative.* We want the input to have a "large, direct and fast effect"
on the output, while we want the disturbance to have a "small, indirect and slow
effect". By "direct" we mean without any delay or inverse response. This leads to the
following conclusion. For both feedback and feedforward control we want k and τ_d
large, and τ, θ and k_d small. For feedforward control we also want θ_d large (we then
have more time to react), but for feedback the value of θ_d does not matter; it translates
time, but otherwise has no effect. Clearly, we want θ_m small for feedback control (it
is not used for feedforward), and we want θ_{md} small for feedforward control (it is not
used for feedback).

(b) *Quantitative.* To stay within the input constraints ($|u| < 1$) we must from Rule 4
require $|G(j\omega)| > |G_d(j\omega)|$ for frequencies $\omega < \omega_d$. Specifically, for both feedback
and feedforward control

$$\boxed{k > k_d; \quad k/\tau > k_d/\tau_d} \tag{5.80}$$

Now consider performance where the results for feedback and feedforward control
differ. (i) First consider *feedback control*. From Rule 1 we need for acceptable
performance ($|e| < 1$) with disturbances

$$\omega_d \approx k_d/\tau_d < \omega_c \tag{5.81}$$

On the other hand, from Rule 5 we require for stability and performance

$$\omega_c < 1/\theta_{tot} \tag{5.82}$$

where $\theta_{tot} = \theta + \theta_m$ is the total delay around the loop. The combination of (5.81)
and (5.82) yields the following requirement for controllability

$$\boxed{\text{Feedback:} \quad \theta + \theta_m < \tau_d/k_d} \tag{5.83}$$

(ii) For *feedforward control*, any delay for the disturbance itself yields a smaller
"net delay", and to have $|e| < 1$ we need "only" require

$$\boxed{\text{Feedforward:} \quad \theta + \theta_{md} - \theta_d < \tau_d/k_d} \tag{5.84}$$

Proof of (5.84): Introduce $\widehat{\theta} = \theta + \theta_{md} - \theta_d$, and consider first the case with $\widehat{\theta} \leq 0$ (so (5.84) is clearly satisfied). In this case perfect control is possible using the controller (5.77),

$$K_d^{\text{perfect}} = -G^{-1}G_dG_{md}^{-1} = -\frac{k_d}{k}\frac{1+\tau s}{1+\tau_d s}e^{\widehat{\theta}s} \tag{5.85}$$

so we can even achieve $e = 0$. Next, consider $\widehat{\theta} > 0$. Perfect control is not possible, so instead we use the "ideal" controller obtained by deleting the prediction $e^{\widehat{\theta}s}$,

$$K_d^{\text{ideal}} = -\frac{k_d}{k}\frac{1+\tau s}{1+\tau_d s} \tag{5.86}$$

From (5.76) the response with this controller is

$$e = (GK_d^{\text{ideal}}G_{md} + G_d)d = \frac{k_d e^{-\theta_d s}}{1+\tau_d s}(1 - e^{-\widehat{\theta}s})d \tag{5.87}$$

and to achieve $|e|/|d| < 1$ we must require $\frac{k_d}{\tau_d}\widehat{\theta} < 1$ (using asymptotic values and $1-e^{-x} \approx x$ for small x) which is equivalent to (5.84). □

5.16.2 Application: Room heating

Consider the problem of maintaining a room at constant temperature, as discussed in Section 1.5. Let y be the room temperature, u the heat input and d the outdoor temperature. Feedback control should be used. Let the measurement delay for temperature (y) be $\theta_m = 100$ s.

1. Is the plant controllable with respect to disturbances?
2. Is the plant controllable with respect to setpoint changes of magnitude $R = 3$ (± 3 K) when the desired response time for setpoint changes is $\tau_r = 1000$ s (17 min) ?

Solution. A critical part of controllability analysis is scaling. A model in terms of scaled variables was derived in (1.26)

$$G(s) = \frac{20}{1000s+1}; \quad G_d(s) = \frac{10}{1000s+1} \tag{5.88}$$

The frequency responses of $|G|$ and $|G_d|$ are shown in Figure 5.18.

1. Disturbances. From Rule 1 feedback control is necessary up to the frequency $\omega_d = 10/1000 = 0.01$ rad/s, where $|G_d|$ crosses 1 in magnitude ($\omega_c > \omega_d$). This is exactly the same frequency as the upper bound given by the delay, $1/\theta = 0.01$ rad/s ($\omega_c < 1/\theta$). We therefore conclude that the system is barely controllable for this disturbance. From Rule 3 no problems with input constraints are expected since $|G| > |G_d|$ at all frequencies. These conclusions are supported by the closed-loop simulation in Figure 5.19(a) for a unit step disturbance (corresponding to a sudden 10 K increase in the outdoor temperature) using a PID-controller of the form in (5.66) with $K_c = 0.4$ (scaled variables), $\tau_I = 200$ s and $\tau_D = 60$ s. The output error

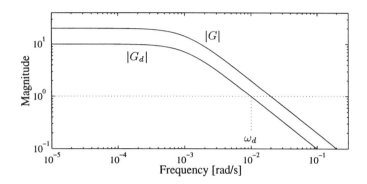

Figure 5.18: Frequency responses for room heating example

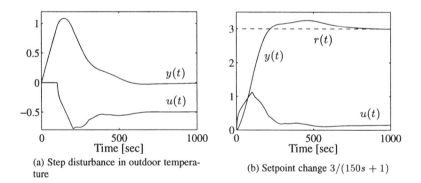

(a) Step disturbance in outdoor temperature

(b) Setpoint change $3/(150s + 1)$

Figure 5.19: PID feedback control of room heating example

exceeds its allowed value of 1 for a very short time after about 100 s, but then returns quite quickly to zero. The input goes down to about -0.8 and thus remains within its allowed bound of ±1.

2. *Setpoints.* The plant is controllable with respect to the desired setpoint changes. First, the delay is 100 s which is much smaller than the desired response time of 1000 s, and thus poses no problem. Second, $|G(j\omega)| \geq R = 3$ up to about $\omega_1 = 0.007$ [rad/s] which is seven times higher than the required $\omega_r = 1/\tau_r = 0.001$ [rad/s]. This means that input constraints pose no problem. In fact, we should be able to achieve response times of about $1/\omega_1 = 150$ s without reaching the input constraints. This is confirmed by the simulation in Figure 5.19(b) for a desired setpoint change $3/(150s + 1)$ using the same PID controller as above.

5.16.3   Application: Neutralization process

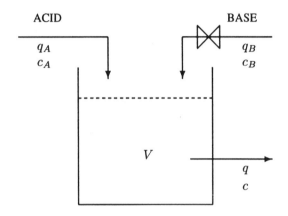

Figure 5.20: Neutralization process with one mixing tank

The following application is interesting in that it shows how the controllability analysis tools may assist the engineer in redesigning the process to make it controllable.

Problem statement. Consider the process in Figure 5.20, where a strong acid with pH= -1 (yes, a negative pH is possible — it corresponds to $c_{H+} = 10$ mol/l) is neutralized by a strong base (pH=15) in a mixing tank with volume $V = 10\text{m}^3$. We want to use feedback control to keep the pH in the product stream (output y) in the range 7 ± 1 ("salt water") by manipulating the amount of base, q_B (input u) in spite of variations in the flow of acid, q_A (disturbance d). The delay in the pH-measurement is $\theta_m = 10$ s.

To achieve the desired product with pH=7 one must exactly balance the inflow of acid (the disturbance) by addition of base (the manipulated input). Intuitively, one might expect that the main control problem is to adjust the base accurately by means of a very accurate valve. However, as we will see this "feedforward" way of thinking is misleading, and the main hurdle to good control is the need for very fast response times.

We take the controlled output to be the excess of acid, c [mol/l], defined as $c = c_{H+} - c_{OH-}$. In terms of this variable the control objective is to keep $|c| \leq c_{max} = 10^{-6}$ mol/l, and the plant is a simple mixing process modelled by

$$\frac{d}{dt}(Vc) = q_A c_A + q_B c_B - qc \qquad (5.89)$$

The nominal values for the acid and base flows are $q_A^* = q_B^* = 0.005$ [m^3/s] resulting in a product flow $q^* = 0.01$ [m^3/s]$= 10$ [l/s]. Here superscript * denotes the steady-state value. Divide each variable by its maximum deviation to get the following scaled variables

$$y = \frac{c}{10^{-6}}; \quad u = \frac{q_B}{q_B^*}; \quad d = \frac{q_A}{0.5 q_A^*} \qquad (5.90)$$

Then the appropriately scaled linear model for one tank becomes

$$G_d(s) = \frac{k_d}{1 + \tau_h s}; \quad G(s) = \frac{-2 k_d}{1 + \tau_h s}; \quad k_d = 2.5 \cdot 10^6 \qquad (5.91)$$

where $\tau_h = V/q = 1000$ s is the residence time for the liquid in the tank. Note that the steady-state gain in terms of scaled variables is more than a million so the output is extremely sensitive to both the input and the disturbance. The reason for this high gain is the much higher concentration in the two feed streams, compared to that desired in the product stream. The question is: Can acceptable control be achieved?

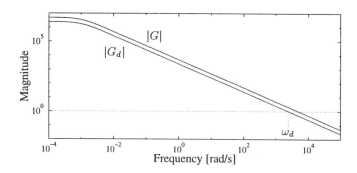

Figure 5.21: Frequency responses for the neutralization process with one mixing tank

Controllability analysis. The frequency responses of $G_d(s)$ and $G(s)$ are shown graphically in Figure 5.21. From Rule 2, input constraints do not pose a problem since $|G| = 2|G_d|$ at all frequencies. The main control problem is the high disturbance sensitivity, and from (5.81) (Rule 1) we find the frequency up to which feedback is needed

$$\omega_d \approx k_d/\tau = 2500 \text{ rad/s} \qquad (5.92)$$

This requires a response time of $1/2500 = 0.4$ milliseconds which is clearly impossible in a process control application, and is in any case much less than the measurement delay of 10 s.

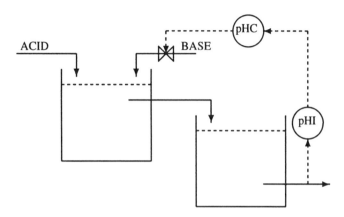

Figure 5.22: Neutralization process with two tanks and one controller

Design change: Multiple tanks. The only way to improve controllability is to modify the process. This is done in practice by performing the neutralization in several steps as illustrated in Figure 5.22 for the case of two tanks. This is similar to playing golf where it is often necessary to use several strokes to get to the hole. With n equal mixing tanks in series the transfer function for the effect of the disturbance becomes

$$G_d(s) = k_d h_n(s); \quad h_n(s) = \frac{1}{\left(\frac{\tau_h}{n}s + 1\right)^n} \qquad (5.93)$$

where $k_d = 2.5 \cdot 10^6$ is the gain for the mixing process, $h_n(s)$ is the transfer function of the mixing tanks, and τ_h is the total residence time, V_{tot}/q. The magnitude of $h_n(s)$ as a function of frequency is shown in Figure 5.23 for one to four equal tanks in series.

From controllability Rules 1 and 5, we must at least require for acceptable disturbance rejection that

$$\boxed{|G_d(j\omega_\theta)| \le 1} \quad \omega_\theta \triangleq 1/\theta \qquad (5.94)$$

where θ is the delay in the feedback loop. Thus, one purpose of the mixing tanks $h_n(s)$ is to reduce the effect of the disturbance by a factor $k_d (= 2.5 \cdot 10^6)$ at the frequency $\omega_\theta (= 0.1 \text{ [rad/s]})$, i.e. $|h_n(j\omega_\theta)| \le 1/k_d$. With $\tau_h = V_{tot}/q$ we obtain the following

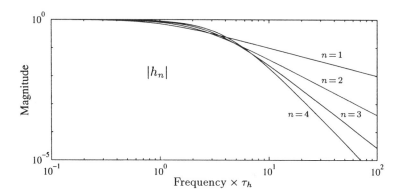

Figure 5.23: Frequency responses for n tanks in series with the same total residence time τ_h; $h_n(s) = 1/(\frac{\tau_h}{n}s + 1)^n$, $n = 1, 2, 3, 4$

minimum value for the total volume for n equal tanks in series

$$V_{tot} = q\theta n\sqrt{(k_d)^{2/n} - 1} \tag{5.95}$$

where $q = 0.01$ m^3/s. With $\theta = 10$ s we then find that the following designs have the same controllability with respect to disturbance rejection:

No. of tanks n	Total volume V_{tot} [m^3]	Volume each tank [m^3]
1	250000	250000
2	316	158
3	40.7	13.6
4	15.9	3.98
5	9.51	1.90
6	6.96	1.16
7	5.70	0.81

With one tank we need a volume corresponding to that of a supertanker to get acceptable controllability. The minimum total volume is obtained with 18 tanks of about 203 litres each — giving a total volume of 3.662 m^3. However, taking into account the additional cost for extra equipment such as piping, mixing, measurements and control, we would probably select a design with 3 or 4 tanks for this example.

Control system design. We are not quite finished yet. The condition $|G_d(j\omega_\theta)| \leq 1$ in (5.94), which formed the basis for redesigning the process, may be optimistic because it only ensures that we have $|S| < 1/|G_d|$ at the crossover frequency

$\omega_B \approx \omega_c \approx \omega_\theta$. However, from Rule 1 we also require that $|S| < 1/|G_d|$, or approximately $|L| > |G_d|$, at frequencies lower than w_c, and this may be difficult to achieve since $G_d(s) = k_d h(s)$ is of high order. The problem is that this requires $|L|$ to drop steeply with frequency, which results in a large negative phase for L, whereas for stability and performance the slope of $|L|$ at crossover should not be steeper than -1, approximately (see Section 2.6.2).

Thus, the control system in Figure 5.22 with a single feedback controller will *not* achieve the desired performance. The solution is to install a *local feedback* control system on each tank and to add base in each tank as shown in Figure 5.24. This is

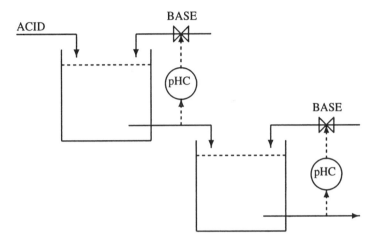

Figure 5.24: Neutralization process with two tanks and two controllers.

another *plant design change* since it requires an additional measurement and actuator for each tank. Consider the case of n tanks in series. With n controllers the overall closed-loop response from a disturbance into the first tank to the pH in the last tank becomes

$$y = G_d \prod_{i=1}^{n} (\frac{1}{1 + L_i}) d \approx \frac{G_d}{L} d, \quad L \triangleq \prod_{i=1}^{n} L_i \qquad (5.96)$$

where $G_d = \prod_{i=1}^{n} G_i$ and $L_i = G_i K_i$, and the approximation applies at low frequencies where feedback is effective.

In this case, we can design each loop $L_i(s)$ with a slope of -1 and bandwidth $\omega_c \approx \omega_\theta$, such that the overall loop transfer function L has slope $-n$ and achieves $|L| > |G_d|$ at all frequencies lower than ω_d (the size of the tanks are selected as before such that $\omega_d \approx \omega_\theta$). Thus, our analysis confirms the usual recommendation of adding base gradually and having one pH-controller for each tank (McMillan, 1984, p. 208). It seems unlikely that any other control strategy can achieve a sufficiently high roll-off for $|L|$.

In summary, this application has shown how a simple controllability analysis may be used to make decisions on both the appropriate size of the equipment, and the selection of actuators and measurements for control. Our conclusions are in agreement with what is used in industry. Importantly, we arrived at these conclusions, without having to design any controllers or perform any simulations. Of course, as a final test, the conclusions from the controllability analysis should be verified by simulations using a nonlinear model.

Exercise 5.8 Comparison of local feedback and cascade control. *Explain why a cascade control system with two measurements (pH in each tank) and only one manipulated input (the base flow into the first tank) will not achieve as good performance as the control system in Figure 5.24 where we use local feedback with two manipulated inputs (one for each tank). (Hint: Show first that the closed-loop response for the cascade control system is as in (5.96) but with*

$$L_1 = G_1 K_1, \ L_2 = G_1 G_2 K_1 K_2, \ \ldots, \ L_i = \prod_{j=1}^{i} G_j K_j$$

rather than $L_i = G_i K_i$.)

The following exercise considers the use of buffer tanks for reducing quality (concentration, temperature) disturbances in chemical processes.

Exercise 5.9 *(a) The effect of a concentration disturbance must be reduced by a factor of 100 at the frequency 0.5 rad/min. The disturbances should be dampened by use of buffer tanks and the objective is to minimize the total volume. How many tanks in series should one have? What is the total residence time?*

(b) The feed to a distillation column has large variations in concentration and the use of one buffer tank is suggested to dampen these. The effect of the feed concentration d on the product composition y is given by (scaled variables, time in minutes)

$$G_d(s) = e^{-s}/3s$$

That is, after a step in d the output y will, after an initial delay of 1 min, increase in a ramp-like fashion and reach its maximum allowed value (which is 1) after another 3 minutes. Feedback control should be used and there is an additional measurement delay of 5 minutes. What should be the residence time in the tank?

(c) Show that in terms of minimizing the total volume for buffer tanks in series, it is optimal to have buffer tanks of equal size.

(d) Is there any reason to have buffer tanks in parallel (they must not be of equal size because then one may simply combine them)?

(e) What about parallel pipes in series (pure delay). Is this a good idea?

Buffer tanks are also used in chemical processes to dampen liquid flowrate disturbances (or gas pressure disturbances). This is the topic of the following exercise.

Exercise 5.10 *Let $d_1 = q_{in}$ [m^3/s] denote a flowrate which acts as a disturbance to the process. We add a buffer tank (with liquid volume V [m^3]), and use a "slow" level controller*

K such that the outflow $d_2 = q_{out}$ (the "new" disturbance) is smoother than the inflow q_{in} (the "original" disturbance). The idea is to temporarily increase or decrease the liquid volume in the tank to avoid sudden changes in q_{out}. Note that the steady-state value of q_{out} must equal that of q_{in}.

A material balance yields $V(s) = (q_{in}(s) - q_{out}(s))/s$ and with a level controller $q_{out}(s) = K(s)V(s)$ we find that

$$d_2(s) = \underbrace{\frac{K(s)}{s + K(s)}}_{h(s)} d_1(s) \qquad (5.97)$$

The design of a buffer tank for a flowrate disturbance then consists of two steps:

1. Design the level controller $K(s)$ such that $h(s)$ has the desired shape (e.g. determined by a controllability analysis of how d_2 affects the remaining process; note that we must always have $h(0) = 1$).
2. Design the size of the tank (determine its volume V_{max}) such that the tanks does not overflow or go empty for the expected disturbances in $d_1 = q_{in}$.

Problem. (a) Assume the inflow varies in the range $q_{in}^* \pm 100\%$ where q_{in}^* is the nominal value, and apply this stepwise procedure to two cases:

(i) The desired transfer function is $h(s) = 1/(\tau s + 1)$.
(ii) The desired transfer function is $h(s) = 1/(\tau_2 s + 1)^2$.

(b) Explain why it is usually not recommended to have integral action in $K(s)$.

(c) In case (ii) one could alternatively use two tanks in series with controllers designed as in (i). Explain why this is most likely not a good solution. (Solution: The required total volume is the same, but the cost of two smaller tanks is larger than one large tank).

5.16.4 Additional exercises

Exercise 5.11 What information about a plant is important for controller design, and in particular, in which frequency range is it important to know the model well? To answer this problem you may think about the following sub-problems:

(a) Explain what information about the plant is used for Ziegler-Nichols tuning of a SISO PID-controller.

(b) Is the steady-state plant gain $G(0)$ important for controller design? (As an example consider the plant $G(s) = \frac{1}{s+a}$ with $|a| \leq 1$ and design a P-controller $K(s) = K_c$ such that $\omega_c = 100$. How does the controller design and the closed-loop response depend on the steady-state gain $G(0) = 1/a$?)

Exercise 5.12 Let $H(s) = K_1 e^{-\theta_1 s}$, $G(s) = K_2 e^{-0.5s} \frac{1}{(30s+1)(Ts+1)}$, and $G_d(s) = G(s)H(s)$. The measurement device for the output has transfer function $G_m(s) = e^{-\theta_2 s}$. The unit for time is seconds. The nominal parameter values are: $K_1 = 0.24$, $\theta_1 = 1$ [s], $K_2 = 38$, $\theta_2 = 5$ [s], and $T = 2$ [s].

(a) Assume all variables have been appropriately scaled. Is the plant input-output controllable?

(b) What is the effect on controllability of changing one model parameter at a time in the following ways:

1. θ_1 *is reduced to* 0.1 *[s].*
2. θ_2 *is reduced to* 2 *[s].*
3. K_1 *is reduced to* 0.024.
4. K_2 *is reduced to* 8.
5. T *is increased to* 30 *[s].*

Exercise 5.13 *A heat exchanger is used to exchange heat between two streams; a coolant with flowrate q* $(1 \pm 1$ *kg/s) is used to cool a hot stream with inlet temperature* T_0 *(*$100 \pm 10° C$*) to the outlet temperature T (which should be* $60 \pm 10° C$*). The measurement delay for T is* 3*s. The main disturbance is on* T_0*. The following model in terms of deviation variables is derived from heat balances*

$$T(s) = \frac{8}{(60s+1)(12s+1)}q(s) + \frac{0.6(20s+1)}{(60s+1)(12s+1)}T_0(s) \qquad (5.98)$$

where T and T_0 *are in* $°C$*, q is in kg/s, and the unit for time is seconds. Derive the scaled model. Is the plant controllable with feedback control? (Solution: The delay poses no problem (performance), but the effect of the disturbance is a bit too large at high frequencies (input saturation), so the plant is not controllable).*

5.17   Conclusion

The chapter has presented a frequency domain controllability analysis for scalar systems applicable to both feedback and feedforward control. We summarized our findings in terms of eight controllability rules; see page 197. These rules are necessary conditions ("minimum requirements") to achieve acceptable control performance. They are not sufficient since among other things they only consider one effect at a time. The rules may be used to determine whether or not a given plant is controllable. The method has been applied to a pH neutralization process, and it is found that the heuristic design rules given in the literature follow directly. The key steps in the analysis are to consider disturbances and to scale the variables properly.

The tools presented in this chapter may also be used to study the effectiveness of adding extra manipulated inputs or extra measurements (cascade control). They may also be generalized to multivariable plants where directionality becomes a further crucial consideration. Interestingly, a direct generalization to decentralized control of multivariable plants is rather straightforward and involves the CLDG and the PRGA; see page 443 in Chapter 10.

6

LIMITATIONS ON PERFORMANCE IN MIMO SYSTEMS

In this chapter, we generalize the results of Chapter 5 to MIMO systems. We first discuss fundamental limitations on the sensitivity and complementary sensitivity functions imposed by the presence of RHP-zeros. We then consider separately the issues of functional controllability, RHP-zeros, RHP-poles, disturbances, input constraints and uncertainty. Finally, we summarize the main steps in a procedure for analyzing the input-output controllability of MIMO plants.

6.1 Introduction

In a MIMO system, disturbances, the plant, RHP-zeros, RHP-poles and delays each have directions associated with them. This makes it more difficult to consider their effects separately, as we did in the SISO case where we were able to reformulate the imposed limitations in terms of bounds on the loop gain, $|L|$, and its crossover frequency, ω_c. For example, a multivariable plant may have a RHP-zero and a RHP-pole at the same location, but their effects may not interact if they are in completely different parts of the system; recall (4.72).

We will quantify the directionality of the various effects in G and G_d by their *output* directions:

- y_z: output direction of a RHP-zero, see (4.67)
- y_p: output direction of a RHP-pole, see (4.68)
- y_d: output direction of a disturbance, see (6.30)
- u_i: i'th output direction (singular vector) of the plant, see (3.34)[1]

All these are $l \times 1$ vectors where l is the number of outputs. y_z and y_p are fixed complex vectors, while $y_d(s)$ and $u_i(s)$ are frequency-dependent (s may here be

---

[1] Note that u_i here is the i'th output singular vector, and *not* the i'th input.

viewed as a generalized complex frequency; in most cases $s = j\omega$). The vectors are here normalized such that they have Euclidean length 1,

$$\|y_z\|_2 = 1, \quad \|y_p\|_2 = 1, \quad \|y_d(s)\|_2 = 1, \quad \|u_i(s)\|_2 = 1$$

We may also consider the associated input directions of G. However, these directions are usually of less interest since we are primarily concerned with the performance at the output of the plant.

The angles between the various output directions can be quantified using their inner products: $|y_z^H y_p|$, $|y_z^H y_d|$, etc. The inner product gives a number between 0 and 1, and from this we can define the angle in the first quadrant, see (A.113). For example, the output angle between a pole and a zero is

$$\phi = \frac{cos^{-1}|y_z^H y_p|}{\|y_z\|_2 \|y_p\|_2} = cos^{-1}|y_z^H y_p|$$

We assume throughout this chapter that the models have been scaled as outlined in Section 1.4. The scaling procedure is the same as that for SISO systems, except that the scaling factors D_u, D_d, D_r and D_e are *diagonal matrices* with elements equal to the maximum change in each variable u_i, d_i, r_i and e_i. The control error in terms of scaled variables is then

$$e = y - r = Gu + G_d d - R\tilde{r}$$

where at each frequency we have $\|u(\omega)\|_{\max} \leq 1$, $\|d(\omega)\|_{\max} \leq 1$ and $\|\tilde{r}(\omega)\|_{\max} \leq 1$, and the control objective is to achieve $\|e\|_{\max}(\omega) < 1$.

Remark 1 Here $\| \cdot \|_{\max}$ is the vector infinity-norm, that is, the largest element in the vector. This norm is sometimes denoted $\| \cdot \|_\infty$, but this is not used here to avoid confusing it with the $\mathcal{H}_\infty$ norm of the transfer function (where the ∞ denotes the maximum over frequency rather than the maximum over the elements of the vector).

Remark 2 As for SISO systems, we see that reference changes may be analyzed as a special case of disturbances by replacing G_d by $-R$.

Remark 3 Whether various disturbances and reference changes should be considered separately or simultaneously is a matter of design philosophy. In this chapter, we mainly consider their effects separately, on the grounds that it is unlikely for several disturbances to attain their worst values simultaneously. This leads to necessary conditions for acceptable performance, which involve the elements of different matrices rather than matrix norms.

6.2   Constraints on S and T

6.2.1 S plus T is the identity matrix

From the identity $S + T = I$ and (A.49), we get

$$|1 - \bar{\sigma}(S)| \leq \bar{\sigma}(T) \leq 1 + \bar{\sigma}(S) \tag{6.1}$$

$$|1 - \bar{\sigma}(T)| \leq \bar{\sigma}(S) \leq 1 + \bar{\sigma}(T) \qquad (6.2)$$

This shows that we cannot have both S and T small simultaneously and that $\bar{\sigma}(S)$ is large if and only if $\bar{\sigma}(T)$ is large.

6.2.2 Sensitivity integrals

For SISO systems we presented several integral constraints on sensitivity (the waterbed effects). These may be generalized to MIMO systems by using the determinant or the singular values of S, see Boyd and Barratt (1991) and Freudenberg and Looze (1988). For example, the generalization of the Bode sensitivity integral in (5.6) may be written

$$\int_0^\infty \ln |\det S(j\omega)| d\omega = \sum_j \int_0^\infty \ln \sigma_j(S(j\omega)) d\omega = \pi \cdot \sum_{i=1}^{N_p} Re(p_i) \qquad (6.3)$$

For a stable $L(s)$, the integrals are zero. Other generalizations are also available, see Zhou et al. (1996). However, although these relationships are interesting, it seems difficult to derive any concrete bounds on achievable performance from them.

6.2.3 Interpolation constraints

RHP-zero. If $G(s)$ has a RHP-zero at z with output direction y_z, then for internal stability of the feedback system the following interpolation constraints must apply:

$$\boxed{y_z^H T(z) = 0; \quad y_z^H S(z) = y_z^H} \qquad (6.4)$$

In words, (6.4) says that T must have a RHP-zero in the same direction as G, and that $S(z)$ has an eigenvalue of 1 corresponding to the left eigenvector y_z.

Proof of (6.4): From (4.67) there exists an output direction y_z such that $y_z^H G(z) = 0$. For internal stability, the controller cannot cancel the RHP-zero and it follows that $L = GK$ has a RHP-zero in the same direction, i.e. $y_z^H L(z) = 0$. Now $S = (I + L)^{-1}$ is stable and has no RHP-pole at $s = z$. It then follows from $T = LS$ that $y_z^H T(z) = 0$ and $y_z^H(I - S) = 0$. $\quad\square$

RHP-pole. If $G(s)$ has a RHPpole at p with output direction y_p, then for internal stability the following interpolation constraints apply

$$\boxed{S(p)y_p = 0; \quad T(p)y_p = y_p} \qquad (6.5)$$

Proof of (6.5): The square matrix $L(p)$ has a RHP-pole at $s = p$, and if we assume that $L(s)$ has no RHP-zeros at $s = p$ then $L^{-1}(p)$ exists and from (4.70) there exists an output pole direction y_p such that

$$L^{-1}(p)y_p = 0 \qquad (6.6)$$

Since T is stable, it has no RHP-pole at $s = p$, so $T(p)$ is finite. It then follows, from $S = TL^{-1}$, that $S(p)y_p = T(p)L^{-1}(p)y_p = 0$. $\quad\square$

Similar constraints apply to L_I, S_I and T_I, but these are in terms of the input zero and pole directions, u_z and u_p.

6.2.4 Sensitivity peaks

Based on the above interpolation constraints we here derive lower bounds on the weighted sensitivity functions. The results show that a peak on $\bar{\sigma}(S)$ larger than 1 is unavoidable if the plant has a RHP-zero, and that a peak on $\bar{\sigma}(T)$ larger 1 is unavoidable if the plant has a RHP-pole. In particular, the peaks may be large if the plant has both RHP-zeros and RHP-poles.

The first result, originally due to Zames (1981), directly generalizes the SISO condition in (5.16).

Theorem 6.1 Weighted sensitivity. *Suppose the plant $G(s)$ has a RHP-zero at $s = z$. Let $w_P(s)$ be any stable scalar weight. Then for closed-loop stability the weighted sensitivity function must satisfy*

$$\|w_P S(s)\|_\infty = \max_\omega \bar{\sigma}(w_P S(j\omega)) \geq |w_P(z)| \qquad (6.7)$$

Proof: Introduce the scalar function $f(s) = y_z^H w_P(s) S(s) y_z$ which is analytic in the RHP. We then have

$$\|w_P S(s)\|_\infty \geq \|f(s)\|_\infty \geq |f(z)| = |w_P(z)| \qquad (6.8)$$

The first inequality follows because the singular value measures the maximum gain of a matrix independent of direction, and so $\bar{\sigma}(A) \geq \|Aw\|_2$ and $\bar{\sigma}(A) \geq \|wA\|_2$ (see (A.111)) for any vector w with $\|w\|_2 = 1$. The second inequality follows from the maximum modulus principle as in the SISO case. The final equality follows since $w_P(s)$ is a scalar and from the interpolation constraint (6.4) we get $y_z^H S(z) y_z = y_z^H y_z = 1$. □

The next theorem generalizes the SISO-condition in (5.17).

Theorem 6.2 Weighted complementary sensitivity. *Suppose the plant $G(s)$ has a RHP-pole at $s = p$. Let $w_T(s)$ be any stable scalar weight. Then for closed-loop stability the weighted complementary sensitivity function must satisfy*

$$\|w_T(s) T(s)\|_\infty = \max_\omega \bar{\sigma}(w_T T(j\omega)) \geq |w_T(p)| \qquad (6.9)$$

Proof: Introduce the scalar function $f(s) = y_p^H w_T(s) T(s) y_p$ which is analytic in the RHP since $w_T T(s)$ is stable. We then have

$$\|w_T T(s)\|_\infty \geq \|f(s)\|_\infty \geq |f(p)| = |w_T(p)| \qquad (6.10)$$

The first inequality follows because the singular value measures the maximum gain of a matrix independent of direction and $\|y_p\|_2 = 1$. The second inequality follows from the maximum modulus principle. The final equality follows since $w_T(s)$ is a scalar and from (6.5) we get $y_p^H T(p) y_p = y_p^H y_p = 1$. □

The third theorem, which is a direct extension of the SISO result in (5.18), generalizes the two theorems above. Consider a plant $G(s)$ with RHP-poles p_i and RHP-zeros z_j, and factorize $G(s)$ in terms of *Blaschke products* as follows

$$G(s) = B_p(s)G_p(s), \quad G(s) = B_z^{-1}(s)G_z(s)$$

where $B_p(s)$ and $B_z(s)$ are stable all-pass transfer matrices (all singular values are 1 for $s = j\omega$) containing the RHP-poles and RHP-zeros, respectively. $B_p(s)$ is obtained by factorizing to the output one RHP-pole at a time, starting with $G(s) = B_{p1}(s)G_{p1}(s)$ where $B_{p1}(s) = I + \frac{2\mathrm{Re}p_1}{s-p_1}\widehat{y}_{p1}\widehat{y}_{p1}^H$ where $\widehat{y}_{p1} = y_{p1}$ is the output pole direction for p_1. This procedure may be continued to factor out p_2 from $G_{p1}(s)$ where $\widehat{y}_{p2}$ is the output pole direction of G_{p1} (which need not coincide with y_{p2}, the pole direction of G_p), and so on. A similar procedure may be used for the RHP-zeros. We get

$$B_p(s) = \prod_{i=1}^{N_p}(I + \frac{2\mathrm{Re}(p_i)}{s - p_i}\widehat{y}_{pi}\widehat{y}_{pi}^H), \quad B_z(s) = \prod_{j=1}^{N_z}(I + \frac{2\mathrm{Re}(z_j)}{s - z_j}\widehat{y}_{zj}\widehat{y}_{zj}^H) \quad (6.11)$$

Remark. State-space realizations are provided by Zhou et al. (1996, p.145). Note that the realizations may be complex.

With this factorization we have the following theorem.

Theorem 6.3 MIMO sensitivity peak. *Suppose that $G(s)$ has N_z RHP-zeros z_j with output directions y_{zj}, and N_p RHP-poles p_i with output directions y_{pi}. Define the all-pass transfer matrices given in (6.11) and compute the real constants*

$$c_{1j} = \|y_{zj}^H B_p(z_j)\|_2 \geq 1; \quad c_{2i} = \|B_z(p_i)y_{pi}\|_2 \geq 1 \quad (6.12)$$

Then for closed-loop stability the weighted sensitivity function must satisfy for each z_j

$$\|w_P S\|_\infty \geq c_{1j}|w_P(z_j)| \quad (6.13)$$

and the weighted complementary sensitivity function must satisfy for each p_i

$$\|w_T T\|_\infty \geq c_{2i}|w_T(p_i)| \quad (6.14)$$

Proof of c_{1j} in (6.12): Consider here a RHP-zero z with direction y_z (the subscript j is omitted). Since G has RHP-poles at p_i, S must have RHP-zeros at p_i, such that $T = SGK$ is stable. We may factorize $S = TL^{-1} = S_1 B_p^{-1}(s)$ and introduce the scalar function $f(s) = y_z^H w_P(s)S_1(s)y$ which is analytic (stable) in the RHP. y is a vector of unit length which can be chosen freely. We then have

$$\|w_P S(s)\|_\infty = \|w_P S_1\|_\infty \geq \|f(s)\|_\infty \geq |f(z)| = |w_P(z)| \cdot |y_z^H B_p(z)y| \quad (6.15)$$

The final equality follows since w_P is a scalar and $y_z^H S_1(z) = y_z^H S(z)B_p(z) = y_z^H B_p(z)$. We finally select y such that the lower bound is as large as possible and derive c_1. To prove that $c_1 \geq 1$, we follow Chen (1995) and introduce the matrix V_i whose columns together with $\hat{y}_{pi}$ form an orthonormal basis for $C^{l \times l}$. Then, $I = \hat{y}_{pi} \hat{y}_{pi}^H + V_i V_i^H$, and

$$B_{pi}(s) = I + \frac{2\text{Re}(p_i)}{s - p_i} \hat{y}_{pi} \hat{y}_{pi}^H = \frac{s + \bar{p}_i}{s - p_i} \hat{y}_{pi} \hat{y}_{pi}^H + V_i V_i^H = [\hat{y}_{pi} \quad V_i] \begin{bmatrix} \frac{s + \bar{p}_i}{s - p_i} & 0 \\ 0 & I \end{bmatrix} \begin{bmatrix} \hat{y}_{pi}^H \\ V_i^H \end{bmatrix} \tag{6.16}$$

and we see that all singular values of $B_{pi}(z)$ are equal to 1, except for one which is $|z + \bar{p}_i|/|z - p_i| \geq 1$ (since z and p_i are both in the RHP). Thus all singular values of $B_p(z)$ are 1 or larger, so $B_p(z)$ is greater than or equal to 1 in all directions and hence $c_1 \geq 1$. The proof of c_{2i} is similar. $\qquad\square$

Lower bound on $\|S\|_\infty$ and $\|T\|_\infty$. From Theorem 6.3 we get by selecting $w_P(s) = 1$ and $w_T(s) = 1$

$$\|S\|_\infty \geq \max_{\text{zeros } z_j} c_{1j}; \quad \|T\|_\infty \geq \max_{\text{poles } p_i} c_{2i} \tag{6.17}$$

One RHP-pole and one RHP-zero. For the case with one RHP-zero z and one RHP-pole p we derive from (6.12)

$$c_1 = c_2 = \sqrt{\sin^2 \phi + \frac{|z + \bar{p}|^2}{|z - p|^2} \cos^2 \phi} \tag{6.18}$$

where $\phi = cos^{-1} |y_z^H y_p|$ is the angle between the output directions of the pole and zero. We then get that if the pole and zero are aligned in the same direction such that $y_z = y_p$ and $\phi = 0$, then (6.18) simplifies to give the SISO-conditions in (5.18) and (5.19) with $c_1 = c_2 = \frac{|z + \bar{p}|}{|z - p|} \geq 1$. Conversely, if the pole and zero are orthogonal to each other, then $\phi = 90°$ and $c_1 = c_2 = 1$ and there is no additional penalty for having both a RHP-pole and a RHP-zero.

Proof of (6.18): From (6.12) $c_1 = \|y_z^H B_p(z)\|_2$. From (6.16) the projection of y_z in the direction of the largest singular value of $B_p(z)$ has magnitude $|z + \bar{p}|/|z - p| \cos\phi$, and the projection onto the remaining subspace is $1 \cdot \sin \phi$, and (6.18) follows. The result was first proved by Boyd and Desoer (1985) and an alternative proof is given by Chen (1995) who presents a slightly improved bound. $\qquad\square$

Later in this chapter we discuss the implications of these results and provide some examples.

6.3 Functional controllability

Consider a plant $G(s)$ with l outputs and let r denote the normal rank of $G(s)$. In order to control all outputs independently we must require $r = l$, that is, the plant must be functionally controllable. This term was introduced by Rosenbrock (1970, p. 70) for square systems, and related concepts are "right invertibility" and "output realizability". We will use the following definition:

Definition 6.1 Functional controllability. *An* m-*input* l-*output system* $G(s)$ *is functionally controllable if the normal rank of* $G(s)$ *is equal to the number of outputs* $(r = l)$, *that is, if* $G(s)$ *has full row rank. A system is functionally uncontrollable if* $r < l$.

The normal rank of $G(s)$ is the rank of $G(s)$ at all values of s except at a finite number of singularities (which are the zeros of $G(s)$).

Remark 1 The only example of a SISO system which is functionally uncontrollable is the system $G(s) = 0$. A square MIMO system is functional uncontrollable if and only if $\det G(s) = 0, \forall s$.

Remark 2 A plant is functionally uncontrollable if and only if $\sigma_l(G(j\omega)) = 0, \forall\omega$. As a measure of how close a plant is to being functionally uncontrollable we may therefore consider $\sigma_l(G(j\omega))$, which for the interesting case when there is at least as many inputs as outputs, $m \geq l$, is the minimum singular value, $\underline{\sigma}(G(j\omega))$.

Remark 3 In most cases functional uncontrollability is a *structural* property of the system, that is, it does not depend on specific parameter values, and it may often be evaluated from cause-and-effect graphs. A typical example of this is when none of the inputs u_i affect a particular output y_j which would be the case if one of the rows in $G(s)$ was identically zero. Another example is when there are fewer inputs than outputs.

Remark 4 For strictly proper systems, $G(s) = C(sI - A)^{-1}B$, we have that $G(s)$ is *functionally uncontrollable* if rank$(B) < l$ (the system is input deficient), or if rank$(C) < l$ (the system is output deficient), or if rank$(sI - A) < l$ (fewer states than outputs). This follows since the rank of a product of matrices is less than or equal to the minimum rank of the individual matrices, see (A.35).

If the plant is *not* functionally controllable, i.e. $r < l$, then there are $l - r$ output directions, denoted y_0, which cannot be affected. These directions will vary with frequency, and we have (analogous to the concept of a zero direction)

$$y_0^H(j\omega)G(j\omega) = 0 \qquad (6.19)$$

From an SVD of $G(j\omega) = U\Sigma V^H$, the *uncontrollable output directions* $y_0(j\omega)$ are the last $l - r$ columns of $U(j\omega)$. By analyzing these directions, an engineer can then decide on whether it is acceptable to keep certain output combinations uncontrolled, or if additional actuators are needed to increase the rank of $G(s)$.

Example 6.1 *The following plant is singular and thus not functionally controllable*

$$G(s) = \begin{bmatrix} \frac{1}{s+1} & \frac{2}{s+1} \\ \frac{2}{s+2} & \frac{4}{s+2} \end{bmatrix}$$

This is easily seen since column 2 of $G(s)$ *is two times column 1. The uncontrollable output directions at low and high frequencies are, respectively*

$$y_0(0) = \frac{1}{\sqrt{2}}\begin{bmatrix} 1 \\ -1 \end{bmatrix} \quad y_0(\infty) = \frac{1}{\sqrt{5}}\begin{bmatrix} 2 \\ -1 \end{bmatrix}$$

6.4   Limitations imposed by time delays

Time delays pose limitations also in MIMO systems. Specifically, let θ_{ij} denote the time delay in the ij'th element of $G(s)$. Then a lower bound on the time delay for output i is given by the smallest delay in row i of $G(s)$, that is,

$$\theta_i^{\min} = \min_j \theta_{ij}$$

This bound is obvious since $\theta_i^{\min}$ is the minimum time for any input to affect output i, and $\theta_i^{\min}$ can be regarded as a delay pinned to output i.

Holt and Morari (1985a) have derived additional bounds, but their usefulness is sometimes limited since they assume a *decoupled* closed-loop response (which is usually not desirable in terms of overall performance) and also assume infinite power in the inputs.

For MIMO systems we have the surprising result that an increased time delay may sometimes improve the achievable performance. As a simple example, consider the plant

$$G(s) = \begin{bmatrix} 1 & 1 \\ e^{-\theta s} & 1 \end{bmatrix} \tag{6.20}$$

With $\theta = 0$, the plant is singular (not functionally controllable), and controlling the two outputs independently is clearly impossible. On the other hand, for $\theta > 0$, effective feedback control is possible, provided the bandwidth is larger than about $1/\theta$. That is, for this example, control is easier the larger θ is. To illustrate this, we may compute the magnitude of the RGA (or the condition number) of G as a function of frequency, and note that it is infinite at low frequencies, but drops to 1 at about frequency $1/\theta$. In words, the presence of the delay decouples the initial (high-frequency) response, so we can obtain tight control if the controller reacts within this initial time period.

Exercise 6.1 *To further illustrate the above arguments, determine the sensitivity function S for the plant (6.20) using a simple diagonal controller $K = \frac{k}{s}I$. Use the approximation $e^{-\theta s} \approx 1 - \theta s$ to show that at low frequencies the elements of $S(s)$ are of magnitude $k/(2k + \theta)$. How large must k be to have acceptable performance (less than 10% offset at low frequencies)? What is the corresponding bandwidth? (Answer: Need $k > 8/\theta$. Bandwidth is equal to k.)*

Remark 1 The observant reader may have noticed that $G(s)$ in (6.20) is singular at $s = 0$ (even with θ non-zero) and thus has a zero at $s = 0$. Therefore, a controller with integral action which cancels this zero, yields an internally unstable system, (e.g. the transfer function KS contains an integrator). This means that although the conclusion that the time delay helps is correct, the derivations given in Exercise 6.1 are not strictly correct. To "fix" the results we may assume that the plant is only going to be controlled over a limited time so that internal instability is not an issue. Alternatively, we may assume, for example, that $e^{-\theta s}$ is replaced by $0.99e^{-\theta s}$ so that the plant is not singular at steady-state (but it is close to singular).

Exercise 6.2 *Repeat Exercise 6.1 numerically, with $e^{-\theta s}$ replaced by $0.99(1 - \frac{\theta}{2n}s)^n/(1 + \frac{\theta}{2n}s)^n$ (where $n = 5$ is the order of the Padé approximation), and plot the elements of $S(j\omega)$ as functions of frequency for $k = 0.1/\theta$, $k = 1/\theta$ and $k = 8/\theta$.*

6.5 Limitations imposed by RHP-zeros

RHP-zeros are common in many practical multivariable problems. The limitations they impose are similar to those for SISO systems, although often not quite so serious as they only apply in particular directions.

For ideal ISE optimal control (the "cheap" LQR problem), the SISO result ISE = $2/z$ from Section 5.4 can be generalized, see Qiu and Davison (1993). They show for a MIMO plant with RHP-zeros at z_i that the ideal ISE-value (the "cheap" LQR cost function) for a step disturbance or reference is directly related to $\sum_i 2/z_i$. Thus, as for SISO systems, RHP-zeros close to the origin imply poor control performance.

The limitations of a RHP-zero located at z may also be derived from the bound

$$\|w_P S(s)\|_\infty = \max_\omega |w_P(j\omega)|\bar{\sigma}(S(j\omega)) \geq |w_P(z)| \tag{6.21}$$

in (6.7) where $w_P(s)$ is a scalar weight. All the results derived in Section 5.6.4 for SISO systems, therefore generalize if we consider the "worst" direction corresponding to the maximum singular value, $\bar{\sigma}(S)$. For instance, by selecting the weight $w_P(s)$ such that we require tight control at low frequencies and a peak for $\bar{\sigma}(S)$ less than 2, we derive from (5.34) that the bandwidth (in the "worst" direction) must for a real RHP-zero satisfy $\omega_B^* < z/2$. Alternatively, if we require tight control at high frequencies, then we must from (5.38) satisfy $\omega_B^* > 2z$.

Remark 1 The use of a scalar weight $w_P(s)$ in (6.21) is somewhat restrictive. However, the assumption is less restrictive if one follows the scaling procedure in Section 1.4 and scales all outputs by their allowed variations such that their magnitudes are of approximately equal importance.

Remark 2 Note that condition (6.21) involves the maximum singular value (which is associated with the "worst" direction), and therefore the RHP-zero may not be a limitation in other directions. Furthermore, we may to some extent choose the worst direction. This is discussed next.

6.5.1 Moving the effect of a RHP-zero to a specific output

In MIMO systems one can often move the deteriorating effect of a RHP-zero to a given output, which may be less important to control well. This is possible because, although the interpolation constraint $y_z^H T(z) = 0$ imposes a certain relationship between the elements within each column of $T(s)$, the columns of $T(s)$ may still be selected independently. Let us first consider an example to motivate the results that follow. Most of the results in this section are from Holt and Morari (1985b) where further extensions can also be found.

Example 6.2 *Consider the plant*

$$G(s) = \frac{1}{(0.2s + 1)(s + 1)} \begin{bmatrix} 1 & 1 \\ 1 + 2s & 2 \end{bmatrix}$$

which has a RHP-zero at $s = z = 0.5$. *This is the same plant considered in Section 3.8 where we performed some* $\mathcal{H}_\infty$ *controller designs. The output zero direction satisfies* $y_z^H G(z) = 0$ *and we find*

$$y_z = \frac{1}{\sqrt{5}} \begin{bmatrix} 2 \\ -1 \end{bmatrix} = \begin{bmatrix} 0.89 \\ -0.45 \end{bmatrix}$$

Any allowable $T(s)$ *must satisfy the interpolation constraint* $y_z^H T(z) = 0$ *in (6.4), and this imposes the following relationships between the column elements of* $T(s)$:

$$2t_{11}(z) - t_{21}(z) = 0; \quad 2t_{12}(z) - t_{22}(z) = 0 \qquad (6.22)$$

We will consider reference tracking $y = Tr$ *and examine three possible choices for* T: T_0 *diagonal (a decoupled design),* T_1 *with output 1 perfectly controlled, and* T_2 *with output 2 perfectly controlled. Of course, we cannot achieve perfect control in practice, but we make the assumption to simplify our argument. In all three cases, we require perfect tracking at steady-state, i.e.* $T(0) = I$.

A decoupled design has $t_{12}(s) = t_{21}(s) = 0$, *and to satisfy (6.22) we then need* $t_{11}(z) = 0$ *and* $t_{22}(z) = 0$, *so the RHP-zero must be contained in both diagonal elements. One possible choice, which also satisfies* $T(0) = I$, *is*

$$T_0(s) = \begin{bmatrix} \frac{-s+z}{s+z} & 0 \\ 0 & \frac{-s+z}{s+z} \end{bmatrix}$$

For the two designs with one output perfectly controlled we choose

$$T_1(s) = \begin{bmatrix} 1 & 0 \\ \frac{\beta_1 s}{s+z} & \frac{-s+z}{s+z} \end{bmatrix} \qquad T_2(s) = \begin{bmatrix} \frac{-s+z}{s+z} & \frac{\beta_2 s}{s+z} \\ 0 & 1 \end{bmatrix}$$

The basis for the last two selections is as follows. For the output which is not perfectly controlled, the diagonal element must have a RHP-zero to satisfy (6.22), and the off-diagonal element must have an s *term in the numerator to give* $T(0) = I$. *To satisfy (6.22), we must then require for the two designs*

$$\beta_1 = 4, \quad \beta_2 = 1$$

The RHP-zero has no effect on output 1 for design $T_1(s)$, *and no effect on output 2 for design* $T_2(s)$. *We therefore see that it is indeed possible to move the effect of the RHP-zero to a particular output. However, we must pay for this by having to accept some interaction. We note that the magnitude of the interaction, as expressed by* β_k, *is largest for the case where output 1 is perfectly controlled. This is reasonable since the zero output direction* $y_z = [0.89 \quad -0.45]^T$ *is mainly in the direction of output 1, so we have to "pay more" to push its effect to output 2. This was also observed in the controller designs in Section 3.8, see Figure 3.10.*

We see from the above example that by requiring a decoupled response from r to y, as in design $T_0(s)$, we have to accept that the multivariable RHP-zero appears as a RHP-zero in each of the diagonal elements of $T(s)$. In other words, requiring a decoupled response generally leads to the introduction of additional RHP-zeros in $T(s)$ which are not present in the plant $G(s)$.

We also see that we can move the effect of the RHP-zero to a particular output, but we then have to accept some interaction. This is stated more exactly in the following Theorem.

Theorem 6.4 *Assume that $G(s)$ is square, functionally controllable and stable and has a single RHP-zero at $s = z$ and no RHP-pole at $s = z$. Then if the k'th element of the output zero direction is non-zero, i.e. $y_{zk} \neq 0$, it is possible to obtain "perfect" control on all outputs $j \neq k$ with the remaining output exhibiting no steady-state offset. Specifically, T can be chosen of the form*

$$T(s) = \begin{bmatrix} 1 & 0 & \cdots & 0 & 0 & 0 & \cdots & 0 \\ 0 & 1 & \cdots & 0 & 0 & 0 & \cdots & 0 \\ \vdots & \vdots & & & & & & \\ \frac{\beta_1 s}{s+z} & \frac{\beta_2 s}{s+z} & \cdots & \frac{\beta_{k-1} s}{s+z} & \frac{-s+z}{s+z} & \frac{\beta_{k+1} s}{s+z} & \cdots & \frac{\beta_n s}{s+z} \\ \vdots & & \ddots & & & & & \vdots \\ 0 & 0 & \cdots & 0 & 0 & 0 & \cdots & 1 \end{bmatrix} \quad (6.23)$$

where

$$\beta_j = -2\frac{y_{zj}}{y_{zk}} \text{ for } j \neq k \quad (6.24)$$

Proof: It is clear that (6.23) satisfies the interpolation constraint $y_z^H T(z) = 0$; see also Holt and Morari (1985b). □

The effect of moving completely the effect of a RHP-zero to output k is quantified by (6.24). We see that if the zero is not "naturally" aligned with this output, i.e. if $|y_{zk}|$ is much smaller than 1, then the interactions will be significant, in terms of yielding some $\beta_j = -2y_{zj}/y_{zk}$ much larger than 1 in magnitude. In particular, we *cannot* move the effect of a RHP-zero to an output corresponding to a zero element in y_z, which occurs frequently if we have a RHP-zero pinned to a subset of the outputs.

Exercise 6.3 *Consider the plant*

$$G(s) = \begin{bmatrix} \alpha & 1 \\ \frac{1}{s+1} & \alpha \end{bmatrix} \quad (6.25)$$

(a) Find the zero and its output direction. (Answer: $z = \frac{1}{\alpha^2} - 1$ and $y_z = [-\alpha \quad 1]^T$).
(b) Which values of α yield a RHP-zero, and which of these values is best/worst in terms of

achievable performance? (Answer: We have a RHP-zero for $|\alpha| < 1$. Best for $\alpha = 0$ with zero at infinity; if control at steady-state is required then worst for $\alpha = 1$ with zero at $s = 0$.)
(c) Suppose $\alpha = 0.1$. Which output is the most difficult to control? Illustrate your conclusion using Theorem 6.4. (Answer: Output 2 is the most difficult since the zero is mainly in that direction; we get strong interaction with $\beta = 20$ if we want to control y_2 perfectly.)

Exercise 6.4 *Repeat the above exercise for the plant*

$$G(s) = \frac{1}{s+1} \begin{bmatrix} s - \alpha & 1 \\ (\alpha+2)^2 & s - \alpha \end{bmatrix} \qquad (6.26)$$

6.6 Limitations imposed by RHP-poles

From the bound $\|w_T(s)T(s)\|_\infty \geq |w_T(p)|$ in (6.9) we find that a RHP-pole p imposes restrictions on $\bar{\sigma}(T)$ which are identical to those derived on $|T|$ for SISO systems in Section 5.8. Thus, we need feedback to stabilize an unstable plant and must require that $\bar{\sigma}(T(j\omega))$ is about 1 or larger up to the frequency $2|p|$, approximately.

6.7 RHP-poles combined with RHP-zeros

For SISO systems we found that performance is poor if the plant has a RHP-pole located close to a RHP-zero. This is also the case in MIMO systems provided that the directions coincide. This was quantified in Theorem 6.3. For example, for a MIMO plant with single RHP-zero z and single RHP-pole p we derive from (6.18) and (6.17)

$$\|S\|_\infty \geq c; \quad \|T\|_\infty \geq c; \quad c = \sqrt{\sin^2\phi + \frac{|z+p|^2}{|z-p|^2}\cos^2\phi} \qquad (6.27)$$

where $\phi = \cos^{-1}|y_z^H y_p|$ is the angle between the RHP-zero and RHP-pole. We next consider an example which demonstrates the importance of the directions as expressed by the angle ϕ.

Example 6.3 *Consider the plant*

$$G_\alpha(s) = \begin{bmatrix} \frac{1}{s-p} & 0 \\ 0 & \frac{1}{s+p} \end{bmatrix} \underbrace{\begin{bmatrix} \cos\alpha & -\sin\alpha \\ \sin\alpha & \cos\alpha \end{bmatrix}}_{U_\alpha} \begin{bmatrix} \frac{s-z}{0.1s+1} & 0 \\ 0 & \frac{s+z}{0.1s+1} \end{bmatrix}; \quad z = 2, p = 3 \quad (6.28)$$

which has for all values of α a RHP-zero at $z = 2$ and a RHP-pole at $p = 3$.
For $\alpha = 0°$ the rotation matrix $U_\alpha = I$, and the plant consists of two decoupled subsystems

$$G_0(s) = \begin{bmatrix} \frac{s-z}{(0.1s+1)(s-p)} & 0 \\ 0 & \frac{s+z}{(0.1s+1)(s+p)} \end{bmatrix}$$

Here the subsystem g_{11} has both a RHP-pole and a RHP-zero, and closed-loop performance is expected to be poor. On the other hand, there are no particular control problems related to the subsystem g_{22}. Next, consider $\alpha = 90°$ for which we have

$$U_\alpha = \begin{bmatrix} 0 & -1 \\ 1 & 0 \end{bmatrix}, \quad \text{and} \quad G_{90}(s) = \begin{bmatrix} 0 & \frac{s+z}{(0.1s+1)(s-p)} \\ -\frac{s-z}{(0.1s+1)(s+p)} & 0 \end{bmatrix}$$

and we again have two decoupled subsystems, but this time in the off-diagonal elements. The main difference, however, is that there is no interaction between the RHP-pole and RHP-zero in this case, so we expect this plant to be easier to control. For intermediate values of α we do not have decoupled subsystems, and there will be some interaction between the RHP-pole and RHP-zero.

Since in (6.28) the RHP-pole is located at the output of the plant, its output direction is fixed and we find $y_p = \begin{bmatrix} 1 & 0 \end{bmatrix}^T$ for all values of α. On the other hand, the RHP-zero output direction changes from $\begin{bmatrix} 1 & 0 \end{bmatrix}^T$ for $\alpha = 0°$ to $\begin{bmatrix} 0 & 1 \end{bmatrix}^T$ for $\alpha = 90°$. Thus, the angle ϕ between the pole and zero direction also varies between $0°$ and $90°$, but ϕ and α are not equal. This is seen from the Table below, where we also give c in (6.27), for four rotation angles, $\alpha = 0°, 30°, 60°$ and $90°$.

α	$0°$	$30°$	$60°$	$90°$
y_z	$\begin{matrix}1\\0\end{matrix}$	$\begin{matrix}0.33\\-0.94\end{matrix}$	$\begin{matrix}0.11\\-0.99\end{matrix}$	$\begin{matrix}0\\1\end{matrix}$
$\phi = \cos^{-1}\lvert y_z^H y_p \rvert$	$0°$	$70.9°$	$83.4°$	$90°$
c	5.0	1.89	1.15	1.0
$\lVert S \rVert_\infty$	7.00	2.60	1.59	1.98
$\lVert T \rVert_\infty$	7.40	2.76	1.60	1.31
$\gamma(S/KS)$	9.55	3.53	2.01	1.59

The Table also shows the values of $\lVert S \rVert_\infty$ and $\lVert T \rVert_\infty$ obtained by an $\mathcal{H}_\infty$ optimal S/KS design using the following weights

$$W_u = I; \quad W_P = \left(\frac{s/M + \omega_B^*}{s} \right) I; \quad M = 2, \omega_B^* = 0.5 \tag{6.29}$$

The weight W_P indicates that we require $\lVert S \rVert_\infty$ less than 2, and require tight control up to a frequency of about $\omega_B^* = 0.5\,rad/s$. The minimum $\mathcal{H}_\infty$ norm for the overall S/KS problem is given by the value of γ in Table 6.3. The corresponding responses to a step change in the reference, $r = \begin{bmatrix} 1 & -1 \end{bmatrix}$, are shown in Figure 6.1.

Several things about the example are worth noting:

1. *We see from the simulation for $\phi = \alpha = 0°$ in Figure 6.1 that the response for y_1 is very poor. This is as expected because of the closeness of the RHP-pole and zero ($z = 2, p = 3$).*
2. *The bound c on $\lVert S \rVert_\infty$ in (6.27) is tight in this case. This can be shown numerically by selecting $W_u = 0.01I$, $\omega_B^* = 0.01$ and $M = 1$ (W_u and ω_B are small so the main objective is to minimize the peak of S). We find with these weights that the $\mathcal{H}_\infty$ designs for the four angles yield $\lVert S \rVert_\infty = 5.04, 1.905, 1.155, 1.005$, which are very close to c.*

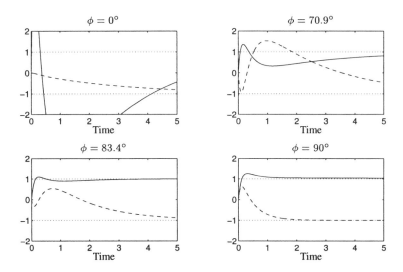

Figure 6.1: MIMO plant with angle ϕ between RHP-pole and RHP-zero. Response to step in reference $r = [1\ -1]^T$ with $\mathcal{H}_\infty$ controller for four different values of ϕ. Solid line: y_1; Dashed line: y_2.

3. *The angle ϕ between the pole and zero, is quite different from the rotation angle α at intermediate values between $0°$ and $90°$. This is because of the influence of the RHP-pole in output 1, which yields a strong gain in this direction, and thus tends to push the zero direction towards output 2.*

4. *For $\alpha = 0°$ we have $c = 5$ so $\|S\|_\infty \geq 5$ and $\|T\|_\infty \geq 5$ and it is clearly impossible to get $\|S\|_\infty$ less than 2, as required by the performance weight W_P.*

5. *The $\mathcal{H}_\infty$ optimal controller is unstable for $\alpha = 0°$ and $30°$. This is not altogether surprising, because for $\alpha = 0°$ the plant becomes two SISO systems one of which needs an unstable controller to stabilize it since $p > z$ (see Section 5.9).*

In conclusion, pole and zero directions provide useful information about a plant, as does the values of c in (6.27). However, the output pole and zero directions do depend on the relative scaling of the outputs, which must therefore be done appropriately prior to any analysis.

6.8   Performance requirements imposed by disturbances

For SISO systems we found that large and "fast" disturbances require tight control and a large bandwidth. The same results apply to MIMO systems, but again the issue of directions is important.

Definition 6.2 Disturbance direction. *Consider a single (scalar) disturbance and let the vector g_d represent its effect on the outputs ($y = g_d d$), The disturbance direction is defined as*

$$y_d = \frac{1}{\|g_d\|_2} g_d \qquad (6.30)$$

The associated disturbance condition number is defined as

$$\gamma_d(G) = \bar{\sigma}(G)\,\bar{\sigma}(G^\dagger y_d) \qquad (6.31)$$

Here $G^\dagger$ is the pseudo-inverse which is G^{-1} for a non-singular G.

Remark. We here use g_d (rather than G_d) to show that we consider a single disturbance, i.e. g_d is a vector. For a plant with many disturbances g_d is a column of the matrix G_d.

The disturbance condition number provides a measure of how a disturbance is aligned with the plant. It may vary between 1 (for $y_d = \bar{u}$) if the disturbance is in the "good" direction, and the condition number $\gamma(G) = \bar{\sigma}(G)\bar{\sigma}(G^\dagger)$ (for $y_d = \underline{u}$) if it is in the "bad" direction. Here $\bar{u}$ and $\underline{u}$ are the output directions in which the plant has its largest and smallest gain; see Chapter 3.

In the following, let $r = 0$ and assume that the disturbance has been scaled such that at each frequency the worst-case disturbance may be selected as $|d(\omega)| = 1$. Also assume that the outputs have been scaled such that the performance objective is that at each frequency the 2-norm of the error should be less than 1, i.e. $\|e(\omega)\|_2 < 1$. With feedback control $e = Sg_d d$ and the performance objective is then satisfied if

$$\|Sg_d\|_2 = \bar{\sigma}(Sg_d) < 1 \,\forall\omega \quad \Leftrightarrow \quad \|Sg_d\|_\infty < 1 \qquad (6.32)$$

For SISO systems, we used this to derive tight bounds on the sensitivity function and the loop gain; $|S| < 1/|G_d|$ and $|1 + L| > |G_d|$. A similar derivation is complicated for MIMO systems because of directions. To see this, we can use (6.30) to get the following requirement, which is equivalent to (6.32),

$$\|Sy_d\|_2 < 1/\|g_d\|_2 \,\forall\omega \qquad (6.33)$$

which shows that the S must be less than $1/\|g_d\|_2$ only in the direction of y_d. We can also derive bounds in terms of the singular values of S. Since g_d is a vector we have from (3.42)

$$\underline{\sigma}(S)\|g_d\|_2 \leq \|Sg_d\|_2 \leq \bar{\sigma}(S)\|g_d\|_2 \qquad (6.34)$$

Now $\underline{\sigma}(S) = 1/\bar{\sigma}(I + L)$ and $\bar{\sigma}(S) = 1/\underline{\sigma}(I + L)$, and we therefore have:

- For acceptable performance ($\|Sg_d\|_2 < 1$) we must *at least* require that $\bar{\sigma}(I + L)$ is larger than $\|g_d\|_2$ and we *may* have to require that $\underline{\sigma}(I + L)$ is larger than $\|g_d\|_2$.

Plant with RHP-zero. If $G(s)$ has a RHP-zero at $s = z$ then the performance may be poor if the disturbance is aligned with the output direction of this zero. To see this apply the maximum modulus principle to $f(s) = y_z^H S G_d$ to get

$$\|S g_d\|_\infty \geq |y_z^H g_d(z)| = |y_z^H y_d| \cdot \|g_d(z)\|_2 \qquad (6.35)$$

To satisfy $\|S g_d\|_\infty < 1$, we must then for a given disturbance d at least require

$$\boxed{|y_z^H g_d(z)| < 1} \qquad (6.36)$$

where y_z is the direction of the RHP-zero. This provides a generalization of the SISO-condition $|G_d(z)| < 1$ in (5.53).

Remark. In the above development we consider at each frequency performance in terms of $\|e\|_2$ (the 2-norm). However, the scaling procedure presented in Section 1.4 leads naturally to the vector max-norm as the way to measure signals and performance. Fortunately, this difference is not too important, and we will neglect it in the following. The reason is that for an $m \times 1$ vector a we have $\|a\|_{\max} \leq \|a\|_2 \leq \sqrt{m} \, \|a\|_{\max}$ (see (A.94)), so the values of max- and 2-norms are at most a factor $\sqrt{m}$ apart.

Example 6.4 *Consider the following plant and disturbance models*

$$G(s) = \frac{1}{s + 2} \begin{bmatrix} s - 1 & 4 \\ 4.5 & 2(s - 1) \end{bmatrix}, \quad g_d(s) = \frac{6}{s + 2} \begin{bmatrix} k \\ 1 \end{bmatrix}, \quad |k| \leq 1 \qquad (6.37)$$

It is assumed that the disturbance and outputs have been appropriately scaled, and the question is whether the plant is input-output controllable, i.e. whether we can achieve $\|S g_d\|_\infty < 1$, for any value of $|k| \leq 1$. $G(s)$ has a RHP-zero $z = 4$ and in Example 4.11 on page 133 we have already computed the zero direction. From this we get

$$|y_z^H g_d(z)| = |\,[\,0.83 \quad -0.55\,] \cdot \begin{bmatrix} k \\ 1 \end{bmatrix}\,| = |0.83k - 0.55|$$

and from (6.36) we conclude that the plant is not input-output controllable if $|0.83k - 0.55| > 1$, i.e. if $k < -0.54$. We cannot really conclude that the plant is controllable for $k > -0.54$ since (6.36) is only a necessary (and not sufficient) condition for acceptable performance, and there may also be other factors that determine controllability, such as input constraints which are discussed next.

Exercise 6.5 *Show that the disturbance condition number may be interpreted as the ratio between the actual input for disturbance rejection and the input that would be needed if the same disturbance was aligned with the "best" plant direction.*

6.9   Limitations imposed by input constraints

Constraints on the manipulated variables can limit the ability to reject disturbances and track references. As was done for SISO plants in Chapter 5, we will consider the case of perfect control ($e = 0$) and then of acceptable control ($\|e\| \leq 1$). We derive the results for disturbances, and the corresponding results for reference tracking are obtained by replacing G_d by $-R$. The results in this section apply to both feedback and feedforward control.

Remark. For MIMO systems the choice of vector norm, $\| \cdot \|$, to measure the vector signal magnitudes at each frequency makes some difference. The vector max-norm (largest element) is the most natural choice when considering input saturation and is also the most natural in terms of our scaling procedure. However, for mathematical convenience we will also consider the vector 2-norm (Euclidean norm). In most cases the difference between these two norms is of little practical significance.

6.9.1 Inputs for perfect control

We here consider the question: can the disturbances be rejected perfectly while maintaining $\|u\| \leq 1$? To answer this, we must quantify the set of possible disturbances and the set of allowed input signals. We will consider both the max-norm and 2-norm.

Max-norm and square plant. For a square plant the input needed for perfect disturbance rejection is $u = -G^{-1}G_d d$ (as for SISO systems). Consider a *single disturbance* (g_d is a vector). Then the worst-case disturbance is $|d(\omega)| = 1$, and we get that input saturation is avoided ($\|u\|_{\max} \leq 1$) if all elements in the vector $G^{-1}g_d$ are less than 1 in magnitude, that is,

$$\|G^{-1}g_d\|_{\max} \leq 1, \forall \omega$$

For *simultaneous disturbances* (G_d is a matrix) the corresponding requirement is

$$\|G^{-1}G_d\|_{i\infty} \leq 1, \forall \omega \tag{6.38}$$

where $\| \cdot \|_{i\infty}$ is the induced max-norm (induced ∞-norm, maximum row sum, see (A.105)). However, it is usually recommended in a preliminary analysis to consider one disturbance at a time, for example, by plotting as a function of frequency the individual elements of the matrix $G^{-1}G_d$. This yields more information about which particular input is most likely to saturate and which disturbance is the most problematic.

Two-norm. We here measure both the disturbance and the input in terms of the 2-norm. Assume that G has full row rank so that the outputs can be perfectly controlled. Then the smallest inputs (in terms of the 2-norm) needed for perfect disturbance rejection are

$$u = -G^{\dagger}G_d d \tag{6.39}$$

where $G^{\dagger} = G^H(GG^H)^{-1}$ is the pseudo-inverse from (A.63). Then with a single disturbance we must require $\|G^{\dagger}g_d\|_2 \leq 1$. With combined disturbances we must require $\bar{\sigma}(G^{\dagger}G_d) \leq 1$, that is, the induced 2-norm is less than 1, see (A.106).

For *combined reference changes*, $\|\tilde{r}(\omega)\|_2 \leq 1$, the corresponding condition for perfect control with $\|u\|_2 \leq 1$ becomes $\bar{\sigma}(G^\dagger R) \leq 1$, or equivalently (see (A.61))

$$\underline{\sigma}(R^{-1}G) \geq 1, \ \forall \omega \leq \omega_r \tag{6.40}$$

where ω_r is the frequency up to which reference tracking is required. Usually R is diagonal with all elements larger than 1, and we must at least require

$$\underline{\sigma}(G(j\omega)) \geq 1, \forall \omega \leq \omega_r \tag{6.41}$$

or, more generally, we want $\underline{\sigma}(G(j\omega))$ large.

6.9.2   Inputs for acceptable control

Let $r = 0$ and consider the response $e = Gu + G_d d$ to a disturbance d. The question we want to answer in this subsection is: is it possible to achieve $\|e\| < 1$ for any $\|d\| \leq 1$ using inputs with $\|u\| \leq 1$? We use here the max-norm, $\| \cdot \|_{\max}$ (the vector infinity-norm), for the vector signals.

We consider this problem frequency-by-frequency. This means that we neglect the issue of causality which is important for plants with RHP-zeros and time delays. The resulting conditions are therefore only necessary (i.e. minimum requirements) for achieving $\|e\|_{\max} < 1$.

Exact conditions

Mathematically, the problem can be formulated in several different ways; by considering the maximum allowed disturbance, the minimum achievable control error or the minimum required input; e.g. see Skogestad and Wolff (1992). We here use the latter approach. To simplify the problem, and also to provide more insight, we consider one disturbance at a time, i.e. d is a scalar and g_d a vector. The worst-case disturbance is then $|d| = 1$ and the problem is at each frequency is to compute

$$U_{\min} \triangleq \min_{u} \|u\|_{\max} \text{ such that } \|Gu + g_d d\|_{\max} \leq 1, |d| = 1 \tag{6.42}$$

A necessary condition for avoiding input saturation (for each disturbance) is then

$$U_{\min} < 1, \forall \omega \tag{6.43}$$

If G and g_d are real (i.e. at steady-state) then (6.42) can be formulated as a linear programming (LP) problem, and in the general case as a convex optimization problem.

For SISO systems we have an analytical solution of (6.42); from the proof of (5.59) we get $U_{\min} = (|g_d| - 1)/|G|$. A necessary condition for avoiding input saturation (for each disturbance) is then

$$\boxed{\text{SISO}: \quad |G| > |g_d| - 1, \text{ at frequencies where } |g_d| > 1} \tag{6.44}$$

We would like to generalize this result to MIMO systems. Unfortunately, we do not have an exact analytical result, but by making the approximation in (6.46) below, a nice approximate generalization is available.

Approximate conditions in terms of the SVD

At each frequency the singular value decomposition of the plant (possibly non-square) is $G = U\Sigma V^H$. Introduce the rotated control error and rotated input

$$\widehat{e} = U^H e, \quad \widehat{u} = V^H u \tag{6.45}$$

and assume that the max-norm is approximately unchanged by these rotations

$$\|\widehat{e}\|_{\max} \approx \|e\|_{\max}, \quad \|\widehat{u}\|_{\max} \approx \|u\|_{\max} \tag{6.46}$$

From (A.124) this would be an equality for the 2-norm, so from (A.94) the error by using the approximation for the max-norm is at most a factor $\sqrt{m}$ where m is the number of elements in the vector. We then find that each singular value of G, $\sigma_i(G)$, must approximately satisfy

$$\boxed{\text{MIMO}: \quad \sigma_i(G) \geq |u_i^H g_d| - 1, \text{ at frequencies where } |u_i^H g_d| > 1} \tag{6.47}$$

where u_i is the i'th output singular vector of G, and g_d is a vector since we consider a single disturbance. More precisely, (6.47) is a necessary condition for achieving acceptable control ($\|e\|_{\max} < 1$) for a single disturbance ($|d| = 1$) with $\|u\|_{\max} \leq 1$, assuming that (6.46) holds.

Condition (6.47) provides a nice generalization of (6.44). $u_i^H g_d$ may be interpreted as the projection of g_d onto the i'th output singular vector of the plant.

Proof of (6.47): Let $r = 0$ and consider the response $e = Gu + g_d d$ to a single disturbance d. We have

$$\widehat{e} = U^H e = U^H(Gu + g_d d) = \Sigma\widehat{u} + U^H g_d d \tag{6.48}$$

where the last equality follows since $U^H G = \Sigma V^H$. For the worst-case disturbance ($|d| = 1$), we want to find the smallest possible input such that $\|e\|_{\max} \approx \|\widehat{e}\|_{\max}$ is less than 1. This is equivalent to requiring $|\widehat{e}_i| \leq 1, \forall i$, where from (6.48) $\widehat{e}_i = \sigma_i(G)\widehat{u}_i + U_i^H g_d d$. Note: u_i (a vector) is the i'th column of U, whereas $\widehat{u}_i$ (a scalar) is the i'th rotated plant input. This is a scalar problem similar to that for the SISO-case in (5.59), and if we assume $|u_i^H g_d d| > 1$ (otherwise we may simply set $\widehat{u}_i = 0$ and achieve $|\widehat{e}_i| < 1$) then the smallest $|\widehat{u}_i|$ is achieved when the right hand side is "lined up" to make $|\widehat{e}_i| = 1$. Thus, the minimum input is

$$|\widehat{u}_i| = (|u_i^H g_d| - 1)/\sigma_i(G) \tag{6.49}$$

and (6.47) follows by requiring that $\|u\|_{\max} \approx \|\widehat{u}\|_{\max}$ is less than 1. $\square$

Based on (6.47) we can find out:

1. For which disturbances and at which frequencies input constraints may cause problems. This may give ideas on which disturbances should be reduced, for example by redesign or use of feedforward control.
2. In which direction i the plant gain is too small. By looking at the corresponding input singular vector, v_i, one can determine which actuators should be redesigned (to get more power in certain directions) and by looking at the corresponding output singular vector, u_i, one can determine on which outputs we may have to reduce our performance requirements.

Several disturbances. For combined disturbances, one requires the i'th row sum of $U^H G_d$ to be less than $\sigma_i(G)$ (at frequencies where the i'th row sum is larger than 1). However, usually we derive more insight by considering one disturbance at a time.

Reference commands. Similar results are derived for references by replacing G_d by $-R$.

Example 6.5 Distillation process *Consider a 2×2 plant with two disturbances. The appropriately scaled steady-state model is*

$$G = 0.5 \begin{bmatrix} 87.8 & -86.4 \\ 108.2 & -109.6 \end{bmatrix}, \ G_d = \begin{bmatrix} 7.88 & 8.81 \\ 11.72 & 11.19 \end{bmatrix} \qquad (6.50)$$

This is a model of a distillation column with product compositions as outputs, reflux and boilup as inputs, and feed rate (20% change) and feed composition (20% change) as disturbances. The elements in G are scaled by a factor 0.5 compared to (3.45) because the allowed input changes are a factor 2 smaller. From an SVD of G we have $\bar{\sigma}(G) = 98.6$ and $\underline{\sigma}(G) = 0.70$. Some immediate observations:

1. *The elements of the matrix G_d are larger than 1 so control is needed to reject disturbances.*
2. *Since $\underline{\sigma}(G) = 0.7$ we are able to perfectly track reference changes of magnitude 0.7 (in terms of the 2-norm) without reaching input constraints.*
3. *The elements in G are about 5 times larger than those in G_d, which suggests that there should be no problems with input constraints. On the other hand, $\underline{\sigma}(G) = 0.7$ is much less than the elements in G_d, so input constraints may be an issue after all.*
4. *The disturbance condition numbers, $\gamma_d(G)$, for the two disturbances, are 11.75 and 1.48, respectively. This indicates that the direction of disturbance 1 is less favourable than that of disturbance 2.*
5. *The condition number $\gamma(G) = \bar{\sigma}(G)/\underline{\sigma}(G) = 141.7$ is large, but this does not by itself imply control problems. In this case, the large value of the condition number is not caused by a small $\underline{\sigma}(G)$ (which would be a problem), but rather by a large $\bar{\sigma}(G)$.*

We will now analyze whether the disturbance rejection requirements will cause input saturation by considering separately the two cases of perfect control ($e = 0$) and acceptable control ($\|e\|_{\max} \le 1$).

1. Perfect control. *The inputs needed for perfect disturbance rejection are $u = G^{-1}G_d$ where*

$$G^{-1}G_d = \begin{bmatrix} -1.09 & -0.009 \\ -1.29 & -0.213 \end{bmatrix}$$

We note that perfect rejection of disturbance $d_1 = 1$ requires an input $u = \begin{bmatrix} -1.09 & -1.29 \end{bmatrix}^T$ which is larger than 1 in magnitude. Thus, perfect control of disturbance 1 is not possible *without violating input constraints. However, perfect rejection of disturbance* $d_2 = 1$ *is possible as it requires a much smaller input* $u = [\,-0.009 \quad -0.213\,]^T$.

2. Approximate result for acceptable control. *We will use the approximate requirement (6.47) to evaluate the inputs needed for acceptable control. We have*

$$U^H G_d = \begin{bmatrix} 14.08 & 14.24 \\ 1.17 & 0.11 \end{bmatrix} \qquad \begin{array}{l} \sigma_1(G) = 98.6 \\ \sigma_2(G) = 0.70 \end{array}$$

and the magnitude of each element in the i*'th row of* $U^H G_d$ *should be less than* $\sigma_i(G) + 1$ *to avoid input constraints. In the high-gain direction this is easily satisfied since 14.08 and 14.24 are both much less than* $\sigma_1(G) + 1 = 99.6$*, and from (6.49) the required input in this direction is thus only about* $|\widehat{u}_1| = (14 - 1)/98.6 = 0.13$ *for both disturbances which is much less than 1. The requirement is also satisfied in the low-gain direction since 1.17 and 0.11 are both less than* $\sigma_2(G) + 1 = 1.7$*, but we note that the margin is relatively small for disturbance 1. More precisely, in the low-gain direction disturbance 1 requires an input magnitude of approximately* $|\widetilde{u}_2| = (1.17 - 1)/0.7 = 0.24$*, whereas disturbance 2 requires no control (as its effect is 0.11 which is less than 1).*

In conclusion, we find that the results based on perfect control are misleading, *as acceptable control is indeed possible. Again we find disturbance 1 to be more difficult, but the difference is much smaller than with perfect control. The reason is that we only need to reject about 12%* $(1.13 - 1/1.13)$ *of disturbance 1 in the low-gain direction.*

However, this changes drastically if disturbance 1 is larger, since then a much larger fraction of it must be rejected. The fact that disturbance 1 is more difficult is confirmed in Section 10.10 on page 445 where we also present closed-loop responses.

3. Exact numerical result for acceptable control. *The exact values of the minimum inputs needed to achieve acceptable control are* $\|u\|_{\max} = 0.098$ *for disturbance 1 and* $\|u\|_{\max} = 0.095$ *for disturbance 2, which confirms that input saturation is not a problem.*

However, the values of $\|u\|_{\max} \approx 0.10$ *indicate that the two disturbances are about equally difficult. This seems inconsistent with the above approximate results, where we found disturbance 1 to be more difficult. However, the results are consistent if for both disturbances control is only needed in the high-gain direction, for which the approximate results gave the same value of 0.13 for both disturbances. (The approximate results indicated that some control was needed for disturbance 1 in the low-gain direction, since 1.17 was just above 1, but apparently this is inaccurate).*

The discussion at the end of the example illustrates an advantage of the approximate method in (6.47); namely that we can easily see whether we are close to a borderline value where control may be needed in some direction. On the other hand, no such "warning" is provided by the exact numerical method.

From the example we conclude that it is difficult to judge, simply by looking at the magnitude of the elements in G_d, whether a disturbance is difficult to reject or not. In the above example, it would appear from the column vectors of G_d in (6.50) that the two disturbances have almost identical effects. However, we found that disturbance 1 may be much more difficult to reject because it has a component of 1.17 in the low-gain direction of G which is about 10 times larger than the value of 0.11 for disturbance 2. This can be seen from the second row of $U^H G_d$.

Exercise 6.6 *Consider again the plant in (6.37). Let $k = 1$ and compute, as a function of frequency, the required inputs $G^{-1}g_d(j\omega)$ for perfect control. You will find that both inputs are about 2 in magnitude at low frequency, so if the inputs and disturbances have been appropriately scaled, we conclude that perfect control is not possible. Next, evaluate $G(j\omega) = U\Sigma V^H$, and compute $U^H g_d(j\omega)$ as a function of frequency and compare with the elements of $\Sigma(j\omega) + I$ to see whether "acceptable control" ($|e_i(j\omega)| < 1$) is possible.*

6.10   Limitations imposed by uncertainty

The presence of uncertainty requires the use of feedback, rather than simply feedforward control, to get acceptable performance. Sensitivity reduction with respect to uncertainty is achieved with high-gain feedback, but for any real system we have a crossover frequency range where the loop gain has to drop below 1, and the presence of uncertainty in this frequency range may result in poor performance or even instability. These issues are the same for SISO and MIMO systems.

However, with MIMO systems there is an additional problem in that there is also uncertainty associated with the plant directionality. The main objective of this section is to introduce some simple tools, like the RGA and the condition number, which are useful in picking out plants for which one might expect sensitivity to directional uncertainty.

Remark. In Chapter 8, we discuss more exact methods for analyzing performance with almost any kind of uncertainty and a given controller. This involves analyzing robust performance by use of the structured singular value. However, in this section the treatment is kept at a more elementary level and we are looking for results which depend on the plant only.

6.10.1   Input and output uncertainty

In practice the difference between the true perturbed plant G' and the plant model G is caused by a number of different sources. In this section, we focus on input uncertainty and output uncertainty. In a multiplicative (relative) form, the output and input uncertainties (as in Figure 6.2) are given by

$$\text{Output uncertainty:}\quad G' = (I + E_O)G \quad \text{or} \quad E_O = (G' - G)G^{-1} \quad (6.51)$$

$$\text{Input uncertainty:}\quad G' = G(I + E_I) \quad \text{or} \quad E_I = G^{-1}(G' - G) \quad (6.52)$$

These forms of uncertainty may seem similar, but we will show that their implications for control may be very different. If all the elements in the matrices E_I or E_O are non-zero, then we have *full block* ("unstructured") uncertainty. However, in many cases the source of uncertainty is in the individual input or output channels, and we have that E_I or E_O are diagonal matrices, for example,

$$E_I = \text{diag}\{\epsilon_1, \epsilon_2, \dots\} \quad (6.53)$$

It is important to stress that this *diagonal input uncertainty* is *always* present in real systems.

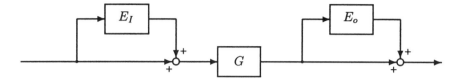

Figure 6.2: Plant with multiplicative input and output uncertainty

6.10.2   Effect of uncertainty on feedforward control

Consider a feedforward controller $u = K_r r$ for the case with no disturbances ($d = 0$). We assume that the plant G is invertible so that we can select

$$K_r = G^{-1}$$

and achieve *perfect* control, $e = y - r = Gu - r = GK_r r - r = 0$, for the nominal case with no uncertainty. However, for the actual plant G' (with uncertainty) the actual control error becomes $e' = y' - r = G'G^{-1}r - r$. We then get for the two sources of uncertainty

Output uncertainty:	$e' = E_O r$	(6.54)
Input uncertainty:	$e' = GE_I G^{-1}r$	(6.55)

For output uncertainty, we see that (6.54) is identical to the result in (5.70) for SISO systems. That is, the worst-case relative control error $\|e'\|_2/\|r\|_2$ is equal to the magnitude of the relative output uncertainty $\bar{\sigma}(E_O)$. However, for input uncertainty the sensitivity may be much larger because the elements in the matrix $GE_I G^{-1}$ can be much larger than the elements in E_I. In particular, for diagonal input uncertainty the elements of $GE_I G^{-1}$ are directly related to the RGA, see (A.80):

$$\text{Diagonal uncertainty:} \qquad \boxed{[GE_I G^{-1}]_{ii} = \sum_{j=1}^{n} \lambda_{ij}(G)\epsilon_j} \qquad (6.56)$$

The RGA-matrix is scaling independent, which makes the use of condition (6.56) attractive. Since diagonal input uncertainty is *always* present we can conclude that

- if the plant has large RGA elements within the frequency range where effective control is desired, then it is *not* possible to achieve good reference tracking with feedforward control because of strong sensitivity to diagonal input uncertainty.

The reverse statement is *not* true, that is, if the RGA has small elements we *cannot* conclude that the sensitivity to input uncertainty is small. This is seen from the following expression for the 2×2 case

$$GE_I G^{-1} = \begin{bmatrix} \lambda_{11}\epsilon_1 + \lambda_{12}\epsilon_2 & -\frac{g_{12}}{g_{22}}\lambda_{11}(\epsilon_1 - \epsilon_2) \\ \frac{g_{21}}{g_{11}}\lambda_{11}(\epsilon_1 - \epsilon_2) & \lambda_{21}\epsilon_1 + \lambda_{22}\epsilon_2 \end{bmatrix} \qquad (6.57)$$

For example, consider a triangular plant with $g_{12} = 0$. In this case the RGA is $\Lambda = I$ so the diagonal elements of $GE_I G^{-1}$ are ϵ_1 and ϵ_2. Still, the system may be sensitive to input uncertainty, since from (6.57) the 2, 1-element of $GE_I G^{-1}$ may be large if g_{21}/g_{11} is large.

6.10.3   Uncertainty and the benefits of feedback

To illustrate the benefits of feedback control in reducing the sensitivity to uncertainty, we consider the effect of output uncertainty on reference tracking. As a basis for comparison we first consider feedforward control.

Feedforward control. Let the nominal transfer function with feedforward control be $y = T_r r$ where $T_r = GK_r$ and K_r denotes the feedforward controller. Ideally, $T_r = I$. With model error $T'_r = G'K_r$, and the change in response is $y' - y = (T'_r - T_r)r$ where

$$T'_r - T_r = (G' - G)G^{-1}T_r = E_O T_r \qquad (6.58)$$

Thus, $y' - y = E_O T_r r = E_O y$, and with feedforward control the relative control error caused by the uncertainty is equal to the relative output uncertainty.

Feedback control. With one degree-of-freedom feedback control the nominal transfer function is $y = Tr$ where $T = L(I + L)^{-1}$ is the complementary sensitivity function. Ideally, $T = I$. The change in response with model error is $y' - y = (T' - T)r$ where from (A.144)

$$T' - T = S'E_O T \qquad (6.59)$$

Thus, $y' - y = S'E_O Tr = S'E_O y$, and we see that

- with feedback control the effect of the uncertainty is reduced by a factor S' relative to that with feedforward control.

Thus, feedback control is much less sensitive to uncertainty than feedforward control at frequencies where feedback is effective and the elements in S' are small. However, the opposite may be true in the crossover frequency range where S' may have elements larger than 1; see Section 6.10.4.

Remark 1 For square plants, $E_O = (G' - G)G^{-1}$ and (6.59) becomes

$$\Delta T \cdot T^{-1} = S' \cdot \Delta G \cdot G^{-1} \qquad (6.60)$$

where $\Delta T = T' - T$ and $\Delta G = G' - G$. Equation (6.60) provides a generalization of Bode's differential relationship (2.23) for SISO systems. To see this, consider a SISO system and let $\Delta G \to 0$. Then $S' \to S$ and we have from (6.60)

$$\frac{dT}{T} = S\frac{dG}{G} \tag{6.61}$$

Remark 2 Alternative expressions showing the benefits of feedback control are derived by introducing the inverse output multiplicative uncertainty $G' = (I - E_{iO})^{-1}G$. We then get (Horowitz and Shaked, 1975).

Feedforward control:	$T'_r - T_r = E_{iO}T'_r$	(6.62)
Feedback control:	$T' - T = SE_{iO}T'$	(6.63)

(Simple proof for square plants: switch G and G' in (6.58) and (6.59) and use $E_{iO} = (G' - G)G'^{-1}$).

Remark 3 Another form of (6.59) is (Zames, 1981)

$$T' - T = S'(L' - L)S \tag{6.64}$$

Conclusion. From (6.59), (6.63) and (6.64) we see that with feedback control $T' - T$ is small at frequencies where feedback is effective (i.e. S and S' are small). This is usually at low frequencies. At higher frequencies we have for real systems that L is small, so T is small, and again $T' - T$ is small. Thus with feedback, uncertainty only has a significant effect in the crossover region where S and T both have norms around 1.

6.10.4   Uncertainty and the sensitivity peak

We demonstrated above how feedback may reduce the effect of uncertainty, but we also pointed out that uncertainty may pose limitations on achievable performance, especially at crossover frequencies. The objective in the following is to investigate how the magnitude of the sensitivity, $\bar{\sigma}(S')$, is affected by the multiplicative output uncertainty and input uncertainty given as (6.51) and (6.52). We will derive *upper bounds* on $\bar{\sigma}(S')$ which involve the plant and controller condition numbers

$$\gamma(G) = \frac{\bar{\sigma}(G)}{\underline{\sigma}(G)}, \qquad \gamma(K) = \frac{\bar{\sigma}(K)}{\underline{\sigma}(K)} \tag{6.65}$$

and the following minimized condition numbers of the plant and the controller

$$\gamma_I^*(G) = \min_{D_I} \gamma(GD_I), \qquad \gamma_O^*(K) = \min_{D_O} \gamma(D_OK) \tag{6.66}$$

where D_I and D_O are diagonal scaling matrices. These minimized condition numbers may be computed using (A.74) and (A.75). Similarly, we state a *lower bound* on $\bar{\sigma}(S')$ for an inverse-based controller in terms of the RGA-matrix of the plant.

The following factorizations of S' in terms of the nominal sensitivity S (see Appendix A.6) form the basis for the development:

Output uncertainty:
$$S' = S(I + E_O T)^{-1} \qquad (6.67)$$

Input uncertainty:
$$S' = S(I + G E_I G^{-1} T)^{-1} = SG(I + E_I T_I)^{-1} G^{-1} \quad (6.68)$$
$$S' = (I + T K^{-1} E_I K)^{-1} S = K^{-1}(I + T_I E_I)^{-1} KS \quad (6.69)$$

We assume that G and G' are stable. We also assume closed-loop stability, so that both S and S' are stable. We then get that $(I + E_O T)^{-1}$ and $(I + E_I T_I)^{-1}$ are stable. In most cases we assume that the magnitude of the multiplicative (relative) uncertainty at each frequency can be bounded in terms of its singular value

$$\bar{\sigma}(E_I) \leq |w_I|, \qquad \bar{\sigma}(E_O) \leq |w_O| \qquad (6.70)$$

where $w_I(s)$ and $w_O(s)$ are scalar weights. Typically the uncertainty bound, $|w_I|$ or $|w_O|$, is 0.2 at low frequencies and exceeds 1 at higher frequencies.

We first state some upper bounds on $\bar{\sigma}(S')$. These are based on identities (6.67)-(6.69) and singular value inequalities (see Appendix A.3.4) of the kind

$$\bar{\sigma}((I + E_I T_I)^{-1}) = \frac{1}{\underline{\sigma}(I + E_I T_I)} \leq \frac{1}{1 - \bar{\sigma}(E_I T_I)} \leq \frac{1}{1 - \bar{\sigma}(E_I)\bar{\sigma}(T_I)} \leq \frac{1}{1 - |w_I|\bar{\sigma}(T_I)}$$

Of course these inequalities only apply if we assume $\bar{\sigma}(E_I T_I) < 1$, $\bar{\sigma}(E_I)\bar{\sigma}(T_I) < 1$ and $|w_I|\bar{\sigma}(T_I) < 1$. For simplicity, we will not state these assumptions each time.

Upper bound on $\bar{\sigma}(S')$ for output uncertainty

From (6.67) we derive

$$\bar{\sigma}(S') \leq \bar{\sigma}(S)\bar{\sigma}((I + E_O T)^{-1}) \leq \frac{\bar{\sigma}(S)}{1 - |w_O|\bar{\sigma}(T)} \qquad (6.71)$$

From (6.71) we see that output uncertainty, be it diagonal or full block, poses no particular problem when performance is measured at the plant output. That is, if we have a reasonable margin to stability ($\|(I + E_O T)^{-1}\|_\infty$ is not too much larger than 1), then the nominal and perturbed sensitivities do not differ very much.

Upper bounds on $\bar{\sigma}(S')$ for input uncertainty

The sensitivity function can be much more sensitive to input uncertainty than output uncertainty.

1. General case (full block or diagonal input uncertainty and any controller). From (6.68) and (6.69) we derive

$$\bar{\sigma}(S') \leq \gamma(G)\bar{\sigma}(S)\bar{\sigma}((I + E_I T_I)^{-1}) \leq \gamma(G)\frac{\bar{\sigma}(S)}{1 - |w_I|\bar{\sigma}(T_I)} \qquad (6.72)$$

$$\bar{\sigma}(S') \leq \gamma(K)\bar{\sigma}(S)\bar{\sigma}((I + T_I E_I)^{-1}) \leq \gamma(K)\frac{\bar{\sigma}(S)}{1 - |w_I|\bar{\sigma}(T_I)} \qquad (6.73)$$

From (6.73) we have the important result that if we use a "round" controller, meaning that $\gamma(K)$ is close to 1, then the sensitivity function is *not* sensitive to input uncertainty. In many cases (6.72) and (6.73) are not very useful because they yield unnecessarily large upper bounds. To improve on this conservativeness we present below some bounds for special cases, where we either restrict the uncertainty to be diagonal or restrict the controller to be of a particular form.

2. Diagonal uncertainty and decoupling controller. Consider a decoupling controller in the form $K(s) = D(s)G^{-1}(s)$ where $D(s)$ is a diagonal matrix. In this case, KG is diagonal so $T_I = KG(I + KG)^{-1}$ is diagonal, and $E_I T_I$ is diagonal. The second identity in (6.68) may then be written $S' = S(GD_I)(I + E_I T_I)^{-1}(GD_I)^{-1}$ where D_I is freely chosen, and we get

$$\bar{\sigma}(S') \leq \gamma_I^*(G)\bar{\sigma}(S)\bar{\sigma}((I + E_I T_I)^{-1}) \leq \gamma_I^*(G)\frac{\bar{\sigma}(S)}{1 - |w_I|\bar{\sigma}(T_I)} \tag{6.74}$$

$$\bar{\sigma}(S') \leq \gamma_O^*(K)\bar{\sigma}(S)\bar{\sigma}((I + T_I E_I)^{-1}) \leq \gamma_O^*(K)\frac{\bar{\sigma}(S)}{1 - |w_I|\bar{\sigma}(T_I)} \tag{6.75}$$

The bounds (6.74) and (6.75) apply to any decoupling controller in the form $K = DG^{-1}$. In particular, they apply to inverse-based control, $D = l(s)I$, which yields input-output decoupling with $T_I = T = t \cdot I$ where $t = \frac{l}{1+l}$.

Remark. A diagonal controller has $\gamma_O^*(K) = 1$, so from (6.77) below we see that (6.75) applies to both a diagonal and decoupling controller. Another bound which does apply to any controller is given in (6.77).

3. Diagonal uncertainty (Any controller). From the first identity in (6.68) we get $S' = S(I + (GD_I)E_I(GD_I)^{-1}T)^{-1}$ and we derive by singular value inequalities

$$\bar{\sigma}(S') \leq \frac{\bar{\sigma}(S)}{1 - \gamma_I^*(G)|w_I|\bar{\sigma}(T)} \tag{6.76}$$

$$\bar{\sigma}(S') \leq \frac{\bar{\sigma}(S)}{1 - \gamma_O^*(K)|w_I|\bar{\sigma}(T)} \tag{6.77}$$

Note that $\gamma_O^*(K) = 1$ for a diagonal controller so (6.77) shows that diagonal uncertainty does not pose much of a problem when we use decentralized control.

Lower bound on $\bar{\sigma}(S')$ for input uncertainty

Above we derived upper bounds on $\bar{\sigma}(S')$; we will next derive a lower bound. *A lower bound is more useful because it allows us to make definite conclusions about when the plant is not input-output controllable.*

Theorem 6.5 Input uncertainty and inverse-based control. *Consider a controller* $K(s) = l(s)G^{-1}(s)$ *which results in a nominally decoupled response with sensitivity*

*S = s · I and complementary sensitivity T = t · I where t(s) = 1 − s(s). Suppose
the plant has diagonal input uncertainty of relative magnitude $|w_I(j\omega)|$ in each input
channel. Then there exists a combination of input uncertainties such that at each
frequency*

$$\bar{\sigma}(S') \geq \bar{\sigma}(S) \left(1 + \frac{|w_I t|}{1 + |w_I t|} \|\Lambda(G)\|_{i\infty} \right) \tag{6.78}$$

where $\|\Lambda(G)\|_{i\infty}$ is the maximum row sum of the RGA and $\bar{\sigma}(S) = |s|$.

The proof is given below. From (6.78) we see that with an inverse based controller the
worst case sensitivity will be much larger than the nominal at frequencies where the
plant has large RGA-elements. At frequencies where control is effective ($|s|$ is small
and $|t| \approx 1$) this implies that control is not as good as expected, but it may still be
acceptable. However, at crossover frequencies where $|s|$ and $|t| = |1 − s|$ are both
close to 1, we find that $\bar{\sigma}(S')$ in (6.78) may become much larger than 1 if the plant
has large RGA-elements at these frequencies. The bound (6.78) applies to diagonal
input uncertainty and therefore also to full-block input uncertainty (since it is a lower
bound).

Worst-case uncertainty. It is useful to know which combinations of input errors
give poor performance. For an inverse-based controller a good indicator results if we
consider $GE_I G^{-1}$, where $E_I = \text{diag}\{\epsilon_k\}$. If all ϵ_k have the same magnitude $|w_I|$,
then the largest possible magnitude of any diagonal element in $GE_I G^{-1}$ is given by
$|w_I| \cdot \|\Lambda(G)\|_{i\infty}$. To obtain this value one may select the phase of each ϵ_k such that
$\angle\epsilon_k = -\angle\lambda_{ik}$ where i denotes the row of $\Lambda(G)$ with the largest elements. Also, if
$\Lambda(G)$ is real (e.g. at steady-state), the signs of the ϵ_k's should be the same as those in
the row of $\Lambda(G)$ with the largest elements.

Proof of Theorem 6.5: (From Skogestad and Havre (1996) and Gjøsæter (1995)). Write the
sensitivity function as

$$S' = (I + G'K)^{-1} = SG \underbrace{(I + E_I T_I)^{-1}}_{D} G^{-1}, \quad E_I = \text{diag}\{\epsilon_k\}, \quad S = sI \tag{6.79}$$

Since D is a diagonal matrix, we have from (6.56) that the diagonal elements of S' are given
in terms of the RGA of the plant G as

$$s'_{ii} = s \sum_{k=1}^{n} \lambda_{ik} d_k; \qquad d_k = \frac{1}{1 + t\epsilon_k}; \qquad \Lambda = G \times (G^{-1})^T \tag{6.80}$$

(Note that s here is a scalar sensitivity function and not the Laplace variable.) The singular
value of a matrix is larger than any of its elements, so $\bar{\sigma}(S') \geq \max_i |s'_{ii}|$, and the objective
in the following is to choose a combination of input errors ϵ_k such that the worst-case $|s'_{ii}|$ is
as large as possible. Consider a given output i and write each term in the sum in (6.80) as

$$\lambda_{ik} d_k = \frac{\lambda_{ik}}{1 + t\epsilon_k} = \lambda_{ik} - \frac{\lambda_{ik} t\epsilon_k}{1 + t\epsilon_k} \tag{6.81}$$

We choose all ϵ_k to have the same magnitude $|w_I(j\omega)|$, so we have $\epsilon_k(j\omega) = |w_I|e^{j\angle\epsilon_k}$. We also assume that $|t\epsilon_k| < 1$ at all frequencies[2], such that the phase of $1 + t\epsilon_k$ lies between $-90°$ and $90°$. It is then always possible to select $\angle\epsilon_k$ (the phase of ϵ_k) such that the last term in (6.81) is real and negative, and we have at each frequency, with these choices for ϵ_k,

$$
\begin{aligned}
\frac{s'_{ii}}{s} &= \sum_{k=1}^{n} \lambda_{ik} d_k = 1 + \sum_{k=1}^{n} \frac{|\lambda_{ik}| \cdot |t\epsilon_k|}{|1 + t\epsilon_k|} \\
&\geq 1 + \sum_{k=1}^{n} \frac{|\lambda_{ik}| \cdot |w_I t|}{1 + |w_I t|} = 1 + \frac{|w_I t|}{1 + |w_I t|} \sum_{k=1}^{n} |\lambda_{ik}|
\end{aligned}
\tag{6.82}
$$

where the first equality makes use of the fact that the row-elements of the RGA sum to 1, $(\sum_{k=1}^{n} \lambda_{ik} = 1)$. The inequality follows since $|\epsilon_k| = |w_I|$ and $|1 + t\epsilon_k| \leq 1 + |t\epsilon_k| = 1 + |w_I t|$. This derivation holds for any i (but only for one at a time), and (6.78) follows by selecting i to maximize $\sum_{k=1}^{n} |\lambda_{ik}|$ (the maximum row-sum of the RGA of G). $\qquad \square$

We next consider three examples. In the first two, we consider feedforward and feedback control of a plant with large RGA-elements. In the third, we consider feedback control of a plant with a large condition number, but with small RGA-elements. The first two are sensitive to diagonal input uncertainty, whereas the third is not.

Example 6.6 Feedforward control of distillation process. *Consider the distillation process in (3.82).*

$$
G(s) = \frac{1}{75s + 1} \begin{bmatrix} 87.8 & -86.4 \\ 108.2 & -109.6 \end{bmatrix}, \quad \Lambda(G) = \begin{bmatrix} 35.1 & -34.1 \\ -34.1 & 35.1 \end{bmatrix}
\tag{6.83}
$$

With $E_I = \mathrm{diag}\{\epsilon_1, \epsilon_2\}$ we get for all frequencies

$$
GE_I G^{-1} = \begin{bmatrix} 35.1\epsilon_1 - 34.1\epsilon_2 & -27.7\epsilon_1 + 27.7\epsilon_2 \\ 43.2\epsilon_1 - 43.2\epsilon_2 & -34.1\epsilon_1 + 35.1\epsilon_2 \end{bmatrix}
\tag{6.84}
$$

We note as expected from (6.57) that the RGA-elements appear on the diagonal elements in the matrix $GE_I G^{-1}$. The elements in the matrix $GE_I G^{-1}$ are largest when ϵ_1 and ϵ_2 have opposite signs. With a 20% error in each input channel we may select $\epsilon_1 = 0.2$ and $\epsilon_2 = -0.2$ and find

$$
GE_I G^{-1} = \begin{bmatrix} 13.8 & -11.1 \\ 17.2 & -13.8 \end{bmatrix}
\tag{6.85}
$$

Thus with an "ideal" feedforward controller and 20% input uncertainty, we get from (6.55) that the relative tracking error at all frequencies, including steady-state, may exceed 1000%. This demonstrates the need for feedback control. However, applying feedback control is also difficult for this plant as seen in Example 6.7.

---

[2] The assumption $|t\epsilon_k| < 1$ is not included in the theorem since it is actually needed for robust stability, so if it does not hold we may have $\bar{\sigma}(S')$ infinite for some allowed uncertainty, and (6.78) clearly holds.

Example 6.7 Feedback control of distillation process. *Consider again the distillation process $G(s)$ in (6.83) for which we have $\|\Lambda(G(j\omega))\|_{i\infty} = 69.1$ and $\gamma(G) \approx \gamma_I^*(G) \approx 141.7$ at all frequencies.*

1. Inverse based feedback controller. *Consider the controller $K(s) = (0.7/s)G^{-1}(s)$ corresponding to the nominal sensitivity function*

$$S(s) = \frac{s}{s + 0.7}I$$

The nominal response is excellent, but we found from simulations in Figure 3.12 that the closed-loop response with 20% input gain uncertainty was extremely poor (we used $\epsilon_1 = 0.2$ and $\epsilon_2 = -0.2$). The poor response is easily explained from the lower RGA-bound on $\bar{\sigma}(S')$ in (6.78). With the inverse-based controller we have $l(s) = k/s$ which has a nominal phase margin of PM= $90°$ so from (2.42) we have, at frequency ω_c, that $|s(j\omega_c)| = |t(j\omega_c)| = 1/\sqrt{2} = 0.707$. With $|w_I| = 0.2$, we then get from (6.78) that

$$\bar{\sigma}(S'(j\omega_c)) \geq 0.707 \left(1 + \frac{0.707 \cdot 0.2 \cdot 69.1}{1.14}\right) = 0.707 \cdot 9.56 = 6.76 \qquad (6.86)$$

(This is close to the peak value in (6.78) of 6.81 at frequency 0.79 rad/min.) Thus, we have that with 20% input uncertainty we may have $\|S'\|_\infty \geq 6.81$ and this explains the observed poor closed-loop performance. For comparison, the actual peak value of $\bar{\sigma}(S')$, with the inverse-based controller and uncertainty $E_I = \text{diag}\{\epsilon_1, \epsilon_2\} = \text{diag}\{0.2, -0.2\}$, is computed numerically (using skewed-μ as discussed below) to be

$$\|S'\|_\infty = \left\| \left(I + \frac{0.7}{s}G\begin{bmatrix} 1.2 & \\ & 0.8 \end{bmatrix}G^{-1}\right)^{-1} \right\|_\infty = 14.2$$

and occurs at 0.69 rad/min. The difference between the values 6.81 and 14.2 illustrates that the bound in terms of the RGA is generally not tight, but it is nevertheless very useful.

Next, we look at the upper bounds. Unfortunately, in this case $\gamma_I^*(G) = \gamma_O^*(K) \approx 141.7$, so the upper bounds on $\bar{\sigma}(S')$ in (6.74) and (6.75) are not very tight (they are of magnitude 141.7 at high frequencies).

2. Diagonal (decentralized) feedback controller. *Consider the controller*

$$K_{\text{diag}}(s) = \frac{k_2(\tau s + 1)}{s}\begin{bmatrix} 1 & 0 \\ 0 & -1 \end{bmatrix}, \qquad k_2 = 2.4 \cdot 10^{-2} \; [\text{min}^{-1}]$$

The peak value for the upper bound on $\bar{\sigma}(S')$ in (6.77) is 1.26, so we are guaranteed $\|S'\|_\infty \leq 1.26$, even with 20% gain uncerainty. For comparison, the actual peak in the perturbed sensitivity function with $E_I = \text{diag}\{0.2, -0.2\}$ is $\|S'\|_\infty = 1.05$. Of course, the problem with the simple diagonal controller is that (although it is robust) even the nominal response is poor.

The following example demonstrates that a large plant condition number, $\gamma(G)$, does not necessarily imply sensitivity to uncertainty even with an inverse-based controller.

Example 6.8 Feedback control of distillation process, DV-model. *In this example we consider the following distillation model given by Skogestad et al. (1988) (it is the same system as studied above but with the DV- rather than the LV-configuration for the lower control levels, see Example 10.5):*

$$G = \frac{1}{75s+1} \begin{bmatrix} -87.8 & 1.4 \\ -108.2 & -1.4 \end{bmatrix}, \quad \Lambda(G) = \begin{bmatrix} 0.448 & 0.552 \\ 0.552 & 0.448 \end{bmatrix} \quad (6.87)$$

We have that $\|\Lambda(G(j\omega))\|_{i\infty} = 1$, $\gamma(G) \approx 70.76$ *and* $\gamma_I^*(G) \approx 1.11$ *at all frequencies. Since both the RGA-elements and* $\gamma_I^*(G)$ *are small we do not expect problems with input uncertainty and an inverse based controller. Consider an inverse-based controller* $K_{inv}(s) = (0.7/s)G^{-1}(s)$ *which yields* $\gamma(K) = \gamma(G)$ *and* $\gamma_O^*(K) = \gamma_I^*(G)$. *In Figure 6.3, we show the lower bound (6.78) given in terms of* $\|\Lambda\|_{i\infty}$ *and the two upper bounds (6.74) and (6.76) given in terms of* $\gamma_I^*(G)$ *for two different uncertainty weights* w_I. *From these curves we see that the bounds are close, and we conclude that the plant in (6.87) is robust against input uncertainty even with a decoupling controller.*

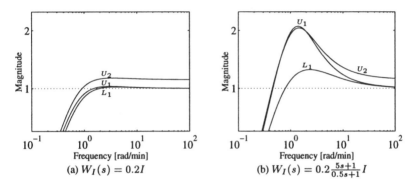

(a) $W_I(s) = 0.2I$ (b) $W_I(s) = 0.2\frac{5s+1}{0.5s+1}I$

Figure 6.3: Bounds on the sensitivity function for the distillation column with the DV configuration: lower bound L_1 from (6.78), upper bounds U_1 from (6.76) and U_2 from (6.74)

Remark. Relationship with the structured singular value: skewed-μ. To analyze exactly the worst-case sensitivity with a given uncertainty $|w_I|$ we may compute skewed-μ (μ^s). With reference to Section 8.11, this involves computing $\mu_{\tilde{\Delta}}(N)$ with $\tilde{\Delta} = \text{diag}(\Delta_I, \Delta_P)$ and $N = \begin{bmatrix} w_I T_I & w_I KS \\ SG/\mu^s & S/\mu^s \end{bmatrix}$ and varying μ^s until $\mu(N) = 1$. The worst-case performance at a given frequency is then $\bar{\sigma}(S') = \mu^s(N)$.

Example 6.9 *Consider the plant*

$$G(s) = \begin{bmatrix} 1 & 100 \\ 0 & 1 \end{bmatrix}$$

for which at all frequencies $\Lambda(G) = I$, $\gamma(G) = 10^4$, $\gamma^*(G) = 1.00$ *and* $\gamma_I^*(G) = 200$. *The RGA-matrix is the identity, but since* $g_{12}/g_{11} = 100$ *we expect from (6.57) that this*

plant will be sensitive to diagonal input uncertainty if we use inverse-based feedback control,
$K = \frac{c}{s}G^{-1}$. *This is confirmed if we compute the worst-case sensitivity function S' for $G' = G(I + w_I \Delta_I)$ where Δ_I is diagonal and $|w_I| = 0.2$. We find by computing $\mu^s(N_1)$ that the peak of $\bar{\sigma}(S')$ is $\|S'\|_\infty = 20.43$.*

Note that the peak is independent of the controller gain c in this case since $G(s)$ is a constant matrix. Also note that with full-block ("unstructured") input uncertainty (Δ_I is a full matrix) the worst-case sensitivity is $\|S'\|_\infty = 1021.7$

Conclusions on input uncertainty and feedback control

The following statements apply to the frequency range around crossover. By "small', we mean about 2 or smaller. By "large" we mean about 10 or larger.

1. Condition number $\gamma(G)$ or $\gamma(K)$ small: robust performance to both diagonal and full-block input uncertainty; see (6.72) and (6.73).
2. Minimized condition numbers $\gamma_I^*(G)$ or $\gamma_O^*(K)$ small: robust performance to diagonal input uncertainty; see (6.76) and (6.77). Note that a diagonal controller always has $\gamma_O^*(K) = 1$.
3. RGA(G) has large elements: inverse-based controller is *not* robust to diagonal input uncertainty; see (6.78). Since diagonal input uncertainty is unavoidable in practice, the rule is never to use a decoupling controller for a plant with large RGA-elements. Furthermore, a diagonal controller will most likely yield poor nominal performance for a plant with large RGA-elements, so we conclude that plants with large RGA-elements are fundamentally difficult to control.
4. $\gamma_I^*(G)$ is large while at the same time the RGA has small elements: cannot make any definite conclusion about the sensitivity to input uncertainty based on the bounds in this section. However, as found in Example 6.9 we may expect there to be problems.

6.10.5   Element-by-element uncertainty

Consider any complex matrix G and let λ_{ij} denote the ij'th element in the RGA-matrix of G. The following result holds (Yu and Luyben, 1987; Hovd and Skogestad, 1992):

Theorem 6.6 *The (complex) matrix G becomes singular if we make a relative change $-1/\lambda_{ij}$ in its ij-th element, that is, if a single element in G is perturbed from g_{ij} to $g_{pij} = g_{ij}(1 - \frac{1}{\lambda_{ij}})$.*

The theorem is proved in Appendix A.4. Thus, the RGA-matrix is a direct measure of sensitivity to element-by-element uncertainty and matrices with large RGA-values become singular for small relative errors in the elements.

Example 6.10 *The matrix G in (6.83) is non-singular. The $1,2$-element of the RGA is $\lambda_{12}(G) = -34.1$. Thus, the matrix G becomes singular if $g_{12} = -86.4$ is perturbed to*

$$g_{p12} = -86.4(1 - 1/(-34.1)) = -88.9 \qquad (6.88)$$

The above result is an important algebraic property of the RGA, but it also has important implications for improved control:

1) **Identification.** Models of multivariable plants, $G(s)$, are often obtained by identifying one element at a time, for example, using step responses. From Theorem 6.6 it is clear that this simple identification procedure will most likely give meaningless results (e.g. the wrong sign of the steady-state RGA) if there are large RGA-elements within the bandwidth where the model is intended to be used.

2) **RHP-zeros.** Consider a plant with transfer function matrix $G(s)$. If the relative uncertainty in an element at a given frequency is larger than $|1/\lambda_{ij}(j\omega)|$ then the plant may be singular at this frequency, implying that the uncertainty allows for a RHP-zero on the $j\omega$-axis. This is of course detrimental to performance both in terms of feedforward and feedback control.

Remark. Theorem 6.6 seems to "prove" that plants with large RGA-elements are fundamentally difficult to control. However, although the statement may be true (see the conclusions on page 244), we cannot draw this conclusion from Theorem 6.6. This is because the assumption of element-by-element uncertainty is often unrealistic from a physical point of view, since the elements are usually *coupled* in some way. For example, this is the case for the distillation column process where we know that the elements are coupled due to an underlying physical constraint in such a way that the model (6.83) never becomes singular even for large changes in the transfer function matrix elements.

6.10.6 Steady-state condition for integral control

Feedback control reduces the sensitivity to model uncertainty at frequencies where the loop gains are large. With integral action in the controller we can achieve zero steady-state control error, even with large model errors, provided the sign of the plant, as expressed by $\det G(0)$, does not change. The statement applies for stable plants, or more generally for cases where the number of unstable poles in the plant does not change. The conditions are stated more exactly in the following theorem.

Theorem 6.7 *Let the number of open-loop unstable poles (excluding poles at $s = 0$) of $G(s)K(s)$ and $G'(s)K(s)$ be P and P', respectively. Assume that the controller K is such that GK has integral action in all channels, and that the transfer functions GK and $G'K$ are strictly proper. Then if*

$$\det G'(0)/\det G(0) \begin{cases} < 0 & \text{for } P - P' \text{ even, including zero} \\ > 0 & \text{for } P - P' \text{ odd} \end{cases} \qquad (6.89)$$

at least one of the following instabilities will occur: a) The negative feedback closed-loop system with loop gain GK is unstable. b) The negative feedback closed-loop system with loop gain $G'K$ is unstable.

Proof: For stability of both $(I + GK)^{-1}$ and $(I + G'K)^{-1}$ we have from Lemma A.5 in Appendix A.6.3 that $\det(I + E_O T(s))$ needs to encircle the origin $P - P'$ times as s traverses the Nyquist D-contour. Here $T(0) = I$ because of the requirement for integral action in all channels of GK. Also, since GK and $G'K$ are strictly proper, $E_O T$ is strictly proper, and hence $E_O(s)T(s) \to 0$ as $s \to \infty$. Thus, the map of $\det(I + E_O T(s))$ starts at $\det G'(0)/\det G(0)$ (for $s = 0$) and ends at 1 (for $s = \infty$). A more careful analysis of the Nyquist plot of $\det(I + E_O T(s))$ reveals that the number of encirclements of the origin will be even for $\det G'(0)/\det G(0) > 0$, and odd for $\det G'(0)/\det G(0) < 0$. Thus, if this parity (odd or even) does not match that of $P - P'$ we will get instability, and the theorem follows. $\square$

Example 6.11 *Suppose the true model of a plant is given by $G(s)$, and that by careful identification we obtain a model $G_1(s)$,*

$$G = \frac{1}{75s + 1}\begin{bmatrix} 87.8 & -86.4 \\ 108.2 & -109.6 \end{bmatrix}, \quad G_1(s) = \frac{1}{75s + 1}\begin{bmatrix} 87 & -88 \\ 109 & -108 \end{bmatrix}$$

At first glance, the identified model seems very good, but it is actually useless for control purposes since $\det G_1(0)$ has the wrong sign; $\det G(0) = -274.4$ and $\det G_1(0) = 196$ (also the RGA-elements have the wrong sign; the $1,1$-element in the RGA is -47.9 instead of $+35.1$). From Theorem 6.7 we then get that any controller with integral action designed based on the model G_1, will yield instability when applied to the plant G.

6.11 Input-output controllability

We now summarize the main findings of this chapter in an analysis procedure for input-output controllability of a MIMO plant. The presence of directions in MIMO systems makes it more difficult to give a precise description of the procedure in terms of a set of rules as was done in the SISO case.

6.11.1 Controllability analysis procedure

The following procedure assumes that we have made a decision on the plant inputs and plant outputs (manipulations and measurements), and we want to analyze the model G to find out what control performance can be expected.

The procedure can also be used to assist in control structure design (the selection of inputs, outputs and control configuration), but it must then be repeated for each candidate set of inputs and outputs. In some cases the number of possibilities is so large that such an approach becomes prohibitive. In this case some pre-screening is required, for example, based on physical insight or by analyzing the "large" model, G_{all}, with all the candidate inputs and outputs included. This is briefly discussed in Section 10.4.

A typical MIMO controllability analysis may proceed as follows:

1. Scale all variables (inputs u, outputs y, disturbances d, references, r) to obtain a scaled model, $y = G(s)u + G_d(s)d$, $r = R\tilde{r}$; see Section 1.4.
2. Obtain a minimal realization.
3. Check functional controllability. To be able to control the outputs independently, we first need at least as many inputs u as outputs y. Second, we need the rank of $G(s)$ to be equal to the number of outputs, l, i.e. the minimum singular value of $G(j\omega)$, $\underline{\sigma}(G) = \sigma_l(G)$, should be non-zero (except at possible $j\omega$-axis zeros). If the plant is not functionally controllable then compute the output direction where the plant has no gain, see (6.19), to obtain insight into the source of the problem.
4. Compute the poles. For RHP (unstable) poles obtain their locations and associated directions; see (6.5). "Fast" RHP-poles far from the origin are bad.
5. Compute the zeros. For RHP-zeros obtain their locations and associated directions. Look for zeros pinned into certain outputs. "Small" RHP-zeros (close to the origin) are bad!
6. Obtain the frequency response $G(j\omega)$ and compute the RGA matrix, $\Lambda = G \times (G^\dagger)^T$. Plants with large RGA-elements at crossover frequencies are difficult to control and should be avoided. For more details about the use of the RGA see Section 3.6, page 86.
7. From now on scaling is critical. Compute the singular values of $G(j\omega)$ and plot them as a function of frequency. Also consider the associated input and output singular vectors.
8. The minimum singular value, $\underline{\sigma}(G(j\omega))$, is a particularly useful controllability measure. It should generally be as large as possible at frequencies where control is needed. If $\underline{\sigma}(G(j\omega)) < 1$ then we cannot at frequency ω make independent output changes of unit magnitude by using inputs of unit magnitude.
9. For disturbances, consider the elements of the matrix G_d. At frequencies where one or more elements is larger than 1, we need control. We get more information by considering one disturbance at a time (the columns g_d of G_d). We must require for each disturbance that S is less than $1/\|g_d\|_2$ in the disturbance direction y_d, i.e. $\|Sy_d\|_2 \leq 1/\|g_d\|_2$; see (6.33). Thus, we must at least require $\underline{\sigma}(S) \leq 1/\|g_d\|_2$ and we may have to require $\bar{\sigma}(S) \leq 1/\|g_d\|_2$; see (6.34).

Remark. If feedforward control is already used, then one may instead analyze $\widehat{G}_d(s) = GK_dG_{md} + G_d$ where K_d denotes the feedforward controller, see (5.78).

10. Disturbances and input saturation:

    First step. Consider the input magnitudes needed for perfect control by computing the elements in the matrix $G^\dagger G_d$. If all elements are less than 1 at all frequencies then input saturation is not expected to be a problem. If some elements of $G^\dagger G_d$ are larger than 1, then perfect control ($e = 0$) cannot be achieved at this frequency, but "acceptable" control ($\|e\|_2 < 1$) may be possible, and this may be tested in the second step.

    Second step. Check condition (6.47), that is, consider the elements of $U^H G_d$ and make sure that the elements in the i'th row are smaller than $\sigma_i(G) + 1$, at all frequencies.

11. Are the requirements compatible? Look at disturbances, RHP-poles and RHP-zeros and their associated locations and directions. For example, we must require for each disturbance and each RHP-zero that $|y_z^H g_d(z)| \leq 1$; see (6.35). For combined RHP-zeros and RHP-poles see (6.13) and (6.14).

12. Uncertainty. If the condition number $\gamma(G)$ is small then we expect no particular problems with uncertainty. If the RGA-elements are large, we expect strong sensitivity to uncertainty. For a more detailed analysis see the conclusion on page 244.

13. If decentralized control (diagonal controller) is of interest see the summary on page 443.

14. The use of the condition number and RGA are summarized separately in Section 3.6, page 86.

A controllability analysis may also be used to obtain initial performance weights for controller design. After a controller design one may analyze the controller by plotting, for example, its elements, singular values, RGA and condition number as a function of frequency.

6.11.2 Plant design changes

If a plant is not input-output controllable, then it must somehow be modified. Some possible modifications are listed below.

Controlled outputs. Identify the output(s) which cannot be controlled satisfactorily. Can the specifications for these be relaxed?

Manipulated inputs. If input constraints are encountered then consider replacing or moving actuators. For example, this could mean replacing a control valve with a larger one, or moving it closer to the controlled output.

If there are RHP-zeros which cause control problems then the zeros may often be eliminated by adding another input (possibly resulting in a non-square plant). This may not be possible if the zero is pinned to a particular output.

Extra Measurements. If the effect of disturbances, or uncertainty, is large, and the dynamics of the plant are such that acceptable control cannot be achieved, then consider adding "fast local loops" based on extra measurements which are located close to the inputs and disturbances; see Section 10.8.2 and the example on page 208.

Disturbances. If the effect of disturbances is too large, then see whether the disturbance itself may be reduced. This may involve adding extra equipment to dampen the disturbances, such as a buffer tank in a chemical process or a spring in a mechanical system. In other cases this may involve improving or changing the control of another part of the system, e.g. we may have a disturbance which is actually the manipulated input for another part of the system.

Plant dynamics and time delays. In most cases, controllability is improved by making the plant dynamics faster and by reducing time delays. An exception to this is a strongly interactive plant, where an increased dynamic lag or time delay, may be helpful if it somehow "delays" the effect of the interactions; see (6.20). Another more obvious exception is for feedforward control of a measured disturbance, where a delay for the disturbance's effect on the outputs is an advantage.

Example 6.12 Removing zeros by adding inputs. *Consider a stable 2 × 2 plant*

$$G_1(s) = \frac{1}{(s+2)^2} \begin{bmatrix} s+1 & s+3 \\ 1 & 2 \end{bmatrix}$$

which has a RHP-zero at $s = 1$ which limits achievable performance. The zero is not pinned to a particular output, so it will most likely disappear if we add a third manipulated input. Suppose the new plant is

$$G_2(s) = \frac{1}{(s+2)^2} \begin{bmatrix} s+1 & s+3 & s+6 \\ 1 & 2 & 3 \end{bmatrix}$$

which indeed has no zeros. It is interesting to note that each of the three 2 × 2 sub-plants of $G_2(s)$ has a RHP-zero (located at $s = 1$, $s = 1.5$ and $s = 3$, respectively).

Remark. Adding outputs. It has also been argued that it is possible to remove multivariable zeros by adding extra outputs. To some extent this is correct. For example, it is well-known that there are no zeros if we use all the states as outputs, see Example 4.13. However, to control all the states independently we need as many inputs as there are states. Thus, by adding outputs to remove the zeros, we generally get a plant which is not functionally controllable.

6.11.3 Additional exercises

The reader will be better prepared for some of these exercises following an initial reading of Chapter 10 on decentralized control. In all cases the variables are assumed to be scaled as outlined in Section 1.4.

Exercise 6.7 *Analyze input-output controllability for*

$$G(s) = \frac{1}{s^2 + 100} \begin{bmatrix} \frac{1}{0.01s+1} & 1 \\ \frac{s+0.1}{s+1} & 1 \end{bmatrix}$$

Compute the zeros and poles, plot the RGA as a function of frequency, etc.

Exercise 6.8 *Analyze input-output controllability for*

$$G(s) = \frac{1}{(\tau s + 1)(\tau s + 1 + 2\alpha)} \begin{bmatrix} \tau s + 1 + \alpha & \alpha \\ \alpha & \tau s + 1 + \alpha \end{bmatrix}$$

where $\tau = 100$; consider two cases: (a) $\alpha = 20$, and (b) $\alpha = 2$.

   Remark. *This is a simple "two-mixing-tank" model of a heat exchanger where $u = \begin{bmatrix} T_{1in} \\ T_{2in} \end{bmatrix}$, $y = \begin{bmatrix} T_{1out} \\ T_{2out} \end{bmatrix}$ and α is the number of heat transfer units.*

Exercise 6.9 *Let*

$$
A = \begin{bmatrix} -10 & 0 \\ 0 & -1 \end{bmatrix}, B = I, C = \begin{bmatrix} 10 & 1.1 \\ 10 & 0 \end{bmatrix}, D = \begin{bmatrix} 0 & 0 \\ 0 & 1 \end{bmatrix}
$$

(a) Perform a controllability analysis of $G(s)$.
(b) Let $\dot{x} = Ax + Bu + d$ and consider a unit disturbance $d = [\, z_1 \quad z_2 \,]^T$. Which direction (value of z_1/z_2) gives a disturbance that is most difficult to reject (consider both RHP-zeros and input saturation)?
(c) Discuss decentralized control of the plant. How would you pair the variables?

Exercise 6.10 *Consider the following two plants. Do you expect any control problems? Could decentralized or inverse-based control be used? What pairing would you use for decentralized control?*

$$
G_a(s) = \frac{1}{1.25(s+1)(s+20)} \begin{bmatrix} s-1 & s \\ -42 & s-20 \end{bmatrix}
$$

$$
G_b(s) = \frac{1}{(s^2+0.1)} \begin{bmatrix} 1 & 0.1(s-1) \\ 10(s+0.1)/s & (s+0.1)/s \end{bmatrix}
$$

Exercise 6.11 *Order the following three plants in terms of their expected ease of controllability*

$$
G_1(s) = \begin{bmatrix} 100 & 95 \\ 100 & 100 \end{bmatrix}, G_2(s) = \begin{bmatrix} 100e^{-s} & 95e^{-s} \\ 100 & 100 \end{bmatrix}, G_3(s) = \begin{bmatrix} 100 & 95e^{-s} \\ 100 & 100 \end{bmatrix}
$$

Remember to also consider the sensitivity to input gain uncertainty.

Exercise 6.12 *Analyze input-output controllability for*

$$
G(s) = \begin{bmatrix} \frac{5000s}{(5000s+1)(2s+1)} & \frac{2(-5s+1)}{100s+1} \\ \frac{3}{5s+1} & \frac{3}{5s+1} \end{bmatrix}
$$

Exercise 6.13 *Analyze input-output controllability for*

$$
G(s) = \begin{bmatrix} 100 & 102 \\ 100 & 100 \end{bmatrix}, \quad g_{d1}(s) = \begin{bmatrix} \frac{10}{s+1} \\ \frac{10}{s+1} \end{bmatrix}; \quad g_{d2} = \begin{bmatrix} \frac{1}{s+1} \\ \frac{-1}{s+1} \end{bmatrix}
$$

Which disturbance is the worst?

Exercise 6.14 *(a) Analyze input-output controllability for the following three plants each of which has 2 inputs and 1 output: $G(s) = (g_1(s) \quad g_2(s))$*
   (i) $g_1(s) = g_2(s) = \frac{s-2}{s+2}$.
   (ii) $g_1(s) = \frac{s-2}{s+2}$, $g_2(s) = \frac{s-2.1}{s+2.1}$.
   (iii) $g_1(s) = \frac{s-2}{s+2}$, $g_2(s) = \frac{s-20}{s+20}$.
   (b) Design controllers and perform closed-loop simulations of reference tracking to complement your analysis. Consider also the input magnitudes.

Exercise 6.15 *Find the poles and zeros and analyze input-output controllability for*

$$G(s) = \begin{bmatrix} c + (1/s) & 1/s \\ c + (1/s) & 1/s \end{bmatrix}$$

Remark. *A similar model form is encountered for distillation columns controlled with the DB-configuration. In which case the physical reason for the model being singular at steady-state is that the sum of the two manipulated inputs is fixed at steady-state, $D + B = F$.*

Exercise 6.16 Controllability of an FCC process. *Consider the following 3×3 plant*

$$\begin{bmatrix} y_1 \\ y_2 \\ y_3 \end{bmatrix} = G(s) \begin{bmatrix} u_1 \\ u_2 \\ u_3 \end{bmatrix}; \quad f(s) = \frac{1}{(18.8s + 1)(75.8s + 1)}$$

$$G(s) = f(s) \begin{bmatrix} 16.8(920s^2 + 32.4s + 1) & 30.5(52.1s + 1) & 4.30(7.28s + 1) \\ -16.7(75.5s + 1) & 31.0(75.8s + 1)(1.58s + 1) & -1.41(74.6s + 1) \\ 1.27(-939s + 1) & 54.1(57.3s + 1) & 5.40 \end{bmatrix}$$

Acceptable control of this 3×3 plant can be achieved with partial control of two outputs with input 3 in manual (not used). That is, we have a 2×2 control problem. Consider three options for the controlled outputs:

$$Y_1 = \begin{bmatrix} y_1 \\ y_2 \end{bmatrix}; \quad Y_2 = \begin{bmatrix} y_2 \\ y_3 \end{bmatrix}; \quad Y_3 = \begin{bmatrix} y_1 \\ y_2 - y_3 \end{bmatrix}$$

In all three cases the inputs are u_1 and u_2. Assume that the third input is a disturbance ($d = u_3$).

(a) Based on the zeros of the three 2×2 plants, $G_1(s)$, $G_2(s)$ and $G_3(s)$, which choice of outputs do you prefer? Which seems to be the worst?

It may be useful to know that the zero polynomials:

a	$5.75 \cdot 10^7 s^4 + 3.92 \cdot 10^7 s^3 + 3.85 \cdot 10^6 s^2 + 1.22 \cdot 10^5 s + 1.03 \cdot 10^3$
b	$4.44 \cdot 10^6 s^3 - 1.05 \cdot 10^6 s^2 - 8.61 \cdot 10^4 s - 9.43 \cdot 10^2$
c	$5.75 \cdot 10^7 s^4 - 8.75 \cdot 10^6 s^3 - 5.66 \cdot 10^5 s^2 + 6.35 \cdot 10^3 s + 1.60 \cdot 10^2$

have the following roots:

a	−0.570	−0.0529	−0.0451	−0.0132
b		0.303	−0.0532	−0.0132
c	0.199	−0.0532	0.0200	−0.0132

(b) For the preferred choice of outputs in (a) do a more detailed analysis of the expected control performance (compute poles and zeros, sketch RGA_{11}, comment on possible problems with input constraints (assume the inputs and outputs have been properly scaled), discuss the effect of the disturbance, etc.). What type of controller would you use? What pairing would you use for decentralized control?

(c) Discuss why the 3×3 plant may be difficult to control.

Remark. *This is actually a model of a fluid catalytic cracking (FCC) reactor where $u = (F_s \ F_a \ k_c)^T$ represents the circulation, airflow and feed composition, and $y = (T_1 \ T_{cy} \ T_{rg})^T$ represents three temperatures. $G_1(s)$ is called the Hicks control structure and $G_3(s)$ the conventional structure. More details are found in Hovd and Skogestad (1993).*

6.13   Conclusion

We have found that most of the insights into the performance limitations of SISO systems developed in Chapter 5 carry over to MIMO systems. For RHP-zeros, RHP-poles and disturbances, the issue of directions usually makes the limitations less severe for MIMO than for SISO systems. However, the situation is usually the opposite with model uncertainty because for MIMO systems there is also uncertainty associated with plant directionality. This is an issue which is unique to MIMO systems.

   We summarized on page 246 the main steps involved in an analysis of input-output controllability of MIMO plants.

7

UNCERTAINTY AND ROBUSTNESS FOR SISO SYSTEMS

In this chapter, we show how to represent uncertainty by real or complex perturbations and we analyze robust stability (RS) and robust performance (RP) for SISO systems using elementary methods. Chapter 8 is devoted to a more general treatment.

7.1 Introduction to robustness

A control system is robust if it is insensitive to differences between the actual system and the model of the system which was used to design the controller. These differences are referred to as model/plant mismatch or simply model uncertainty. The key idea in the $\mathcal{H}_\infty$ robust control paradigm we use is to check whether the design specifications are satisfied even for the "worst-case" uncertainty.

Our approach is then as follows:

1. Determine the uncertainty set: find a mathematical representation of the model uncertainty ("clarify what we know about what we don't know").
2. Check Robust stability (RS): determine whether the system remains stable for all plants in the uncertainty set.
3. Check Robust performance (RP): if RS is satisfied, determine whether the performance specifications are met for all plants in the uncertainty set.

This approach may not always achieve optimal performance. In particular, if the worst-case plant rarely or never occurs, other approaches, such as optimizing some average performance or using adaptive control, may yield better performance. Nevertheless, the linear uncertainty descriptions presented in this book are very useful in many practical situations.

It should also be appreciated that model uncertainty is not the only concern when it comes to robustness. Other considerations include sensor and actuator failures, physical constraints, changes in control objectives, the opening and closing of loops, etc. Furthermore, if a control design is based on an optimization, then robustness problems may also be caused by the mathematical objective function not properly describing the real control problem. Also, the numerical design algorithms themselves may not be robust. However, when we refer to robustness in this book, we mean robustness with respect to model uncertainty, and assume that a fixed (linear) controller is used.

To account for model uncertainty we will assume that the dynamic behaviour of a plant is described not by a single linear time invariant model but by a set Π of possible linear time invariant models, sometimes denoted the "uncertainty set". We adopt the following notation:

Π – a set of possible perturbed plant models.

$G(s) \in \Pi$ – nominal plant model (with no uncertainty).

$G_p(s) \in \Pi$ and $G'(s) \in \Pi$ – particular perturbed plant models.

Sometimes G_p is used rather than Π to denote the uncertainty set, whereas G' always refers to a specific uncertain plant. The subscript p stands for *perturbed* or *possible* or Π (pick your choice). This should not be confused with the subscript capital P, e.g. in w_P, which denotes *performance*.

We will use a "norm-bounded uncertainty description" where the set Π is generated by allowing $\mathcal{H}_\infty$ norm-bounded stable perturbations to the nominal plant $G(s)$. This corresponds to a continuous description of the model uncertainty, and there will be an infinite number of possible plants G_p in the set Π. We let E denote a perturbation which is not normalized, and let Δ denote a normalized perturbation with $\mathcal{H}_\infty$ norm less than 1.

Remark. Another strategy for dealing with model uncertainty is to approximate its effect on the feedback system by adding fictitious disturbances or noise. For example, this is the only way of handling model uncertainty within the so-called LQG approach to optimal control (see Chapter 9). Is this an acceptable strategy? In general, the answer is *no*. This is easily illustrated for linear systems where the addition of disturbances does *not* affect system stability, whereas model uncertainty combined with feedback may easily create instability.

For example, consider a plant with a nominal model $y = Gu + G_d d$, and let the perturbed plant model be $G_p = G + E$ where E represents additive model uncertainty. Then the output of the perturbed plant is

$$y = G_p u + G_d d = Gu + d_1 + d_2 \tag{7.1}$$

where y is different from what we ideally expect (namely Gu) for two reasons:

1. Uncertainty in the model ($d_1 = Eu$)
2. Signal uncertainty ($d_2 = G_d d$)

In LQG control we set $w_d = d_1 + d_2$ where w_d is assumed to be an independent variable such as white noise. Then in the design problem we may make w_d large by selecting appropriate weighting functions, but its presence will never cause instability. However, in reality $w_d =$

$Eu + d_2$, so w_d depends on the signal u and this may cause instability in the presence of feedback when u depends on y. Specifically, the closed-loop system $(I + (G + E)K)^{-1}$ may be unstable for some $E \neq 0$. In conclusion, it may be important to explicitly take into account model uncertainty when studying feedback control.

We will next discuss some sources of model uncertainty and outline how to represent these mathematically.

7.2   Representing uncertainty

Uncertainty in the plant model may have several origins:

1. There are always parameters in the linear model which are only known approximately or are simply in error.
2. The parameters in the linear model may vary due to nonlinearities or changes in the operating conditions.
3. Measurement devices have imperfections. This may even give rise to uncertainty on the manipulated inputs, since the actual input is often measured and adjusted in a cascade manner. For example, this is often the case with valves used to measure flow. In other cases limited valve resolution may cause input uncertainty.
4. At high frequencies even the structure and the model order is unknown, and the uncertainty will always exceed 100% at some frequency.
5. Even when a very detailed model is available we may choose to work with a simpler (low-order) nominal model and represent the neglected dynamics as "uncertainty".
6. Finally, the controller implemented may differ from the one obtained by solving the synthesis problem. In this case one may include uncertainty to allow for controller order reduction and implementation inaccuracies.

The various sources of model uncertainty mentioned above may be grouped into two main classes:

1. **Parametric uncertainty.** Here the structure of the model (including the order) is known, but some of the parameters are uncertain.
2. **Neglected and unmodelled dynamics uncertainty.** Here the model is in error because of missing dynamics, usually at high frequencies, either through deliberate neglect or because of a lack of understanding of the physical process. Any model of a real system will contain this source of uncertainty.

Parametric uncertainty will be quantified by assuming that each uncertain parameter is bounded within some region $[\alpha_{\min}, \alpha_{\max}]$. That is, we have parameter sets of the form

$$\alpha_p = \bar{\alpha}(1 + r_\alpha \Delta)$$

where $\bar{\alpha}$ is the mean parameter value, $r_\alpha = (\alpha_{\max} - \alpha_{\min})/(\alpha_{\max} + \alpha_{\min})$ is the relative uncertainty in the parameter, and Δ is any real scalar satisfying $|\Delta| \leq 1$.

Neglected and unmodelled dynamics uncertainty is somewhat less precise and thus more difficult to quantify, but it appears that the frequency domain is particularly well suited for this class. This leads to complex perturbations which we normalize such that $\|\Delta\|_\infty \leq 1$. In this chapter, we will deal mainly with this class of perturbations.

For completeness one may consider a third class of uncertainty (which is really a combination of the other two):

3. **Lumped uncertainty.** Here the uncertainty description represents one or several sources of parametric and/or unmodelled dynamics uncertainty combined into a single lumped perturbation of a chosen structure.

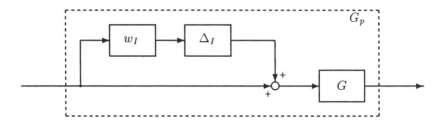

Figure 7.1: Plant with multiplicative uncertainty

The frequency domain is also well suited for describing lumped uncertainty. In most cases we prefer to lump the uncertainty into a *multiplicative uncertainty* of the form

$$\Pi_I : \quad G_p(s) = G(s)(1 + w_I(s)\Delta_I(s)); \quad \underbrace{|\Delta_I(j\omega)| \leq 1 \; \forall \omega}_{\|\Delta_I\|_\infty \leq 1} \qquad (7.2)$$

which may be represented by the block diagram in Figure 7.1. Here $\Delta_I(s)$ is *any* stable transfer function which at each frequency is less than or equal to one in magnitude. Some examples of allowable $\Delta_I(s)$'s with $\mathcal{H}_\infty$ norm less than one, $\|\Delta_I\|_\infty \leq 1$, are

$$\frac{s - z}{s + z}, \quad \frac{1}{\tau s + 1}, \quad \frac{1}{(5s + 1)^3}, \quad \frac{0.1}{s^2 + 0.1s + 1}$$

Remark 1 The stability requirement on $\Delta_I(s)$ may be removed if one instead assumes that the number of RHP poles in $G(s)$ and $G_p(s)$ remains unchanged. However, in order to simplify the stability proofs we will in this book assume that the perturbations are stable.

Remark 2 The subscript I denotes "input", but for SISO systems it doesn't matter whether we consider the perturbation at the input or output of the plant, since

$$G(1 + w_I\Delta_I) = (1 + w_O\Delta_O)G \quad \text{with } \Delta_I(s) = \Delta_O(s) \text{ and } w_I(s) = w_O(s)$$

Another uncertainty form, which is better suited for representing pole uncertainty, is the *inverse multiplicative uncertainty*

$$\Pi_{iI}: \quad G_p(s) = G(s)(1 + w_{iI}(s)\Delta_{iI}(s))^{-1}; \quad |\Delta_{iI}(j\omega)| \le 1 \; \forall\omega \qquad (7.3)$$

Even with a stable $\Delta_{iI}(s)$ this form allows for uncertainty in the location of an unstable pole, and it also allows for poles crossing between the left- and right-half planes.

Parametric uncertainty is sometimes called *structured uncertainty* as it models the uncertainty in a structured manner. Analogously, lumped dynamics uncertainty is sometimes called *unstructured uncertainty*. However, one should be careful about using these terms because there can be several levels of structure, especially for MIMO systems.

Remark. Alternative approaches for describing uncertainty and the resulting performance may be considered. One approach for parametric uncertainty is to assume a probabilistic (e.g. normal) distribution of the parameters, and to consider the "average" response. This stochastic uncertainty is, however, difficult to analyze exactly.

Another approach is the multi-model approach in which one considers a finite set of alternative models. This approach is well suited for parametric uncertainty as it eases the burden of the engineer in representing the uncertainty. Performance may be measured in terms of the worst-case or some average of these models' responses. The multi-model approach can also be used when there is unmodelled dynamics uncertainty. A problem with the multi-model approach is that it is not clear how to pick the set of models such that they represent the limiting ("worst-case") plants.

To summarize, there are many ways to define uncertainty, from stochastic uncertainty to differential sensitivity (local robustness) and multi-models. Weinmann (1991) gives a good overview. In particular, there are several ways to handle parametric uncertainty, and of these the $\mathcal{H}_\infty$ frequency-domain approach, used in this book, may not be the best or the simplest, but it can handle most situations as we will see. In addition, the frequency-domain is excellent for describing neglected or unknown dynamics, and it is very well suited when it comes to making simple yet realistic lumped uncertainty descriptions.

7.3   Parametric uncertainty

In spite of what is sometimes claimed, parametric uncertainty may also be represented in the $\mathcal{H}_\infty$ framework, at least if we restrict the perturbations Δ to be real. This is discussed in more detail in Section 7.7. Here we provide just two simple examples.

Example 7.1 Gain uncertainty. *Let the set of possible plants be*

$$G_p(s) = k_p G_0(s); \quad k_{\min} \le k_p \le k_{\max} \qquad (7.4)$$

where k_p is an uncertain gain and $G_0(s)$ is a transfer function with no uncertainty. By writing

$$k_p = \bar{k}(1 + r_k\Delta), \quad r_k \triangleq \frac{k_{max} - k_{min}}{k_{max} + k_{min}}, \quad \bar{k} \triangleq \frac{k_{min} + k_{max}}{2} \tag{7.5}$$

where r_k is the relative magnitude of the gain uncertainty and $\bar{k}$ is the average gain, (7.4) may be rewritten as multiplicative uncertainty

$$G_p(s) = \underbrace{\bar{k}G_0(s)}_{G(s)}(1 + r_k\Delta), \quad |\Delta| \leq 1 \tag{7.6}$$

where Δ is a real scalar and $G(s)$ is the nominal plant. We see that the uncertainty in (7.6) is in the form of (7.2) with a constant multiplicative weight $w_I(s) = r_k$. The uncertainty description in (7.6) can also handle cases where the gain changes sign ($k_{min} < 0$ and $k_{max} > 0$) corresponding to $r_k > 1$. The usefulness of this is rather limited, however, since it is impossible to get any benefit from control for a plant where we can have $G_p = 0$, at least with a linear controller.

Example 7.2 **Time constant uncertainty.** *Consider a set of plants, with an uncertain time constant, given by*

$$G_p(s) = \frac{1}{\tau_p s + 1}G_0(s); \quad \tau_{min} \leq \tau_p \leq \tau_{max} \tag{7.7}$$

By writing $\tau_p = \bar{\tau}(1 + r_\tau\Delta)$, *similar to (7.5) with* $|\Delta| < 1$, *the model set (7.7) can be rewritten as*

$$G_p(s) = \frac{G_0}{1 + \bar{\tau}s + r_\tau\bar{\tau}s\Delta} = \underbrace{\frac{G_0}{1 + \bar{\tau}s}}_{G(s)} \frac{1}{1 + w_{iI}(s)\Delta}; \quad w_{iI}(s) = \frac{r_\tau\bar{\tau}s}{1 + \bar{\tau}s} \tag{7.8}$$

which is in the inverse multiplicative form of (7.3). Note that it does not make physical sense for τ_p to change sign, because a value $\tau_p = 0^-$ corresponds to a pole at infinity in the RHP, and the corresponding plant would be impossible to stabilize. To represent cases in which a pole may cross between the half planes, one should instead consider parametric uncertainty in the pole itself, $1/(s + p)$, as described in (7.85).

As shown by the above examples one can represent parametric uncertainty in the $\mathcal{H}_\infty$ framework. However, parametric uncertainty is often avoided for the following reasons:

1. It usually requires a large effort to model parametric uncertainty.
2. A parametric uncertainty model is somewhat deceiving in the sense that it provides a very detailed and accurate description, even though the underlying assumptions about the model and the parameters may be much less exact.
3. The exact model structure is required and so unmodelled dynamics cannot be dealt with.
4. Real perturbations are required, which are more difficult to deal with mathematically and numerically, especially when it comes to controller synthesis.

Therefore, parametric uncertainty is often represented by complex perturbations. For example, we may simply replace the real perturbation, $-1 \leq \Delta \leq 1$ by a complex perturbation with $|\Delta(j\omega)| \leq 1$. This is of course conservative as it introduces possible plants that are not present in the original set. However, if there are several real perturbations, then the conservatism is often reduced by *lumping* these perturbations into a *single* complex perturbation. Typically, a complex multiplicative perturbation is used, e.g. $G_p = G(I + w_I \Delta)$.

How is it possible that we can reduce conservatism by lumping together several real perturbations? This will become clearer from the examples in the next section, but simply stated the answer is that with several uncertain parameters the true uncertainty region is often quite "disk-shaped", and may be more accurately represented by a single complex perturbation.

7.4 Representing uncertainty in the frequency domain

In terms of quantifying unmodelled dynamics uncertainty the frequency-domain approach ($\mathcal{H}_\infty$) does not seem to have much competition (when compared with other norms). In fact, Owen and Zames (1992) make the following observation:

> The design of feedback controllers in the presence of non-parametric and unstructured uncertainty ... is the *raison d'être* for $\mathcal{H}_\infty$ feedback optimization, for if disturbances and plant models are clearly parameterized then $\mathcal{H}_\infty$ methods seem to offer no clear advantages over more conventional state-space and parametric methods.

7.4.1 Uncertainty regions

To illustrate how parametric uncertainty translates into frequency domain uncertainty, consider in Figure 7.2 the Nyquist plots (or regions) generated by the following set of plants

$$G_p(s) = \frac{k}{\tau s + 1} e^{-\theta s}, \quad 2 \leq k, \theta, \tau \leq 3 \qquad (7.9)$$

Step 1. At each frequency, a *region* of complex numbers $G_p(j\omega)$ is generated by varying the three parameters in the ranges given by (7.9). In general, these *uncertainty regions* have complicated shapes and complex mathematical descriptions, and are cumbersome to deal with in the context of control system design.

Step 2. We therefore approximate such complex regions as discs (circles) as shown in Figure 7.3, resulting in a (complex) additive uncertainty description as discussed next.

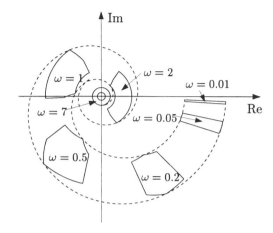

Figure 7.2: Uncertainty regions of the Nyquist plot at given frequencies. Data from (7.9)

Remark 1 There is no conservatism introduced in the first step when we go from a parametric uncertainty description as in (7.9) to an uncertainty region description as in Figure 7.2. This is somewhat surprising since the uncertainty regions in Figure 7.2 seem to allow for more uncertainty. For example, they allow for "jumps" in $G_p(j\omega)$ from one frequency to the next (e.g. from one corner of a region to another). Nevertheless, we derive in this and the next chapter necessary and sufficient frequency-by-frequency conditions for robust stability based on uncertainty regions. Thus, the only conservatism is in the second step where we approximate the original uncertainty region by a larger disc-shaped region as shown in Figure 7.3.

Remark 2 Exact methods do exist (using complex region mapping, e.g. see Laughlin et al. (1986)) which avoid the second conservative step. However, as already mentioned these methods are rather complex, and although they may be used in analysis, at least for simple systems, they are not really suitable for controller synthesis and will not be pursued further in this book.

Remark 3 From Figure 7.3 we see that the radius of the disc may be reduced by moving the center (selecting another nominal model). This is discussed in Section 7.4.4.

7.4.2 Representing uncertainty regions by complex perturbations

We will use disc-shaped regions to represent uncertainty regions as illustrated by the Nyquist plots in Figures 7.3 and 7.4. These disc-shaped regions may be generated by additive complex norm-bounded perturbations (additive uncertainty) around a nominal plant G

$$\Pi_A: \quad G_p(s) = G(s) + w_A(s)\Delta_A(s); \quad |\Delta_A(j\omega)| \leq 1 \; \forall \omega \tag{7.10}$$

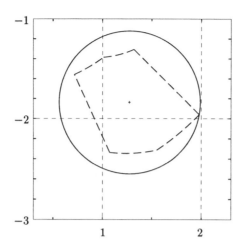

Figure 7.3: Disc approximation (solid line) of the original uncertainty region (dashed line). Plot corresponds to $\omega = 0.2$ in Figure 7.2

where $\Delta_A(s)$ is *any* stable transfer function which at each frequency is no larger than one in magnitude. How is this possible? If we consider all possible Δ_A's, then at each frequency $\Delta_A(j\omega)$ "generates" a disc-shaped region with radius 1 centred at 0, so $G(j\omega) + w_A(j\omega)\Delta_A(j\omega)$ generates at each frequency a disc-shaped region of radius $|w_A(j\omega)|$ centred at $G(j\omega)$ as shown in Figure 7.4.

In most cases $w_A(s)$ is a rational transfer function (although this need not always be the case).

One may also view $w_A(s)$ as a weight which is introduced in order to normalize the perturbation to be less than 1 in magnitude at each frequency. Thus only the magnitude of the weight matters, and in order to avoid unnecessary problems we always choose $w_A(s)$ to be stable and minimum phase (this applies to all weights used in this book).

The disk-shaped regions may alternatively be represented by a *multiplicative uncertainty* description as in (7.2),

$$\Pi_I : \quad G_p(s) = G(s)(1 + w_I(s)\Delta_I(s)); \quad |\Delta_I(j\omega)| \leq 1, \forall \omega \qquad (7.11)$$

By comparing (7.10) and (7.11) we see that for SISO systems the additive and multiplicative uncertainty descriptions are equivalent if at each frequency

$$|w_I(j\omega)| = |w_A(j\omega)|/|G(j\omega)| \qquad (7.12)$$

However, multiplicative (relative) weights are often preferred because their numerical value is more informative. At frequencies where $|w_I(j\omega)| > 1$ the uncertainty exceeds 100% and the Nyquist curve may pass through the origin. This follows since, as illustrated in Figure 7.5, the radius of the discs in the Nyquist plot,

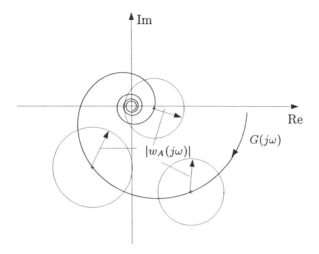

Figure 7.4: Disc-shaped uncertainty regions generated by complex additive uncertainty, $G_p = G + w_A \Delta$

$|w_A(j\omega)| = |G(j\omega)w_I(j\omega)|$, then exceeds the distance from $G(j\omega)$ to the origin. At these frequencies we do not know the phase of the plant, and we allow for zeros crossing from the left to the right-half plane. To see this, consider a frequency ω_0 where $|w_I(j\omega_0)| \geq 1$. Then there exists a $|\Delta_I| \leq 1$ such that $G_p(j\omega_0) = 0$ in (7.11), that is, there exists a possible plant with zeros at $s = \pm j\omega_0$. For this plant at frequency ω_0 the input has no effect on the output, so control has no effect. It then follows that *tight control is not possible at frequencies where* $|w_I(j\omega)| \geq 1$ (this condition is derived more rigorously in (7.33)).

7.4.3 Obtaining the weight for complex uncertainty

Consider a set Π of possible plants resulting, for example, from parametric uncertainty as in (7.9). We now want to describe this set of plants by a single (lumped) complex perturbation, Δ_A or Δ_I. This complex (disk-shaped) uncertainty description may be generated as follows:

1. Select a nominal model $G(s)$.
2. *Additive uncertainty*. At each frequency find the smallest radius $l_A(\omega)$ which includes all the possible plants Π:

$$l_A(\omega) = \max_{G_P \in \Pi} |G_p(j\omega) - G(j\omega)| \qquad (7.13)$$

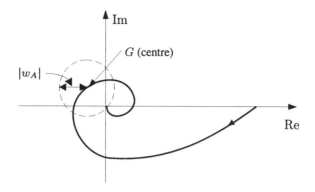

Figure 7.5: The set of possible plants includes the origin at frequencies where $|w_A(j\omega)| \geq |G(j\omega)|$, or equivalently $|w_I(j\omega)| \geq 1$

If we want a rational transfer function weight, $w_A(s)$ (which may not be the case if we only want to do analysis), then it must be chosen to cover the set, so

$$|w_A(j\omega)| \geq l_A(\omega) \quad \forall \omega \tag{7.14}$$

Usually $w_A(s)$ is of low order to simplify the controller design. Furthermore, an objective of frequency-domain uncertainty is usually to represent uncertainty in a simple straightforward manner.

3. *Multiplicative (relative) uncertainty.* This is often the preferred uncertainty form, and we have

$$l_I(\omega) = \max_{G_p \in \Pi} \left| \frac{G_p(j\omega) - G(j\omega)}{G(j\omega)} \right| \tag{7.15}$$

and with a rational weight

$$|w_I(j\omega)| \geq l_I(\omega), \forall \omega \tag{7.16}$$

Example 7.3 Multiplicative weight for parametric uncertainty. *Consider again the set of plants with parametric uncertainty given in (7.9)*

$$\Pi: \quad G_p(s) = \frac{k}{\tau s + 1} e^{-\theta s}, \quad 2 \leq k, \theta, \tau \leq 3 \tag{7.17}$$

We want to represent this set using multiplicative uncertainty with a rational weight $w_I(s)$. To simplify subsequent controller design we select a delay-free nominal model

$$G(s) = \frac{\bar{k}}{\bar{\tau} s + 1} = \frac{2.5}{2.5s + 1} \tag{7.18}$$

To obtain $l_I(\omega)$ in (7.15) we consider three values (2, 2.5 and 3) for each of the three parameters (k, θ, τ). (This is not, in general, guaranteed to yield the worst case as the worst

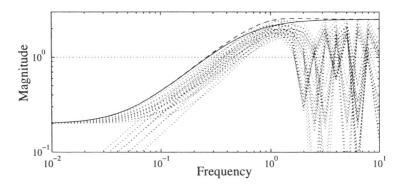

Figure 7.6: Relative errors for 27 combinations of k, τ and θ with delay-free nominal plant (dotted lines). Solid line: First-order weight $|w_{I1}|$ in (7.19). Dashed line: Third-order weight $|w_I|$ in (7.20)

case may be at the interior of the intervals.) The corresponding relative errors $|(G_p - G)/G|$ are shown as functions of frequency for the $3^3 = 27$ resulting G_p's in Figure 7.6. The curve for $l_I(\omega)$ must at each frequency lie above all the dotted lines, and we find that $l_I(\omega)$ is 0.2 at low frequencies and 2.5 at high frequencies. To derive $w_I(s)$ we first try a simple first-order weight that matches this limiting behaviour:

$$w_{I1}(s) = \frac{Ts + 0.2}{(T/2.5)s + 1}, \quad T = 4 \tag{7.19}$$

As seen from the solid line in Figure 7.6, this weight gives a good fit of $l_I(\omega)$, except around $\omega = 1$ where $|w_{I1}(j\omega)|$ is slightly too small, and so this weight does not include all possible plants. To change this so that $|w_I(j\omega)| \geq l_I(\omega)$ at all frequencies, we can multiply w_{I1} by a correction factor to lift the gain slightly at $\omega = 1$. The following works well

$$w_I(s) = w_{I1}(s) \frac{s^2 + 1.6s + 1}{s^2 + 1.4s + 1} \tag{7.20}$$

as is seen from the dashed line in Figure 7.6. The magnitude of the weight crosses 1 at about $\omega = 0.26$. This seems reasonable since we have neglected the delay in our nominal model, which by itself yields 100% uncertainty at a frequency of about $1/\theta_{max} = 0.33$ (see Figure 7.8(a) below).

An uncertainty description for the same parametric uncertainty, but with a mean-value nominal model (with delay), is given in Exercise 7.8. Parametric gain and delay uncertainty (without time constant uncertainty) is discussed further in Example 7.5.

Remark. Pole uncertainty. In the example we represented pole (time constant) uncertainty by a multiplicative perturbation, Δ_I. We *may* even do this for unstable plants, provided the poles do not shift between the half planes and one allows $\Delta_I(s)$ to be unstable. However, if the pole uncertainty is large, and in particular if poles can cross form the LHP to the RHP, then one should use an inverse ("feedback") uncertainty representation as in (7.3).

7.4.4 Choice of nominal model

With parametric uncertainty there are three main options for the choice of nominal model:

1. A simplified model, e.g. a low-order, delay-free model.
2. A model of mean parameter values, $G(s) = \bar{G}(s)$.
3. The central plant obtained from a Nyquist plot (yielding the smallest discs).

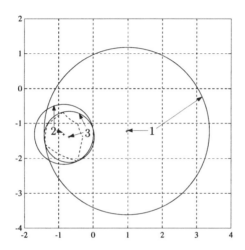

Figure 7.7: Nyquist plot of $G_p(j\omega)$ at frequency $\omega = 0.5$ (dashed region) showing complex disc approximations using three options for the nominal model:
1. Simplified nominal model with no time delay
2. Mean parameter values
3. Nominal model corresponding to the smallest radius

Option 1 usually yields the largest uncertainty region, but the model is simple and this facilitates controller design in later stages. Option 2 is probably the most straightforward choice. Option 3 yields the smallest region, but in this case a significant effort may be required to obtain the nominal model, which is usually not a rational transfer function and a rational approximation could be of very high order.

Example 7.4 *Consider again the uncertainty set (7.17) used in Example 7.3. The nominal models selected for options 1 and 2 are*

$$G_1(s) = \frac{\bar{k}}{\bar{\tau}s + 1}, \quad G_2(s) = \frac{\bar{k}}{\bar{\tau}s + 1}e^{-\bar{\theta}s}$$

For option 3 the nominal model is not rational. The Nyquist plot of the three resulting discs at frequency $\omega = 0.5$ are shown in Figure 7.7.

A similar example was studied by Wang et al. (1994), who obtained the best controller designs with option 1, although the uncertainty region is clearly much larger in this case. The reason for this is that the "worst-case region" in the Nyquist plot in Figure 7.7 corresponds quite closely to those plants with the most negative phase (at coordinates about $(-1.5, -1.5)$). Thus, the additional plants included in the largest region (option 1) are generally easier to control and do not really matter when evaluating the worst-case plant with respect to stability or performance. In conclusion, at least for SISO plants, we find that for plants with an uncertain time delay, it is simplest and sometimes best (!) to use a delay-free nominal model, and to represent the nominal delay as additional uncertainty.

Remark. The choice of nominal model is only an issue since we are lumping several sources of parametric uncertainty into a single complex perturbation. Of course, if we use a parametric uncertainty description, based on multiple real perturbations, then we should always use the mean parameter values in the nominal model.

7.4.5   Neglected dynamics represented as uncertainty

We saw above that one advantage of frequency domain uncertainty descriptions is that one can choose to work with a simple nominal model, and represent neglected dynamics as uncertainty. We will now consider this in a little more detail. Consider a set of plants

$$G_p(s) = G_0(s) f(s)$$

where $G_0(s)$ is fixed (and certain). We want to neglect the term $f(s)$ (which may be fixed or may be an uncertain set Π_f), and represent G_p by multiplicative uncertainty with a nominal model $G = G_0$. From (7.15) we get that the magnitude of the relative uncertainty caused by neglecting the dynamics in $f(s)$ is

$$l_I(\omega) = \max_{G_p} \left| \frac{G_p - G}{G} \right| = \max_{f(s) \in \Pi_f} |f(j\omega) - 1| \tag{7.21}$$

Two simple examples illustrate the procedure.

1. Neglected delay. Let $f(s) = e^{-\theta_p s}$, where $0 \le \theta_p \le \theta_{\max}$. We want to represent $G_p = G_0(s) e^{-\theta_p s}$ by a delay-free plant $G_0(s)$ and multiplicative uncertainty. Let us first consider the maximum delay, for which the relative error $|1 - e^{-j\omega\theta_{\max}}|$ is shown as a function of frequency in Figure 7.8(a). The relative uncertainty crosses 1 in magnitude at about frequency $1/\theta_{\max}$, reaches 2 at frequency $\pi/\theta_{\max}$ (since at this frequency $e^{j\omega\theta_{\max}} = -1$), and oscillates between 0 and 2 at higher frequencies (which corresponds to the Nyquist plot of $e^{-j\omega\theta_{\max}}$ going around and around the unit circle). Similar curves are generated for smaller values of the delay, and they also oscillate between 0 and 2 but at higher frequencies. It then follows that if we consider all $\theta \in [0, \theta_{\max}]$ then the relative error bound is 2 at frequencies above $\pi/\theta_{\max}$, and we have

$$l_I(\omega) = \begin{cases} |1 - e^{-j\omega\theta_{\max}}| & \omega < \pi/\theta_{\max} \\ 2 & \omega \ge \pi/\theta_{\max} \end{cases} \tag{7.22}$$

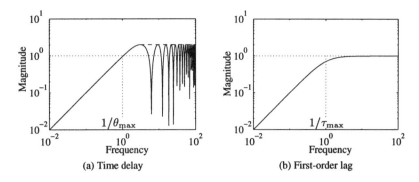

(a) Time delay   (b) First-order lag

Figure 7.8: Multiplicative uncertainty resulting from neglected dynamics

Rational approximations of (7.22) are given in (7.26) and (7.27) with $r_k = 0$.

2. Neglected lag. Let $f(s) = 1/(\tau_p s + 1)$, where $0 \leq \tau_p \leq \tau_{max}$. In this case the resulting $l_I(\omega)$, which is shown in Figure 7.8(b), can be represented by a rational transfer function with $|w_I(j\omega)| = l_I(\omega)$ where

$$w_I(s) = 1 - \frac{1}{\tau_{max} s + 1} = \frac{\tau_{max} s}{\tau_{max} s + 1}$$

This weight approaches 1 at high frequency, and the low-frequency asymptote crosses 1 at frequency $1/\tau_{max}$.

Example 7.5 Multiplicative weight for gain and delay uncertainty. *Consider the following set of plants*

$$G_p(s) = k_p e^{-\theta_p s} G_0(s); \quad k_p \in [k_{min}, k_{max}], \quad \theta_p \in [\theta_{min}, \theta_{max}] \tag{7.23}$$

which we want to represent by multiplicative uncertainty and a delay-free nominal model, $G(s) = \bar{k} G_0(s)$. Lundström (1994) derived the following exact expression for the relative uncertainty weight

$$l_I(\omega) = \begin{cases} \sqrt{r_k^2 + 2(1 + r_k)(1 - \cos(\theta_{max}\omega))} & \text{for } \omega < \pi/\theta_{max} \\ 2 + r_k & \text{for } \omega \geq \pi/\theta_{max} \end{cases} \tag{7.24}$$

where r_k is the relative uncertainty in the gain. This bound is irrational. To derive a rational weight we first approximate the delay by a first-order Padé approximation to get

$$k_{max} e^{-\theta_{max} s} - \bar{k} \approx \bar{k}(1 + r_k) \frac{1 - \frac{\theta_{max}}{2} s}{1 + \frac{\theta_{max}}{2} s} - \bar{k} = \bar{k} \frac{-\left(1 + \frac{r_k}{2}\right)\theta_{max} s + r_k}{\frac{\theta_{max}}{2} s + 1} \tag{7.25}$$

Since only the magnitude matters this may be represented by the following first-order weight

$$w_I(s) = \frac{(1 + \frac{r_k}{2})\theta_{max} s + r_k}{\frac{\theta_{max}}{2} s + 1} \tag{7.26}$$

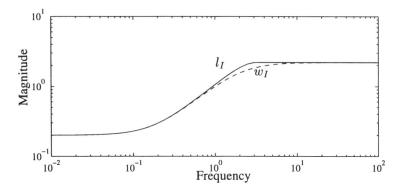

Figure 7.9: Multiplicative weight for gain and delay uncertainty in (7.23)

However, as seen from Figure 7.9 by comparing the dotted line (representing w_I) with the solid line (representing l_I), this weight w_I is somewhat optimistic (too small), especially around frequencies $1/\theta_{max}$. To make sure that $|w_I(j\omega)| \geq l_I(\omega)$ at all frequencies we apply a correction factor and get a third-order weight

$$w_I(s) = \frac{(1 + \frac{r_k}{2})\theta_{max}s + r_k}{\frac{\theta_{max}}{2}s + 1} \cdot \frac{\left(\frac{\theta_{max}}{2.363}\right)^2 s^2 + 2 \cdot 0.838 \cdot \frac{\theta_{max}}{2.363}s + 1}{\left(\frac{\theta_{max}}{2.363}\right)^2 s^2 + 2 \cdot 0.685 \cdot \frac{\theta_{max}}{2.363}s + 1} \qquad (7.27)$$

The improved weight $w_I(s)$ in (7.27) is not shown in Figure 7.9, but it would be almost indistinguishable from the exact bound given by the solid curve. In practical applications, it is suggested that one starts with a simple weight as in (7.26), and if it later appears important to eke out a little extra performance then one can try a higher-order weight as in (7.27).

7.4.6   Unmodelled dynamics uncertainty

Although we have spent a considerable amount of time on modelling uncertainty and deriving weights, we have not yet addressed the most important reason for using frequency domain ($\mathcal{H}_\infty$) uncertainty descriptions and complex perturbations, namely the incorporation of unmodelled dynamics. Of course, *unmodelled dynamics* is close to *neglected dynamics* which we have just discussed, but it is not quite the same. In unmodelled dynamics we also include unknown dynamics of unknown or even infinite order. To represent unmodelled dynamics we usually use a simple multiplicative weight of the form

$$w_I(s) = \frac{\tau s + r_0}{(\tau/r_\infty)s + 1} \qquad (7.28)$$

where r_0 is the relative uncertainty at steady-state, $1/\tau$ is (approximately) the frequency at which the relative uncertainty reaches 100%, and r_∞ is the magnitude of the weight at high frequency (typically, $r_\infty \geq 2$). Based on the above examples and discussions it is hoped that the reader has now accumulated the necessary insight to select reasonable values for the parameters r_0, r_∞ and τ for a specific application. The following exercise provides further support and gives a good review of the main ideas.

Exercise 7.1 *Suppose that the nominal model of a plant is*

$$G(s) = \frac{1}{s+1}$$

and the uncertainty in the model is parameterized by multiplicative uncertainty with the weight

$$w_I(s) = \frac{0.125s + 0.25}{(0.125/4)s + 1}$$

Call the resulting set Π. *Now find the extreme parameter values in each of the plants (a)-(g) below so that each plant belongs to the set* Π. *All parameters are assumed to be positive. One approach is to plot* $l_I(\omega) = |G^{-1}G' - 1|$ *in (7.15) for each* G' *(G_a, G_b, etc.) and adjust the parameter in question until* l_I *just touches* $|w_I(j\omega)|$.
   (a) Neglected delay: Find the largest θ *for* $G_a = Ge^{-\theta s}$ *(Answer: 0.13).*
   (b) Neglected lag: Find the largest τ *for* $G_b = G\frac{1}{\tau s+1}$ *(Answer: 0.15).*
   (c) Uncertain pole: Find the range of a *for* $G_c = \frac{1}{s+a}$ *(Answer: 0.8 to 1.33).*
   (d) Uncertain pole (time constant form): Find the range of T *for* $G_d = \frac{1}{Ts+1}$ *(Answer: 0.7 to 1.5).*
   (e) Neglected resonance: Find the range of ζ *for* $G_e = G\frac{1}{(s/70)^2 + 2\zeta(s/70)+1}$ *(Answer: 0.02 to 0.8).*
   (f) Neglected dynamics: Find the largest integer m *for* $G_f = G\left(\frac{1}{0.01s+1}\right)^m$ *(Answer: 13).*
   (g) Neglected RHP-zero: Find the largest τ_z *for* $G_g = G\frac{-\tau_z s+1}{\tau_z s+1}$ *(Answer: 0.07). These results imply that a control system which meets given stability and performance requirements for all plants in* Π, *is also guaranteed to satisfy the same requirements for the above plants* $G_a,$ $G_b, \ldots, G_g$.
   (h) Repeat the above with a new nominal plant $G = 1/(s-1)$ *(and with everything else the same except* $G_d = 1/(Ts-1)$). *(Answer: Same as above).*

Exercise 7.2 *Repeat Exercise 7.1 with a new weight,*

$$w_I(s) = \frac{s + 0.3}{(1/3)s + 1}$$

We end this section with a couple of remarks on uncertainty modelling:

1. We can usually get away with just one source of complex uncertainty for SISO systems.
2. With an $\mathcal{H}_\infty$ uncertainty description, it is possible to represent time delays (corresponding to an infinite-dimensional plant) and unmodelled dynamics of infinite order , using a nominal model and associated weights with finite order.

7.5   SISO Robust stability

We have so far discussed how to represent the uncertainty mathematically. In this section, we derive conditions which will ensure that the system remains stable for all perturbations in the uncertainty set, and then in the subsequent section we study robust performance.

7.5.1   RS with multiplicative uncertainty

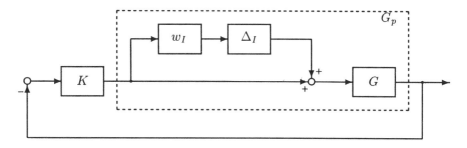

Figure 7.10: Feedback system with multiplicative uncertainty

We want to determine the stability of the uncertain feedback system in Figure 7.10 when there is multiplicative (relative) uncertainty of magnitude $|w_I(j\omega)|$. With uncertainty the loop transfer function becomes

$$L_p = G_p K = GK(1 + w_I \Delta_I) = L + w_I L \Delta_I, \quad |\Delta_I(j\omega)| \leq 1, \forall \omega \quad (7.29)$$

As always, we assume (by design) stability of the nominal closed-loop system (i.e. with $\Delta_I = 0$). For simplicity, we also assume that the loop transfer function L_p is stable. We now use the Nyquist stability condition to test for robust stability of the closed-loop system. We have

$$\text{RS} \quad \overset{\text{def}}{\Leftrightarrow} \quad \text{System stable } \forall L_p$$

$$\Leftrightarrow L_p \text{ should not encircle the point } -1, \quad \forall L_p \quad (7.30)$$

1. Graphical derivation of RS-condition. Consider the Nyquist plot of L_p as shown in Figure 7.11. Convince yourself that $|-1-L| = |1+L|$ is the distance from the point -1 to the centre of the disc representing L_p, and that $|w_I L|$ is the radius of the disc. Encirclements are avoided if none of the discs cover -1, and we get from Figure 7.11

$$\text{RS} \quad \Leftrightarrow \quad |w_I L| < |1 + L|, \quad \forall \omega \quad (7.31)$$

$$\Leftrightarrow \quad \left|\frac{w_I L}{1 + L}\right| < 1, \forall \omega \quad \Leftrightarrow \quad |w_I T| < 1, \forall \omega \quad (7.32)$$

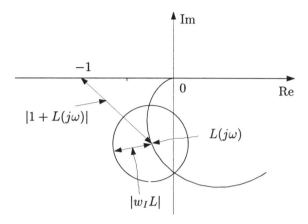

Figure 7.11: Nyquist plot of L_p for robust stability

$$\overset{\text{def}}{\Leftrightarrow} \quad \|w_I T\|_\infty < 1 \tag{7.33}$$

Note that for SISO systems $w_I = w_O$ and $T = T_I = GK(1 + GK)^{-1}$, so the condition could equivalently be written in terms of $w_I T_I$ or $w_O T$. Thus, the requirement of robust stability for the case with multiplicative uncertainty gives an upper bound on the complementary sensitivity:

$$\boxed{\text{RS} \Leftrightarrow |T| < 1/|w_I|, \quad \forall\omega} \tag{7.34}$$

We see that we have to detune the system (i.e. make T small) at frequencies where the relative uncertainty $|w_I|$ exceeds 1 in magnitude. Condition (7.34) is *exact* (necessary and sufficient) provided there exist uncertain plants such that at each frequency all perturbations satisfying $|\Delta(j\omega)| \leq 1$ are possible. If this is not the case, then (7.34) is only *sufficient* for RS, e.g. this is the case if the perturbation is restricted to be real, as for the parametric gain uncertainty in (7.6).

Example 7.6 *Consider the following nominal plant and PI-controller*

$$G(s) = \frac{3(-2s + 1)}{(5s + 1)(10s + 1)} \quad K(s) = K_c \frac{12.7s + 1}{12.7s}$$

Recall that this is the inverse response process from Chapter 2. Initially, we select $K_c = K_{c1} = 1.13$ as suggested by the Ziegler-Nichols' tuning rule. It results in a nominally stable closed-loop system. Suppose that one "extreme" uncertain plant is $G'(s) = 4(-3s + 1)/(4s + 1)^2$. For this plant the relative error $|(G' - G)/G|$ is 0.33 at low frequencies; it is 1 at about 0.1 rad/s, and it is 5.25 at high frequencies. Based on this and (7.28) we choose the following uncertainty weight

$$w_I(s) = \frac{10s + 0.33}{(10/5.25)s + 1}$$

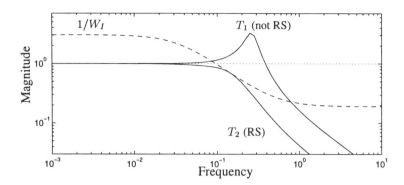

Figure 7.12: Checking robust stability with multiplicative uncertainty

which closely matches this relative error. We now want to evaluate whether the system remains stable for all possible plants as given by $G_p = G(1 + w_I \Delta_I)$ where $\Delta_I(s)$ is any perturbation satisfying $\|\Delta\|_\infty \leq 1$. This is not *the case as seen from Figure 7.12 where we see that the magnitude of the nominal complementary sensitivity function $T_1 = GK_1/(1+GK_1)$ exceeds the bound $1/|w_I|$ from about 0.1 to 1 rad/s, so (7.34) is not satisfied. To achieve robust stability we need to reduce the controller gain. By trial and error we find that reducing the gain to $K_{c2} = 0.31$ just achieves RS, as is seen from the curve for $T_2 = GK_2/(1 + GK_2)$ in Figure 7.12.*

Remark. *For the "extreme" plant $G'(s)$ we find as expected that the closed-loop system is unstable with $K_{c1} = 1.13$. However, with $K_{c2} = 0.31$ the system is stable with reasonable margins (and not at the limit of instability as one might have expected); we can increase the gain by almost a factor of two to $K_c = 0.58$ before we get instability. This illustrates that condition (7.34) is only a* sufficient *condition for stability, and a violation of this bound does not imply instability for a specific plant G'. However, with $K_{c2} = 0.31$ there exists an allowed Δ_I and a corresponding $G_p = G(1 + w_I\Delta_I)$ which yields $T_{2p} = \frac{G_p K_2}{1+G_p K_2}$ on the limit of instability.*

2. Algebraic derivation of RS-condition. Since L_p is assumed stable, and the nominal closed-loop is stable, the nominal loop transfer function $L(j\omega)$ does not encircle -1. Therefore, since the set of plants is norm-bounded, it then follows that if some L_{p1} in the uncertainty set encircles -1, then there must be another L_{p2} in the uncertainty set which goes exactly through -1 at some frequency. Thus,

$$\text{RS} \quad \Leftrightarrow \quad |1 + L_p| \neq 0, \quad \forall L_p, \forall \omega \tag{7.35}$$

$$\Leftrightarrow \quad |1 + L_p| > 0, \quad \forall L_p, \forall \omega \tag{7.36}$$

$$\Leftrightarrow \quad |1 + L + w_I L \Delta_I| > 0, \quad \forall |\Delta_I| \leq 1, \forall \omega \tag{7.37}$$

At each frequency the last condition is most easily violated (the worst case) when the complex number $\Delta_I(j\omega)$ is selected with $|\Delta_I(j\omega)| = 1$ and with phase such that the terms $(1 + L)$ and $w_I L \Delta_I$ have opposite signs (point in the opposite direction). Thus

$$\text{RS} \Leftrightarrow |1 + L| - |w_I L| > 0, \quad \forall \omega \quad \Leftrightarrow |w_I T| < 1, \quad \forall \omega \quad (7.38)$$

and we have rederived (7.33).

Remark. Unstable plants. The stability condition (7.33) also applies to the case when L and L_p are unstable as long as the number of RHP-poles remains the same for each plant in the uncertainty set. This follows since the nominal closed-loop system is assumed stable, so we must make sure that the perturbation does not change the number of encirclements, and (7.33) is the condition which guarantees this.

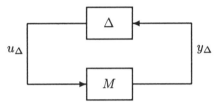

Figure 7.13: $M\Delta$-structure

3. $M\Delta$-structure derivation of RS-condition. This derivation is a preview of a general analysis presented in the next chapter. The reader should not be too concerned if he or she does not fully understand the details at this point. The derivation is based on applying the Nyquist stability condition to an alternative "loop transfer function" $M\Delta$ rather than L_p. The argument goes as follows. If the nominal ($\Delta_I = 0$) feedback system is stable then the stability of the system in Figure 7.10 is equivalent to stability of the system in Figure 7.13, where $\Delta = \Delta_I$ and

$$M = w_I K (1 + GK)^{-1} G = w_I T \quad (7.39)$$

is the transfer function from the output of Δ_I to the input of Δ_I. Notice that the only source of instability in Figure 7.10 is the new feedback loop created by Δ_I. We now apply the Nyquist stability condition to the system in Figure 7.13. We assume that Δ and $M = w_I T$ are stable; the former implies that G and G_p must have the same unstable poles, the latter is equivalent to assuming nominal stability of the closed-loop system. The Nyquist stability condition then determines RS if and only if the "loop transfer function" $M\Delta$ does not encircle -1 for all Δ. Thus,

$$\text{RS} \quad \Leftrightarrow \quad |1 + M\Delta| > 0, \quad \forall \omega, \forall |\Delta| \le 1 \quad (7.40)$$

The last condition is most easily violated (the worst case) when Δ is selected at each frequency such that $|\Delta| = 1$ and the terms $M\Delta$ and 1 have opposite signs (point in the opposite direction). We therefore get

$$\text{RS} \quad \Leftrightarrow \quad 1 - |M(j\omega)| > 0, \quad \forall\omega \tag{7.41}$$

$$\Leftrightarrow \quad |M(j\omega)| < 1, \quad \forall\omega \tag{7.42}$$

which is the same as (7.33) and (7.38) since $M = w_I T$. The $M\Delta$-structure provides a very general way of handling robust stability, and we will discuss this at length in the next chapter where we will see that (7.42) is essentially a clever application of the small gain theorem where we avoid the usual conservatism since any phase in $M\Delta$ is allowed.

7.5.2   Comparison with gain margin

By what factor, $k_{\max}$, can we multiply the loop gain, $L_0 = G_0 K$, before we get instability? In other words, given

$$L_p = k_p L_0; \quad k_p \in [1, k_{\max}] \tag{7.43}$$

find the largest value of $k_{\max}$ such that the closed-loop system is stable.

1. Real perturbation. The exact value of $k_{\max}$, which is obtained with Δ real, is the gain margin (GM) from classical control. We have (recall (2.33))

$$k_{\max,1} = GM = \frac{1}{|L_0(j\omega_{180})|} \tag{7.44}$$

where ω_{180} is the frequency where $\angle L_0 = -180°$.

2. Complex perturbation. Alternatively, represent the gain uncertainty as complex multiplicative uncertainty,

$$L_p = k_p L_0 = \bar{k} L_0 (1 + r_k \Delta) \tag{7.45}$$

where

$$\bar{k} = \frac{k_{\max} + 1}{2}, \quad r_k = \frac{k_{\max} - 1}{k_{\max} + 1} \tag{7.46}$$

Note that the nominal $L = \bar{k} L_0$ is not fixed, but depends on $k_{\max}$. The robust stability condition $\|w_I T\|_\infty < 1$ (which is derived for complex Δ) with $w_I = r_k$ then gives

$$\left\| r_k \frac{\bar{k} L_0}{1 + \bar{k} L_0} \right\|_\infty < 1 \tag{7.47}$$

Here both r_k and $\bar{k}$ depend on $k_{\max}$, and (7.47) must be solved iteratively to find $k_{\max,2}$. Condition (7.47) would be exact if Δ were complex, but since it is not we expect $k_{\max,2}$ to be somewhat smaller than GM.

Example 7.7 *To check this numerically consider a system with $L_0 = \frac{1}{s}\frac{-s+2}{s+2}$. We find $\omega_{180} = 2$ [rad/s] and $|L_0(j\omega_{180})| = 0.5$, and the exact factor by which we can increase the loop gain is, from (7.44), $k_{\max,1} = GM = 2$. On the other hand, use of (7.47) yields $k_{\max,2} = 1.78$, which as expected is less than GM=2. This illustrates the conservatism involved in replacing a real perturbation by a complex one.*

Exercise 7.3 *Represent the gain uncertainty in (7.43) as multiplicative complex uncertainty with nominal model $G = G_0$ (rather than $G = \bar{k}G_0$ used above).*

(a) Find w_I and use the RS-condition $\|w_I T\|_\infty < 1$ to find $k_{\max,3}$. Note that no iteration is needed in this case since the nominal model and thus $T = T_0$ is independent of $k_{\max}$.

(b) One expects $k_{\max,3}$ to be even more conservative than $k_{\max,2}$ since this uncertainty description is not even tight when Δ is real. Show that this is indeed the case using the numerical values from Example 7.7.

7.5.3   RS with inverse multiplicative uncertainty

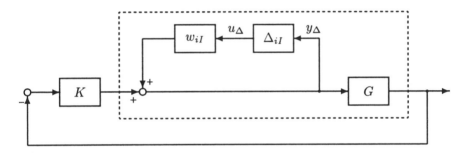

Figure 7.14: Feedback system with inverse multiplicative uncertainty

We will derive a corresponding RS-condition for a feedback system with inverse multiplicative uncertainty (see Figure 7.14) in which

$$G_p = G(1 + w_{iI}(s)\Delta_{iI})^{-1} \qquad (7.48)$$

Algebraic derivation. Assume for simplicity that the loop transfer function L_p is stable, and assume stability of the nominal closed-loop system. Robust stability is then guaranteed if encirclements by $L_p(j\omega)$ of the point -1 are avoided, and since L_p is in a norm-bounded set we have

$$\text{RS} \quad \Leftrightarrow \quad |1 + L_p| > 0, \quad \forall L_p, \forall \omega \qquad (7.49)$$

$$\Leftrightarrow \quad |1 + L(1 + w_{iI}\Delta_{iI})^{-1}| > 0, \quad \forall|\Delta_{iI}| \le 1, \forall \omega \qquad (7.50)$$

$$\Leftrightarrow \quad |1 + w_{iI}\Delta_{iI} + L| > 0, \quad \forall|\Delta_{iI}| \le 1, \forall \omega \qquad (7.51)$$

The last condition is most easily violated (the worst case) when Δ_{iI} is selected at each frequency such that $|\Delta_{iI}| = 1$ and the terms $1 + L$ and $w_{iI}\Delta_{iI}$ have opposite signs (point in the opposite direction). Thus

$$\text{RS} \quad \Leftrightarrow \quad |1 + L| - |w_{iI}| > 0, \quad \forall\omega \tag{7.52}$$

$$\Leftrightarrow \quad |w_{iI}S| < 1, \quad \forall\omega \tag{7.53}$$

Remark. In this derivation we have assumed that L_p is stable, but this is not necessary as one may show by deriving the condition using the $M\Delta$-structure. Actually, the RS-condition (7.53) applies even when the number of RHP-poles of G_p can change.

Control implications. From (7.53) we find that the requirement of robust stability for the case with inverse multiplicative uncertainty gives an upper bound on the sensitivity,

$$\boxed{\text{RS} \quad \Leftrightarrow \quad |S| < 1/|w_{iI}|, \quad \forall\omega} \tag{7.54}$$

We see that we need tight control and have to make S small at frequencies where the uncertainty is large and $|w_{iI}|$ exceeds 1 in magnitude. This may be somewhat surprising since we intuitively expect to have to detune the system (and make $S \approx 1$) when we have uncertainty, while this condition tells us to do the opposite. The reason is that this uncertainty represents pole uncertainty, and at frequencies where $|w_{iI}|$ exceeds 1 we allow for poles crossing from the left to the right-half plane (G_p becoming unstable), and we then know that we need feedback ($|S| < 1$) in order to stabilize the system.

However, $|S| < 1$ may not always be possible. In particular, assume that the plant has a RHP-zero at $s = z$. Then we have the interpolation constraint $S(z) = 1$ and we must as a prerequisite for RS, $\|w_{iI}S\|_\infty < 1$, require that $w_{iI}(z) \leq 1$ (recall the maximum modulus theorem, see (5.15)). Thus, we cannot have large pole uncertainty with $|w_{iI}(j\omega)| > 1$ (and hence the possibility of instability) at frequencies where the plant has a RHP-zero. This is consistent with the results we obtained in Section 5.9.

7.6   SISO Robust performance

7.6.1   SISO nominal performance in the Nyquist plot

Consider performance in terms of the weighted sensitivity function as discussed in Section 2.7.2. The condition for nominal performance (NP) is then

$$\text{NP} \quad \Leftrightarrow \quad |w_P S| < 1 \;\; \forall\omega \quad \Leftrightarrow \quad |w_P| < |1 + L| \;\; \forall\omega \tag{7.55}$$

Now $|1 + L|$ represents at each frequency the distance of $L(j\omega)$ from the point -1 in the Nyquist plot, so $L(j\omega)$ must be at least a distance of $|w_P(j\omega)|$ from -1. This is illustrated graphically in Figure 7.15, where we see that for NP, $L(j\omega)$ must stay outside a disc of radius $|w_P(j\omega)|$ centred on -1.

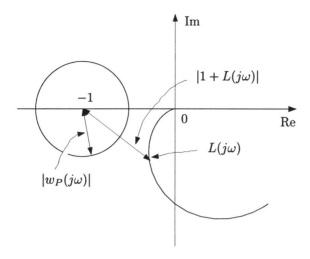

Figure 7.15: Nyquist plot illustration of nominal performance condition $|w_P| < |1 + L|$

7.6.2 Robust performance

For robust performance we require the performance condition (7.55) to be satisfied for *all* possible plants, that is, including the worst-case uncertainty.

$$\text{RP} \quad \overset{\text{def}}{\Leftrightarrow} \quad |w_P S_p| < 1 \quad \forall S_p, \forall \omega \tag{7.56}$$

$$\Leftrightarrow \quad |w_P| < |1 + L_p| \quad \forall L_p, \forall \omega \tag{7.57}$$

This corresponds to requiring $|\widehat{y}/d| < 1 \ \forall \Delta_I$ in Figure 7.16, where we consider multiplicative uncertainty, and the set of possible loop transfer functions is

$$L_p = G_p K = L(1 + w_I \Delta_I) = L + w_I L \Delta_I \tag{7.58}$$

1. Graphical derivation of RP-condition. Condition (7.57) is illustrated graphically by the Nyquist plot in Figure 7.17. For RP we must require that all possible

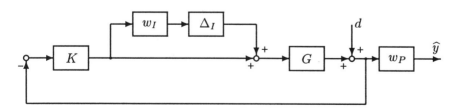

Figure 7.16: Diagram for robust performance with multiplicative uncertainty

$L_p(j\omega)$ stay outside a disc of radius $|w_P(j\omega)|$ centred on -1. Since L_p at each frequency stays within a disc of radius $w_I L$ centred on L, we see from Figure 7.17 that the condition for RP is that the two discs, with radii $|w_P|$ and $|w_I L|$, do not overlap. Since their centres are located a distance $|1 + L|$ apart, the RP-condition becomes

$$\text{RP} \quad \Leftrightarrow \quad |w_P| + |w_I L| < |1 + L|, \quad \forall\omega \tag{7.59}$$

$$\Leftrightarrow \quad |w_P(1 + L)^{-1}| + |w_I L(1 + L)^{-1}| < 1, \quad \forall\omega \tag{7.60}$$

or in other words

$$\boxed{\text{RP} \quad \Leftrightarrow \quad \max_\omega \left(|w_P S| + |w_I T|\right) < 1} \tag{7.61}$$

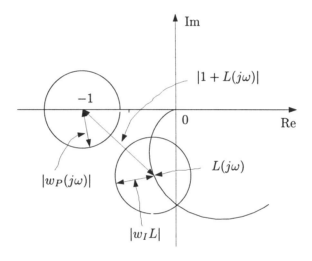

Figure 7.17: Nyquist plot illustration of robust performance condition $|w_P| < |1 + L_p|$

2. Algebraic derivation of RP-condition. From the definition in (7.56) we have that RP is satisfied if the worst-case (maximum) weighted sensitivity at each frequency is less than 1, that is,

$$\text{RP} \quad \Leftrightarrow \quad \max_{S_p} |w_P S_p| < 1, \quad \forall\omega \tag{7.62}$$

(strictly speaking, max should be replaced by sup, the supremum). The perturbed sensitivity is $S_p = (I + L_p)^{-1} = 1/(1 + L + w_I L \Delta_I)$, and the worst-case (maximum) is obtained at each frequency by selecting $|\Delta_I| = 1$ such that the terms $(1 + L)$ and $w_I L \Delta_I$ (which are complex numbers) point in opposite directions. We get

$$\max_{S_p} |w_P S_p| = \frac{|w_P|}{|1 + L| - |w_I L|} = \frac{|w_P S|}{1 - |w_I T|} \tag{7.63}$$

and by substituting (7.63) into (7.62) we rederive the RP-condition in (7.61).

Remarks on RP-condition (7.61).

1. The RP-condition (7.61) for this problem is closely approximated by the following mixed sensitivity $\mathcal{H}_\infty$ condition:

$$\left\| \begin{matrix} w_P S \\ w_I T \end{matrix} \right\|_\infty = \max_\omega \sqrt{|w_P S|^2 + |w_I T|^2} < 1 \qquad (7.64)$$

To be more precise, we find from (A.95) that condition (7.64) is within a factor of at most $\sqrt{2}$ to condition (7.61). This means that for SISO systems we can closely approximate the RP-condition in terms of an $\mathcal{H}_\infty$ problem, so there is little need to make use of the structured singular value. However, we will see in the next chapter that the situation can be very different for MIMO systems.

2. The RP-condition (7.61) can be used to derive bounds on the loop shape $|L|$. At a given frequency we have that $|w_P S| + |w_I T| < 1$ (RP) is satisfied if (see Exercise 7.4)

$$|L| > \frac{1 + |w_P|}{1 - |w_I|}, \quad \text{(at frequencies where } |w_I| < 1) \qquad (7.65)$$

or if

$$|L| < \frac{1 - |w_P|}{1 + |w_I|}, \quad \text{(at frequencies where } |w_P| < 1) \qquad (7.66)$$

Conditions (7.65) and (7.66) may be combined over different frequency ranges. Condition (7.65) is most useful at low frequencies where generally $|w_I| < 1$ and $|w_P| > 1$ (tight performance requirement) and we need $|L|$ large. Conversely, condition (7.66) is most useful at high frequencies where generally $|w_I| > 1$, (more than 100% uncertainty), $|w_P| < 1$ and we need L small. The loop-shaping conditions (7.65) and (7.66) may in the general case be obtained numerically from μ-conditions as outlined in Remark 13 on page 317. This is discussed by Braatz et al. (1996) who derive bounds also in terms of S and T, and furthermore derive necessary bounds for RP in addition to the sufficient bounds in (7.65) and (7.66); see also Exercise 7.5.

3. The term $\mu(N_{RP}) = |w_P S| + |w_I T|$ in (7.61) is the *structured singular value* (μ) for RP for this particular problem; see (8.128). We will discuss μ in much more detail in the next chapter.

4. The structured singular value μ is *not* equal to the worst-case weighted sensitivity, $\max_{S_p} |w_P S_p|$, given in (7.63) (although many people seem to think it is). The worst-case weighted sensitivity is equal to skewed-μ (μ^s) with fixed uncertainty; see Section 8.10.3. Thus, in summary we have for this particular robust performance problem:

$$\mu = |w_P S| + |w_I T|, \quad \mu^s = \frac{|w_P S|}{1 - |w_I T|} \qquad (7.67)$$

Note that μ and μ^s are closely related since $\mu \le 1$ if and only if $\mu^s \le 1$.

Exercise 7.4 *Derive the loop-shaping bounds in (7.65) and (7.66) which are sufficient for* $|w_P S| + |w_I T| < 1$ *(RP).* **Hint:** *Start from the RP-condition in the form* $|w_P| + |w_I L| < |1 + L|$ *and use the facts that* $|1 + L| \ge 1 - |L|$ *and* $|1 + L| \ge |L| - 1$.

Exercise 7.5 *Also derive, from $|w_P S| + |w_I T| < 1$, the following necessary bounds for RP (which must be satisfied)*

$$|L| > \frac{|w_P| - 1}{1 - |w_I|}, \quad (\text{for } |w_P| > 1 \text{ and } |w_I| < 1)$$

$$|L| < \frac{1 - |w_P|}{|w_I| - 1}, \quad (\text{for } |w_P| < 1 \text{ and } |w_I| > 1)$$

Hint: *Use* $|1 + L| \leq 1 + |L|$.

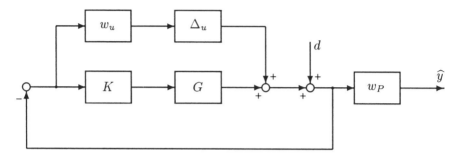

Figure 7.18: Diagram for robust performance in Example 7.8

Example 7.8 Robust performance problem. *Consider robust performance of the SISO system in Figure 7.18, for which we have*

$$\text{RP} \overset{\text{def}}{\Leftrightarrow} \left| \frac{\widehat{y}}{d} \right| < 1, \ \forall |\Delta_u| \leq 1, \ \forall \omega; \quad w_P(s) = 0.25 + \frac{0.1}{s}; \ w_u(s) = r_u \frac{s}{s+1} \quad (7.68)$$

(a) Derive a condition for robust performance (RP).

(b) For what values of r_u is it impossible to satisfy the robust performance condition?

(c) Let $r_u = 0.5$. Consider two cases for the nominal loop transfer function: 1) $GK_1(s) = 0.5/s$ and 2) $GK_2(s) = \frac{0.5}{s} \frac{1-s}{1+s}$. For each system, sketch the magnitudes of S and its performance bound as a function of frequency. Does each system satisfy robust performance?

Solution. *(a) The requirement for RP is $|w_P S_p| < 1$, $\forall S_p, \forall \omega$, where the possible sensitivities are given by*

$$S_p = \frac{1}{1 + GK + w_u \Delta_u} = \frac{S}{1 + w_u \Delta_u S} \quad (7.69)$$

The condition for RP then becomes

$$\text{RP} \quad \Leftrightarrow \quad \left| \frac{w_P S}{1 + w_u \Delta_u S} \right| < 1, \quad \forall \Delta_u, \forall \omega \quad (7.70)$$

A simple analysis shows that the worst case corresponds to selecting Δ_u with magnitude 1 such that the term $w_u \Delta_u S$ is purely real and negative, and hence we have

$$\text{RP} \quad \Leftrightarrow \quad |w_P S| < 1 - |w_u S|, \quad \forall \omega \quad (7.71)$$

$$\Leftrightarrow \quad |w_P S| + |w_u S| < 1, \quad \forall \omega \quad (7.72)$$

$$\Leftrightarrow \quad |S(jw)| < \frac{1}{|w_P(jw)| + |w_u(jw)|}, \quad \forall \omega \quad (7.73)$$

(b) Since any real system is strictly proper we have $|S| = 1$ at high frequencies and therefore we must require $|w_u(j\omega)| + |w_P(j\omega)| < 1$ as $\omega \to \infty$. With the weights in (7.68) this is equivalent to $r_u + 0.25 < 1$. Therefore, we must at least require $r_u < 0.75$ for RP, so RP cannot be satisfied if $r_u \geq 0.75$.

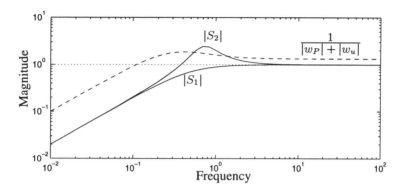

Figure 7.19: Robust performance test

(c) Design S_1 yields RP, while S_2 does not. This is seen by checking the RP-condition (7.73) graphically as shown in Figure 7.19; $|S_1|$ has a peak of 1 while $|S_2|$ has a peak of about 2.45.

7.6.3   The relationship between NP, RS and RP

Consider a SISO system with multiplicative uncertainty, and assume that the closed-loop is nominally stable (NS). The conditions for nominal performance (NP), robust stability (RS) and robust performance (RP) can then be summarized as follows

$$\text{NP} \quad \Leftrightarrow \quad |w_P S| < 1, \forall \omega \tag{7.74}$$

$$\text{RS} \quad \Leftrightarrow \quad |w_I T| < 1, \forall \omega \tag{7.75}$$

$$\text{RP} \quad \Leftrightarrow \quad |w_P S| + |w_I T| < 1, \forall \omega \tag{7.76}$$

From this we see that a prerequisite for RP is that we satisfy NP and RS. This applies in general, both for SISO and MIMO systems and for any uncertainty. In addition, for SISO systems, if we satisfy both RS and NP, then we have at each frequency

$$|w_P S| + |w_I T| \leq 2 \max\{|w_P S|, |w_I T|\} < 2 \tag{7.77}$$

It then follows that, within a factor of at most 2, we will automatically get RP when the subobjectives of NP and RS are satisfied. Thus, RP is not a "big issue" for SISO systems, and this is probably the main reason why there is little discussion about robust performance in the classical control literature. On the other hand, as we will see in the next chapter, for MIMO systems we may get very poor RP even though the subobjectives of NP and RS are individually satisfied.

To satisfy RS we generally want T small, whereas to satisfy NP we generally want S small. However, we cannot make *both* S and T small at the same frequency because of the identity $S + T = 1$. This has implications for RP, since $|w_P||S| + |w_I||T| \geq \min\{|w_P|, |w_I|\}(|S| + |T|)$, where $|S| + |T| \geq |S + T| = 1$, and we derive at each frequency

$$|w_P S| + |w_I T| \geq \min\{|w_P|, |w_I|\} \qquad (7.78)$$

We conclude that we *cannot have both* $|w_P| > 1$ *(i.e. good performance) and* $|w_I| > 1$ *(i.e. more than* 100% *uncertainty) at the same frequency.* One explanation for this is that at frequencies where $|w_I| > 1$ the uncertainty will allow for RHP-zeros, and we know that we cannot have tight performance in the presence of RHP-zeros.

7.6.4   The similarity between RS and RP

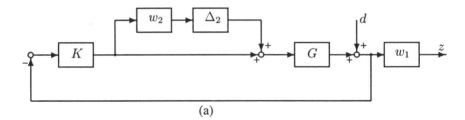

(a)

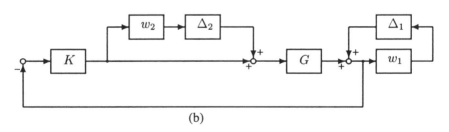

(b)

Figure 7.20: (a) Robust performance with multiplicative uncertainty
(b) Robust stability with combined multiplicative and inverse multiplicative uncertainty

Robust performance may be viewed as a special case of robust stability (with muptiple perturbations). To see this consider the following two cases as illustrated in Figure 7.20:

(a) RP with multiplicative uncertainty
(b) RS with combined multiplicative and inverse multiplicative uncertainty

As usual the uncertain perturbations are normalized such that $\|\Delta_1\|_\infty \leq 1$ and $\|\Delta_2\|_\infty \leq 1$. Since we use the $\mathcal{H}_\infty$ norm to define both uncertainty and performance and since the weights in Figures 7.20(a) and (b) are the same, the tests for RP and RS in cases (a) and (b), respectively, are identical. This may be argued from the block diagrams, or by simply evaluating the conditions for the two cases as shown below.

(a) The condition for RP with multiplicative uncertainty was derived in (7.61), but with w_1 replaced by w_P and with w_2 replaced by w_I. We found

$$\text{RP} \quad \Leftrightarrow \quad |w_1 S| + |w_2 T| < 1, \quad \forall \omega \qquad (7.79)$$

(b) We will now derive the RS-condition for the case where L_p is stable (this assumption may be relaxed if the more general $M\Delta$-structure and the structured singular value are used). We want the system to be closed-loop stable for all possible Δ_1 and Δ_2. RS is equivalent to avoiding encirclements of -1 by the Nyquist plot of L_p. That is, the distance between L_p and -1 must be larger than zero, i.e. $|1 + L_p| > 0$, and therefore

$$\text{RS} \quad \Leftrightarrow \quad |1 + L_p| > 0 \quad \forall L_p, \forall \omega \qquad (7.80)$$
$$\Leftrightarrow \quad |1 + L(1 + w_2\Delta_2)(1 - w_1\Delta_1)^{-1}| > 0, \quad \forall \Delta_1, \forall \Delta_2, \forall \omega \quad (7.81)$$
$$\Leftrightarrow \quad |1 + L + Lw_2\Delta_2 - w_1\Delta_1| > 0, \quad \forall \Delta_1, \forall \Delta_2, \forall \omega \qquad (7.82)$$

Here the worst case is obtained when we choose Δ_1 and Δ_2 with magnitudes 1 such that the terms $Lw_2\Delta_2$ and $w_1\Delta_1$ are in the opposite direction of the term $1 + L$. We get

$$\text{RS} \quad \Leftrightarrow \quad |1 + L| - |Lw_2| - |w_1| > 0, \quad \forall \omega \qquad (7.83)$$
$$\Leftrightarrow \quad |w_1 S| + |w_2 T| < 1, \quad \forall \omega \qquad (7.84)$$

which is the same condition as found for RP.

7.7 Examples of parametric uncertainty

We now provide some further examples of how to represent parametric uncertainty. The perturbations Δ must be real to exactly represent parametric uncertainty.

7.7.1 Parametric pole uncertainty

Consider uncertainty in the parameter a in a state space model, $\dot{y} = ay + bu$, corresponding to the uncertain transfer function $G_p(s) = b/(s - a_p)$. More generally, consider the following set of plants

$$G_p(s) = \frac{1}{s - a_p} G_0(s); \quad a_{\min} \leq a_p \leq a_{\max} \qquad (7.85)$$

If $a_{\min}$ and $a_{\max}$ have different signs then this means that the plant can change from stable to unstable with the pole crossing through the origin (which happens in some applications). This set of plants can be written as

$$G_p = \frac{G_0(s)}{s - \bar{a}(1 + r_a\Delta)}; \quad -1 \le \Delta \le 1 \tag{7.86}$$

which can be exactly described by inverse multiplicative uncertainty as in (7.118) with nominal model $G = G_0(s)/(s - \bar{a})$ and

$$w_{iI}(s) = \frac{r_a\bar{a}}{s - \bar{a}} \tag{7.87}$$

The magnitude of the weight $w_{iI}(s)$ is equal to r_a at low frequencies. If r_a is larger than 1 then the plant can be both stable and unstable. As seen from the RS-condition in (7.53), a value of $|w_{iI}|$ larger than 1 means that $|S|$ must be less than 1 at the same frequency, which is consistent with the fact that we need feedback (S small) to stabilize an unstable plant.

Time constant form. It is also interesting to consider another form of pole uncertainty, namely that associated with the time constant:

$$G_p(s) = \frac{1}{\tau_p s + 1} G_0(s); \quad \tau_{\min} \le \tau_p \le \tau_{\max} \tag{7.88}$$

This results in uncertainty in the pole location, but the set of plants is entirely different from that in (7.85). The reason is that in (7.85) the uncertainty affects the model at low frequency, whereas in (7.88) the uncertainty affects the model at high frequency. The corresponding uncertainty weight as derived in (7.8) is

$$w_{iI}(s) = \frac{r_\tau \bar{\tau} s}{1 + \bar{\tau} s} \tag{7.89}$$

This weight is zero at $\omega = 0$ and approaches r_τ at high frequency, whereas the weight w_{iI} in (7.87) is r_a at $\omega = 0$ and approaches zero at high frequencies.

7.7.2 Parametric zero uncertainty

Consider zero uncertainty in the "time constant" form, as in

$$G_p(s) = (1 + \tau_p s) G_0(s); \quad \tau_{\min} \le \tau_p \le \tau_{\max} \tag{7.90}$$

where the remaining dynamics $G_0(s)$ are as usual assumed to have no uncertainty. For example, let $-1 \le \tau_p \le 3$. Then the possible zeros $z_p = -1/\tau_p$ cross from the LHP to the RHP through infinity: $z_p \le -1/3$ (in LHP) and $z_p \ge 1$ (in RHP). The set of plants in (7.90) may be written as multiplicative (relative) uncertainty with

$$w_I(s) = r_\tau \bar{\tau} s / (1 + \bar{\tau} s) \tag{7.91}$$

The magnitude $|w_I(j\omega)|$ is small at low frequencies, and approaches r_τ (the relative uncertainty in τ) at high frequencies. For cases with $r_\tau > 1$ we allow the zero to cross from the LHP to the RHP (through infinity).

Exercise 7.6 Parametric zero uncertainty in zero form. *Consider the following alternative form of parametric zero uncertainty*

$$G_p(s) = (s + z_p)G_0(s); \quad z_{\min} \le z_p \le z_{\max} \tag{7.92}$$

which caters for zeros crossing from the LHP to the RHP through the origin (corresponding to a sign change in the steady-state gain). Show that the resulting multiplicative weight is $w_I(s) = r_z \bar{z}/(s + \bar{z})$ and explain why the set of plants given by (7.92) is entirely different from that with the zero uncertainty in "time constant" form in (7.90). Explain what the implications are for control if $r_z > 1$.

Remark. Both of the two zero uncertainty forms, (7.90) and (7.92), can occur in practice. An example of the zero uncertainty form in (7.92), which allows for changes in the steady-state gain, is given in Example 7.10.

7.7.3 Parametric state-space uncertainty

We here introduce the reader to parametric state-space uncertainty by way of a few examples. A general procedure for handling this kind of uncertainty is given by Packard (1988). Consider an uncertain state-space model

$$\dot{x} = A_p x + B_p u \tag{7.93}$$
$$y = C_p x + D_p u \tag{7.94}$$

or equivalently

$$G_p(s) = C_p(sI - A_p)^{-1}B_p + D_p \tag{7.95}$$

Assume that the underlying cause for the uncertainty is uncertainty in some real parameters $\delta_1, \delta_2, \ldots$ (these could be temperature, mass, volume, etc.), and assume in the simplest case that the state-space matrices depend linearly on these parameters i.e.

$$A_p = A + \sum \delta_i A_i, \; B_p = B + \sum \delta_i B_i \; C_p = C + \sum \delta_i C_i, \; D_p = D + \sum \delta_i D_i \tag{7.96}$$

where A, B, C and D model the nominal system. This description has multiple perturbations, so it cannot be represented by a single perturbation, but it should be fairly clear that we can separate out the perturbations affecting A, B, C and D, and then collect them in a large diagonal matrix Δ with the real δ_i's along its diagonal. Some of the δ_i's may have to be repeated. For example, we may write

$$A_p = A + \sum \delta_i A_i = A + W_2 \Delta W_1 \tag{7.97}$$

where Δ is diagonal with the δ_i's along its diagonal. Introduce $\Phi(s) \triangleq (sI - A)^{-1}$, and we get

$$(sI - A_p)^{-1} = (sI - A - W_2 \Delta W_1)^{-1} = (I - \Phi(s)W_2 \Delta W_1)^{-1}\Phi(s) \quad (7.98)$$

This is illustrated in the block diagram of Figure 7.21, which is in the form of an inverse additive perturbation (see Figure 8.5(d)).

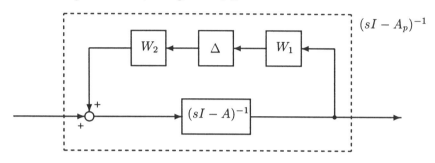

Figure 7.21: Uncertainty in state space A-matrix

Example 7.9 *Suppose A_p is a function of two parameters $k_p = 1 + w_1 \delta_1$ $(-1 \leq \delta_1 \leq 1)$ and $\alpha_p = 3 + w_2 \delta_2$ $(-1 \leq \delta_2 \leq 1)$ as follows:*

$$A_p = \begin{bmatrix} -2 - k_p & k_p - \alpha_p \\ k_p + 2\alpha_p & -k_p \end{bmatrix} \quad (7.99)$$

Then

$$A_p = \underbrace{\begin{bmatrix} -3 & -2 \\ 7 & -1 \end{bmatrix}}_{A} + \delta_1 \underbrace{\begin{bmatrix} -w_1 & w_1 \\ w_1 & -w_1 \end{bmatrix}}_{A_1} + \delta_2 \underbrace{\begin{bmatrix} 0 & -w_2 \\ 2w_2 & 0 \end{bmatrix}}_{A_2} \quad (7.100)$$

$$= A + \underbrace{\begin{bmatrix} -w_1 & 0 & -w_2 \\ w_1 & 2w_2 & 0 \end{bmatrix}}_{W_2} \underbrace{\begin{bmatrix} \delta_1 & 0 & 0 \\ 0 & \delta_2 & 0 \\ 0 & 0 & \delta_2 \end{bmatrix}}_{\Delta} \underbrace{\begin{bmatrix} 1 & -1 \\ 1 & 0 \\ 0 & 1 \end{bmatrix}}_{W_1} \quad (7.101)$$

Note that δ_1 appears only once in Δ, whereas δ_2 needs to be repeated. This is related to the ranks of the matrices A_1 (which has rank 1) and A_2 (which has rank 2).

Additional repetitions of the parameters δ_i may be necessary if we also have uncertainty in B, C and D. It can be shown that the minimum number of repetitions of each δ_i in the overall Δ-matrix is equal to the rank of each matrix $\begin{bmatrix} A_i & B_i \\ C_i & D_i \end{bmatrix}$ (Packard, 1988; Zhou et al., 1996). Also, note that seemingly nonlinear parameter dependencies may be rewritten in our standard linear block diagram form, for example, we can handle δ_1^2 (which would need δ_1 repeated), $\frac{\alpha + w_1 \delta_1 \delta_2}{1 + w_2 \delta_2}$, etc. This is illustrated next by an example.

Example 7.10 Parametric uncertainty and repeated perturbations. *This example illustrates how most forms of parametric uncertainty can be represented in terms of the Δ-representation using linear fractional transformations (LFTs). Consider the following state space description of a SISO plant*[1]

$$\dot{x} = A_p x + B_p u, \quad y = Cx \tag{7.102}$$

$$A_p = \begin{bmatrix} -(1+k) & 0 \\ 1 & -(1+k) \end{bmatrix}, \quad B_p = \begin{bmatrix} \frac{1-k}{k} \\ -1 \end{bmatrix}, \quad C = \begin{bmatrix} 0 & 1 \end{bmatrix} \tag{7.103}$$

The constant $k > 0$ may vary during operation, so the above description generates a set of possible plants. Assume that $k = 0.5 \pm 0.1$, which may be written as

$$k = 0.5 + 0.1\delta, \quad |\delta| \le 1 \tag{7.104}$$

Note that the parameter k enters into the plant model in several places, and we will need to use repeated perturbations in order to rearrange the model to fit our standard formulation with the uncertainty represented as a block-diagonal Δ-matrix.

Let us first consider the input gain uncertainty for state 1, that is, the variations in $b_{p1} = (1-k)/k$. Even though b_{p1} is a nonlinear function of δ, it has a block-diagram representation and may thus be written as a linear fractional transformation (LFT). We have

$$b_{p1}(\delta) = \frac{1-k}{k} = \frac{0.5 - 0.1\delta}{0.5 + 0.1\delta} = \frac{1 - 0.2\delta}{1 + 0.2\delta} \tag{7.105}$$

which may be written as a scalar LFT

$$b_{p1}(\delta) = F_u(N, \delta) = n_{22} + n_{12}\delta(1 - n_{11}\delta)^{-1}n_{21} \tag{7.106}$$

with $n_{22} = 1$, $n_{11} = -0.2$, $n_{12}n_{21} = -0.4$. Next consider the pole uncertainty caused by variations in the A-matrix, which may be written as

$$A_p = \begin{bmatrix} -1.5 & 0 \\ 1 & -1.5 \end{bmatrix} + \begin{bmatrix} -0.1 & 0 \\ 0 & -0.1 \end{bmatrix} \begin{bmatrix} \delta & 0 \\ 0 & \delta \end{bmatrix} \tag{7.107}$$

---

[1] This is actually a simple model of a chemical reactor (CSTR) where u is the feed flowrate, x_1 is the concentration of reactant A, $y = x_2$ is the concentration of intermediate product B and $k = q^*$ is the steady-state value of the feed flowrate. Component balances yield

$$V\dot{c}_A = qc_{Af} - qc_A - k_1 c_A V \quad [\text{mol A/s}]$$

$$V\dot{c}_B = -qc_B + k_1 c_A V - k_2 c_B V \quad [\text{mol B/s}]$$

where V is the reactor volume. Linearization and introduction of deviation variables, $x_1 = \Delta c_A$, $x_2 = \Delta c_B$, and $u = \Delta q$, yields, with $k_1 = 1, k_2 = 1, V = 1$ and $c_A^* = c_B^* = 1$,

$$\dot{x}_1 = -(1 + q^*)x_1 + (c_{Af}^* - 1)u$$

$$\dot{x}_2 = x_1 - (1 + q^*)x_2 - u$$

where the superscript * signifies a steady-state value. The values of q^* and c_{Af}^* depend on the operating point, and it is given that at steady-state we always have $q^* c_{Af}^* = 1$ (physically, we may have an upstream mixing tank where a fixed amount of A is fed in). By introducing $k = q^*$ we get the model in (7.103).

For our specific example with uncertainty in both B and A, the plant uncertainty may be represented as shown in Figure 7.22 where $K(s)$ is a scalar controller. Consequently, we may pull out the perturbations and collect them in a 3×3 diagonal Δ-block with the scalar perturbation δ repeated three times,

$$\Delta = \begin{bmatrix} \delta & & \\ & \delta & \\ & & \delta \end{bmatrix} \qquad (7.108)$$

and we may then obtain the interconnection matrix P by rearranging the block diagram of Figure 7.22 to fit Figure 3.21. It is rather tedious to do this by hand, but it is relatively straightforward with the appropriate software tools.

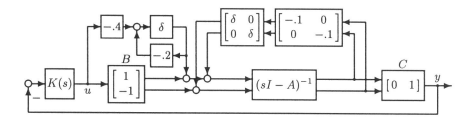

Figure 7.22: Block diagram of parametric uncertainty

Remark. The above example is included in order to show that quite complex uncertainty representations can be captured by the general framework of block-diagonal perturbations. It is *not* suggested, however, that such a complicated description should be used in practice for this example. A little more analysis will show why. The transfer function for this plant is

$$G_p(s) = \frac{-\left(s + \frac{(k+2.414)(k-0.414)}{k}\right)}{(s+1+k)^2} \qquad (7.109)$$

and we note that it has a RHP-zero for $k < 0.414$, and that the steady state gain is zero for $k = 0.414$. The three plants corresponding to $k = 0.5, 0.4$ and 0.6 are

$$G(s) = -\frac{(s+0.5)}{(s+1.5)^2}, \quad G_{p1}(s) = -\frac{(s-0.1)}{(s+1.4)^2}, \quad G_{p2}(s) = -\frac{(s+0.93)}{(s+1.6)^2} \qquad (7.110)$$

From a practical point of view the pole uncertainty therefore seems to be of very little importance and we may use a simplified uncertainty description with zero uncertainty only, e.g.

$$g_p(s) = -\frac{(s+z_p)}{(s+1.5)^2}; \quad -0.1 \le z_p \le 0.93 \qquad (7.111)$$

In any case, we know that because of the RHP-zero crossing through the origin, the performance at low frequencies will be very poor for this plant.

7.8 Additional exercises

Exercise 7.7 *Consider a "true" plant*

$$G'(s) = \frac{3e^{-0.1s}}{(2s+1)(0.1s+1)^2}$$

(a) Derive and sketch the additive uncertainty weight when the nominal model is $G(s) = 3/(2s+1)$.

(b) Derive the corresponding robust stability condition.

(c) Apply this test for the controller $K(s) = k/s$ and find the values of k that yield stability. Is this condition tight?

Exercise 7.8 **Uncertainty weight for a first-order model with delay.** *Laughlin et al. (1987) considered the following parametric uncertainty description*

$$G_p(s) = \frac{k_p}{\tau_p s + 1} e^{-\theta_p s}; \quad k_p \in [k_{\min}, k_{\max}], \ \tau_p \in [\tau_{\min}, \tau_{\max}], \ \theta_p \in [\theta_{\min}, \theta_{\max}]$$

$$(7.112)$$

where all parameters are assumed positive. They chose the mean parameter values as $(\bar{k}, \bar{\theta}, \bar{\tau})$ giving the nominal model

$$G(s) = \bar{G}(s) \triangleq \frac{\bar{k}}{\bar{\tau}s + 1} e^{-\bar{\theta}s} \qquad (7.113)$$

and suggested use of the following multiplicative uncertainty weight

$$w_{IL}(s) = \frac{k_{\max}}{\bar{k}} \cdot \frac{\bar{\tau}s+1}{\tau_{\min}s+1} \cdot \frac{Ts+1}{-Ts+1} - 1; \quad T = \frac{\theta_{\max} - \theta_{\min}}{4} \qquad (7.114)$$

(a) Show that the resulting stable and minimum phase weight corresponding to the uncertainty description in (7.17) is

$$w_{IL}(s) = (1.25s^2 + 1.55s + 0.2)/(2s+1)(0.25s+1) \qquad (7.115)$$

Note that this weight cannot be compared with (7.19) or (7.20) since the nominal plant is different.

(b) Plot the magnitude of w_{IL} as a function of frequency. Find the frequency where the weight crosses 1 in magnitude, and compare this with $1/\theta_{\max}$. Comment on your answer.

(c) Find $l_I(j\omega)$ using (7.15) and compare with $|w_{IL}|$. Does the weight (7.115) and the uncertainty model (7.2) include all possible plants? (Answer: No, not quite around frequency $\omega = 5$).

Exercise 7.9 *Consider again the system in Figure 7.18. What kind of uncertainty might w_u and Δ_u represent?*

Exercise 7.10 **Neglected dynamics.** *Assume we have derived the following detailed model*

$$G_{detail}(s) = \frac{3(-0.5s+1)}{(2s+1)(0.1s+1)^2} \qquad (7.116)$$

and we want to use the simplified nominal model $G(s) = 3/(2s+1)$ with multiplicative uncertainty. Plot $l_I(\omega)$ and approximate it by a rational transfer function $w_I(s)$.

Exercise 7.11 Parametric gain uncertainty. *We showed in Example 7.1 how to represent scalar parametric gain uncertainty $G_p(s) = k_p G_0(s)$ where*

$$k_{\min} \leq k_p \leq k_{\max} \qquad (7.117)$$

as multiplicative uncertainty $G_p = G(1 + w_I \Delta_I)$ with nominal model $G(s) = \bar{k} G_0(s)$ and uncertainty weight $w_I = r_k = (k_{\max} - k_{\min})/(k_{\max} + k_{\min})$. Δ_I is here a real scalar, $-1 \leq \Delta_I \leq 1$. Alternatively, we can represent gain uncertainty as inverse multiplicative uncertainty:

$$\Pi_{iI} : \quad G_p(s) = G(s)(1 + w_{iI}(s)\Delta_{iI})^{-1}; \quad -1 \leq \Delta_{iI} \leq 1 \qquad (7.118)$$

with $w_{iI} = r_k$ and $G(s) = k_i G$ where

$$k_i = 2 \frac{k_{\min} k_{\max}}{k_{\max} + k_{\min}} \qquad (7.119)$$

(a) Derive (7.118) and (7.119). (Hint: The gain variation in (7.117) can be written exactly as $k_p = k_i/(1 - r_k \Delta)$.)

(b) Show that the form in (7.118) does not allow for $k_p = 0$.

(c) Discuss why (b) may be a possible advantage.

Exercise 7.12 *The model of an industrial robot arm is as follows*

$$G(s) = \frac{250(as^2 + 0.0001s + 100)}{s(as^2 + 0.0001(500a + 1)s + 100(500a + 1))}$$

where $a \in [0.0002, 0.002]$. Sketch the Bode plot for the two extreme values of a. What kind of control performance do you expect? Discuss how you may best represent this uncertainty.

7.9   Conclusion

In this chapter we have shown how model uncertainty for SISO systems can be represented in the frequency domain using complex norm-bounded perturbations, $\|\Delta\|_\infty \leq 1$. At the end of the chapter we also discussed how to represent parametric uncertainty using real perturbations.

We showed that the requirement of robust stability for the case of multiplicative complex uncertainty imposes an upper bound on the allowed complementary sensitivity, $|w_I T| < 1, \forall \omega$. Similarly, the inverse multiplicative uncertainty imposes an upper bound on the sensitivity, $|w_{iI} S| < 1, \forall \omega$. We also derived a condition for robust performance with multiplicative uncertainty, $|w_P S| + |w_I T| < 1, \forall \omega$.

The approach in this chapter was rather elementary, and to extend the results to MIMO systems and to more complex uncertainty descriptions we need to make use of the structured singular value, μ. This is the theme of the next chapter, where we find that $|w_I T|$ and $|w_{iI} S|$ are the structured singular values for evaluating robust stability for the two sources of uncertainty in question, whereas $|w_P S| + |w_I T|$ is the structured singular value for evaluating robust performance with multiplicative uncertainty.

8

ROBUST STABILITY AND PERFORMANCE ANALYSIS

The objective of this chapter is to present a general method for analyzing robust stability and robust performance of MIMO systems with multiple perturbations. Our main analysis tool will be the *structured singular value*, μ. We also show how the "optimal" robust controller, in terms of minimizing μ, can be designed using DK-iteration. This involves solving a sequence of scaled $\mathcal{H}_\infty$ problems.

8.1 General control configuration with uncertainty

For useful notation and an introduction to model uncertainty the reader is referred to Sections 7.1 and 7.2. The starting point for our robustness analysis is a system representation in which the uncertain perturbations are "pulled out" into a block-diagonal matrix,

$$\Delta = \text{diag}\{\Delta_i\} = \begin{bmatrix} \Delta_1 & & & \\ & \ddots & & \\ & & \Delta_i & \\ & & & \ddots \end{bmatrix} \tag{8.1}$$

where each Δ_i represents a specific source of uncertainty, e.g. input uncertainty, Δ_I, or parametric uncertainty, δ_i, where δ_i is real. If we also pull out the controller K, we get the generalized plant P, as shown in Figure 8.1. This form is useful for controller synthesis. Alternatively, if the controller is given and we want to analyze the uncertain system, we use the $N\Delta$-structure in Figure 8.2.

In Section 3.8.8, we discussed how to find P and N for cases without uncertainty. The procedure with uncertainty is similar and is demonstrated by examples below; see Section 8.3. To illustrate the main idea, consider Figure 8.4 where it is shown how to pull out the perturbation blocks to form Δ and the nominal system N. As shown in (3.112), N is related to P and K by a lower LFT

$$N = F_l(P, K) \triangleq P_{11} + P_{12}K(I - P_{22}K)^{-1}P_{21} \tag{8.2}$$

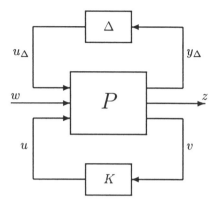

Figure 8.1: General control configuration (for controller synthesis)

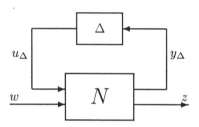

Figure 8.2: $N\Delta$-structure for robust performance analysis

Similarly, the uncertain closed-loop transfer function from w to z, $z = Fw$, is related to N and Δ by an upper LFT (see (3.113)),

$$F = F_u(N, \Delta) \triangleq N_{22} + N_{21}\Delta(I - N_{11}\Delta)^{-1}N_{12} \qquad (8.3)$$

To analyze robust stability of F, we can then rearrange the system into the $M\Delta$-structure of Figure 8.3 where $M = N_{11}$ is the transfer function from the output to the input of the perturbations.

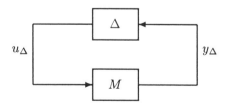

Figure 8.3: $M\Delta$-structure for robust stability analysis

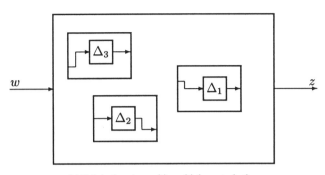

(a) Original system with multiple perturbations

$\Downarrow$

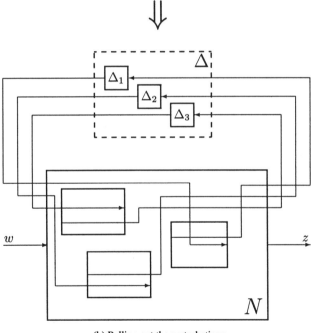

(b) Pulling out the perturbations

Figure 8.4: Rearranging an uncertain system into the $N\Delta$-structure

8.2 Representing uncertainty

As usual, each individual perturbation is assumed to be stable and is normalized,

$$\bar{\sigma}(\Delta_i(j\omega)) \leq 1 \; \forall\omega \tag{8.4}$$

For a complex scalar perturbation we have $|\delta_i(j\omega)| \leq 1$, $\forall\omega$, and for a real scalar perturbation $-1 \leq \delta_i \leq 1$. Since from (A.47) the maximum singular value of a block diagonal matrix is equal to the largest of the maximum singular values of the individual blocks, it then follows for $\Delta = \text{diag}\{\Delta_i\}$ that

$$\bar{\sigma}(\Delta_i(j\omega)) \leq 1 \; \forall\omega, \; \forall i \quad \Leftrightarrow \quad \boxed{\|\Delta\|_\infty \leq 1} \tag{8.5}$$

Note that Δ has *structure*, and therefore in the robustness analysis we do *not* want to allow all Δ such that (8.5) is satisfied. Only the subset which has the block-diagonal structure in (8.1) should be considered. In some cases the blocks in Δ may be repeated or may be real, that is, we have additional structure. For example, as shown in Section 7.7.3, repetition is often needed to handle parametric uncertainty.

Remark. The assumption of a stable Δ may be relaxed, but then the resulting robust stability and performance conditions will be harder to derive and more complex to state. Furthermore, if we use a suitable form for the uncertainty and allow for multiple perturbations, then we can always generate the desired class of plants with stable perturbations, so assuming Δ stable is not really a restriction.

8.2.1 Differences between SISO and MIMO systems

The main difference between SISO and MIMO systems is the concept of directions which is only relevant in the latter. As a consequence MIMO systems may experience much larger sensitivity to uncertainty than SISO systems. The following example illustrates that for MIMO systems it is sometimes critical to represent the coupling between uncertainty in different transfer function elements.

Example 8.1 **Coupling between transfer function elements.** *Consider a distillation process where at steady-state*

$$G = \begin{bmatrix} 87.8 & -86.4 \\ 108.2 & -109.6 \end{bmatrix}, \quad \Lambda = \text{RGA}(G) = \begin{bmatrix} 35.1 & -34.1 \\ -34.1 & 35.1 \end{bmatrix} \tag{8.6}$$

From the large RGA-elements we know that G becomes singular for small relative changes in the individual elements. For example, from (6.88) we know that perturbing the 1, 2-element from -86.4 to -88.9 makes G singular. Since variations in the steady-state gains of $\pm 50\%$ or more may occur during operation of the distillation process, this seems to indicate that independent control of both outputs is impossible. However, this conclusion is incorrect since, for a distillation process, G never becomes singular. The reason is that the transfer function elements are coupled due to underlying physical constraints (e.g. the material balance).

Specifically, for the distillation process a more reasonable description of the gain uncertainty is (Skogestad et al., 1988)

$$G_p = G + w \begin{bmatrix} \delta & -\delta \\ -\delta & \delta \end{bmatrix}, \quad |\delta| \leq 1 \tag{8.7}$$

where w in this case is a real constant, e.g. $w = 50$. For the numerical data above $\det G_p = \det G$ irrespective of δ, so G_p is never singular for this uncertainty. (Note that $\det G_p = \det G$ is not generally true for the uncertainty description given in (8.7)).

Exercise 8.1 *The uncertain plant in (8.7) may be represented in the additive uncertainty form $G_p = G + W_2 \Delta_A W_1$ where $\Delta_A = \delta$ is a single scalar perturbation. Find W_1 and W_2.*

8.2.2 Parametric uncertainty

The representation of parametric uncertainty, as discussed in Chapter 7 for SISO systems, carries straight over to MIMO systems. However, the inclusion of parametric uncertainty may be more significant for MIMO plants because it offers a simple method of representing the coupling between uncertain transfer function elements. For example, the simple uncertainty description used in (8.7) originated from a parametric uncertainty description of the distillation process.

8.2.3 Unstructured uncertainty

Unstructured perturbations are often used to get a simple uncertainty model. We here define *unstructured* uncertainty as the use of a "full" complex perturbation matrix Δ, usually with dimensions compatible with those of the plant, where at each frequency any $\Delta(j\omega)$ satisfying $\bar{\sigma}(\Delta(j\omega)) \leq 1$ is allowed.

Six common forms of unstructured uncertainty are shown in Figure 8.5. In Figure 8.5(a), (b) and (c) are shown three *feedforward* forms; additive uncertainty, multiplicative input uncertainty and multiplicative output uncertainty:

$$\Pi_A: \quad G_p = G + E_A; \quad E_a = w_A \Delta_a \tag{8.8}$$

$$\Pi_I: \quad G_p = G(I + E_I); \quad E_I = w_I \Delta_I \tag{8.9}$$

$$\Pi_O: \quad G_p = (I + E_O)G; \quad E_O = w_O \Delta_O \tag{8.10}$$

In Figure 8.5(d), (e) and (f) are shown three *feedback* or *inverse* forms; inverse additive uncertainty, inverse multiplicative input uncertainty and inverse multiplicative output uncertainty:

$$\Pi_{iA}: \quad G_p = G(I - E_{iA}G)^{-1}; \quad E_{iA} = w_{iA} \Delta_{iA} \tag{8.11}$$

$$\Pi_{iI}: \quad G_p = G(I - E_{iI})^{-1}; \quad E_{iI} = w_{iI} \Delta_{iI} \tag{8.12}$$

$$\Pi_{iO}: \quad G_p = (I - E_{iO})^{-1}G; \quad E_{iO} = w_{iO} \Delta_{iO} \tag{8.13}$$

The negative sign in front of the E's does not really matter here since we assume that Δ can have any sign. Δ denotes the normalized perturbation and E the "actual"

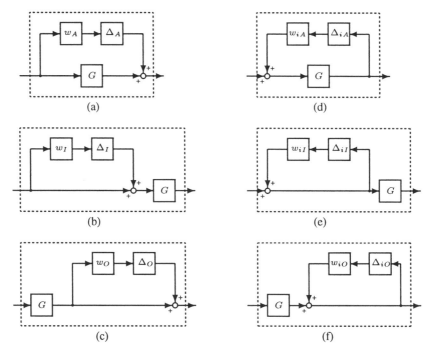

Figure 8.5: (a) Additive uncertainty, (b) Multiplicative input uncertainty, (c) Multiplicative output uncertainty, (d) Inverse additive uncertainty, (e) Inverse multiplicative input uncertainty, (f) Inverse multiplicative output uncertainty

perturbation. We have here used scalar weights w, so $E = w\Delta = \Delta w$, but sometimes one may want to use matrix weights, $E = W_2 \Delta W_1$ where W_1 and W_2 are given transfer function matrices.

Another common form of unstructured uncertainty is coprime factor uncertainty discussed later in Section 8.6.2.

Remark. In practice, one can have several perturbations which themselves are unstructured. For example, we may have Δ_I at the input and Δ_O at the output, which may be combined into a larger perturbation, $\Delta = \text{diag}\{\Delta_I, \Delta_O\}$. However, this Δ is a block-diagonal matrix and is therefore no longer truly unstructured.

Lumping uncertainty into a single perturbation

For SISO systems we usually lump multiple sources of uncertainty into a single complex perturbation; often in multiplicative form. This may also be done for MIMO systems, but then it makes a difference whether the perturbation is at the input or the output.

Since output uncertainty is frequently less restrictive than input uncertainty in terms of control performance (see Section 6.10.4), we first attempt to lump the uncertainty at the output. For example, a set of plants Π may be represented by *multiplicative output uncertainty* with a scalar weight $w_O(s)$ using

$$G_p = (I + w_O \Delta_O)G, \quad \|\Delta_O\|_\infty \leq 1 \tag{8.14}$$

where, similar to (7.15),

$$l_O(\omega) = \max_{G_p \in \Pi} \bar{\sigma}\left((G_p - G)G^{-1}(j\omega)\right); \quad |w_O(j\omega)| \geq l_O(\omega) \quad \forall \omega \tag{8.15}$$

(we can use the pseudo-inverse if G is singular). If the resulting uncertainty weight is reasonable (i.e. it must at least be less than 1 in the frequency range where we want control), and the subsequent analysis shows that robust stability and performance may be achieved, then this lumping of uncertainty at the output is fine. If this is not the case, then one may try to lump the uncertainty at the input instead, using *multiplicative input uncertainty* with a scalar weight,

$$G_p = G(I + w_I \Delta_I), \quad \|\Delta_I\|_\infty \leq 1 \tag{8.16}$$

where, similar to (7.15),

$$l_I(\omega) = \max_{G_p \in \Pi} \bar{\sigma}\left(G^{-1}(G_p - G)(j\omega)\right); \quad |w_I(j\omega)| \geq l_I(\omega) \quad \forall \omega \tag{8.17}$$

However, in many cases this approach of lumping uncertainty either at the output or the input does not work well. This is because one cannot in general shift a perturbation from one location in the plant (say at the input) to another location (say the output) without introducing candidate plants which were not present in the original set. In particular, one should be careful when the plant is ill-conditioned. This is discussed next.

Moving uncertainty from the input to the output

For a scalar plant, we have $G_p = G(1 + w_I \Delta_I) = (1 + w_O \Delta_O)G$ and we may simply "move" the multiplicative uncertainty from the input to the output without changing the value of the weight, i.e. $w_I = w_O$. However, for multivariable plants we usually need to multiply by the condition number $\gamma(G)$ as is shown next.

Suppose the true uncertainty is represented as unstructured input uncertainty (E_I is a full matrix) on the form

$$G_p = G(I + E_I) \tag{8.18}$$

Then from (8.17) the magnitude of multiplicative input uncertainty is

$$l_I(\omega) = \max_{E_I} \bar{\sigma}(G^{-1}(G_p - G)) = \max_{E_I} \bar{\sigma}(E_I) \tag{8.19}$$

On the other hand, if we want to represent (8.18) as multiplicative output uncertainty, then from (8.15)

$$l_O(\omega) = \max_{E_I} \bar{\sigma}((G_p - G)G^{-1}) = \max_{E_I} \bar{\sigma}(GE_IG^{-1}) \tag{8.20}$$

which is much larger than $l_I(\omega)$ if the condition number of the plant is large. To see this, write $E_I = w_I \Delta_I$ where we allow any $\Delta_I(j\omega)$ satisfying $\bar{\sigma}(\Delta_I(j\omega)) \leq 1, \forall \omega$. Then at a given frequency

$$l_O(\omega) = |w_I| \max_{\Delta_I} \bar{\sigma}(G\Delta_IG^{-1}) = |w_I(j\omega)| \, \gamma(G(j\omega)) \tag{8.21}$$

Proof of (8.21): Write at each frequency $G = U\Sigma V^H$ and $G^{-1} = \tilde{U}\tilde{\Sigma}\tilde{V}^H$. Select $\Delta_I = V\tilde{U}^H$ (which is a unitary matrix with all singular values equal to 1). Then $\bar{\sigma}(G\Delta_IG^{-1}) = \bar{\sigma}(U\Sigma\tilde{\Sigma}V^H) = \bar{\sigma}(\Sigma\tilde{\Sigma}) = \bar{\sigma}(G)\bar{\sigma}(G^{-1}) = \gamma(G)$.   □

Example 8.2 *Assume the relative input uncertainty is 10%, that is, $w_I = 0.1$, and the condition number of the plant is 141.7. Then we must select $l_0 = w_O = 0.1 \times 141.7 = 14.2$ in order to represent this as multiplicative output uncertainty (this is larger than 1 and therefore not useful for controller design).*

Also for diagonal uncertainty (E_I diagonal) we may have a similar situation. For example, if the plant has large RGA-elements then the elements in GE_IG^{-1} will be much larger than those of E_I, see (A.80), making it impractical to move the uncertainty from the input to the output.

Example 8.3 *Let Π be the set of plants generated by the additive uncertainty in (8.7) with $w = 10$ (corresponding to about 10% uncertainty in each element). Then from (8.7) one plant G' in this set (corresponding to $\delta = 1$) has*

$$G' = G + \begin{bmatrix} 10 & -10 \\ -10 & 10 \end{bmatrix} \tag{8.22}$$

for which we have $l_I = \bar{\sigma}(G^{-1}(G' - G)) = 14.3$. Therefore, to represent G' in terms of input uncertainty we would need a relative uncertainty of more than 1400%. This would imply that the plant could become singular at steady-state and thus impossible to control, which we know is incorrect. Fortunately, we can instead represent this additive uncertainty as multiplicative output uncertainty (which is also generally preferable for a subsequent controller design) with $l_O = \bar{\sigma}((G' - G)G^{-1}) = 0.10$. Therefore output uncertainty works well for this particular example.

Conclusion. Ideally, we would like to lump several sources of uncertainty into a single perturbation to get a simple uncertainty description. Often an unstructured multiplicative output perturbations is used. However, from the above discussion we have learnt that we should be careful about doing this, at least for plants with a large condition number. In such cases we may have to represent the uncertainty as it occurs physically (at the input, in the elements, etc.) thereby generating several perturbations. For uncertainty associated with unstable plant poles, we should use one of the *inverse* forms in Figure 8.5.

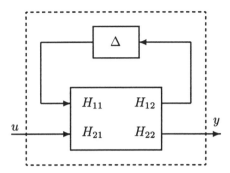

Figure 8.6: Uncertain plant, $y = G_p u$, represented by LFT, see (8.23)

Exercise 8.2 *A fairly general way of representing an uncertain plant G_p is in terms of a linear fractional transformation (LFT) of Δ as shown in Figure 8.6. Here*

$$G_p = F_u \left(\begin{bmatrix} H_{11} & H_{12} \\ H_{21} & H_{22} \end{bmatrix}, \Delta \right) = H_{22} + H_{21}\Delta(I - H_{11}\Delta)^{-1}H_{12} \qquad (8.23)$$

where $G = H_{22}$ is the nominal plant model. Obtain H for each of the six uncertainty forms in (8.8)-(8.13) using $E = W_2\Delta W_1$ (Hint for the inverse forms: $(I - W_1\Delta W_2)^{-1} = I + W_1\Delta(I - W_2W_1\Delta)^{-1}W_2$, see (3.7)-(3.9)).

Exercise 8.3 *Obtain H in Figure 8.6 for the uncertain plant in Figure 7.20(b).*

Exercise 8.4 *Obtain H in Figure 8.6 for the uncertain plant in Figure 7.22.*

8.2.4 Diagonal uncertainty

By "diagonal uncertainty" we mean that the perturbation is a *complex diagonal* matrix

$$\Delta(s) = \text{diag}\{\delta_i(s)\}; \quad |\delta_i(j\omega)| \leq 1, \forall \omega, \forall i \qquad (8.24)$$

(usually of the same size as the plant). For example, this is the case if Δ is diagonal in any of the six uncertainty forms in Figure 8.5. Diagonal uncertainty usually arises from a consideration of uncertainty or neglected dynamics in the individual input channels (actuators) or in the individual output channels (sensors). This type of diagonal uncertainty is *always* present, and since it has a scalar origin it may be represented using the methods presented in Chapter 7.

To make this clearer, let us consider uncertainty in the input channels. With each input u_i there is associated a separate physical system (amplifier, signal converter, actuator, valve, etc.) which based on the *controller output* signal, u_i, generates a *physical plant input m_i*

$$m_i = h_i(s)u_i \qquad (8.25)$$

The scalar transfer function $h_i(s)$ is often absorbed into the plant model $G(s)$, but for representing the uncertainty it is important to notice that it originates at the input. We can represent this actuator uncertainty as multiplicative (relative) uncertainty given by

$$h_{pi}(s) = h_i(s)(1 + w_{Ii}(s)\delta_i(s)); \quad |\delta_i(j\omega)| \leq 1, \forall \omega \qquad (8.26)$$

which after combining all input channels results in *diagonal input uncertainty* for the plant

$$G_p(s) = G(1 + W_I\Delta_I); \quad \Delta_I = \text{diag}\{\delta_i\}, W_I = \text{diag}\{w_{Ii}\} \qquad (8.27)$$

Normally we would represent the uncertainty in each input or output channel using a simple weight in the form given in (7.28), namely

$$w(s) = \frac{\tau s + r_0}{(\tau/r_\infty)s + 1} \qquad (8.28)$$

where r_0 is the relative uncertainty at steady-state, $1/\tau$ is (approximately) the frequency where the relative uncertainty reaches 100%, and r_∞ is the magnitude of the weight at higher frequencies. Typically, the uncertainty $|w|$, associated with each input, is at least 10% at steady-state ($r_0 \geq 0.1$), and it increases at higher frequencies to account for neglected or uncertain dynamics (typically, $r_\infty \geq 2$).

Remark 1 The diagonal uncertainty in (8.27) originates from independent scalar uncertainty in each input channel. If we choose to represent this as *unstructured* input uncertainty (Δ_I is a full matrix) then we must realize that this will introduce non-physical couplings at the input to the plant, resulting in a set of plants which is too large, and the resulting robustness analysis may be conservative (meaning that we may incorrectly conclude that the system may not meet its specifications).

Remark 2 The claim is often made that one can easily reduce the static input gain uncertainty to significantly less than 10%, but this is probably not true in most cases. Consider again (8.25). A commonly suggested method to reduce the uncertainty is to measure the actual input (m_i) and employ local feedback (cascade control) to readjust u_i. As a simple example, consider a bathroom shower, in which the input variables are the flows of hot and cold water. One can then imagine measuring these flows and using cascade control so that each flow can be adjusted more accurately. However, even in this case there will be uncertainty related to the accuracy of each measurement. Note that it is *not* the absolute measurement error that yields problems, but rather the error in the sensitivity of the measurement with respect to changes (i.e. the "gain" of the sensor). For example, assume that the nominal flow in our shower is 1 l/min and we want to increase it to 1.1 l/min, that is, in terms of deviation variables we want $u = 0.1$ [l/min]. Suppose the vendor guarantees that the measurement error is less than 1%. But, even with this small absolute error, the actual flow rate may have increased from 0.99 l/min (measured value of 1 l/min is 1% too high) to 1.11 l/min (measured value of 1.1 l/min is 1% too low), corresponding to a change $u' = 0.12$ [l/min], and an input gain uncertainty of 20%.

In conclusion, diagonal input uncertainty, as given in (8.27), should always be considered because:

1. It is *always* present and a system which is sensitive to this uncertainty will not work in practice.
2. It often restricts achievable performance with multivariable control.

8.3   Obtaining P, N and M

We will now illustrate, by way of an example, how to obtain the interconnection matrices P, N and M in a given situation.

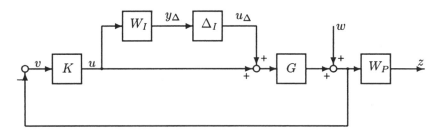

Figure 8.7: System with multiplicative input uncertainty and performance measured at the output

Example 8.4 System with input uncertainty. *Consider a feedback system with multiplicative input uncertainty Δ_I as shown in Figure 8.7. Here W_I is a normalization weight for the uncertainty and W_P is a performance weight. We want to derive the generalized plant P in Figure 8.1 which has inputs $[\, u_\Delta \quad w \quad u \,]^T$ and outputs $[\, y_\Delta \quad z \quad v \,]^T$. By writing down the equations (e.g. see Example 3.13) or simply by inspecting Figure 8.7 (remember to break the loop before and after K) we get*

$$P = \begin{bmatrix} 0 & 0 & W_I \\ W_P G & W_P & W_P G \\ -G & -I & -G \end{bmatrix} \tag{8.29}$$

It is recommended that the reader carefully derives P (as instructed in Exercise 8.5). Note that the transfer function from u_Δ to y_Δ (upper left element in P) is 0 because u_Δ has no direct effect on y_Δ (except through K). Next, we want to derive the matrix N corresponding to Figure 8.2. First, partition P to be compatible with K, i.e.

$$P_{11} = \begin{bmatrix} 0 & 0 \\ W_P G & W_P \end{bmatrix}, \; P_{12} = \begin{bmatrix} W_I \\ W_P G \end{bmatrix} \tag{8.30}$$

$$P_{21} = [-G \quad -I], \; P_{22} = -G \tag{8.31}$$

and then find $N = F_l(P, K)$ using (8.2). We get (see Exercise 8.7)

$$N = \begin{bmatrix} -W_I K G(I + KG)^{-1} & -W_I K(I + GK)^{-1} \\ W_P G(I + KG)^{-1} & W_P(I + GK)^{-1} \end{bmatrix} \tag{8.32}$$

Alternatively, we can derive N directly from Figure 8.7 by evaluating the closed-loop transfer function from inputs $\begin{bmatrix} u_\Delta \\ w \end{bmatrix}$ *to outputs* $\begin{bmatrix} y_\Delta \\ z \end{bmatrix}$ *(without breaking the loop before and after K).*
For example, to derive N_{12}, which is the transfer function from w to y_Δ, we start at the output (y_Δ) and move backwards to the input (w) using the MIMO Rule in Section 3.2 (we first meet W_I, then $-K$ and we then exit the feedback loop and get the term $(I + GK)^{-1}$).
The upper left block, N_{11}, in (8.32) is the transfer function from u_Δ to y_Δ. This is the transfer function M needed in Figure 8.3 for evaluating robust stability. Thus, we have $M = -W_I KG(I + KG)^{-1} = -W_I T_I$.

Exercise 8.5 *Show in detail how P in (8.29) is derived.*

Exercise 8.6 *For the system in Figure 8.7 we see easily from the block diagram that the uncertain transfer function from w to z is $F = W_P(I + G(I + W_I\Delta_I)K)^{-1}$. Show that this is identical to $F_u(N, \Delta)$ evaluated using (8.35) where from (8.32) we have $N_{11} = -W_I T_I$, $N_{12} = -W_I KS$, $N_{21} = W_P SG$ and $N_{22} = W_P S$.*

Exercise 8.7 *Derive N in (8.32) from P in (8.29) using the lower LFT in (8.2). You will note that the algebra is quite tedious, and that it is much simpler to derive N directly from the block diagram as described above.*

Remark. Of course, deriving N from P is straightforward using available software. For example, in the MATLAB μ-toolbox we can evaluate $N = F_l(P, K)$ using the command `N=starp(P,K)`, and with a specific Δ the perturbed transfer function $F_u(N, \Delta)$ from w to z is obtained with the command `F=starp(delta,N)`.

Exercise 8.8 *Derive P and N for the case when the multiplicative uncertainty is at the output rather than the input.*

Exercise 8.9 *Find P for the uncertain system in Figure 7.18.*

Exercise 8.10 *Find P for the uncertain plant G_p in (8.23) when $w = r$ and $z = y - r$.*

Exercise 8.11 *Find the interconnection matrix N for the uncertain system in Figure 7.18. What is M?*

Exercise 8.12 *Find the transfer function $M = N_{11}$ for studying robust stability for the uncertain plant G_p in (8.23).*

Exercise 8.13 *$M\Delta$-structure for combined input and output uncertainties. Consider the block diagram in Figure 8.8 where we have both input and output multiplicative uncertainty blocks. The set of possible plants is given by*

$$G_p = (I + W_{2O}\Delta_O W_{1O})G(I + W_{2I}\Delta_I W_{1I}) \tag{8.33}$$

where $\|\Delta_I\|_\infty \le 1$ and $\|\Delta_O\|_\infty \le 1$. Collect the perturbations into $\Delta = \mathrm{diag}\{\Delta_I, \Delta_O\}$ and rearrange Figure 8.8 into the $M\Delta$-structure in Figure 8.3 Show that

$$M = \begin{bmatrix} W_{1I} & 0 \\ 0 & W_{1O} \end{bmatrix} \begin{bmatrix} -T_I & -KS \\ SG & -T \end{bmatrix} \begin{bmatrix} W_{2I} & 0 \\ 0 & W_{2O} \end{bmatrix} \tag{8.34}$$

Exercise 8.14 *Find μ for the uncertain system in Figure 7.20(b).*

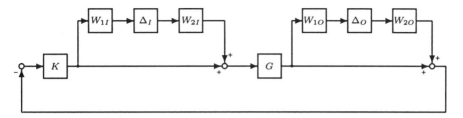

Figure 8.8: System with input and output multiplicative uncertainty

8.4 Definitions of robust stability and robust performance

We have discussed how to represent an uncertain set of plants in terms of the $N\Delta$-structure in Figure 8.2. The next step is to check whether we have stability and acceptable performance for all plants in the set:

1. *Robust stability (RS) analysis*: with a given controller K we determine whether the system remains stable for all plants in the uncertainty set.
2. *Robust performance (RP) analysis*: if RS is satisfied, we determine how "large" the transfer function from exogenous inputs w to outputs z may be for all plants in the uncertainty set.

Before proceeding, we need to define performance more precisely. In Figure 8.2, w represents the exogenous inputs (normalized disturbances and references), and z the exogenous outputs (normalized errors). We have $z = F(\Delta)w$, where from (8.3)

$$F = F_u(N, \Delta) \triangleq N_{22} + N_{21}\Delta(I - N_{11}\Delta)^{-1}N_{12} \qquad (8.35)$$

We here use the $\mathcal{H}_\infty$ norm to define performance and require for RP that $\|F(\Delta)\|_\infty \leq 1$ for all allowed Δ's. A typical choice is $F = w_P S_p$ (the weighted sensitivity function), where w_P is the performance weight (capital P for performance) and S_p represents the set of perturbed sensitivity functions (lower-case p for perturbed).

In terms of the $N\Delta$-structure in Figure 8.2 our requirements for stability and performance can then be summarized as follows

$$\text{NS} \overset{\text{def}}{\Leftrightarrow} N \text{ is internally stable} \qquad (8.36)$$

$$\text{NP} \overset{\text{def}}{\Leftrightarrow} \|N_{22}\|_\infty < 1; \quad \text{and NS} \qquad (8.37)$$

$$\text{RS} \overset{\text{def}}{\Leftrightarrow} F = F_u(N, \Delta) \text{ is stable } \forall\Delta, \|\Delta\|_\infty \leq 1; \quad \text{and NS} \qquad (8.38)$$

$$\text{RP} \overset{\text{def}}{\Leftrightarrow} \|F\|_\infty < 1, \quad \forall\Delta, \|\Delta\|_\infty \leq 1; \quad \text{and NS} \qquad (8.39)$$

Remark 1 Important. As a prerequisite for nominal performance (NP), robust stability (RS) and robust performance (RP), we must first satisfy nominal stability (NS). This is because the frequency-by-frequency conditions can also be satisfied for unstable systems.

Remark 2 The definitions of RS and RP are useful only if we can test them in an efficient manner, that is, without having to search through the infinite set of allowable perturbations Δ. We will show how this can be done by introducing the structured singular value, μ, as our analysis tool. At the end of the chapter we also discuss how to synthesize controllers such that we have "optimal robust performance" by minimizing μ over the set of stabilizing controllers.

Remark 3 Convention for inequalities. In this book we use the convention that the perturbations are bounded such that they are less than *or equal* to one. This results in a stability condition with a strict inequality, for example, RS $\forall \|\Delta\|_\infty \leq 1$ if $\|M\|_\infty < 1$. (We could alternatively have bounded the uncertainty with a strict inequality, yielding the equivalent condition RS $\forall \|\Delta\|_\infty < 1$ if $\|M\|_\infty \leq 1$.)

Remark 4 Allowed perturbations. For simplicity below we will use the shorthand notation

$$\forall \Delta \quad \text{and} \quad \max_\Delta \qquad (8.40)$$

to mean "for all Δ's in the set of allowed perturbations", and "maximizing over all Δ's in the set of allowed perturbations". By *allowed perturbations* we mean that the $\mathcal{H}_\infty$ norm of Δ is less or equal to 1, $\|\Delta\|_\infty \leq 1$, and that Δ has a specified block-diagonal structure where certain blocks may be restricted to be real. To be mathematically exact, we should replace Δ in (8.40) by $\Delta \in \mathbf{B_\Delta}$, where

$$\mathbf{B_\Delta} = \{\Delta \in \mathbf{\Delta} : \|\Delta\|_\infty \leq 1\}$$

is the set of unity norm-bounded perturbations with a given structure $\mathbf{\Delta}$. The allowed structure should also be defined, for example by

$$\mathbf{\Delta} = \{\text{diag}\,[\delta_1 I_{r1}, \ldots, \delta_S I_{rS}, \Delta_1, \ldots, \Delta_F] : \delta_i \in \mathcal{R}, \Delta_j \in \mathcal{C}^{m_j \times m_j}\}$$

where in this case S denotes the number of real scalars (some of which may be repeated), and F the number of complex blocks. This gets rather involved. Fortunately, this amount of detail is rarely required as it is usually clear what we mean by "for all allowed perturbations" or "$\forall \Delta$".

8.5   Robust stability of the $M\Delta$-structure

Consider the uncertain $N\Delta$-system in Figure 8.2 for which the transfer function from w to z is, as in (8.35), given by

$$F_u(N, \Delta) = N_{22} + N_{21}\Delta(I - N_{11}\Delta)^{-1}N_{12} \qquad (8.41)$$

Suppose that the system is nominally stable (with $\Delta = 0$), that is, N is stable (which means that the whole of N, and not only N_{22} must be stable). We also assume that Δ is stable. We then see directly from (8.41) that the only possible source of instability is the feedback term $(I - N_{11}\Delta)^{-1}$. Thus, when we have nominal stability (NS), the stability of the system in Figure 8.2 is equivalent to the stability of the $M\Delta$-structure in Figure 8.3 where $M = N_{11}$.

We thus need to derive conditions for checking the stability of the $M\Delta$-structure. The next theorem follows from the generalized Nyquist Theorem 4.7. It applies to

$\mathcal{H}_\infty$ norm-bounded Δ-perturbations, but as can be seen from the statement it also applies to any other *convex* set of perturbations (e.g. sets with other structures or sets bounded by different norms).

Theorem 8.1 Determinant stability condition (Real or complex perturbations).
Assume that the nominal system $M(s)$ and the perturbations $\Delta(s)$ are stable. Consider the convex set of perturbations Δ, such that if Δ' is an allowed perturbation then so is $c\Delta'$ where c is any __real__ scalar such that $|c| \leq 1$. Then the $M\Delta$-system in Figure 8.3 is stable for all allowed perturbations (we have RS) if and only if

$$\text{Nyquist plot of } \det (I - M\Delta(s)) \text{ does not encircle the origin, } \forall\Delta \quad (8.42)$$

$$\Leftrightarrow \quad \boxed{\det (I - M\Delta(j\omega)) \neq 0, \quad \forall\omega, \forall\Delta} \quad (8.43)$$

$$\Leftrightarrow \quad \lambda_i(M\Delta) \neq 1, \quad \forall i, \forall\omega, \forall\Delta \quad (8.44)$$

Proof: Condition (8.42) is simply the generalized Nyquist Theorem (page 146) applied to a positive feedback system with a stable loop transfer function $M\Delta$.

(8.42) $\Rightarrow$ (8.43): This is obvious since by "encirclement of the origin" we also include the origin itself.

(8.42) $\Leftarrow$ (8.43) is proved by proving not(8.42) $\Rightarrow$ not(8.43): First note that with $\Delta = 0$, $\det(I - M\Delta) = 1$ at all frequencies. Assume there exists a perturbation Δ' such that the image of $\det(I - M\Delta'(s))$ encircles the origin as s traverses the Nyquist D-contour. Because the Nyquist contour and its map is closed, there then exists another perturbation in the set, $\Delta'' = \epsilon\Delta'$ with $\epsilon \in [0, 1]$, and an ω' such that $\det(I - M\Delta''(j\omega')) = 0$.

(8.44) is equivalent to (8.43) since $\det(I - A) = \prod_i \lambda_i(I - A)$ and $\lambda_i(I - A) = 1 - \lambda_i(A)$ (see Appendix A.2.1). $\qquad\square$

The following is a special case of Theorem 8.1 which applies to complex perturbations.

Theorem 8.2 Spectral radius condition for complex perturbations. *Assume that the nominal system $M(s)$ and the perturbations $\Delta(s)$ are stable. Consider the class of perturbations, Δ, such that if Δ' is an allowed perturbation then so is $c\Delta'$ where c is any __complex__ scalar such that $|c| \leq 1$. Then the $M\Delta$- system in Figure 8.3 is stable for all allowed perturbations (we have RS) if and only if*

$$\rho(M\Delta(j\omega)) < 1, \quad \forall\omega, \forall\Delta \quad (8.45)$$

or equivalently

$$\text{RS} \quad \Leftrightarrow \quad \max_\Delta \rho(M\Delta(j\omega)) < 1, \quad \forall\omega \quad (8.46)$$

Proof: (8.45) $\Rightarrow$ (8.43) ($\Leftrightarrow$ RS) is "obvious": It follows from the definition of the spectral radius ρ, and applies also to real Δ's.

(8.43) $\Rightarrow$ (8.45) is proved by proving not(8.45) $\Rightarrow$ not(8.43): Assume there exists a perturbation Δ' such that $\rho(M\Delta') = 1$ at some frequency. Then $|\lambda_i(M\Delta')| = 1$ for some eigenvalue i, and there always exists another perturbation in the set, $\Delta'' = c\Delta'$ where c is a *complex* scalar with $|c| = 1$, such that $\lambda_i(M\Delta'') = +1$ (real and positive) and therefore $\det(I - M\Delta'') = \prod_i \lambda_i(I - M\Delta'') = \prod_i(1 - \lambda_i(M\Delta'')) = 0$. Finally, the equivalence between (8.45) and (8.46) is simply the definition of $\max_\Delta$. $\square$

Remark 1 The proof of (8.45) relies on adjusting the phase of $\lambda_i(Mc\Delta')$ using the complex scalar c and thus requires the perturbation to be complex.

Remark 2 In words, Theorem 8.2 tells us that we have stability if and only if the spectral radius of $M\Delta$ is less than 1 at all frequencies and for all allowed perturbations, Δ. The main problem here is of course that we have to test the condition for an infinite set of Δ's, and this is difficult to check numerically.

Remark 3 Theorem 8.1, which applies to both real and complex perturbations, forms the basis for the general definition of the structured singular value in (8.76).

8.6   RS for complex unstructured uncertainty

In this section, we consider the special case where $\Delta(s)$ is allowed to be *any* (full) complex transfer function matrix satisfying $\|\Delta\|_\infty \leq 1$. This is often referred to as *unstructured uncertainty* or as *full-block complex perturbation uncertainty*.

Lemma 8.3 *Let Δ be the set of* all *complex matrices such that $\bar\sigma(\Delta) \leq 1$. Then the following holds*

$$\max_\Delta \rho(M\Delta) = \max_\Delta \bar\sigma(M\Delta) = \max_\Delta \bar\sigma(\Delta)\bar\sigma(M) = \bar\sigma(M) \qquad (8.47)$$

Proof: In general, the spectral radius (ρ) provides a lower bound on the spectral norm ($\bar\sigma$) (see (A.116)), and we have

$$\max_\Delta \rho(M\Delta) \leq \max_\Delta \bar\sigma(M\Delta) \leq \max_\Delta \bar\sigma(\Delta)\bar\sigma(M) = \bar\sigma(M) \qquad (8.48)$$

where the second inequality in (8.48) follows since $\bar\sigma(AB) \leq \bar\sigma(A)\bar\sigma(B)$. Now, we need to show that we actually have equality. This will be the case if for any M there exists an allowed Δ' such that $\rho(M\Delta') = \bar\sigma(M)$. Such a Δ' does indeed exist if we allow Δ' to be a full matrix such that all directions in Δ' are allowed: Select $\Delta' = VU^H$ where U and V are matrices of the left and right singular vectors of $M = U\Sigma V^H$. Then $\bar\sigma(\Delta') = 1$ and $\rho(M\Delta') = \rho(U\Sigma V^H V U^H) = \rho(U\Sigma U^H) = \rho(\Sigma) = \bar\sigma(M)$. The second to last equality follows since $U^H = U^{-1}$ and the eigenvalues are invariant under similarity transformations. $\square$

Lemma 8.3 together with Theorem 8.2 directly yield the following theorem:

Theorem 8.4 RS for unstructured ("full") perturbations. *Assume that the nominal system $M(s)$ is stable (NS) and that the perturbations $\Delta(s)$ are stable. Then the*

M Δ-system in Figure 8.3 is stable for all perturbations Δ satisfying $\|\Delta\|_\infty \leq 1$ (i.e. we have RS) if and only if

$$\boxed{\bar{\sigma}(M(j\omega)) < 1 \quad \forall w} \qquad \Leftrightarrow \qquad \boxed{\|M\|_\infty < 1} \qquad (8.49)$$

Remark 1 Condition (8.49) may be rewritten as

$$\text{RS} \Leftrightarrow \bar{\sigma}(M(j\omega))\,\bar{\sigma}(\Delta(j\omega)) < 1, \quad \forall w, \forall \Delta, \qquad (8.50)$$

The sufficiency of (8.50) ($\Rightarrow$) also follows directly from the small gain theorem by choosing $L = M\Delta$. The small gain theorem applies to any operator norm satisfying $\|AB\| \leq \|A\| \cdot \|B\|$.

Remark 2 An important reason for using the $\mathcal{H}_\infty$ norm to describe model uncertainty, is that the stability condition in (8.50) is both necessary and sufficient. In contrast, use of the $\mathcal{H}_2$ norm yields neither necessary nor sufficient conditions for stability. We do not get sufficiency since the $\mathcal{H}_2$ norm does *not* in general satisfy $\|AB\| \leq \|A\| \cdot \|B\|$.

8.6.1 Application of the unstructured RS-condition

We will now present necessary and sufficient conditions for robust stability (RS) for each of the six single unstructured perturbations in Figure 8.5. with

$$E = W_2 \Delta W_1, \quad \|\Delta\|_\infty \leq 1 \qquad (8.51)$$

To derive the matrix M we simply "isolate" the perturbation, and determine the transfer function matrix

$$M = W_1 M_0 W_2 \qquad (8.52)$$

from the output to the input of the perturbation, where M_0 for each of the six cases becomes (disregarding some negative signs which do not affect the subsequent robustness condition) is given by

$$\begin{aligned}
G_p = G + E_A : & \quad M_0 = K(I + GK)^{-1} = KS & (8.53) \\
G_p = G(I + E_I) : & \quad M_0 = K(I + GK)^{-1}G = T_I & (8.54) \\
G_p = (I + E_O)G : & \quad M_0 = GK(I + GK)^{-1} = T & (8.55) \\
G_p = G(I - E_{iA}G)^{-1} : & \quad M_0 = (I + GK)^{-1}G = SG & (8.56) \\
G_p = G(I - E_{iI})^{-1} : & \quad M_0 = (I + KG)^{-1} = S_I & (8.57) \\
G_p = (I - E_{iO})^{-1}G : & \quad M_0 = (I + GK)^{-1} = S & (8.58)
\end{aligned}$$

For example, (8.54) and (8.55) follow from the diagonal elements in the M-matrix in (8.34), and the others are derived in a similar fashion. Note that the sign of M_0 does not matter as it may be absorbed into Δ. Theorem 8.4 then yields

$$\text{RS} \quad \Leftrightarrow \quad \|W_1 M_0 W_2(j\omega)\|_\infty < 1, \forall\, w \qquad (8.59)$$

For instance, from (8.54) and (8.59) we get for multiplicative input uncertainty with a scalar weight:

$$\text{RS } \forall G_p = G(I + w_I \Delta_I), \; \|\Delta_I\|_\infty \leq 1 \; \Leftrightarrow \; \|w_I T_I\|_\infty < 1 \qquad (8.60)$$

Note that the SISO-condition (7.33) follows as a special case of (8.60) Similarly, (7.53) follows as a special case of the inverse multiplicative output uncertainty in (8.58):

$$\text{RS } \forall G_p = (I - w_{iO}\Delta_{iO})^{-1}G, \; \|\Delta_{iO}\|_\infty \leq 1 \; \Leftrightarrow \; \|w_{iO}S\|_\infty < 1 \qquad (8.61)$$

In general, the unstructured uncertainty descriptions in terms of a single perturbation are not "tight" (in the sense that at each frequency all complex perturbations satisfying $\bar{\sigma}(\Delta(j\omega)) \leq 1$ may not be possible in practice). Thus, the above RS-conditions are often conservative. In order to get tighter conditions we must use a tighter uncertainty description in terms of a block-diagonal Δ.

8.6.2   RS for coprime factor uncertainty

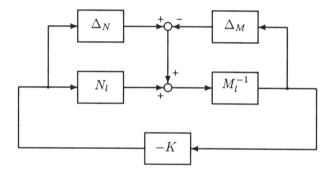

Figure 8.9: Coprime uncertainty

Robust stability bounds in terms of the $\mathcal{H}_\infty$ norm (RS $\Leftrightarrow \|M\|_\infty < 1$) are in general only tight when there is a single full perturbation block. An "exception" to this is when the uncertainty blocks enter or exit from the same location in the block diagram, because they can then be stacked on top of each other or side-by-side, in an overall Δ which is then a full matrix. If we norm-bound the combined (stacked) uncertainty, we then get a tight condition for RS in terms of $\|M\|_\infty$.

One important uncertainty description that falls into this category is the coprime uncertainty description shown in Figure 8.9, for which the set of plants is

$$G_p = (M_l + \Delta_M)^{-1}(N_l + \Delta_N), \quad \|[\Delta_N \quad \Delta_M]\|_\infty \leq \epsilon \qquad (8.62)$$

where $G = M_l^{-1} N_l$ is a left coprime factorization of the nominal plant, see (4.20). This uncertainty description is surprisingly general, it allows both zeros and poles to cross into the right-half plane, and has proved to be very useful in applications (McFarlane and Glover, 1990). Since we have no weights on the perturbations, it is reasonable to use a normalized coprime factorization of the nominal plant; see (4.25). Also note that we choose *not* to normalize the perturbations to be less than 1 in this case. This is because this uncertainty description is most often used in a controller design procedure where the objective is to maximize the magnitude of the uncertainty (ϵ) such that RS is maintained.

In any case, to test for RS we can rearrange the block diagram to match the $M\Delta$-structure in Figure 8.3 with

$$\Delta = [\Delta_N \quad \Delta_M]; \quad M = - \begin{bmatrix} K \\ I \end{bmatrix} (I + GK)^{-1} M_l^{-1} \qquad (8.63)$$

We then get from Theorem 8.4

$$\text{RS } \forall \| \Delta_N \quad \Delta_M \|_\infty \leq \epsilon \quad \Leftrightarrow \quad \|M\|_\infty < 1/\epsilon \qquad (8.64)$$

The above robust stability result is central to the $\mathcal{H}_\infty$ loop-shaping design procedure discussed in Chapter 9.

Remark. In (8.62) we bound the combined (stacked) uncertainty, $\| [\Delta_N \quad \Delta_M] \|_\infty \leq \epsilon$, which is *not* quite the same as bounding the individual blocks, $\|\Delta_N\|_\infty \leq \epsilon$ and $\|\Delta_M\|_\infty \leq \epsilon$. However, from (A.45) we see that these two approaches differ at most by a factor of $\sqrt{2}$, so it is not an important issue from a practical point of view.

Exercise 8.15 *Consider combined multiplicative and inverse multiplicative uncertainty at the output, $G_p = (I - \Delta_{iO} W_{iO})^{-1}(I + \Delta_O W_O)G$, where we choose to norm-bound the combined uncertainty, $\| [\Delta_{iO} \quad \Delta_O] \|_\infty \leq 1$. Make a block diagram of the uncertain plant, and derive a necessary and sufficient condition for robust stability of the closed-loop system.*

8.7 RS with structured uncertainty: Motivation

Consider now the presence of structured uncertainty, where $\Delta = \text{diag}\{\Delta_i\}$ is block-diagonal. To test for robust stability we rearrange the system into the $M\Delta$-structure and we have from (8.49)

$$\text{RS} \quad \text{if} \quad \bar{\sigma}(M(j\omega)) < 1, \forall \omega \qquad (8.65)$$

We have here written "if" rather than "if and only if" since this condition is only sufficient for RS when Δ has "no structure" (full-block uncertainty). The question is whether we can take advantage of the fact that $\Delta = \text{diag}\{\Delta_i\}$ is structured to obtain an RS-condition which is tighter than (8.65). One idea is to make use of the fact that

SAME UNCERTAINTY

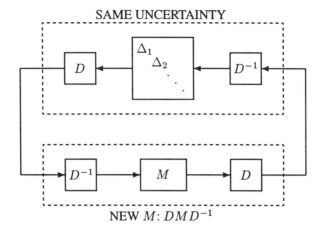

NEW M: DMD^{-1}

Figure 8.10: Use of block-diagonal scalings, $\Delta D = D\Delta$

stability must be independent of scaling. To this effect, introduce the block-diagonal scaling matrix

$$D = \text{diag}\{d_i I_i\} \tag{8.66}$$

where d_i is a scalar and I_i is an identity matrix of the same dimension as the i'th perturbation block, Δ_i. Now rescale the inputs and outputs to M and Δ by inserting the matrices D and D^{-1} on both sides as shown in Figure 8.10. This clearly has no effect on stability. Next, note that with the chosen form for the scalings we have for each perturbation block $\Delta_i = d_i \Delta_i d_i^{-1}$, that is, we have $\Delta = D\Delta D^{-1}$. This means that (8.65) must also apply if we replace M by DMD^{-1} (see Figure 8.10), and we have

$$\text{RS} \quad \text{if} \quad \bar{\sigma}(DMD^{-1}) < 1, \forall \omega \tag{8.67}$$

This applies for any D in (8.66), and therefore the "most improved" (least conservative) RS-condition is obtained by minimizing at each frequency the scaled singular value, and we have

$$\boxed{\text{RS} \quad \text{if} \quad \min_{D(\omega)\in\mathcal{D}} \bar{\sigma}(D(\omega)M(j\omega)D(\omega)^{-1}) < 1, \forall \omega} \tag{8.68}$$

where $\mathcal{D}$ is the set of block-diagonal matrices whose structure is compatible to that of Δ, i.e, $\Delta D = D\Delta$. We will return with more examples of this compatibility later. Note that when Δ is a full matrix, we must select $D = dI$ and we have $\bar{\sigma}(DMD^{-1}) = \bar{\sigma}(M)$, and so as expected (8.68) is identical to (8.65). However, when Δ has structure, we get more degrees of freedom in D, and $\bar{\sigma}(DMD^{-1})$ may be significantly smaller than $\bar{\sigma}(M)$.

Remark 1 Historically, the RS-condition in (8.68) directly motivated the introduction of the structured singular value, $\mu(M)$, discussed in detail in the next section. As one might guess, we have that $\mu(M) \leq \min_D \bar{\sigma}(DMD^{-1})$. In fact, for block-diagonal complex perturbations we generally have that $\mu(M)$ is very close to $\min_D \bar{\sigma}(DMD^{-1})$.

Remark 2 Other norms. Condition (8.68) is essentially a scaled version of the small gain theorem. Thus, a similar condition applies when we use other matrix norms. The $M\Delta$-structure in Figure 8.3 is stable for all block-diagonal Δ's which satisfy $\|\Delta(j\omega)\| \leq 1, \forall w$ if

$$\min_{D(\omega) \in \mathcal{D}} \|D(\omega)M(j\omega)D(\omega)^{-1}\| < 1, \forall w \tag{8.69}$$

where D as before is compatible with the block-structure of Δ. Any matrix norm may be used; for example, the Frobenius norm, $\|M\|_F$, or any induced matrix norm such as $\|M\|_{i1}$ (maximum column sum), $\|M\|_{i\infty}$ (maximum row sum), or $\|M\|_{i2} = \bar{\sigma}(M)$, which is the one we will use. Although in some cases it may be convenient to use other norms, we usually prefer $\bar{\sigma}$ because for this norm we get a necessary and sufficient RS-condition.

8.8   The structured singular value

The structured singular value (denoted Mu, mu, SSV or μ) is a function which provides a generalization of the singular value, $\bar{\sigma}$, and the spectral radius, ρ. We will use μ to get necessary and sufficient conditions for robust stability and also for robust performance. How is μ defined? A simple statement is:

> *Find the smallest structured Δ (measured in terms of $\bar{\sigma}(\Delta)$) which makes* $\det(I - M\Delta) = 0$; *then* $\mu(M) = 1/\bar{\sigma}(\Delta)$.

Mathematically,

$$\mu(M)^{-1} \triangleq \min_\Delta \{\bar{\sigma}(\Delta)| \det(I - M\Delta) = 0 \text{ for structured } \Delta\} \tag{8.70}$$

Clearly, $\mu(M)$ depends not only on M but also on the allowed structure for Δ. This is sometimes shown explicitly by using the notation $\mu_\Delta(M)$.

Remark. For the case where Δ is "unstructured" (a full matrix), the smallest Δ which yields singularity has $\bar{\sigma}(\Delta) = 1/\bar{\sigma}(M)$, and we have $\mu(M) = \bar{\sigma}(M)$. A particular smallest Δ which achieves this is $\Delta = \frac{1}{\sigma_1} v_1 u_1^H$.

Example 8.5 Full perturbation (Δ *is unstructured*). *Consider*

$$M = \begin{bmatrix} 2 & 2 \\ -1 & -1 \end{bmatrix} = \begin{bmatrix} 0.894 & 0.447 \\ -0.447 & 0.894 \end{bmatrix} \begin{bmatrix} 3.162 & 0 \\ 0 & 0 \end{bmatrix} \begin{bmatrix} 0.707 & -0.707 \\ 0.707 & 0.707 \end{bmatrix}^H \tag{8.71}$$

The perturbation

$$\Delta = \frac{1}{\sigma_1} v_1 u_1^H = \frac{1}{3.162} \begin{bmatrix} 0.707 \\ 0.707 \end{bmatrix} [0.894 \quad -0.447] = \begin{bmatrix} 0.200 & 0.200 \\ -0.100 & -0.100 \end{bmatrix} \tag{8.72}$$

with $\bar{\sigma}(\Delta) = 1/\bar{\sigma}(M) = 1/3.162 = 0.316$ *makes* $\det(I - M\Delta) = 0$. *Thus* $\mu(M) = 3.162$ *when* Δ *is a full matrix.*

Note that the perturbation Δ in (8.72) is a full matrix. If we restrict Δ to be diagonal then we need a larger perturbation to make $\det(I - M\Delta) = 0$. This is illustrated next.

Example 8.5 **continued. Diagonal perturbation** (Δ *is structured*). *For the matrix* M *in* (8.71), *the smallest diagonal* Δ *which makes* $\det(I - M\Delta) = 0$ *is*

$$\Delta = \frac{1}{3} \begin{bmatrix} 1 & 0 \\ 0 & -1 \end{bmatrix} \tag{8.73}$$

with $\bar{\sigma}(\Delta) = 0.333$. *Thus* $\mu(M) = 3$ *when* Δ *is a diagonal matrix.*

The above example shows that μ depends on the structure of Δ. The following example demonstrates that μ also depends on whether the perturbation is real or complex.

Example 8.6 μ **of a scalar.** *If* M *is a scalar then in most cases* $\mu(M) = |M|$. *This follows from* (8.70) *by selecting* $|\Delta| = 1/|M|$ *such that* $(1 - M\Delta) = 0$. *However, this requires that we can select the phase of* Δ *such that* $M\Delta$ *is real, which is impossible when* Δ *is real and* M *has an imaginary component, so in this case* $\mu(M) = 0$. *In summary, we have*

$$\Delta \text{ complex}: \qquad \mu(M) = |M| \tag{8.74}$$

$$\Delta \text{ real}: \qquad \mu(M) = \begin{cases} |M| & \text{for real } M \\ 0 & \text{otherwise} \end{cases} \tag{8.75}$$

The definition of μ in (8.70) involves varying $\bar{\sigma}(\Delta)$. However, we prefer to normalize Δ such that $\bar{\sigma}(\Delta) \leq 1$. We can do this by scaling Δ by a factor k_m, and looking for the *smallest* k_m which makes the matrix $I - k_m M\Delta$ singular, and μ is then the reciprocal of this smallest k_m, i.e. $\mu = 1/k_m$. This results in the following alternative definition of μ.

Definition 8.1 Structured Singular Value. *Let* M *be a given complex matrix and let* $\Delta = \text{diag}\{\Delta_i\}$ *denote a set of complex matrices with* $\bar{\sigma}(\Delta) \leq 1$ *and with a given block-diagonal structure (in which some of the blocks may be repeated and some may be restricted to be real). The real non-negative function* $\mu(M)$, *called the structured singular value, is defined by*

$$\mu(M) \triangleq \frac{1}{\min\{k_m | \det(I - k_m M\Delta) = 0 \text{ for structured } \Delta, \bar{\sigma}(\Delta) \leq 1\}} \tag{8.76}$$

If no such structured Δ *exists then* $\mu(M) = 0$.

A value of $\mu = 1$ means that there exists a perturbation with $\bar{\sigma}(\Delta) = 1$ which is just large enough to make $I - M\Delta$ singular. A larger value of μ is "bad" as it means that a smaller perturbation makes $I - M\Delta$ singular, whereas a smaller value of μ is "good".

8.8.1   Remarks on the definition of μ

1. The structured singular value was introduced by Doyle (1982). At the same time (in fact, in the same issue of the same journal) Safonov (1982) introduced the *Multivariable Stability Margin* k_m for a diagonally perturbed system as the inverse of μ, that is, $k_m(M) = \mu(M)^{-1}$. In many respects, this is a more natural definition of a robustness margin. However, $\mu(M)$ has a number of other advantages, such as providing a generalization of the spectral radius, $\rho(M)$, and the spectral norm, $\bar{\sigma}(M)$.

2. The Δ corresponding to the smallest k_m in (8.76) will always have $\bar{\sigma}(\Delta) = 1$, since if $\det(I - k'_m M\Delta') = 0$ for some Δ' with $\bar{\sigma}(\Delta') = c < 1$, then $1/k'_m$ cannot be the structured singular value of M, since there exists a smaller scalar $k_m = k'_m c$ such that $\det(I - k_m M\Delta) = 0$ where $\Delta = \frac{1}{c}\Delta'$ and $\bar{\sigma}(\Delta) = 1$.

3. Note that with $k_m = 0$ we obtain $I - k_m M\Delta = I$ which is clearly non-singular. Thus, one possible way to obtain μ numerically, is to start with $k_m = 0$, and gradually increase k_m until we first find an allowed Δ with $\bar{\sigma}(\Delta) = 1$ such that $(I - k_m M\Delta)$ is singular (this value of k_m is then $1/\mu$). By "allowed" we mean that Δ must have the specified block-diagonal structure and that some of the blocks may have to be real.

4. The sequence of M and Δ in the definition of μ does not matter. This follows from the identity (A.12) which yields

$$\det(I - k_m M\Delta) = \det(I - k_m \Delta M) \tag{8.77}$$

5. In most cases M and Δ are square, but this need not be the case. If they are non-square, then we make use of (8.77) and work with either $M\Delta$ or ΔM (whichever has the lowest dimension).

The remainder of this section deals with the properties and computation of μ. Readers who are primarily interested in the practical use of μ may skip most of this material.

8.8.2   Properties of μ for real and complex Δ

Two properties of μ which hold for both real and complex perturbations are:

1. $\mu(\alpha M) = |\alpha|\mu(M)$ for any *real* scalar α.
2. Let $\Delta = \text{diag}\{\Delta_1, \Delta_2\}$ be a block-diagonal perturbation (in which Δ_1 and Δ_2 may have additional structure) and let M be partitioned accordingly. Then

$$\mu_\Delta(M) \geq \max\{\mu_{\Delta_1}(M_{11}), \mu_{\Delta_2}(M_{22})\} \tag{8.78}$$

Proof: Consider $\det(I - \frac{1}{\mu}M\Delta)$ where $\mu = \mu_\Delta(M)$ and use Schur's formula in (A.14) with $A_{11} = I - \frac{1}{\mu}M_{11}\Delta_1$ and $A_{12} = I - \frac{1}{\mu}M_{22}\Delta_2$.   □

In words, (8.78) simply says that robustness with respect to two perturbations taken together is at least as bad as for the worst perturbation considered alone. This agrees with our intuition that we cannot improve robust stability by including another uncertain perturbation.

In addition, the *upper bounds* given below for complex perturbations, e.g. $\mu_\Delta(M) \leq \min_{D \in \mathcal{D}} \bar{\sigma}(DMD^{-1})$ in (8.87), also hold for real or mixed real/complex perturbations Δ. This follows because complex perturbations include real perturbations as a special case. However, the lower bounds, e.g. $\mu(M) \geq \rho(M)$ in (8.82), generally hold only for complex perturbations.

8.8.3 μ for complex Δ

When all the blocks in Δ are complex, μ may be computed relatively easily. This is discussed below and in more detail in the survey paper by Packard and Doyle (1993). The results are mainly based on the following result, which may be viewed as another definition of μ that applies for complex Δ only.

Lemma 8.5 *For complex perturbations Δ with $\bar{\sigma}(\Delta) \leq 1$:*

$$\boxed{\mu(M) = \max_{\Delta, \bar{\sigma}(\Delta) \leq 1} \rho(M\Delta)} \tag{8.79}$$

Proof: The lemma follows directly from the definition of μ and the equivalence between (8.43) and (8.46).   □

Properties of μ for complex perturbations

Most of the properties below follow easily from (8.79).

1. $\mu(\alpha M) = |\alpha|\mu(M)$ for any (*complex*) scalar α.
2. For a repeated scalar complex perturbation we have

$$\Delta = \delta I \ (\delta \text{ is a complex scalar}) : \quad \mu(M) = \rho(M) \tag{8.80}$$

Proof: Follows directly from (8.79) since there are no degrees-of-freedom for the maximization.   □

3. For a full block complex perturbation we have from (8.79) and (8.47):

$$\Delta \text{ full matrix} : \quad \mu(M) = \bar{\sigma}(M) \tag{8.81}$$

4. μ for complex perturbations is bounded by the spectral radius and the singular value (spectral norm):

$$\boxed{\rho(M) \leq \mu(M) \leq \bar{\sigma}(M)} \tag{8.82}$$

This follows from (8.80) and (8.81), since selecting $\Delta = \delta I$ gives the fewest degrees-of-freedom for the optimization in (8.79), whereas selecting Δ full gives the most degrees-of-freedom.

5. Consider any unitary matrix U with the same structure as Δ. Then

$$\mu(MU) = \mu(M) = \mu(UM) \tag{8.83}$$

Proof: Follows from (8.79) by writing $MU\Delta = M\Delta'$ where $\bar{\sigma}(\Delta') = \bar{\sigma}(U\Delta) = \bar{\sigma}(\Delta)$, and so U may always be absorbed into Δ.   □

6. Consider any matrix D which commutes with Δ, that is, $\Delta D = D\Delta$. Then

$$\mu(DM) = \mu(MD) \quad \text{and} \quad \mu(DMD^{-1}) = \mu(M) \tag{8.84}$$

Proof: $\mu(DM) = \mu(MD)$ follows from

$$\mu_\Delta(DM) = \max_\Delta \rho(DM\Delta) = \max_\Delta \rho(M\Delta D) = \max_\Delta \rho(MD\Delta) = \mu_\Delta(MD) \tag{8.85}$$

The first equality is (8.79). The second equality applies since $\rho(AB) = \rho(BA)$ (by the eigenvalue properties in the Appendix). The key step is the third equality which applies only when $D\Delta = \Delta D$. The fourth equality again follows from (8.79).   □

7. **Improved lower bound.** Define $\mathcal{U}$ as the set of all unitary matrices U with the same block-diagonal structure as Δ. Then for complex Δ

$$\boxed{\mu(M) = \max_{U \in \mathcal{U}} \rho(MU)} \tag{8.86}$$

Proof: The proof of this important result is given by Doyle (1982) and Packard and Doyle (1993). It follows from a generalization of the maximum modulus theorem for rational functions.   □

The result (8.86) is motivated by combining (8.83) and (8.82) to yield

$$\mu(M) \geq \max_{U \in \mathcal{U}} \rho(MU)$$

The surprise is that this is always an equality. Unfortunately, the optimization in (8.86) is not convex and so it may be difficult to use in calculating μ numerically.

8. **Improved upper bound.** Define $\mathcal{D}$ to be the set of matrices D which commute with Δ (i.e. satisfy $D\Delta = \Delta D$). Then it follows from (8.84) and (8.82) that

$$\boxed{\mu(M) \leq \min_{D \in \mathcal{D}} \bar{\sigma}(DMD^{-1})} \tag{8.87}$$

This optimization is convex in D (i.e. has only one minimum, the global minimum) and it may be shown (Doyle, 1982) that the inequality is in fact an equality if there are 3 or fewer blocks in Δ. Furthermore, numerical evidence suggests that the bound is tight (within a few percent) for 4 blocks or more; the worst known example has an upper bound which is about 15% larger than μ (Balas et al., 1993). Some examples of D's which commute with Δ are

$$\Delta = \delta I : \quad D = \text{full matrix} \tag{8.88}$$

$$\Delta = \text{full matrix} : \quad D = dI \tag{8.89}$$

$$\Delta = \begin{bmatrix} \Delta_1(\text{full}) & 0 \\ 0 & \Delta_2(\text{full}) \end{bmatrix} : D = \begin{bmatrix} d_1 I & 0 \\ 0 & d_2 I \end{bmatrix} \tag{8.90}$$

$$\Delta = \text{diag}\{\Delta_1(\text{full}), \delta_2 I, \delta_3, \delta_4\} : D = \text{diag}\{d_1 I, D_2(\text{full}), d_3, d_4\} \tag{8.91}$$

In short, we see that the structures of Δ and D are "opposites".

9. Without affecting the optimization in (8.87), we may assume the blocks in D to be Hermitian positive definite, i.e. $D_i = D_i^H > 0$, and for scalars $d_i > 0$ (Packard and Doyle, 1993).

10. One can always simplify the optimization in (8.87) by fixing one of the scalar blocks in D equal to 1. For example, let $D = \text{diag}\{d_1, d_2, \ldots, d_n\}$, then one may without loss of generality set $d_n = 1$.

Proof: Let $D' = \frac{1}{d_n} D$ and note that $\bar{\sigma}(DMD^{-1}) = \bar{\sigma}(D'MD'^{-1})$. $\qquad\qquad$ □

Similarly, for cases where Δ has one or more scalar blocks, one may simplify the optimization in (8.86) by fixing one of the corresponding unitary scalars in U equal to 1. This follows from Property 1 with $|c| = 1$.

11. The following property is useful for finding $\mu(AB)$ when Δ has a structure similar to that of A or B:

$$\mu_\Delta(AB) \leq \bar{\sigma}(A)\mu_{\Delta A}(B) \qquad (8.92)$$

$$\mu_\Delta(AB) \leq \bar{\sigma}(B)\mu_{B\Delta}(A) \qquad (8.93)$$

Here the subscript "ΔA" denotes the structure of the matrix ΔA, and "$B\Delta$" denotes the structure of $B\Delta$.

Proof: The proof is from (Skogestad and Morari, 1988a). Use the fact that $\mu(AB) = \max_\Delta \rho(\Delta AB) = \max_\Delta \rho(VB)\bar{\sigma}(A)$ where $V = \Delta A/\bar{\sigma}(A)$. When we maximize over Δ then V generates a certain set of matrices with $\bar{\sigma}(V) \leq 1$. Let us *extend* this set by maximizing over *all* matrices V with $\bar{\sigma}(V) \leq 1$ and with the same structure as ΔA. We then get $\mu(AB) \leq \max_V \rho(VB)\bar{\sigma}(A) = \mu_V(B)\bar{\sigma}(A)$. $\qquad\qquad$ □

Some special cases of (8.92):

(a) If A is a full matrix then the structure of ΔA is a full matrix, and we simply get $\mu(AB) \leq \bar{\sigma}(A)\bar{\sigma}(B)$ (which is not a very exciting result since we always have $\mu(AB) \leq \bar{\sigma}(AB) \leq \bar{\sigma}(A)\bar{\sigma}(B)$).

(b) If Δ has the same structure as A (e.g. they are both diagonal) then

$$\boxed{\mu_\Delta(AB) \leq \bar{\sigma}(A)\mu_\Delta(B)} \qquad (8.94)$$

Note: (8.94) is stated incorrectly in Doyle (1982) since it is not specified that Δ must have the same structure as A.

(c) If $\Delta = \delta I$ (i.e. Δ consists of repeated scalars), we get the spectral radius inequality $\rho(AB) \leq \bar{\sigma}(A)\mu_A(B)$. A useful special case of this is

$$\rho(M\Delta) \leq \bar{\sigma}(\Delta)\mu_\Delta(M) \qquad (8.95)$$

12. A generalization of (8.92) and (8.93) is:

$$\mu_\Delta(ARB) \leq \bar{\sigma}(R)\mu_{\widetilde{\Delta}}^2 \begin{bmatrix} 0 & A \\ B & 0 \end{bmatrix} \qquad (8.96)$$

where $\widetilde{\Delta} = \text{diag}\{\Delta, R\}$. The result is proved by (Skogestad and Morari, 1988a).

13. The following is a further generalization of these bounds. Assume that M is an LFT of R: $M = N_{11} + N_{12}R(I - N_{22}R)^{-1}N_{21}$. The problem is to find an upper bound on R, $\bar{\sigma}(R) \leq c$, which guarantees that $\mu_\Delta(M) < 1$ when $\mu_\Delta(N_{11}) < 1$. Skogestad and Morari (1988a) show that the best upper bound is the c which solves

$$\mu_{\tilde{\Delta}} \begin{bmatrix} N_{11} & N_{12} \\ cN_{21} & cN_{22} \end{bmatrix} = 1 \qquad (8.97)$$

where $\tilde{\Delta} = \text{diag}\{\Delta, R\}$, and c is easily computed using skewed-μ. Given the μ-condition $\mu_\Delta(M) < 1$ (for RS or RP), (8.97) may be used to derive a sufficient loop-shaping bound on a transfer function of interest, e.g. R may be S, T, L, L^{-1} or K.

Remark. Above we have used $\min_D$. To be mathematically correct, we should have used $\inf_D$ because the set of allowed D's is not bounded and therefore the exact minimum may not be achieved (although we may get arbitrarily close). The use of $\max_\Delta$ (rather than $\sup_\Delta$) is mathematically correct since the set Δ is closed (with $\bar{\sigma}(\Delta) \leq 1$).

Example 8.7 *Let M and Δ be complex 2×2 matrices. Then*

$$M = \begin{bmatrix} a & a \\ b & b \end{bmatrix}, \quad \mu(M) = \begin{cases} \rho(M) = |a + b| & \text{for } \Delta = \delta I \\ |a| + |b| & \text{for } \Delta = \text{diag}\{\delta_1, \delta_2\} \\ \bar{\sigma}(M) = \sqrt{2|a|^2 + 2|b|^2} & \text{for } \Delta \text{ a full matrix} \end{cases}$$
$$(8.98)$$

Proof: For $\Delta = \delta I$, $\mu(M) = \rho(M)$ and $\rho(M) = |a + b|$ since M is singular and its non-zero eigenvalue is $\lambda_1(M) = \text{tr}(M) = a + b$. For Δ full, $\mu(M) = \bar{\sigma}(M)$ and $\bar{\sigma}(M) = \sqrt{2|a|^2 + 2|b|^2}$ since M is singular and its non-zero singular value is $\bar{\sigma}(M) = \|M\|_F$, see (A.126). For a diagonal Δ, it is interesting to consider three different proofs of the result $\mu(M) = |a| + |b|$:

(a) A direct calculation based on the definition of μ.
(b) Use of the lower "bound" in (8.86) (which is always exact).
(c) Use of the upper bound in (8.87) (which is exact here since we have only two blocks).

We will use approach (a) here and leave (b) and (c) for Exercises 8.16 and 8.17. We have

$$M\Delta = \begin{bmatrix} a & a \\ b & b \end{bmatrix} \begin{bmatrix} \delta_1 & \\ & \delta_2 \end{bmatrix} = \begin{bmatrix} a \\ b \end{bmatrix} [\delta_1 \quad \delta_2] = \tilde{M}\tilde{\Delta}$$

From (8.77) we then get

$$\det(I - M\Delta) = \det(I - \tilde{\Delta}\tilde{M}) = 1 - [\delta_1 \quad \delta_2] \begin{bmatrix} a \\ b \end{bmatrix} = 1 - a\delta_1 - b\delta_2$$

The smallest δ_1 and δ_2 which make this matrix singular, i.e. $1 - a\delta_1 - b\delta_2 = 0$, are obtained when $|\delta_1| = |\delta_2| = |\delta|$ and the phases of δ_1 and δ_2 are adjusted such that $1 - |a| \cdot |\delta| - |b| \cdot |\delta| = 0$. We get $|\delta| = 1/(|a| + |b|)$, and from (8.70) we have that $\mu = 1/|\delta| = |a| + |b|$. □

Example 8.8 *Let M be a partitioned matrix with both diagonal blocks equal to zero. Then*

$$\mu \underbrace{\begin{bmatrix} 0 & A \\ B & 0 \end{bmatrix}}_{M} = \begin{cases} \rho(M) = \sqrt{\rho(AB)} & \text{for } \Delta = \delta I \\ \sqrt{\bar{\sigma}(A)\bar{\sigma}(B)} & \text{for } \Delta = \text{diag}\{\Delta_1, \Delta_2\}, \Delta_i \text{ full} \\ \bar{\sigma}(M) = \max\{\bar{\sigma}(A), \bar{\sigma}(B)\} & \text{for } \Delta \text{ a full matrix} \end{cases}$$

(8.99)

Proof: From the definition of eigenvalues and Schur's formula (A.14) we get $\lambda_i(M) = \sqrt{\lambda_i(AB)}$ and $\rho(M) = \sqrt{\rho(AB)}$ follows. For block-diagonal Δ, $\mu(M) = \sqrt{\bar{\sigma}(A)\bar{\sigma}(B)}$ follows in a similar way using $\mu(M) = \max_{\Delta} \rho(M\Delta) = \max_{\Delta_1, \Delta_2} \rho(A\Delta_2 B\Delta_1)$, and then realizing that we can always select Δ_1 and Δ_2 such that $\rho(A\Delta_2 B\Delta_1) = \bar{\sigma}(A)\bar{\sigma}(B)$ (recall (8.47)). $\bar{\sigma}(M) = \max\{\bar{\sigma}(A), \bar{\sigma}(B)\}$ follows since $\bar{\sigma}(M) = \sqrt{\rho(M^H M)}$ where $M^H M = \text{diag}\{B^H B, A^H A\}$. □

Exercise 8.16 *For M in (8.98) and a diagonal Δ show that $\mu(M) = |a| + |b|$ using the lower "bound" $\mu(M) = \max_U \rho(MU)$ (which is always exact). Hint: Use $U = \text{diag}\{e^{j\phi}, 1\}$ (the blocks in U are unitary scalars, and we may fix one of them equal to 1).*

Exercise 8.17 *For M in (8.98) and a diagonal Δ show that $\mu(M) = |a| + |b|$ using the upper bound $\mu(M) \leq \min_D \bar{\sigma}(DMD^{-1})$ (which is exact in this case since D has two "blocks").*
Solution: Use $D = \text{diag}\{d, 1\}$. Since DMD^{-1} is a singular matrix we have from (A.36) that

$$\bar{\sigma}(DMD^{-1}) = \bar{\sigma} \begin{bmatrix} a & da \\ \frac{1}{d}b & b \end{bmatrix} = \sqrt{|a|^2 + |da|^2 + |b/d|^2 + |b|^2}$$

(8.100)

which we want to minimize with respect to d. The solution is $d = \sqrt{|a|/|b|}$ which gives $\mu(M) = \sqrt{|a|^2 + 2|ab| + |b|^2} = |a| + |b|$.

Exercise 8.18 *Let c be a complex scalar. Show that for*

$$\Delta = \text{diag}\{\Delta_1, \Delta_2\} : \quad \mu \begin{bmatrix} M_{11} & M_{12} \\ M_{21} & M_{22} \end{bmatrix} = \mu \begin{bmatrix} M_{11} & cM_{12} \\ \frac{1}{c}M_{21} & M_{22} \end{bmatrix}$$

(8.101)

Exercise 8.19 *Let a, b, c and d be complex scalars. Show that for*

$$\Delta = \text{diag}\{\delta_1, \delta_2\} : \quad \mu \begin{bmatrix} ab & ad \\ bc & cd \end{bmatrix} = \mu \begin{bmatrix} ab & ab \\ cd & cd \end{bmatrix} = |ab| + |cd|$$

(8.102)

Does this hold when Δ is scalar times identity, or when Δ is full? (Answers: Yes and No).

Exercise 8.20 *Assume A and B are square matrices. Show by a counterexample that $\bar{\sigma}(AB)$ is not in general equal to $\bar{\sigma}(BA)$. Under what conditions is $\mu(AB) = \mu(BA)$? (Hint: Recall (8.84)).*

Exercise 8.21 *If (8.94) were true for any structure of Δ then it would imply $\rho(AB) \leq \bar{\sigma}(A)\rho(B)$. Show by a counterexample that this is not true.*

8.9   Robust stability with structured uncertainty

Consider stability of the $M\Delta$-structure in Figure 8.3 for the case where Δ is a set of norm-bounded block-diagonal perturbations. From the determinant stability condition in (8.43) which applies to both complex and real perturbations we get

$$\text{RS} \quad \Leftrightarrow \quad \det(I - M\Delta(j\omega)) \neq 0, \quad \forall \omega, \forall \Delta, \bar{\sigma}(\Delta(j\omega)) \leq 1 \quad \forall \omega \qquad (8.103)$$

A problem with (8.103) is that it is only a "yes/no" condition. To find the factor k_m by which the system is robustly stable, we scale the uncertainty Δ by k_m, and look for the smallest k_m which yields "borderline instability", namely

$$\det(I - k_m M\Delta) = 0 \qquad (8.104)$$

From the definition of μ in (8.76) this value is $k_m = 1/\mu(M)$, and we obtain the following necessary and sufficient condition for robust stability.

Theorem 8.6 RS for block-diagonal perturbations (real or complex). *Assume that the nominal system M and the perturbations Δ are stable. Then the $M\Delta$-system in Figure 8.3 is stable for all allowed perturbations with $\bar{\sigma}(\Delta) \leq 1, \forall \omega$, if and only if*

$$\boxed{\mu(M(j\omega)) < 1, \quad \forall \omega} \qquad (8.105)$$

Proof: $\mu(M) < 1 \Leftrightarrow k_m > 1$, so if $\mu(M) < 1$ at all frequencies the required perturbation Δ to make $\det(I - M\Delta) = 0$ is larger than 1, and the system is stable. On the other hand, $\mu(M) = 1 \Leftrightarrow k_m = 1$, so if $\mu(M) = 1$ at some frequency there does exist a perturbation with $\bar{\sigma}(\Delta) = 1$ such that $\det(I - M\Delta) = 0$ at this frequency, and the system is unstable. $\square$

Condition (8.105) for robust stability may be rewritten as

$$\text{RS} \quad \Leftrightarrow \quad \mu(M(j\omega))\, \bar{\sigma}(\Delta(j\omega)) < 1, \quad \forall \omega \qquad (8.106)$$

which may be interpreted as a "generalized small gain theorem" that also takes into account the *structure* of Δ.

One may argue whether Theorem 8.6 is really a theorem, or a restatement of the definition of μ. In either case, we see from (8.105) that it is trivial to check for robust stability provided we can compute μ.

Let us consider two examples that illustrate how we use μ to check for robust stability with structured uncertainty. In the first example, the structure of the uncertainty is important, and an analysis based on the $\mathcal{H}_\infty$ norm leads to the incorrect conclusion that the system is not robustly stable. In the second example the structure makes no difference.

Example 8.9 RS with diagonal input uncertainty   *Consider robust stability of the feedback system in Figure 8.7 for the case when the multiplicative input uncertainty is*

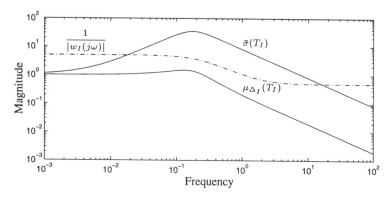

Figure 8.11: Robust stability for diagonal input uncertainty is guaranteed since $\mu_{\Delta_I}(T_I) <$ $1/|w_I|$, $\forall \omega$. The use of unstructured uncertainty and $\bar{\sigma}(T_I)$ is conservative

diagonal. A nominal 2×2 plant and the controller (which represents PI-control of a distillation process using the DV-configuration) is given by

$$G(s) = \frac{1}{\tau s + 1} \begin{bmatrix} -87.8 & 1.4 \\ -108.2 & -1.4 \end{bmatrix}; \quad K(s) = \frac{1 + \tau s}{s} \begin{bmatrix} -0.0015 & 0 \\ 0 & -0.075 \end{bmatrix} \quad (8.107)$$

(time in minutes). The controller results in a nominally stable system with acceptable performance. Assume there is complex multiplicative uncertainty in each manipulated input of magnitude

$$w_I(s) = \frac{s + 0.2}{0.5s + 1} \quad (8.108)$$

This implies a relative uncertainty of up to 20% in the low frequency range, which increases at high frequencies, reaching a value of 1 (100% uncertainty) at about 1 rad/min. The increase with frequency allows for various neglected dynamics associated with the actuator and valve. The uncertainty may be represented as multiplicative input uncertainty as shown in Figure 8.7 where Δ_I is a diagonal complex matrix and the weight is $W_I = w_I I$ where $w_I(s)$ is a scalar. On rearranging the block diagram to match the $M\Delta$-structure in Figure 8.3 we get $M = w_I KG(I + KG)^{-1} = w_I T_I$ (recall (8.32)), and the RS-condition $\mu(M) < 1$ in Theorem 8.6 yields

$$\text{RS} \Leftrightarrow \mu_{\Delta_I}(T_I) < \frac{1}{|w_I(j\omega)|} \quad \forall \omega, \quad \Delta_I = \begin{bmatrix} \delta_1 & \\ & \delta_2 \end{bmatrix} \quad (8.109)$$

This condition is shown graphically in Figure 8.11 and is seen to be satisfied at all frequencies, so the system is robustly stable. Also in Figure 8.11, $\bar{\sigma}(T_I)$ can be seen to be larger than $1/|w_I(j\omega)|$ over a wide frequency range. This shows that the system would be unstable for full-block input uncertainty (Δ_I full). However, full-block uncertainty is not reasonable for this plant, and therefore we conclude that the use of the singular value is conservative in this case. This demonstrates the need for the structured singular value.

Exercise 8.22 *Consider the same example and check for robust stability with full-block multiplicative output uncertainty of the same magnitude. (Solution: RS is satisfied).*

Example 8.10 RS of spinning satellite. *Recall Motivating Example No. 1 from Section 3.7.1 with the plant $G(s)$ given in (3.77) and the controller $K = I$. We want to study how sensitive this design is to multiplicative input uncertainty.*

In this case $T_I = T$, so for RS there is no difference between multiplicative input and multiplicative output uncertainty. In Figure 8.12, we plot $\mu(T)$ as a function of frequency. We find for this case that $\mu(T) = \bar{\sigma}(T)$ irrespective of the structure of the complex multiplicative perturbation (full-block, diagonal or repeated complex scalar). Since $\mu(T)$ crosses 1 at about 10 rad/s, we can tolerate more than 100% uncertainty at frequencies above 10 rad/s. At low frequencies $\mu(T)$ is about 10, so to guarantee RS we can at most tolerate 10% (complex) uncertainty at low frequencies. This confirms the results from Section 3.7.1, where we found

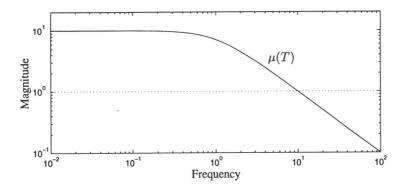

Figure 8.12: μ-plot for robust stability of spinning satellite

that real perturbations $\delta_1 = 0.1$ and $\delta_2 = -0.1$ yield instability. Thus, the use of complex rather than real perturbations is not conservative in this case, at least for Δ_I diagonal.

However, with repeated scalar perturbations (i.e. the uncertainty in each channel is identical) there is a difference between real and complex perturbations. With repeated real perturbations, available software (e.g. using the command mu *with* blk = [-2 0] *in the μ-toolbox in MATLAB) yields a peak μ-value of 1, so we can tolerate a perturbation $\delta_1 = \delta_2$ of magnitude 1 before getting instability (This is confirmed by considering the characteristic polynomial in (3.81), from which we see that $\delta_1 = \delta_2 = -1$ yields instability.) On the other hand, with complex repeated perturbations, we have that $\mu(T) = \rho(T)$ is 10 at low frequencies, so instability may occur with a (non-physical) complex $\delta_1 = \delta_2$ of magnitude 0.1. (Indeed, from (3.81) we see that the non-physical constant perturbation $\delta_1 = \delta_2 = j0.1$ yields instability.)*

8.9.1   What do $\mu \neq 1$ and skewed-μ mean?

A value of $\mu = 1.1$ for robust stability means that *all* the uncertainty blocks must be decreased in magnitude by a factor 1.1 in order to guarantee stability.

But if we want to keep some of the uncertainty blocks fixed, how large can one particular source of uncertainty be before we get instability? We define this value as $1/\mu^s$, where μ^s is called *skewed-μ*. We may view $\mu^s(M)$ as a generalization of $\mu(M)$.

For example, let $\Delta = \text{diag}\{\Delta_1, \Delta_2\}$ and assume we have fixed $\|\Delta_1\| \leq 1$ and we want to find how large Δ_2 can be before we get instability. The solution is to select

$$K_m = \begin{bmatrix} I & 0 \\ 0 & k_m I \end{bmatrix} \tag{8.110}$$

and look at each frequency for the smallest value of k_m which makes $\det(I - K_m M \Delta) = 0$, and we have that skewed-μ is

$$\mu^s(M) \triangleq 1/k_m$$

Note that to compute skewed-μ we must first define which part of the perturbations is to be constant. $\mu^s(M)$ is always further from 1 than $\mu(M)$ is, i.e. $\mu^s \geq \mu$ for $\mu > 1$, $\mu^s = \mu$ for $\mu = 1$, and $\mu^s \leq \mu$ for $\mu < 1$. In practice, with available software to compute μ, we obtain μ^s by iterating on k_m until $\mu(K_m M) = 1$ where K_m may be as in (8.110). This iteration is straightforward since μ increases uniformly with k_m.

8.10 Robust performance

Robust performance (RP) means that the performance objective is satisfied for all possible plants in the uncertainty set, even the worst-case plant. We showed in Chapter 7 that for a SISO system with an $\mathcal{H}_\infty$ performance objective, the RP-condition is identical to a RS-condition with an additional perturbation block.

This also holds for MIMO systems, as illustrated by the step-wise derivation in Figure 8.13. Step B is the key step and the reader is advised to study this carefully in the treatment below. Note that the block Δ_P (where capital P denotes Performance) is always a full matrix. It is a fictitious uncertainty block representing the $\mathcal{H}_\infty$ performance specification.

8.10.1 Testing RP using μ

To test for RP, we first "pull out" the uncertain perturbations and rearrange the uncertain system into the $N\Delta$-form of Figure 8.2. Our RP-requirement, as given in (8.39), is that the $\mathcal{H}_\infty$ norm of the transfer function $F = F_u(N, \Delta)$ remains less than 1 for all allowed perturbations. This may be tested exactly by computing $\mu(N)$ as stated in the following theorem.

Theorem 8.7 Robust performance. *Rearrange the uncertain system into the $N\Delta$-structure of Figure 8.13. Assume nominal stability such that N is (internally) stable. Then*

$$\text{RP} \quad \overset{\text{def}}{\Leftrightarrow} \quad \|F\|_\infty = \|F_u(N, \Delta)\|_\infty < 1, \quad \forall \|\Delta\|_\infty \leq 1 \tag{8.111}$$

$$\Leftrightarrow \qquad \boxed{\mu_{\widehat{\Delta}}(N(j\omega)) < 1, \quad \forall w} \qquad (8.112)$$

where μ is computed with respect to the structure

$$\widehat{\Delta} = \begin{bmatrix} \Delta & 0 \\ 0 & \Delta_P \end{bmatrix} \qquad (8.113)$$

and Δ_P is a full complex perturbation with the same dimensions as F^T.

Below we prove the theorem in two alternative ways, but first a few remarks:

1. Condition (8.112) allows us to test if $\|F\|_\infty < 1$ for all possible Δ's without having to test each Δ individually. Essentially, μ is defined such that it directly addresses the worst case.
2. The μ-condition for RP involves the enlarged perturbation $\widehat{\Delta} = \text{diag}\{\Delta, \Delta_P\}$. Here Δ, which itself may be a block-diagonal matrix, represents the true uncertainty, whereas Δ_P is a *full complex matrix* stemming from the $\mathcal{H}_\infty$ norm performance specification. For example, for the nominal system (with $\Delta = 0$) we get from (8.81) that $\bar{\sigma}(N_{22}) = \mu_{\Delta_P}(N_{22})$, and we see that Δ_P must be a full matrix.
3. Since $\widehat{\Delta}$ always has structure, the use of the $\mathcal{H}_\infty$ norm, $\|N\|_\infty < 1$, is generally conservative for robust performance.
4. From (8.78) we have that

$$\underbrace{\mu_{\widehat{\Delta}}(N)}_{\text{RP}} \geq \max\{\underbrace{\mu_\Delta(N_{11})}_{\text{RS}}, \underbrace{\mu_{\Delta_P}(N_{22})}_{\text{NP}}\} \qquad (8.114)$$

where as just noted $\mu_{\Delta_P}(N_{22}) = \bar{\sigma}(N_{22})$. (8.114) implies that RS ($\mu_\Delta(N_{11}) < 1$) and NP ($\bar{\sigma}(N_{22}) < 1$) are automatically satisfied when RP ($\mu(N) < 1$) is satisfied. However, note that NS (stability of N) is not guaranteed by (8.112) and must be tested separately (Beware! It is a common mistake to get a design with apparently great RP, but which is not nominally stable and thus is actually robustly *unstable*).
5. For a generalization of Theorem 8.7 see the *main loop theorem* of Packard and Doyle (1993); see also Zhou et al. (1996).

Block diagram proof of Theorem 8.7

In the following, let $F = F_u(N, \Delta)$ denote the perturbed closed-loop system for which we want to test RP. The theorem is proved by the equivalence between the various block diagrams in Figure 8.13.

Step A. This is simply the definition of RP; $\|F\|_\infty < 1$.

Step B (the key step). Recall first from Theorem 8.4 that stability of the $M\Delta$-structure in Figure 8.3, where Δ is a *full* complex matrix, is equivalent to $\|M\|_\infty < 1$. From this theorem, we get that the RP-condition $\|F\|_\infty < 1$ is equivalent to RS of the $F\Delta_P$-structure, where Δ_P is a *full* complex matrix.

Step C. Introduce $F = F_u(N, \Delta)$ from Figure 8.2.

Step D. Collect Δ and Δ_P into the block-diagonal matrix $\widehat{\Delta}$. We then have that the original RP-problem is equivalent to RS of the $N\widehat{\Delta}$-structure which from Theorem 8.6 is equivalent to $\mu_{\widehat{\Delta}}(N) < 1$. $\qquad \square$

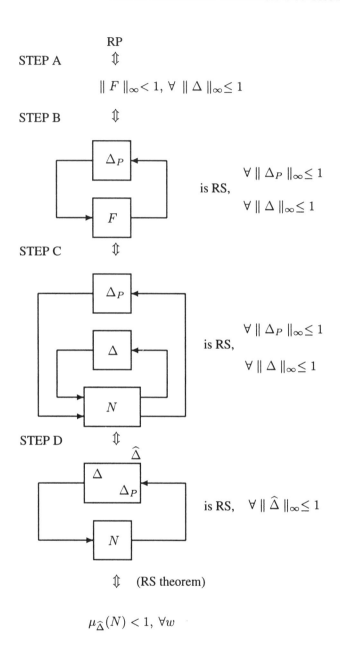

Figure 8.13: RP as a special case of structured RS. $F = F_u(N, \Delta)$

Algebraic proof of Theorem 8.7

The definition of μ gives at each frequency

$$\mu_{\widehat{\Delta}}(N(j\omega)) < 1 \Leftrightarrow \det(I - N(j\omega)\widehat{\Delta}(j\omega)) \neq 0, \forall\widehat{\Delta}, \bar{\sigma}(\widehat{\Delta}_P(j\omega)) \leq 1$$

By Schur's formula in (A.14) we have

$$\det(I - N\widehat{\Delta}) = \det \begin{bmatrix} I - N_{11}\Delta & -N_{12}\Delta_P \\ -N_{21}\Delta & I - N_{22}\Delta_P \end{bmatrix} =$$

$$\det(I - N_{11}\Delta) \cdot \det[I - N_{22}\Delta_P - N_{21}\Delta(I - N_{11}\Delta)^{-1}N_{12}\Delta_P] =$$

$$\det(I - N_{11}\Delta) \cdot \det[I - (N_{22} + N_{21}\Delta(I - N_{11}\Delta)^{-1}N_{12})\Delta_P] =$$

$$\det(I - N_{11}\Delta) \cdot \det(I - F_u(N, \Delta)\Delta_P)$$

Since this expression should not be zero, both terms must be non-zero at each frequency, i.e.

$$\det(I - N_{11}\Delta) \neq 0 \,\forall\Delta \Leftrightarrow \mu_\Delta(N_{11}) < 1, \quad \forall\omega \quad \text{(RS)}$$

and for all Δ

$$\det(I - F\Delta_P) \neq 0 \,\forall\Delta_P \Leftrightarrow \mu_{\Delta_P}(F) < 1 \Leftrightarrow \bar{\sigma}(F) < 1, \quad \forall\omega \quad \text{(RP definition)}$$

Theorem 8.7 is proved by reading the above lines in the opposite direction. Note that it is not necessary to test for RS separately as it follows as a special case of the RP requirement. $\square$

8.10.2 Summary of μ-conditions for NP, RS and RP

Rearrange the uncertain system into the $N\Delta$-structure of Figure 8.2, where the block-diagonal perturbations satisfy $\|\Delta\|_\infty \leq 1$. Introduce

$$F = F_u(N, \Delta) = N_{22} + N_{21}\Delta(I - N_{11}\Delta)^{-1}N_{12}$$

and let the performance requirement (RP) be $\|F\|_\infty \leq 1$ for all allowable perturbations. Then we have:

NS	$\Leftrightarrow$	N (internally) stable	(8.115)
NP	$\Leftrightarrow$	$\bar{\sigma}(N_{22}) = \mu_{\Delta_P} < 1, \forall\omega$, and NS	(8.116)
RS	$\Leftrightarrow$	$\mu_\Delta(N_{11}) < 1, \forall\omega$, and NS	(8.117)
RP	$\Leftrightarrow$	$\mu_{\widetilde{\Delta}}(N) < 1, \forall\omega, \widetilde{\Delta} = \begin{bmatrix} \Delta & 0 \\ 0 & \Delta_P \end{bmatrix}$, and NS	(8.118)

Here Δ is a block-diagonal matrix (its detailed structure depends on the uncertainty we are representing), whereas Δ_P always is a full complex matrix. Note that nominal stability (NS) must be tested separately in all cases.

Although the structured singular value is not a norm, it is sometimes convenient to refer to the peak μ-value as the "Δ-norm". For a stable rational transfer matrix $H(s)$, with an associated block structure Δ, we therefore define

$$\|H(s)\|_\Delta \triangleq \max_\omega \mu_\Delta(H(j\omega))$$

For a nominally stable system we then have

$$\text{NP} \Leftrightarrow \|N_{22}\|_\infty < 1, \quad \text{RS} \Leftrightarrow \|N_{11}\|_\Delta < 1, \quad \text{RP} \Leftrightarrow \|N\|_{\tilde\Delta} < 1$$

8.10.3 Worst-case performance and skewed-μ

Assume we have a system for which the peak μ-value for RP is 1.1. What does this mean? The definition of μ tells us that our RP-requirement would be satisfied exactly if we reduced *both* the performance requirement *and* the uncertainty by a factor of 1.1. So, μ does *not* directly give us the worst-case performance, i.e. $\max_\Delta \bar\sigma(F(\Delta))$, as one might have expected.

To find the worst-case weighted performance for a given uncertainty, one needs to keep the magnitude of the perturbations fixed ($\bar\sigma(\Delta) \leq 1$), that is, we must compute skewed-μ of N as discussed in Section 8.9.1. We have, in this case,

$$\max_{\bar\sigma(\Delta)\leq 1} \bar\sigma(F_l(N,\Delta)(j\omega)) = \mu^s(N(j\omega)) \tag{8.119}$$

To find μ^s numerically, we scale the performance part of N by a factor $k_m = 1/\mu^s$ and iterate on k_m until $\mu = 1$. That is, at each frequency skewed-μ is the value $\mu^s(N)$ which solves

$$\mu(K_m N) = 1, \quad K_m = \begin{bmatrix} I & 0 \\ 0 & 1/\mu^s \end{bmatrix} \tag{8.120}$$

Note that μ underestimates how bad or good the actual worst-case performance is. This follows because $\mu^s(N)$ is always further from 1 than $\mu(N)$.

Remark. The corresponding worst-case perturbation may be obtained as follows: First compute the worst-case performance at each frequency using skewed-μ. At the frequency where $\mu^s(N)$ has its peak, we may extract the corresponding worst-case perturbation generated by the software, and then find a stable, all-pass transfer function that matches this. In the MATLAB μ-toolbox, the single command wcperf combines these three steps: [delwc,mulow,muup] = wcperf(N,blk,1);.

8.11 Application: RP with input uncertainty

We will now consider in some detail the case of multiplicative input uncertainty with performance defined in terms of weighted sensitivity, as illustrated in Figure 8.14.

The performance requirement is then

$$\text{RP} \overset{\text{def}}{\Leftrightarrow} \|w_P(I + G_pK)^{-1}\|_\infty < 1, \quad \forall G_p \tag{8.121}$$

where the set of plants is given by

$$G_p = G(I + w_I\Delta_I), \quad \|\Delta_I\|_\infty \le 1 \tag{8.122}$$

Here $w_P(s)$ and $w_I(s)$ are scalar weights, so the performance objective is the same for all the outputs, and the uncertainty is the same for all inputs. We will mostly assume that Δ_I is diagonal, but we will also consider the case when Δ_I is a full matrix. This problem is excellent for illustrating the robustness analysis of uncertain

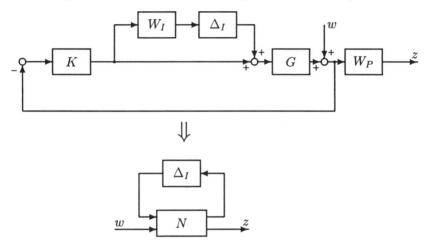

Figure 8.14: Robust performance of system with input uncertainty

multivariable systems. It should be noted, however, that although the problem set-up in (8.121) and (8.122) is fine for analyzing a given controller, it is less suitable for controller synthesis. For example, the problem formulation does not penalize directly the outputs from the controller.

In this section, we will:

1. Find the interconnection matrix N for this problem.
2. Consider the SISO case, so that useful connections can be made with results from the previous chapter.
3. Consider a multivariable distillation process for which we have already seen from simulations in Chapter 3 that a decoupling controller is sensitive to small errors in the input gains. We will find that μ for RP is indeed much larger than 1 for this decoupling controller.
4. Find some simple bounds on μ for this problem and discuss the role of the condition number.
5. Make comparisons with the case where the uncertainty is located at the output.

8.11.1 Interconnection matrix

On rearranging the system into the $N\Delta$-structure, as shown in Figure 8.14, we get, as in (8.32),

$$N = \begin{bmatrix} w_I T_I & w_I KS \\ w_P SG & w_P S \end{bmatrix} \tag{8.123}$$

where $T_I = KG(I + KG)^{-1}$, $S = (I + GK)^{-1}$ and for simplicity we have omitted the negative signs in the 1,1 and 1,2 blocks of N, since $\mu(N) = \mu(UN)$ with unitary $U = \begin{bmatrix} -I & 0 \\ 0 & I \end{bmatrix}$; see (8.83).

For a given controller K we can now test for NS, NP, RS and RP using (8.115)-(8.118). Here $\Delta = \Delta_I$ may be a full or diagonal matrix (depending on the physical situation).

8.11.2 RP with input uncertainty for SISO system

For a SISO system, conditions (8.115)-(8.118) with N as in (8.123) become

$$\text{NS} \quad \Leftrightarrow \quad S, SG, KS \text{ and } T_I \text{ are stable} \tag{8.124}$$

$$\text{NP} \quad \Leftrightarrow \quad |w_P S| < 1, \; \forall\omega \tag{8.125}$$

$$\text{RS} \quad \Leftrightarrow \quad |w_I T_I| < 1, \; \forall\omega \tag{8.126}$$

$$\text{RP} \quad \Leftrightarrow \quad |w_P S| + |w_I T_I| < 1, \; \forall\omega \tag{8.127}$$

The RP-condition (8.127) follows using (8.102); that is

$$\mu(N) = \mu \begin{bmatrix} w_I T_I & w_I KS \\ w_P SG & w_P S \end{bmatrix} = \mu \begin{bmatrix} w_I T_I & w_I T_I \\ w_P S & w_P S \end{bmatrix} = |w_I T_I| + |w_P S| \tag{8.128}$$

where we have used $T_I = KSG$. For SISO systems $T_I = T$, and we see that (8.127) is identical to (7.61), which was derived in Chapter 7 using a simple graphical argument based on the Nyquist plot of $L = GK$.

Robust performance optimization, in terms of weighted sensitivity with multiplicative uncertainty for a SISO system, thus involves minimizing the peak value of $\mu(N) = |w_I T| + |w_P S|$. This may be solved using DK-iteration as outlined later in Section 8.12. A closely related problem, which is easier to solve both mathematically and numerically, is to minimize the peak value ($\mathcal{H}_\infty$ norm) of the mixed sensitivity matrix

$$N_{\text{mix}} = \begin{bmatrix} w_P S \\ w_I T \end{bmatrix} \tag{8.129}$$

From (A.95) we get that at each frequency $\mu(N) = |w_I T| + |w_P S|$ differs from $\bar{\sigma}(N_{\text{mix}}) = \sqrt{|w_I T|^2 + |w_P S|^2}$ by at most a factor $\sqrt{2}$; recall (7.64). Thus, minimizing $\|N_{\text{mix}}\|_\infty$ is close to optimizing robust performance in terms of $\mu(N)$.

8.11.3   Robust performance for 2×2 distillation process

Consider again the distillation process example from Chapter 3 (Motivating Example No. 2) and the corresponding inverse-based controller:

$$G(s) = \frac{1}{75s+1} \begin{bmatrix} 87.8 & -86.4 \\ 108.2 & -109.6 \end{bmatrix}; \quad K(s) = \frac{0.7}{s} G(s)^{-1} \tag{8.130}$$

The controller provides a nominally decoupled system with

$$L = l\,I, \ S = \epsilon I \text{ and } T = tI \tag{8.131}$$

where

$$l = \frac{0.7}{s}, \quad \epsilon = \frac{1}{1+l} = \frac{s}{s+0.7}, \quad t = 1 - \epsilon = \frac{0.7}{s+0.7} = \frac{1}{1.43s+1}$$

We have used ϵ for the nominal sensitivity in each loop to distinguish it from the Laplace variable s. Recall from Figure 3.12 that this controller gave an excellent nominal response, but that the response with 20% gain uncertainty in each input channel was extremely poor. We will now confirm these findings by a μ-analysis. To this effect we use the following weights for uncertainty and performance:

$$w_I(s) = \frac{s+0.2}{0.5s+1}; \quad w_P(s) = \frac{s/2+0.05}{s} \tag{8.132}$$

With reference to (7.26) we see the weight $w_I(s)$ may approximately represent a 20% gain error and a neglected time delay of 0.9 min. $|w_I(j\omega)|$ levels off at 2 (200% uncertainty) at high frequencies. With reference to (2.72) we see that the performance weight $w_P(s)$ specifies integral action, a closed-loop bandwidth of about 0.05 [rad/min] (which is relatively slow in the presence of an allowed time delay of 0.9 min) and a maximum peak for $\bar{\sigma}(S)$ of $M_s = 2$.

We now test for NS, NP, NS and RP. Note that Δ_I is a diagonal matrix in this example.

NS With G and K as given in (8.130) we find that S, SG, KS and T_I are stable, so the system is nominally stable.

NP With the decoupling controller we have

$$\bar{\sigma}(N_{22}) = \bar{\sigma}(w_P S) = \left| \frac{s/2+0.05}{s+0.7} \right|$$

and we see from the dashed-dot line in Figure 8.15 that the NP-condition is easily satisfied: $\bar{\sigma}(w_P S)$ is small at low frequencies $(0.05/0.7 = 0.07$ at $\omega = 0$) and approaches $1/2 = 0.5$ at high frequencies.

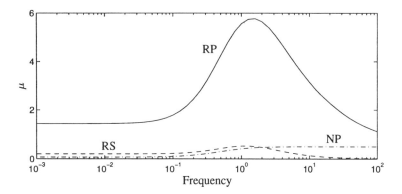

Figure 8.15: μ-plots for distillation process with decoupling controller

RS Since in this case $w_I T_I = w_I T$ is a scalar times the identity matrix, we have, independent of the structure of Δ_I, that

$$\mu_{\Delta_I}(w_I T_I) = |w_I t| = \left| 0.2 \frac{5s + 1}{(0.5s + 1)(1.43s + 1)} \right|$$

and we see from the dashed line in Figure 8.15 that RS is easily satisfied. The peak value of $\mu_{\Delta_I}(M)$ over frequency is $\|M\|_{\Delta_I} = 0.53$. This means that we may increase the uncertainty by a factor of $1/0.53 = 1.89$ before the worst-case uncertainty yields instability, that is, we can tolerate about 38% gain uncertainty and a time delay of about 1.7 min before we get instability.

RP Although our system has good robustness margins (RS easily satisfied) and excellent nominal performance we know from the simulations in Figure 3.12 that the robust performance is poor. This is confirmed by the μ-curve for RP in Figure 8.15 which was computed numerically using $\mu_{\widehat{\Delta}}(N)$ with N as in (8.123), $\widehat{\Delta} = \text{diag}\{\Delta_I, \Delta_P\}$ and $\Delta_I = \text{diag}\{\delta_1, \delta_2\}$. The peak value is close to 6, meaning that even with 6 times less uncertainty, the weighted sensitivity will be about 6 times larger than what we require. The peak of the actual worst-case weighted sensitivity with uncertainty blocks of magnitude 1, which may be computed using skewed-μ, is for comparison 44.93.

The MATLAB μ-toolbox commands to generate Figure 8.15 are given in Table 8.1.

In general, μ with unstructured uncertainty (Δ_I full) is larger than μ with structured uncertainty (Δ_I diagonal). However, for our particular plant and controller in (8.130) it appears from numerical calculations and by use of (8.135) below, that they are the same. Of course, this is not generally true, as is confirmed in the following exercise.

Exercise 8.23 *Consider the plant $G(s)$ in (8.107) which is ill-conditioned with $\gamma(G) = 70.8$ at all frequencies (but note that the RGA-elements of G are all about 0.5). With an inverse-based controller $K(s) = \frac{0.7}{s}G(s)^{-1}$, compute μ for RP with both diagonal and full-block*

Table 8.1: MATLAB program for μ-analysis (generates Figure 8.15)

```
% Uses the Mu toolbox
G0 = [87.8 -86.4; 108.2 -109.6];
dyn = nd2sys(1,[75 1]);
Dyn=daug(dyn,dyn); G=mmult(Dyn,G0);
%
% Inverse-based control.
%
dynk=nd2sys([75 1],[1 1.e-5],0.7);
Dynk=daug(dynk,dynk); Kinv=mmult(Dynk,minv(G0));
%
% Weights.
%
wp=nd2sys([10 1],[10 1.e-5],0.5); Wp=daug(wp,wp);
wi=nd2sys([1 0.2],[0.5 1]); Wi=daug(wi,wi);
%
% Generalized plant P.
%
systemnames = 'G Wp Wi';
inputvar = '[ydel(2); w(2) ; u(2)]';
outputvar = '[Wi; Wp; -G-w]';
input_to_G = '[u+ydel]';
input_to_Wp = '[G+w]'; input_to_Wi = '[u]';
sysoutname = 'P';
cleanupsysic = 'yes'; sysic;
%
N = starp(P,Kinv); omega = logspace(-3,3,61);
Nf = frsp(N,omega);
%
% mu for RP.
%
blk = [1 1; 1 1; 2 2];
[mubnds,rowd,sens,rowp,rowg] = mu(Nf,blk,'c');
muRP = sel(mubnds,':',1); pkvnorm(muRP)              % (ans = 5.7726).
%
% Worst-case weighted sensitivity
%
[delworst,muslow,musup] = wcperf(Nf,blk,1); musup    % (musup = 44.93 for
%                                                    %   delta=1).
% mu for RS.
%
Nrs=sel(Nf,1:2,1:2);
[mubnds,rowd,sens,rowp,rowg]=mu(Nrs,[1 1; 1 1],'c');
muRS = sel(mubnds,':',1); pkvnorm(muRS)              % (ans = 0.5242).
%
% mu for NP (= max. singular value of Nnp).
%
Nnp=sel(Nf,3:4,3:4);
[mubnds,rowd,sens,rowp,rowg]=mu(Nnp,[2 2],'c');
muNP = sel(mubnds,':',1); pkvnorm(muNP)              % (ans = 0.5000).
vplot('liv,m',muRP,muRS,muNP);
```

input uncertainty using the weights in (8.132). The value of μ is much smaller in the former case.

8.11.4 μ and the condition number

In this subsection we consider the relationship between μ for RP and the condition number of the plant or of the controller. We consider *unstructured* multiplicative input uncertainty (i.e. Δ_I is a full matrix) and performance measured in terms of weighted sensitivity.

Any controller. Let N be given as in (8.123). Then

$$\overbrace{\mu_{\tilde{\Delta}}(N)}^{RP} \leq [\overbrace{\bar{\sigma}(w_I T_I)}^{RS} + \overbrace{\bar{\sigma}(w_P S)}^{NP}](1 + \sqrt{k}) \qquad (8.133)$$

where k is either the condition number of the plant or the controller (the smallest one should be used):

$$k = \gamma(G) \quad \text{or} \quad k = \gamma(K) \qquad (8.134)$$

Proof of (8.133): Since Δ_I is a full matrix, (8.87) yields

$$\mu(N) = \min_d \bar{\sigma} \begin{bmatrix} N_{11} & dN_{12} \\ d^{-1}N_{21} & N_{22} \end{bmatrix}$$

where from (A.46)

$$\bar{\sigma} \begin{bmatrix} w_I T_I & dw_I KS \\ d^{-1}w_P SG & w_P S \end{bmatrix} \leq \bar{\sigma}(w_I T_I \begin{bmatrix} I & dG^{-1} \end{bmatrix}) + \bar{\sigma}(w_P S \begin{bmatrix} d^{-1}G & I \end{bmatrix})$$

$$\leq \bar{\sigma}(w_I T_I)\underbrace{\bar{\sigma}(I \quad dG^{-1})}_{\leq 1 + |d|\bar{\sigma}(G^{-1})} + \bar{\sigma}(w_P S)\underbrace{\bar{\sigma}(d^{-1}G \quad I)}_{\leq 1 + |d^{-1}|\bar{\sigma}(G)}$$

and selecting $d = \sqrt{\frac{\bar{\sigma}(G)}{\bar{\sigma}(G^{-1})}} = \sqrt{\gamma(G)}$ gives

$$\mu(N) \leq [\bar{\sigma}(w_I T_I) + \bar{\sigma}(w_P S)](1 + \sqrt{\gamma(G)})$$

A similar derivation may be performed using $SG = K^{-1}T_I$ to derive the same expression but with $\gamma(K)$ instead of $\gamma(G)$. $\square$

From (8.133) we see that with a "round" controller, i.e. one with $\gamma(K) = 1$, there is less sensitivity to uncertainty (but it may be difficult to achieve NP in this case). On the other hand, we would expect μ for RP to be large if we used an inverse-based controller for a plant with a large condition number, since then $\gamma(K) = \gamma(G)$ is large. This is confirmed by (8.135) below.

Example 8.11 *For the distillation process studied above, we have $\gamma(G) = \gamma(K) = 141.7$ at all frequencies, and at frequency $w = 1$ rad/min the upper bound given by (8.133) becomes $(0.52 + 0.41)(1 + \sqrt{141.7}) = 13.1$. This is higher than the actual value of $\mu(N)$ which is 5.56, which illustrates that the bound in (8.133) is generally not tight.*

Inverse-based controller. With an inverse-based controller (resulting in the nominal decoupled system (8.131)) and unstructured input uncertainty, it is possible to derive an analytic expression for μ for RP with N as in (8.123):

$$\mu_{\tilde{\Delta}}(N) = \sqrt{|w_P \epsilon|^2 + |w_I t|^2 + |w_P \epsilon| \cdot |w_I t| \left(\gamma(G) + \frac{1}{\gamma(G)} \right)} \qquad (8.135)$$

where ϵ is the nominal sensitivity and $\gamma(G)$ is the condition number of the plant. We see that for plants with a large condition number, μ for RP increases approximately in proportion to $\sqrt{\gamma(G)}$.

Proof of (8.135): The proof originates from Stein and Doyle (1991). The upper μ-bound in (8.87) with $D = \text{diag}\{dI, I\}$ yields

$$
\begin{aligned}
\mu(N) &= \min_d \bar{\sigma} \begin{bmatrix} w_I t I & w_I t (dG)^{-1} \\ w_P \epsilon (dG) & w_P \epsilon I \end{bmatrix} = \min_d \bar{\sigma} \begin{bmatrix} w_I t I & w_I t (d\Sigma)^{-1} \\ w_P \epsilon (d\Sigma) & w_P \epsilon I \end{bmatrix} \\
&= \min_d \max_i \bar{\sigma} \begin{bmatrix} w_I t & w_I t (d\sigma_i)^{-1} \\ w_P \epsilon (d\sigma_i) & w_P \epsilon \end{bmatrix} \\
&= \min_d \max_i \sqrt{|w_P \epsilon|^2 + |w_I t|^2 + |w_P \epsilon d\sigma_i|^2 + |w_I t (d\sigma_i)^{-1}|^2}
\end{aligned}
$$

We have here used the SVD of $G = U\Sigma V^H$ at each frequency, and have used the fact that $\bar{\sigma}$ is unitary invariant. σ_i denotes the i'th singular value of G. The expression is minimized by selecting at each frequency $d = |w_I t|/(|w_P \epsilon|\bar{\sigma}(G)\underline{\sigma}(G))$, see (8.100), and hence the desired result. $\qquad \square$

Example 8.12 *For the distillation column example studied above, we have at frequency $\omega = 1$ rad/min, $|w_P \epsilon| = 0.41$ and $|w_I t| = 0.52$, and since $\gamma(G) = 141.7$ at all frequencies, (8.135) yields $\mu(N) = \sqrt{0.17 + 0.27 + 30.51} = 5.56$ which agrees with the plot in Figure 8.15.*

Worst-case performance (any controller)

We next derive relationships between worst-case performance and the condition number. Suppose that at each frequency the worst-case sensitivity is $\bar{\sigma}(S')$. We then have that the worst-case weighted sensitivity is equal to skewed-μ:

$$\max_{S_p} \bar{\sigma}(w_P S_p) = \bar{\sigma}(w_P S') = \mu^s(N)$$

Now, recall that in Section 6.10.4 we derived a number of upper bounds on $\bar{\sigma}(S')$, and referring back to (6.72) we find

$$\bar{\sigma}(S') \leq \gamma(G) \frac{\bar{\sigma}(S)}{1 - \bar{\sigma}(w_I T_I)} \qquad (8.136)$$

A similar bound involving $\gamma(K)$ applies. We then have

$$\mu^s(N) = \bar{\sigma}(w_P S') \le k \frac{\bar{\sigma}(w_P S)}{1 - \bar{\sigma}(w_I T_I)} \tag{8.137}$$

where k as before denotes either the condition number of the plant *or* the controller (preferably the smallest). Equation (8.137) holds for any controller and for any structure of the uncertainty (including Δ_I unstructured).

Remark 1 In Section 6.10.4 we derived tighter upper bounds for cases when Δ_I is restricted to be diagonal and when we have a decoupling controller. In (6.78) we also derived a lower bound in terms of the RGA.

Remark 2 Since $\mu^s = \mu$ when $\mu = 1$, we may, from (8.133), (8.137) and expressions similar to (6.76) and (6.77), derive the following *sufficient* (conservative) tests for RP ($\mu(N) < 1$) with unstructured input uncertainty (any controller):

$$\text{RP} \quad \Leftarrow \quad [\bar{\sigma}(w_P S) + \bar{\sigma}(w_I T_I)](1 + \sqrt{k}) < 1, \ \forall \omega$$
$$\text{RP} \quad \Leftarrow \quad k\bar{\sigma}(w_P S) + \bar{\sigma}(w_I T_I) < 1, \quad \forall \omega$$
$$\text{RP} \quad \Leftarrow \quad \bar{\sigma}(w_P S) + k\bar{\sigma}(w_I T) < 1, \quad \forall \omega$$

where k denotes the condition number of the plant *or* the controller (the smallest being the most useful).

Example 8.13 *For the distillation process the upper bound given by (8.137) at $\omega = 1$ rad/min is $141.7 \cdot 0.41/(1 - 0.52) = 121$. This is higher than the actual peak value of $\mu^s = \max_{S_p} \bar{\sigma}(w_P S_p)$ which as found earlier is 44.9 (at frequency 1.2 rad/min), and demonstrates that these bounds are not generally tight.*

8.11.5   Comparison with output uncertainty

Consider output multiplicative uncertainty of magnitude $w_O(j\omega)$. In this case we get the interconnection matrix

$$N = \begin{bmatrix} w_O T & w_O T \\ w_P S & w_P S \end{bmatrix} \tag{8.138}$$

and for any structure of the uncertainty $\mu(N)$ is bounded as follows:

$$\bar{\sigma} \begin{bmatrix} w_O T \\ w_P S \end{bmatrix} \le \overbrace{\mu(N)}^{\text{RP}} \le \sqrt{2} \, \bar{\sigma} \underbrace{\overbrace{\begin{bmatrix} w_O T \\ w_P S \end{bmatrix}}^{\text{RS}}}_{\text{NP}} \tag{8.139}$$

This follows since the uncertainty and performance blocks both enter at the output (see Section 8.6.2) and from (A.45) the difference between bounding the combined perturbations, $\bar{\sigma}\begin{bmatrix} \Delta_O & \Delta_P \end{bmatrix}$ and individual perturbations, $\bar{\sigma}(\Delta_O)$ and $\bar{\sigma}(\Delta_P)$, is at most a factor of $\sqrt{2}$. Thus, in this case we "automatically" achieve RP (at least within $\sqrt{2}$) if we have satisfied separately the subobjectives of NP and RS. This confirms our findings from Section 6.10.4 that multiplicative output uncertainty poses no particular problem for performance. It also implies that for practical purposes we may optimize robust performance with output uncertainty by minimizing the $\mathcal{H}_\infty$ norm of the stacked matrix $\begin{bmatrix} w_O T \\ w_P S \end{bmatrix}$.

Exercise 8.24 *Consider the RP-problem with weighted sensitivity and multiplicative output uncertainty. Derive the interconnection matrix N for, 1) the conventional case with $\widehat{\Delta} = \mathrm{diag}\{\Delta, \Delta_P\}$, and 2) the stacked case when $\widehat{\Delta} = \begin{bmatrix} \Delta & \Delta_P \end{bmatrix}$. Use this to prove (8.139).*

8.12 μ-synthesis and DK-iteration

The structured singular value μ is a very powerful tool for the analysis of robust performance with a given controller. However, one may also seek to find the controller that minimizes a given μ-condition: this is the μ-synthesis problem.

8.12.1 DK-iteration

At present there is no direct method to synthesize a μ-optimal controller. However, for complex perturbations a method known as DK-iteration is available. It combines $\mathcal{H}_\infty$-synthesis and μ-analysis, and often yields good results. The starting point is the upper bound (8.87) on μ in terms of the scaled singular value

$$\mu(N) \leq \min_{D \in \mathcal{D}} \bar{\sigma}(DND^{-1})$$

The idea is to find the controller that minimizes the peak value over frequency of this upper bound, namely

$$\min_{K}(\min_{D \in \mathcal{D}} \|DN(K)D^{-1}\|_\infty) \tag{8.140}$$

by alternating between minimizing $\|DN(K)D^{-1}\|_\infty$ with respect to either K or D (while holding the other fixed). To start the iterations, one selects an initial stable rational transfer matrix $D(s)$ with appropriate structure. The identity matrix is often a good initial choice for D provided the system has been reasonably scaled for performance. The DK-iteration then proceeds as follows:

1. **K-step.** Synthesize an $\mathcal{H}_\infty$ controller for the scaled problem, $\min_K \|DN(K)D^{-1}\|_\infty$ with fixed $D(s)$.
2. **D-step.** Find $D(j\omega)$ to minimize at each frequency $\bar{\sigma}(DND^{-1}(j\omega))$ with fixed N.

3. Fit the magnitude of each element of $D(j\omega)$ to a stable and minimum phase transfer function $D(s)$ and go to Step 1.

The iteration may continue until satisfactory performance is achieved, $\|DND^{-1}\|_\infty <$ 1, or until the $\mathcal{H}_\infty$ norm no longer decreases. One fundamental problem with this approach is that although each of the minimization steps (K-step and D-step) are convex, joint convexity is *not* guaranteed. Therefore, the iterations may converge to a local optimum. However, practical experience suggests that the method works well in most cases.

The order of the controller resulting from each iteration is equal to the number of states in the plant $G(s)$ plus the number of states in the weights plus twice the number of states in $D(s)$. For most cases, the true μ-optimal controller is not rational, and will thus be of infinite order, but because we use a finite-order $D(s)$ to approximate the D-scales, we get a controller of finite (but often high) order. The true μ-optimal controller would have a flat μ-curve (as a function of frequency), except at infinite frequency where μ generally has to approach a fixed value independent of the controller (because $L(j\infty) = 0$ for real systems). However, with a finite-order controller we will generally not be able (and it may not be desirable) to extend the flatness to infinite frequencies.

The DK-iteration depends heavily on optimal solutions for Steps 1 and 2, and also on good fits in Step 3, preferably by a transfer function of low order. One reason for preferring a low-order fit is that this reduces the order of the $\mathcal{H}_\infty$ problem, which usually improves the numerical properties of the $\mathcal{H}_\infty$ optimization (Step 1) and also yields a controller of lower order. In some cases the iterations converge slowly, and it may be difficult to judge whether the iterations are converging or not. One may even experience the μ-value increasing. This may be caused by numerical problems or inaccuracies (e.g. the upper bound μ-value in Step 2 being higher than the $\mathcal{H}_\infty$ norm obtained in Step 1), or by a poor fit of the D-scales. In any case, if the iterations converge slowly, then one may consider going back to the initial problem and rescaling the inputs and outputs.

In the K-step (Step 1) where the $\mathcal{H}_\infty$ controller is synthesized, it is often desirable to use a slightly sub-optimal controller (e.g. with an $\mathcal{H}_\infty$ norm, γ, which is 5% higher than the optimal value, $\gamma_{\min}$). This yields a blend of $\mathcal{H}_\infty$ and $\mathcal{H}_2$ optimality with a controller which usually has a steeper high-frequency roll-off than the $\mathcal{H}_\infty$ optimal controller.

8.12.2   Adjusting the performance weight

Recall that if μ at a given frequency is different from 1, then the interpretation is that at this frequency we can tolerate $1/\mu$-times more uncertainty and still satisfy our performance objective with a margin of $1/\mu$. In μ-synthesis, the designer will usually adjust some parameter(s) in the performance or uncertainty weights until the peak μ-value is close to 1. Sometimes the uncertainty is fixed, and we effectively optimize worst-case performance by adjusting a parameter in the performance weight. For example, consider the performance weight

$$w_P(s) = \frac{s/M + \omega_B^*}{s + \omega_B^* A} \tag{8.141}$$

where we want to keep M constant and find the highest achievable bandwidth frequency ω_B^*. The optimization problem becomes

$$\max |\omega_B^*| \quad \text{such that} \quad \mu(N) < 1, \forall \omega \tag{8.142}$$

where N, the interconnection matrix for the RP-problem, depends on ω_B^*. This may be implemented as an outer loop around the DK-iteration.

8.12.3 Fixed structure controller

Sometimes it is desirable to find a low-order controller with a given structure. This may be achieved by numerical optimization where μ is minimized with respect to the controller parameters The problem here is that the optimization is not generally convex in the parameters. Sometimes it helps to switch the optimization between minimizing the peak of μ (i.e. $\|\mu\|_\infty$) and minimizing the integral square deviation of μ away from k (i.e. $\|\mu(j\omega) - k\|_2$) where k usually is close to 1. The latter is an attempt to "flatten out" μ.

8.12.4 Example: μ-synthesis with DK-iteration

We will consider again the case of multiplicative input uncertainty and performance defined in terms of weighted sensitivity, as discussed in detail in Section 8.11. We noted there that this set-up is fine for analysis, but less suitable for controller synthesis, as it does not explicitly penalize the outputs from the controller. Nevertheless we will use it here as an example of μ-synthesis because of its simplicity. The resulting controller will have very large gains at high frequencies and should not be used directly for implementation. In practice, one can add extra roll-off to the controller (which should work well because the system should be robust with respect to uncertain high-frequency dynamics), or one may consider a more complicated problem set-up (see Section 12.4).

With this caution in mind, we proceed with the problem description. Again, we use the model of the simplified distillation process

$$G(s) = \frac{1}{75s + 1} \begin{bmatrix} 87.8 & -86.4 \\ 108.2 & -109.6 \end{bmatrix} \tag{8.143}$$

The uncertainty weight $w_I I$ and performance weight $w_P I$ are given in (8.132), and are shown graphically in Figure 8.16. The objective is to minimize the peak value of $\mu_{\tilde{\Delta}}(N)$, where N is given in (8.123) and $\tilde{\Delta} = \mathrm{diag}\{\Delta_I, \Delta_P\}$. We will consider diagonal input uncertainty (which is always present in any real problem), so Δ_I is a 2×2 diagonal matrix. Δ_P is a full 2×2 matrix representing the performance specification. Note that we have only three complex uncertainty blocks, so $\mu(N)$ is equal to the upper bound $\min_D \bar{\sigma}(DND^{-1})$ in this case.

We will now use DK-iteration in attempt to obtain the μ-optimal controller for this example. The appropriate commands for the MATLAB μ-toolbox are listed in Table 8.2.

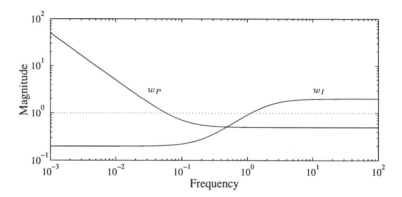

Figure 8.16: Uncertainty and performance weights. Notice that there is a frequency range ("window") where both weights are less than one in magnitude.

First the generalized plant P as given in (8.29) is constructed. It includes the plant model, the uncertainty weight and the performance weight, but not the controller which is to be designed (note that $N = F_l(P, K)$). Then the block-structure is defined; it consists of two 1×1 blocks to represent Δ_I and a 2×2 block to represent Δ_P. The scaling matrix D for DND^{-1} then has the structure $D = \mathrm{diag}\{d_1, d_2, d_3 I_2\}$ where I_2 is a 2×2 identity matrix, and we may set $d_3 = 1$. As initial scalings we select $d_1^0 = d_2^0 = 1$. P is then scaled with the matrix $\mathrm{diag}\{D, I_2\}$ where I_2 is associated with the inputs and outputs from the controller (we do not want to scale the controller).

Iteration No. 1. Step 1: With the initial scalings, $D^0 = I$, the $\mathcal{H}_\infty$ software produced a 6 state controller (2 states from the plant model and 2 from each of the weights) with an $\mathcal{H}_\infty$ norm of $\gamma = 1.1823$. Step 2: The upper μ-bound gave the μ-curve shown as curve "Iter. 1" in Figure 8.17, corresponding to a peak value of $\mu = 1.1818$. Step 3: The frequency-dependent $d_1(\omega)$ and $d_2(\omega)$ from Step 2 were each fitted using a 4th order transfer function. $d_1(w)$ and the fitted 4th-order transfer function (dotted line) are shown in Figure 8.18 and labelled "Iter. 1". The fit is very good so it is hard to distinguish the two curves. d_2 is not shown because it was found that $d_1 \approx d_2$ (indicating that the worst-case full-block Δ_I is in fact diagonal).

Iteration No. 2. Step 1: With the 8 state scaling $D^1(s)$ the $\mathcal{H}_\infty$ software gave a 22

Table 8.2: MATLAB program to perform DK-iteration

```
% Uses the Mu toolbox
G0 = [87.8 -86.4; 108.2 -109.6];                          % Distillation
dyn = nd2sys(1,[75 1]); Dyn = daug(dyn,dyn);              % process.
G = mmult(Dyn,G0);
%
% Weights.
%
wp = nd2sys([10 1],[10 1.e-5],0.5);                       % Approximated
wi = nd2sys([1 0.2],[0.5 1]);                             % integrator.
Wp = daug(wp,wp); Wi = daug(wi,wi);
%
% Generalized plant P. %
systemnames = 'G Wp Wi';
inputvar = '[ydel(2); w(2) ; u(2)]';
outputvar = '[Wi; Wp; -G-w]';
input_to_G = '[u+ydel]';
input_to_Wp = '[G+w]'; input_to_Wi = '[u]';
sysoutname = 'P'; cleanupsysic = 'yes';
sysic;
%
% Initialize.
%
omega = logspace(-3,3,61);
blk = [1 1; 1 1; 2 2];
nmeas=2; nu=2; gmin=0.9; gamma=2; tol=0.01; d0 = 1;
dsysl = daug(d0,d0,eye(2),eye(2)); dsysr=dsysl;
%
% START ITERATION.
%
% STEP 1: Find H-infinity optimal controller
% with given scalings:
%
DPD = mmult(dsysl,P,minv(dsysr)); gmax=1.05*gamma;
[K,Nsc,gamma] = hinfsyn(DPD,nmeas,nu,gmin,gmax,tol);
Nf=frsp(Nsc,omega);                                       % (Remark:
%                                                         %  Without scaling:
%                                                         %  N=starp(P,K);).
% STEP 2: Compute mu using upper bound:
%
[mubnds,rowd,sens,rowp,rowg] = mu(Nf,blk,'c');
vplot('liv,m',mubnds); murp=pkvnorm(mubnds,inf)
%
% STEP 3: Fit resulting D-scales:
%
[dsysl,dsysr]=musynflp(dsysl,rowd,sens,blk,nmeas,nu);     % choose 4th order.
%
% New Version:
% [dsysL,dsysR]=msf(Nf,mubnds,rowd,sens,blk);             % order: 4, 4, 0.
% dsysl=daug(dsysL,eye(2)); dsysr=daug(dsysR,eye(2));
%
% GOTO STEP 1 (unless satisfied with murp).
%
```

state controller and $\|D^1 N (D^1)^{-1}\|_\infty = 1.0238$. Step 2: This controller gave a peak value of μ of 1.0238. Step 3: The resulting scalings D^2 were only slightly changed from the previous iteration as can be seen from $d_1^2(\omega)$ labelled "Iter. 2" in Figure 8.18.

Iteration No. 3. Step 1: With the scalings $D^2(s)$ the $\mathcal{H}_\infty$ norm was only slightly reduced from 1.024 to 1.019. Since the improvement was very small and since the value was very close to the desired value of 1, it was decided to stop the iterations. The resulting controller with 22 states (denoted K_3 in the following) gives a peak μ-value of 1.019.

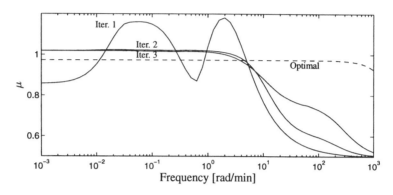

Figure 8.17: Change in μ during DK-iteration

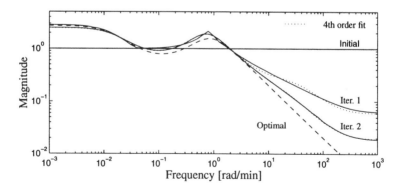

Figure 8.18: Change in D-scale d_1 during DK-iteration

Analysis of μ-"optimal" controller K_3

The final μ-curves for NP, RS and RP with controller K_3 are shown in Figure 8.19. The objectives of RS and NP are easily satisfied. Furthermore, the peak μ-value of

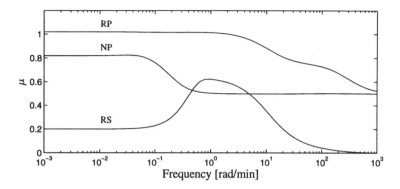

Figure 8.19: μ-plots with μ-"optimal" controller K_3

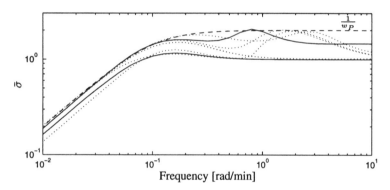

Figure 8.20: Perturbed sensitivity functions $\bar{\sigma}(S')$ using μ-"optimal" controller K_3. Dotted lines: Plants $G'_i, i = 1, 6$. Lower solid line: Nominal plant G. Upper solid line: Worst-case plant G'_{wc}.

1.019 with controller K_3 is only slightly above 1, so the performance specification $\bar{\sigma}(w_P S_p) < 1$ is almost satisfied for all possible plants. To confirm this we considered the nominal plant and six perturbed plants

$$G'_i(s) = G(s)E_{Ii}(s)$$

where $E_{Ii} = I + w_I \Delta_I$ is a diagonal transfer function matrix representing input uncertainty (with nominal $E_{I0} = I$). Recall that the uncertainty weight is

$$w_I(s) = \frac{s + 0.2}{0.5s + 1}$$

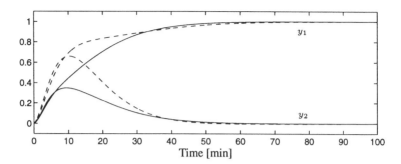

Figure 8.21: Setpoint response for μ-"optimal" controller K_3. Solid line: nominal plant. Dashed line: uncertain plant G_3'

which is 0.2 in magnitude at low frequencies. Thus, the following input gain perturbations are allowable

$$E_{I1} = \begin{bmatrix} 1.2 & 0 \\ 0 & 1.2 \end{bmatrix}, \; E_{I2} = \begin{bmatrix} 0.8 & 0 \\ 0 & 1.2 \end{bmatrix}, \; E_{I3} = \begin{bmatrix} 1.2 & 0 \\ 0 & 0.8 \end{bmatrix}, \; E_{I4} = \begin{bmatrix} 0.8 & 0 \\ 0 & 0.8 \end{bmatrix}$$

These perturbations do not make use of the fact that $w_I(s)$ increases with frequency. Two allowed dynamic perturbations for the diagonal elements in $w_I \Delta_I$ are

$$\epsilon_1(s) = \frac{-s + 0.2}{0.5s + 1}, \quad \epsilon_2(s) = -\frac{s + 0.2}{0.5s + 1}$$

corresponding to elements in E_{I_i} of

$$f_1(s) = 1 + \epsilon_1(s) = 1.2 \frac{-0.417s + 1}{0.5s + 1}, \quad f_2(s) = 1 + \epsilon_2(s) = 0.8 \frac{-0.633s + 1}{0.5s + 1}$$

so let us also consider

$$E_{I5} = \begin{bmatrix} f_1(s) & 0 \\ 0 & f_1(s) \end{bmatrix}, \; E_{I6} = \begin{bmatrix} f_2(s) & 0 \\ 0 & f_1(s) \end{bmatrix}$$

The maximum singular value of the sensitivity, $\bar{\sigma}(S_i')$, is shown in Figure 8.20 for the nominal and six perturbed plants, and is seen to be almost below the bound $1/|w_I(j\omega)|$ for all seven cases ($i = 0, 6$) illustrating that RP is almost satisfied. The sensitivity for the nominal plant is shown by the lower solid line, and the others with dotted lines. At low frequencies the worst-case corresponds closely to a plant with gains 1.2 and 0.8, such as G_2', G_3' or G_6'. Overall, the worst-case of these six plants seems to be $G_6' = GE_{I6}$, which has $\bar{\sigma}(S')$ close to the bound at low frequencies, and has a peak of about 2.02 (above the allowed bound of 2) at 1.6 rad/min.

To find the "true" worst-case performance and plant we used the MATLAB μ-tools command wcperf as explained in Section 8.10.3 on page 326. This gives a worst-case performance of $\max_{S_p} \|w_P S_p\|_\infty = 1.037$, and the sensitivity function for the corresponding worst-case plant $G'_{wc}(s) = G(s)(I + w_I(s)\Delta_{wc}(s))$ found with the software is shown by the upper solid line in Figure 8.20. It has a peak value of $\bar{\sigma}(S')$ of about 2.05 at 0.79 rad/min.

Remark. The "worst-case" plant is not unique, and there are many plants which yield a worst-case performance of $\max_{S_p} \|w_P S_p\|_\infty = 1.037$. For example, it is likely that we could find plants which were more consistently "worse" at all frequencies than the one shown by the upper solid line in Figure 8.20.

The time response of y_1 and y_2 to a filtered setpoint change in $y_1, r_1 = 1/(5s+1)$, is shown in Figure 8.21 both for the nominal case (solid line) and for 20% input gain uncertainty (dashed line) using the plant $G'_3 = GF_3$ (which we know is one of the worst plants). The response is interactive, but shows no strong sensitivity to the uncertainty. The response with uncertainty is seen to be much better than that with the inverse-based controller studied earlier and shown in Figure 3.12.

Remarks on the μ-synthesis example.

1. By trial and error, and many long nights, Petter Lundström was able to reduce the peak μ-value for robust performance for this problem down to about $\mu_{\text{opt}} = 0.974$ (Lundström, 1994). The resulting design produces the curves labelled *optimal* in Figures 8.17 and 8.18. The corresponding controller, K_{opt}, may be synthesized using $\mathcal{H}_\infty$-synthesis with the following third-order D-scales

$$d_1(s) = d_2(s) = 2\frac{(0.001s + 1)(s + 0.25)(s + 0.054)}{((s + 0.67)^2 + 0.56^2)(s + 0.013)}, \quad d_3 = 1 \qquad (8.144)$$

2. Note that the optimal controller K_{opt} for this problem has a SVD-form. That is, let $G = U\Sigma V^H$, then $K_{\text{opt}} = VK_sU^H$ where K_s is a diagonal matrix. This arises because in this example U and V are constant matrices. For more details see Hovd (1992) and Hovd et al. (1994).
3. For this particular plant it appears that the worst-case full-block input uncertainty is a diagonal perturbation, so we might as well have used a full matrix for Δ_I. But this does *not* hold in general.
4. The $\mathcal{H}_\infty$ software may encounter numerical problems if $P(s)$ has poles on the $j\omega$-axis. This is the reason why in the MATLAB code we have moved the integrators (in the performance weights) slightly into the left-half plane.
5. The initial choice of scaling $D = I$ gave a good design for this plant with an $\mathcal{H}_\infty$ norm of about 1.18. This scaling worked well because the inputs and outputs had been scaled to be of unit magnitude. For a comparison, consider the original model in Skogestad et al. (1988) which was in terms of *unscaled* physical variables.

$$G_{\text{unscaled}}(s) = \frac{1}{75s + 1}\begin{bmatrix} 0.878 & -0.864 \\ 1.082 & -1.096 \end{bmatrix} \qquad (8.145)$$

(8.145) has all its elements 100 times smaller than in the scaled model (8.143). Therefore, using this model should give the same optimal μ-value but with controller gains 100 times larger. However, starting the DK-iteration with $D = I$ works very poorly in this case. The first iteration yields an $\mathcal{H}_\infty$ norm of 14.9 (Step 1) resulting in a peak μ-value of 5.2 (Step 2). Subsequent iterations yield with 3rd and 4th order fits of the D-scales the following peak μ-values: 2.92, 2.22, 1.87, 1.67, 1.58, 1.53, 1.49, 1.46, 1.44, 1.42. At this point (after 11 iterations) the μ-plot is fairly flat up to 10 [rad/min] and one may be tempted to stop the iterations. However, we are still far away from the optimal value which we know is less than 1. This demonstrates the importance of good initial D-scales, which is related to scaling the plant model properly.

Exercise 8.25 *Explain why the optimal μ-value would be the same if in the model (8.143) we changed the time constant of 75 [min] to another value. Note that the μ-iteration itself would be affected.*

8.13   Further remarks on μ

8.13.1   Further justification for the upper bound on μ

For complex perturbations, the scaled singular value $\bar{\sigma}(DND^{-1})$ is a tight upper bound on $\mu(N)$ in most cases, and minimizing the upper bound $\|DND^{-1}\|_\infty$ forms the basis for the DK-iteration. However, $\|DND^{-1}\|_\infty$ is also of interest in its own right. The reason for this, is that when all uncertainty blocks are full and complex, the upper bound provides a necessary and sufficient condition for robustness to *arbitrary-slow time-varying linear uncertainty* (Poolla and Tikku, 1995). On the other hand, the use of μ assumes the uncertain perturbations to be *time-invariant*. In some cases, it can be argued that slowly time-varying uncertainty is more useful than constant perturbations, and therefore that it is better to minimize $\|DND^{-1}\|_\infty$ instead of $\mu(N)$. In addition, by considering how $D(\omega)$ varies with frequency, one can find bounds on the allowed time variations in the perturbations.

Another interesting fact is that the use of *constant D-scales* (D is *not* allowed to vary with frequency), provides a necessary and sufficient condition for robustness to *arbitrary-fast time-varying linear uncertainty* (Shamma, 1994). It may be argued that such perturbations are unlikely in a practical situation. Nevertheless, we see that if we can get an acceptable controller design using constant D-scales, then we know that this controller will work very well even for rapid changes in the plant model. Another advantage of constant D-scales is that the computation of μ is then straightforward and may be solved using LMIs, see (8.151) below.

8.13.2   Real perturbations and the mixed μ problem

We have not discussed in any detail the analysis and design problems which arise with real or, more importantly, mixed real and complex perturbations.

The current algorithms, implemented in the MATLAB μ-toolbox, employ a generalization of the upper bound $\bar{\sigma}(DMD^{-1})$, where in addition to D-matrices, which exploit the block-diagonal structure of the perturbations, there are G-matrices, which exploit the structure of the real perturbations. The G-matrices (which should not be confused with the plant transfer function $G(s)$) have real diagonal elements at locations where Δ is real and have zeros elsewhere. The algorithm in the μ-toolbox makes use of the following result from Young et al. (1992): if there exists a $\beta > 0$, a D and a G with the appropriate block-diagonal structure such that

$$\bar{\sigma}\left((I+G^2)^{-\frac{1}{4}}(\frac{1}{\beta}DMD^{-1}-jG)(I+G^2)^{-\frac{1}{4}}\right) \leq 1 \tag{8.146}$$

then $\mu(M) \leq \beta$. For more details, the reader is referred to Young (1993).

There is also a corresponding DGK-iteration procedure for synthesis (Young, 1994). The practical implementation of this algorithm is however difficult, and a very high order fit may be required for the G-scales. An alternative approach which involves solving a series of scaled DK-iterations is given by Tøffner-Clausen et al. (1995).

8.13.3 Computational complexity

It has been established that the computational complexity of computing μ has a combinatoric (non-polynomial, "NP-hard") growth with the number of parameters involved even for purely complex perturbations (Toker and Ozbay, 1995).

This does not mean, however, that practical algorithms are not possible, and we have described practical algorithms for computing upper bounds of μ for cases with complex, real or mixed real/complex perturbations.

As mentioned, the upper bound $\bar{\sigma}(DMD^{-1})$ for complex perturbations is generally tight, whereas the present upper bounds for mixed perturbations (see (8.146)) may be arbitrarily conservative.

There also exists a number of lower bounds for computing μ. Most of these involve generating a perturbation which makes $I - M\Delta$ singular.

8.13.4 Discrete case

It is also possible to use μ for analyzing robust performance of discrete-time systems (Packard and Doyle, 1993). Consider a discrete time system

$$x_{k+1} = Ax_k + Bu_k, \quad y_k = Cx_k + Du_k$$

The corresponding discrete transfer function matrix from u to y is $N(z) = C(zI - A)^{-1}B + D$. First, note that the $\mathcal{H}_\infty$ norm of a discrete transfer function is

$$\|N\|_\infty \triangleq \max_{|z|\geq 1} \bar{\sigma}(C(zI - A)^{-1}B + D)$$

This follows since evaluation on the $j\omega$-axis in the continuous case is equivalent to the unit circle ($|z| = 1$) in the discrete case. Second, note that $N(z)$ may be written as an LFT in terms of $1/z$,

$$N(z) = C(zI - A)^{-1}B + D = F_u(H, \frac{1}{z}I); \quad H = \begin{bmatrix} A & B \\ C & D \end{bmatrix} \quad (8.147)$$

Thus, by introducing $\delta_z = 1/z$ and $\Delta_z = \delta_z I$ we have from the main loop theorem of Packard and Doyle (1993) (which generalizes Theorem 8.7) that $\|N\|_\infty < 1$ (NP) if and only if

$$\mu_{\widehat{\Delta}}(H) < 1, \quad \widehat{\Delta} = \text{diag}\{\Delta_z, \Delta_P\} \quad (8.148)$$

where Δ_z is a matrix of repeated complex scalars, representing the discrete "frequencies", and Δ_P is a full complex matrix, representing the singular value performance specification. Thus, we see that the search over frequencies in the frequency domain is avoided, but at the expense of a complicated μ-calculation. The condition in (8.148) is also referred to as the *state-space μ test*.

Condition (8.148) only considers nominal performance (NP). However, note that in this case nominal stability (NS) follows as a special case (and thus does not need to be tested separately), since when $\mu_{\widehat{\Delta}}(H) \leq 1$ (NP) we have from (8.78) that $\mu_{\Delta_z}(A) = \rho(A) < 1$, which is the well-known stability condition for discrete systems.

We can also generalize the treatment to consider RS and RP. In particular, since the state-space matrices are contained explicitly in H in (8.147), it follows that the discrete time formulation is convenient if we want to consider parametric uncertainty in the state-space matrices. This is discussed by Packard and Doyle (1993). However, this results in real perturbations, and the resulting μ-problem which involves repeated complex perturbations (from the evaluation of z on the unit circle), a full-block complex perturbation (from the performance specification), and real perturbations (from the uncertainty), is difficult to solve numerically both for analysis and in particular for synthesis. For this reason the discrete-time formulation is little used in practical applications.

8.13.5 Relationship to linear matrix inequalities (LMIs)

An example of an LMI problem is the following. For given matrices P, Q, R and S, does there exist an $X = X^H$ (where X may have a block-diagonal structure) such that

$$P^H X P - Q^H X Q + X R + R^H X + S < 0 \quad ? \quad (8.149)$$

Depending on the particular problem, the $<$ may be replaced by $\leq$. These inequality conditions produce a set of solutions which are convex, which make LMIs attractive computationally. Sometimes, the matrices P, Q, R and S are functions of a real parameter β, and we want to know, for example, what is the largest β for which there is no solution.

The upper bound for μ can be rewritten as an LMI:

$$\bar{\sigma}(DMD^{-1}) < \beta \Leftrightarrow \rho(D^{-1}M^H DDMD^{-1}) < \beta^2 \quad (8.150)$$

$$\Leftrightarrow D^{-1}M^H DDMD^{-1} - \beta^2 I < 0 \Leftrightarrow M^H D^2 M - \beta_2 D^2 < 0 \quad (8.151)$$

which is an LMI in $X = D^2 > 0$. To compute the upper bound for μ based on this LMI we need to iterate on β. It has been shown that a number of other problems, including $\mathcal{H}_2$ and $\mathcal{H}_\infty$ optimal control problems, can be reduced to solving LMIs. The reader is referred to Boyd et al. (1994) for more details.

8.14 Conclusion

In this chapter and the last we have discussed how to represent uncertainty and how to analyze its effect on stability (RS) and performance (RP) using the structured singular value μ as our main tool.

To analyze robust stability (RS) of an uncertain system we make use of the $M\Delta$-structure (Figure 8.3) where M represents the transfer function for the "new" feedback part generated by the uncertainty. From the small gain theorem,

$$\text{RS} \quad \Leftarrow \quad \bar{\sigma}(M) < 1 \quad \forall \omega \quad (8.152)$$

which is tight (necessary and sufficient) for the special case where at each frequency *any* complex Δ satisfying $\bar{\sigma}(\Delta) \leq 1$ is allowed. More generally, the tight condition is

$$\text{RS} \Leftrightarrow \mu(M) < 1 \quad \forall \omega \quad (8.153)$$

where $\mu(M)$ is the structured singular value $\mu(M)$. The calculation of μ makes use of the fact that Δ has a given block-diagonal structure, where certain blocks may also be real (e.g. to handle parametric uncertainty).

We defined robust performance (RP) as $\|F_l(N, \Delta)\|_\infty < 1$ for all allowed Δ's. Since we used the $\mathcal{H}_\infty$ norm in both the representation of uncertainty and the definition of performance, we found that RP could be viewed as a special case of RS, and we derived

$$\text{RP} \quad \Leftrightarrow \quad \mu(N) < 1 \quad \forall \omega \quad (8.154)$$

where μ is computed with respect to the block-diagonal structure diag$\{\Delta, \Delta_P\}$. Here Δ represents the uncertainty and Δ_P is a fictitious full uncertainty block representing the $\mathcal{H}_\infty$ performance bound.

It should be noted that there are two main approaches to getting a robust design:

1. We aim to make the system robust to some "general" class of uncertainty which we do not explicitly model. For SISO systems the classical gain and phase margins and the peaks of S and T provide useful general robustness measures. For MIMO systems, normalized coprime factor uncertainty provides a good general class

of uncertainty, and the associated Glover-McFarlane $\mathcal{H}_\infty$ loop-shaping design procedure, see Chapter 10, has proved itself very useful in applications.

2. We explicitly model and quantify the uncertainty in the plant and aim to make the system robust to this specific uncertainty. This second approach has been the focus of the preceding two chapters. Potentially, it yields better designs, but it may require a much larger effort in terms of uncertainty modelling, especially if parametric uncertainty is considered. Analysis and, in particular, synthesis using μ can be very involved.

In applications, it is therefore recommended to start with the first approach, at least for design. The robust stability and performance is then analyzed in simulations and using the structured singular value, for example, by considering first simple sources of uncertainty such as multiplicative input uncertainty. One then iterates between design and analysis until a satisfactory solution is obtained.

Practical μ-analysis

We end the chapter by providing a few recommendations on how to use the structured singular value μ in practice.

1. Because of the effort involved in deriving detailed uncertainty descriptions, and the subsequent complexity in synthesizing controllers, the rule is to "start simple" with a crude uncertainty description, and then to see whether the performance specifications can be met. Only if they can't, should one consider more detailed uncertainty descriptions such as parametric uncertainty (with real perturbations).

2. The use of μ implies a worst-case analysis, so one should be careful about including too many sources of uncertainty, noise and disturbances – otherwise it becomes very unlikely for the worst case to occur, and the resulting analysis and design may be unnecessarily conservative.

3. There is always uncertainty with respect to the inputs and outputs, so it is generally "safe" to include *diagonal* input and output uncertainty. The relative (multiplicative) form is very convenient in this case.

4. μ is most commonly used for analysis. If μ is used for synthesis, then we recommend that you keep the uncertainty fixed and adjust the parameters in the performance weight until μ is close to 1.

9

CONTROLLER DESIGN

In this chapter, we present practical procedures for multivariable controller design which are relatively straightforward to apply and which, in our opinion, have an important role to play in industrial control.

For industrial systems which are either SISO or loosely coupled, the classical loop-shaping approach to control system design as described in Section 2.6 has been successfully applied. But for truly multivariable systems it has only been in the last decade, or so, that reliable generalizations of this classical approach have emerged.

9.1   Trade-offs in MIMO feedback design

The shaping of multivariable transfer functions is based on the idea that a satisfactory definition of gain (range of gain) for a matrix transfer function is given by the singular values of the transfer function. By multivariable transfer function shaping, therefore, we mean the shaping of singular values of appropriately specified transfer functions such as the loop transfer function or possibly one or more closed-loop transfer functions. This methodology for controller design is central to the practical procedures described in this chapter.

In February 1981, the IEEE Transactions on Automatic Control published a Special Issue on Linear Multivariable Control Systems, the first six papers of which were on the use of singular values in the analysis and design of multivariable feedback systems. The paper by Doyle and Stein (1981) was particularly influential: it was primarily concerned with the fundamental question of how to achieve the benefits of feedback in the presence of unstructured uncertainty, and through the use of singular values it showed how the classical loop-shaping ideas of feedback design could be generalized to multivariable systems. To see how this was done, consider the one degree-of-freedom configuration shown in Figure 9.1. The plant G and controller K interconnection is driven by reference commands r, output disturbances d, and measurement noise n. y are the outputs to be controlled, and u are the control signals. In terms of the sensitivity function $S = (I + GK)^{-1}$ and the closed-loop transfer function $T = GK(I + GK)^{-1} = I - S$, we have the following important relationships:

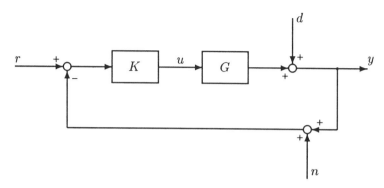

Figure 9.1: One degree-of-freedom feedback configuration

$$y(s) = T(s)r(s) + S(s)d(s) - T(s)n(s) \qquad (9.1)$$
$$u(s) = K(s)S(s)[r(s) - n(s) - d(s)] \qquad (9.2)$$

These relationships determine several closed-loop objectives, in addition to the requirement that K stabilizes G, namely:

1. For *disturbance rejection* make $\bar{\sigma}(S)$ small.
2. For *noise attenuation* make $\bar{\sigma}(T)$ small.
3. For *reference tracking* make $\bar{\sigma}(T) \approx \underline{\sigma}(T) \approx 1$.
4. For *control energy reduction* make $\bar{\sigma}(KS)$ small.

If the unstructured uncertainty in the plant model G is represented by an additive perturbation, i.e. $G_p = G + \Delta$, then a further closed-loop objective is

5. For *robust stability* in the presence of an additive perturbation make $\bar{\sigma}(KS)$ small.

Alternatively, if the uncertainty is modelled by a multiplicative output perturbation such that $G_p = (I + \Delta)G$, then we have:

6. For *robust stability* in the presence of a multiplicative output perturbation make $\bar{\sigma}(T)$ small.

The closed-loop requirements 1 to 6 cannot all be satisfied simultaneously. Feedback design is therefore a trade-off over frequency of conflicting objectives. This is not always as difficult as it sounds because the frequency ranges over which the objectives are important can be quite different. For example, disturbance rejection is typically a low frequency requirement, while noise mitigation is often only relevant at higher frequencies.

In classical *loop shaping*, it is the magnitude of the open-loop transfer function $L = GK$ which is shaped, whereas the above design requirements are all in terms of closed-loop transfer functions. However, recall from (3.51) that

$$\underline{\sigma}(L) - 1 \leq \frac{1}{\bar{\sigma}(S)} \leq \underline{\sigma}(L) + 1 \tag{9.3}$$

from which we see that $\bar{\sigma}(S) \approx 1/\underline{\sigma}(L)$ at frequencies where $\underline{\sigma}(L)$ is much larger than 1. It also follows that at the bandwidth frequency (where $1/\bar{\sigma}(S(j\omega_B)) = \sqrt{2} = 1.41$) we have $\underline{\sigma}(L(j\omega_B))$ between 0.41 and 2.41. Furthermore, from $T = L(I + L)^{-1}$ it follows that $\bar{\sigma}(T) \approx \bar{\sigma}(L)$ at frequencies where $\bar{\sigma}(L)$ is small. Thus, over specified frequency ranges, it is relatively easy to approximate the closed-loop requirements by the following open-loop objectives:

1. For *disturbance rejection* make $\underline{\sigma}(GK)$ large; valid for frequencies at which $\underline{\sigma}(GK) \gg 1$.
2. For *noise attenuation* make $\bar{\sigma}(GK)$ small; valid for frequencies at which $\bar{\sigma}(GK) \ll 1$.
3. For *reference tracking* make $\underline{\sigma}(GK)$ large; valid for frequencies at which $\underline{\sigma}(GK) \gg 1$.
4. For *control energy reduction* make $\bar{\sigma}(K)$ small; valid for frequencies at which $\bar{\sigma}(GK) \ll 1$.
5. For *robust stability to an additive perturbation* make $\bar{\sigma}(K)$ small; valid for frequencies at which $\bar{\sigma}(GK) \ll 1$.
6. For *robust stability to a multiplicative output perturbation* make $\bar{\sigma}(GK)$ small; valid for frequencies at which $\bar{\sigma}(GK) \ll 1$.

Typically, the open-loop requirements 1 and 3 are valid and important at low frequencies, $0 \leq \omega \leq \omega_l \leq \omega_B$, while 2, 4, 5 and 6 are conditions which are valid and important at high frequencies, $\omega_B \leq \omega_h \leq \omega \leq \infty$, as illustrated in Figure 9.2. From this we see that at frequencies where we want high gains (at low frequencies) the "worst-case" direction is related to $\underline{\sigma}(GK)$, whereas at frequencies where we want low gains (at high frequencies) the "worst-case" direction is related to $\bar{\sigma}(GK)$.

Exercise 9.1 *Show that the closed-loop objectives 1 to 6 can be approximated by the open-loop objectives 1 to 6 at the specified frequency ranges.*

From Figure 9.2, it follows that the control engineer must design K so that $\bar{\sigma}(GK)$ and $\underline{\sigma}(GK)$ avoid the shaded regions. That is, for good performance, $\underline{\sigma}(GK)$ must be made to lie above a performance boundary for all ω up to ω_l, and for robust stability $\bar{\sigma}(GK)$ must be forced below a robustness boundary for all ω above ω_h. To shape the singular values of GK by selecting K is a relatively easy task, but to do this in a way which also guarantees closed-loop stability is in general difficult. Closed-loop stability cannot be determined from open-loop singular values.

For SISO systems, it is clear from Bode's work (1945) that closed-loop stability is closely related to open-loop gain and phase near the crossover frequency ω_c, where

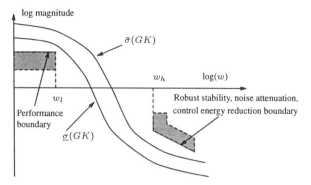

Figure 9.2: Design trade-offs for the multivariable loop transfer function GK

$|GK(j\omega_c)| = 1$. In particular, the roll-off rate from high to low gain at crossover is limited by phase requirements for stability, and in practice this corresponds to a roll-off rate less than 40 $dB/decade$; see Section 2.6.2. An immediate consequence of this is that there is a lower limit to the difference between ω_h and ω_l in Figure 9.2.

For MIMO systems a similar gain-phase relationship holds in the crossover frequency region, but this is in terms of the eigenvalues of GK and results in a limit on the roll-off rate of the magnitude of the eigenvalues of GK, not the singular values (Doyle and Stein, 1981). The stability constraint is therefore even more difficult to handle in multivariable loop-shaping than it is in classical loop-shaping. To overcome this difficulty Doyle and Stein (1981) proposed that the loop-shaping should be done with a controller that was already known to guarantee stability. They suggested that an LQG controller could be used in which the regulator part is designed using a "sensitivity recovery" procedure of Kwakernaak (1969) to give desirable properties (gain and phase margins) in GK. They also gave a dual "robustness recovery" procedure for designing the filter in an LQG controller to give desirable properties in KG. Recall that KG is not in general equal to GK, which implies that stability margins vary from one break point to another in a multivariable system. Both of these loop transfer recovery (LTR) procedures are discussed below after first describing traditional LQG control.

9.2 LQG control

Optimal control, building on the optimal filtering work of Wiener in the 1940's, reached maturity in the 1960's with what we now call linear quadratic gaussian or LQG Control. Its development coincided with large research programs and considerable funding in the United States and the former Soviet Union on space related problems. These were problems, such as rocket manoeuvering with minimum fuel consumption, which could be well defined and easily formulated as optimizations. Aerospace engineers were particularly successful at applying LQG, but when other control engineers attempted to use LQG on everyday industrial problems a different story emerged. Accurate plant models were frequently not available and the assumption of white noise disturbances was not always relevant or meaningful to practising control engineers. As a result LQG designs were sometimes not robust enough to be used in practice. In this section, we will describe the LQG problem and its solution, we will discuss its robustness properties, and we will describe procedures for improving robustness. Many text books consider this topic in far greater detail; we recommend Anderson and Moore (1989) and Kwakernaak and Sivan (1972).

9.2.1 Traditional LQG and LQR problems

In traditional LQG Control, it is assumed that the plant dynamics are linear and known, and that the measurement noise and disturbance signals (process noise) are stochastic with known statistical properties. That is, we have a plant model

$$\dot{x} = Ax + Bu + w_d \tag{9.4}$$

$$y = Cx + w_n \tag{9.5}$$

where w_d and w_n are the disturbance (process noise) and measurement noise inputs respectively, which are usually assumed to be uncorrelated zero-mean Gaussian stochastic processes with constant power spectral density matrices W and V respectively. That is, w_d and w_n are white noise processes with covariances

$$E\left\{w_d(t)w_d(\tau)^T\right\} = W\delta(t - \tau) \tag{9.6}$$

$$E\left\{w_n(t)w_n(\tau)^T\right\} = V\delta(t - \tau) \tag{9.7}$$

and

$$E\left\{w_d(t)w_n(\tau)^T\right\} = 0, \ E\left\{w_n(t)w_d(\tau)^T\right\} = 0 \tag{9.8}$$

where E is the expectation operator and $\delta(t - \tau)$ is a delta function.

The LQG control problem is to find the optimal control $u(t)$ which minimizes

$$J = E\left\{\lim_{T \to \infty} \frac{1}{T} \int_0^T \left[x^T Q x + u^T R u\right] dt\right\} \tag{9.9}$$

where Q and R are appropriately chosen constant weighting matrices (design parameters) such that $Q = Q^T \geq 0$ and $R = R^T > 0$. The name LQG arises from the use of a linear model, an integral Quadratic cost function, and Gaussian white noise processes to model disturbance signals and noise.

The solution to the LQG problem, known as the Separation Theorem or Certainty Equivalence Principle, is surprisingly simple and elegant. It consists of first determining the optimal control to a deterministic linear quadratic regulator (LQR) problem:

namely, the above LQG problem without w_d and w_n. It happens that the solution to this problem can be written in terms of the simple state feedback law

$$u(t) = -K_r x(t) \qquad (9.10)$$

where K_r is a constant matrix which is easy to compute and is clearly independent of W and V, the statistical properties of the plant noise. The next step is to find an optimal estimate $\hat{x}$ of the state x, so that $E\left\{ [x - \hat{x}]^T [x - \hat{x}] \right\}$ is minimized. The optimal state estimate is given by a Kalman filter and is independent of Q and R. The required solution to the LQG problem is then found by replacing x by $\hat{x}$, to give $u(t) = -K_r\hat{x}(t)$. We therefore see that the LQG problem and its solution can be separated into two distinct parts, as illustrated in Figure 9.3.

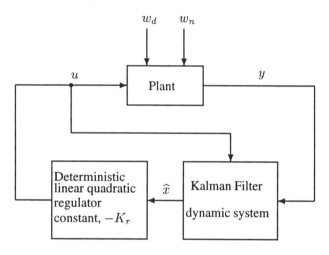

Figure 9.3: The Separation Theorem

We will now give the equations necessary to find the optimal state-feedback matrix K_r and the Kalman filter.

Optimal state feedback. The LQR problem, where all the states are known, is the deterministic initial value problem: given the system $\dot{x} = Ax + Bu$ with a non-zero initial state $x(0)$, find the input signal $u(t)$ which takes the system to the zero state ($x = 0$) in an optimal manner, i.e. by minimizing the deterministic cost

$$J_r = \int_0^\infty (x(t)^T Q x(t) + u(t)^T R u(t)) dt \qquad (9.11)$$

The optimal solution (for any initial state) is $u(t) = -K_r x(t)$, where

$$K_r = R^{-1} B^T X \qquad (9.12)$$

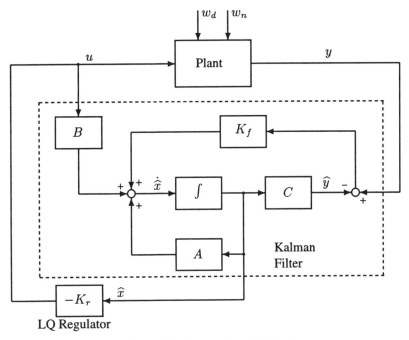

Figure 9.4: The LQG controller and noisy plant

and $X = X^T \geq 0$ is the unique positive-semidefinite solution of the algebraic Riccati equation

$$A^T X + X A - X B R^{-1} B^T X + Q = 0 \qquad (9.13)$$

Kalman filter. The Kalman filter has the structure of an ordinary state-estimator or observer, as shown in Figure 9.4, with

$$\dot{\widehat{x}} = A\widehat{x} + Bu + K_f(y - C\widehat{x}) \qquad (9.14)$$

The optimal choice of K_f, which minimizes $E\left\{[x - \widehat{x}]^T [x - \widehat{x}]\right\}$, is given by

$$K_f = YC^T V^{-1} \qquad (9.15)$$

where $Y = Y^T \geq 0$ is the unique positive-semidefinite solution of the algebraic Riccati equation

$$YA^T + AY - YC^T V^{-1} CY + W = 0 \qquad (9.16)$$

LQG: Combined optimal state estimation and optimal state feedback. The LQG control problem is to minimize J in (9.9). The structure of the LQG controller is illustrated in Figure 9.4; its transfer function, from y to u (i.e. assuming positive feedback), is easily shown to be given by

$$
K_{LQG}(s) \;\overset{s}{=}\; \left[\begin{array}{c|c} A - BK_r - K_fC & K_f \\ \hline -K_r & 0 \end{array}\right]
$$

$$
= \left[\begin{array}{c|c} A - BR^{-1}B^TX - YC^TV^{-1}C & YC^TV^{-1} \\ \hline -R^{-1}B^TX & 0 \end{array}\right] \quad (9.17)
$$

It has the same degree (number of poles) as the plant.

Remark. The optimal gain matrices K_f and K_r exist, and the LQG-controlled system is internally stable, provided the systems with state-space realizations $(A, B, Q^{\frac{1}{2}})$ and $(A, W^{\frac{1}{2}}, C)$ are stabilizable and detectable.

Exercise 9.2 *For the plant and LQG controller arrangement of Figure 9.4, show that the closed-loop dynamics are described by*

$$
\frac{d}{dt}\left[\begin{array}{c} x \\ x - \widehat{x} \end{array}\right] = \left[\begin{array}{cc} A - BK_r & BK_r \\ 0 & A - K_fC \end{array}\right]\left[\begin{array}{c} x \\ x - \widehat{x} \end{array}\right] + \left[\begin{array}{cc} I & 0 \\ I & -K_f \end{array}\right]\left[\begin{array}{c} w_d \\ w_n \end{array}\right]
$$

This shows that the closed-loop poles are simply the union of the poles of the deterministic LQR system (eigenvalues of $A - BK_r$) and the poles of the Kalman filter (eigenvalues of $A - K_fC$). It is exactly as we would have expected from the Separation Theorem.

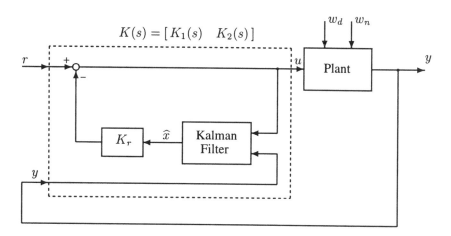

Figure 9.5: A rearranged LQG controller with reference input

Exercise 9.3 *In Figure 9.5, a reference input r is introduced into the LQG controller-plant configuration and the controller is rearranged to illustrate its two degrees-of-freedom structure. Show that the transfer function of the controller $K(s)$ linking $[\,r^T \quad y^T\,]^T$ to u is*

$$
K(s) = [\,K_1(s) \quad K_2(s)\,]
$$
$$
= -K_r(sI - A + BK_r + K_fC)^{-1}[B \quad K_f] + [I \quad 0]
$$

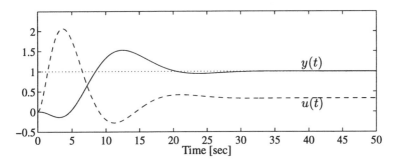

Figure 9.6: LQG design for inverse response process. Closed-loop response to unit step in reference.

If the LQG controller is stable, for which there is no guarantee, then $K_1(s)$ can be implemented as a pre-filter and $K_2(s)$ as the feedback controller. However, since the degrees (number of states) of $K_1(s), K_2(s)$ and $K_{LQG}(s)$ are in general the same, there is an increase in complexity on implementing $K_1(s)$ and $K_2(s)$ separately and so this is not recommended.

Example 9.1 LQG design for inverse response process. *We will here apply LQG to the SISO inverse response process introduced in Chapter 2. Recall from (2.26) that the plant model has a RHP-zero and is given by $G(s) = \frac{3(-2s+1)}{(5s+1)(10s+1)}$. The standard LQG design procedure does not give a controller with integral action, so we augment the plant $G(s)$ with an integrator before starting the design, and then use this integrator as part of the final controller. For the objective function $J = \int (x^T Q x + u^T R u) dt$ we use the weights $Q = q C^T C$, with $q = 1$, and $R = 1$. (Only the ratio between q and R matters and reducing R yields a faster response). Our choice for Q with the term $C^T C$ implies that we are weighting the output y rather than the states. Because we have augmented the plant with an integrator the weight R penalizes the derivative of the input, du/dt. For the noise weights we select $W = wI$, where $w = 1$, (process noise directly on the states) and $V = 1$ (measurement noise). (Only the ratio between w and V matter and reducing V yields a faster response). The MATLAB file in Table 9.1 was used to design the LQG controller.*

The resulting closed-loop response is shown in Figure 9.6. The response is not very good, nor is it easy to improve by adjusting the weights. One reason for this is that in our formulation we have penalized du/dt which makes it difficult for the controller to respond quickly to changes. For a comparison, see the loop-shaping design in Figure 2.17.

9.2.2 Robustness properties

For an LQG-controlled system with a combined Kalman filter and LQR control law there are no guaranteed stability margins. This was brought starkly to the attention of the control community by Doyle (1978). He showed, by example, that there exist LQG combinations with arbitrarily small gain margins.

However, for an LQR-controlled system (i.e. assuming all the states are available and no stochastic inputs) it is well known (Kalman, 1964; Safonov and Athans, 1977)

Table 9.1: **MATLAB commands to generate LQG controller in Example 9.1**

```
% Uses the Robust Control toolbox
G=nd2sys([-2,1],[50 15 1],3);       % original plant.
int = nd2sys(1,[1 0.0001]);         % integrator (moved slightly into LHP for
%                                   % numerical reasons).
Gs = mmult(G,int);                  % augmented plant (with integrator).
[A,B,C,D] = unpck(Gs);              % Augmented Plant is [A,B,C,D].
Q = 1.0*C'*C;                       % weight on outputs.
R = 1;                              % weight on inputs, R small gives faster
%                                   % response.
Kx = lqr(A,B,Q,R);                  % optimal regulator.
Bnoise = eye(size(A));             % process noise directly on states.
W = eye(size(A));                   % process noise.
V = 1;                              % measurement noise, V small gives faster
%                                   % response.
Ke = lqe(A,Bnoise,C,W,V);          % optimal estimator.
[Ac,Bc,Cc,Dc] = reg(A,B,C,D,Kx,Ke); % combine regulator and estimator.
Ks = pck(Ac,Bc,Cc,Dc);
Klqg = mmult(Ks,int,-1);            % include integrator in final controller
%                                   % with positive feedback
```

that, if the weight R is chosen to be diagonal, the sensitivity function $S = (I + K_r (sI - A)^{-1} B)^{-1}$ satisfies

$$\bar{\sigma} (S(j\omega)) \leq 1, \ \forall w \qquad (9.18)$$

From this it can be shown that the system will have a gain margin equal to infinity, a gain reduction margin equal to 0.5, and a (minimum) phase margin of 60° in each plant input control channel. This means that in the LQR-controlled system $u = -K_r x$, a complex perturbation diag $\{k_i e^{j\theta_i}\}$ can be introduced at the plant inputs without causing instability providing

(i) $\theta_i = 0$ and $0.5 \leq k_i \leq \infty, i = 1, 2, \ldots, m$

or

(ii) $k_i = 1$ and $|\theta_i| \leq 60°, i = 1, 2, \ldots, m$

where m is the number of plant inputs. For a single-input plant, the above shows that the Nyquist diagram of the open-loop regulator transfer function $K_r(sI - A)^{-1}B$ will always lie outside the unit circle with centre -1. This was first shown by Kalman (1964), and is illustrated in Example 9.2 below.

Example 9.2 **LQR design of a first order process.** *Consider a first order process* $G(s) = 1/(s - a)$ *with the state-space realization*

$$\dot{x}(t) = ax(t) + u(t), \quad y(t) = x(t)$$

so that the state is directly measured. For a non-zero initial state the cost function to be minimized is

$$J_r = \int_0^\infty (x^2 + Ru^2)dt$$

The algebraic Riccati equation (9.13) becomes ($A = a$, $B = 1$, $Q = 1$)

$$aX + Xa - XR^{-1}X + 1 = 0 \quad \Leftrightarrow \quad X^2 - 2aRX - R = 0$$

which, since $X \geq 0$, gives $X = aR + \sqrt{(aR)^2 + R}$. The optimal control is given by $u = -K_r x$ where from (9.12)

$$K_r = X/R = a + \sqrt{a^2 + 1/R}$$

and we get the closed-loop system

$$\dot{x} = ax + u = -\sqrt{a^2 + 1/R}\, x$$

The closed-loop pole is located at $s = -\sqrt{a^2 + 1/R} < 0$. Thus, the root locus for the optimal closed-loop pole with respect to R starts at $s = -|a|$ for $R = \infty$ (infinite weight on the input) and moves to $-\infty$ along the real axis as R approaches zero. Note that the root locus is identical for stable ($a < 0$) and unstable ($a > 0$) plants $G(s)$ with the same value of $|a|$. In particular, for $a > 0$ we see that the minimum input energy needed to stabilize the plant (corresponding to $R = \infty$) is obtained with the input $u = -2|a|x$, which moves the pole from $s = a$ to its mirror image at $s = -a$.

For R small ("cheap control") the gain crossover frequency of the loop transfer function $L = GK_r = K_r/(s-a)$ is given approximately by $\omega_c \approx \sqrt{1/R}$. Note also that $L(j\omega)$ has a roll-off of -1 at high frequencies, which is a general property of LQR designs. Furthermore, the Nyquist plot of $L(j\omega)$ avoids the unit disc centred on the critical point -1, that is $|S(j\omega)| = 1/|1 + L(j\omega)| \leq 1$ at all frequencies. This is obvious for the stable plant with $a < 0$ since $K_r > 0$ and then the phase of $L(j\omega)$ varies from $0°$ (at zero frequency) to $-90°$ (at infinite frequency). The surprise is that it is also true for the unstable plant with $a > 0$ even though the phase of $L(j\omega)$ varies from $-180°$ to $-90°$.

Consider now the Kalman filter shown earlier in Figure 9.4. Notice that it is itself a feedback system. Arguments dual to those employed for the LQR-controlled system can then be used to show that, if the power spectral density matrix V is chosen to be diagonal, then at the input to the Kalman gain matrix K_f there will be an infinite gain margin, a gain reduction margin of 0.5 and a minimum phase margin of 60°. Consequently, for a single-output plant, the Nyquist diagram of the open-loop filter transfer function $C(sI - A)^{-1}K_f$ will lie outside the unit circle with centre at -1.

An LQR-controlled system has good stability margins at the plant inputs, and a Kalman filter has good stability margins at the inputs to K_f, so why are there no guarantees for LQG control? To answer this, consider the LQG controller arrangement shown in Figure 9.7. The loop transfer functions associated with the labelled points 1 to 4 are respectively

$$
\begin{aligned}
L_1(s) &= K_r\left[\Phi(s)^{-1} + BK_r + K_fC\right]^{-1}K_fC\Phi(s)B \\
&= -K_{LQG}(s)G(s) & (9.19) \\
L_2(s) &= -G(s)K_{LQG}(s) & (9.20) \\
L_3(s) &= K_r\Phi(s)B \quad \text{(regulator transfer function)} & (9.21) \\
L_4(s) &= C\Phi(s)K_f \quad \text{(filter transfer function)} & (9.22)
\end{aligned}
$$

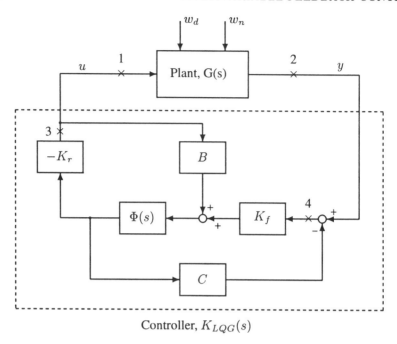

Controller, $K_{LQG}(s)$

Figure 9.7: LQG-controlled plant

where

$$\Phi(s) \stackrel{\triangle}{=} (sI - A)^{-1} \tag{9.23}$$

$K_{LQG}(s)$ is as in (9.17) and $G(s) = C\Phi(s)B$ is the plant model.

Remark. $L_3(s)$ and $L_4(s)$ are surprisingly simple. For $L_3(s)$ the reason is that after opening the loop at point 3 the error dynamics (point 4) of the Kalman filter are not excited by the plant inputs; in fact they are uncontrollable from u.

Exercise 9.4 *Derive the expressions for $L_1(s)$, $L_2(s)$, $L_3(s)$ and $L_4(s)$, and explain why $L_4(s)$ (like $L_3(s)$) has such a simple form.*

At points 3 and 4 we have the guaranteed robustness properties of the LQR system and the Kalman filter respectively. But at the actual input and output of the plant (points 1 and 2) where we are most interested in achieving good stability margins, we have complex transfer functions which in general give no guarantees of satisfactory robustness properties. Notice also that points 3 and 4 are effectively inside the LQG controller which has to be implemented, most likely as software, and so we have good stability margins where they are not really needed and no guarantees where they are.

Fortunately, for a minimum phase plant procedures developed by Kwakernaak (1969) and Doyle and Stein (1979; 1981) show how, by a suitable choice of parameters, either $L_1(s)$ can be made to tend asymptotically to $L_3(s)$ or $L_2(s)$ can be made to approach $L_4(s)$. These procedures are considered next.

9.2.3   Loop transfer recovery (LTR) procedures

For full details of the recovery procedures, we refer the reader to the original communications (Kwakernaak, 1969; Doyle and Stein, 1979; Doyle and Stein, 1981) or to the tutorial paper by Stein and Athans (1987). We will only give an outline of the major steps here, since we will argue later that the procedures are somewhat limited for practical control system design. For a more recent appraisal of LTR, we recommend a Special Issue of the International Journal of Robust and Nonlinear Control, edited by Niemann and Stoustrup (1995).

The LQG loop transfer function $L_2(s)$ can be made to approach $C\Phi(s)K_f$, with its guaranteed stability margins, if K_r in the LQR problem is designed to be large using the sensitivity recovery procedure of Kwakernaak (1969). It is necessary to assume that the plant model $G(s)$ is minimum phase and that it has at least as many inputs as outputs.

Alternatively, the LQG loop transfer function $L_1(s)$ can be made to approach $K_r\Phi(s)B$ by designing K_f in the Kalman filter to be large using the robustness recovery procedure of Doyle and Stein (1979). Again, it is necessary to assume that the plant model $G(s)$ is minimum phase, but this time it must have at least as many outputs as inputs.

The procedures are dual and therefore we will only consider recovering robustness at the plant output. That is, we aim to make $L_2(s) = G(s)K_{\text{LQG}}(s)$ approximately equal to the Kalman filter transfer function $C\Phi(s)K_f$.

First, we design a Kalman filter whose transfer function $C\Phi(s)K_f$ is desirable. This is done, in an iterative fashion, by choosing the power spectral density matrices W and V so that the minimum singular value of $C\Phi(s)K_f$ is large enough at low frequencies for good performance and its maximum singular value is small enough at high frequencies for robust stability, as discussed in section 9.1. Notice that W and V are being used here as design parameters and their associated stochastic processes are considered to be fictitious. In tuning W and V we should be careful to choose V as diagonal and $W = (BS)(BS)^T$, where S is a scaling matrix which can be used to balance, raise, or lower the singular values. When the singular values of $C\Phi(s)K_f$ are thought to be satisfactory, loop transfer recovery is achieved by designing K_r in an LQR problem with $Q = C^T C$ and $R = \rho I$, where ρ is a scalar. As ρ tends to zero $G(s)K_{\text{LQG}}$ tends to the desired loop transfer function $C\Phi(s)K_f$.

Much has been written on the use of LTR procedures in multivariable control system design. But as methods for multivariable loop-shaping they are limited in their applicability and sometimes difficult to use. Their main limitation is to minimum phase plants. This is because the recovery procedures work by cancelling the plant zeros, and a cancelled non-minimum phase zero would lead to instability. The cancellation of lightly damped zeros is also of concern because of undesirable oscillations at these modes during transients. A further disadvantage is that the limiting process ($\rho \rightarrow 0$) which brings about full recovery also introduces high gains which may cause problems with unmodelled dynamics. Because of the above disadvantages, the recovery procedures are not usually taken to their limits ($\rho \rightarrow 0$) to achieve full recovery, but rather a set of designs is obtained (for small ρ) and an acceptable design is selected. The result is a somewhat ad-hoc design procedure in which the singular values of a loop transfer function, $G(s)K_{\mathrm{LQG}}(s)$ or $K_{\mathrm{LQG}}(s)G(s)$, are indirectly shaped. A more direct and intuitively appealing method for multivariable loop-shaping will be given in Section 9.4.

9.3 $\mathcal{H}_2$ and $\mathcal{H}_\infty$ control

Motivated by the shortcomings of LQG control there was in the 1980's a significant shift towards $\mathcal{H}_\infty$ optimization for robust control. This development originated from the influential work of Zames (1981), although an earlier use of $\mathcal{H}_\infty$ optimization in an engineering context can be found in Helton (1976). Zames argued that the poor robustness properties of LQG could be attributed to the integral criterion in terms of the $\mathcal{H}_2$ norm, and he also criticized the representation of uncertain disturbances by white noise processes as often unrealistic. As the $\mathcal{H}_\infty$ theory developed, however, the two approaches of $\mathcal{H}_2$ and $\mathcal{H}_\infty$ control were seen to be more closely related than originally thought, particularly in the solution process; see for example Glover and Doyle (1988) and Doyle et al. (1989). In this section, we will begin with a general control problem formulation into which we can cast all $\mathcal{H}_2$ and $\mathcal{H}_\infty$ optimizations of practical interest. The general $\mathcal{H}_2$ and $\mathcal{H}_\infty$ problems will be described along with some specific and typical control problems. It is not our intention to describe in detail the mathematical solutions, since efficient, commercial software for solving such problems is now so easily available. Rather we seek to provide an understanding of some useful problem formulations which might then be used by the reader, or modified to suit his or her application.

9.3.1   General control problem formulation

There are many ways in which feedback design problems can be cast as $\mathcal{H}_2$ and $\mathcal{H}_\infty$ optimization problems. It is very useful therefore to have a standard problem formulation into which any particular problem may be manipulated. Such a general formulation is afforded by the general configuration shown in Figure 9.8 and discussed earlier in Chapter 3. The system of Figure 9.8 is described by

$$\begin{bmatrix} z \\ v \end{bmatrix} = P(s) \begin{bmatrix} w \\ u \end{bmatrix} = \begin{bmatrix} P_{11}(s) & P_{12}(s) \\ P_{21}(s) & P_{22}(s) \end{bmatrix} \begin{bmatrix} w \\ u \end{bmatrix} \tag{9.24}$$

$$u = K(s)v \tag{9.25}$$

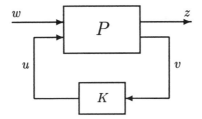

Figure 9.8: General control configuration

with a state-space realization of the generalized plant P given by

$$P \stackrel{s}{=} \left[\begin{array}{c|cc} A & B_1 & B_2 \\ \hline C_1 & D_{11} & D_{12} \\ C_2 & D_{21} & D_{22} \end{array} \right] \qquad (9.26)$$

The signals are: u the control variables, v the measured variables, w the exogenous signals such as disturbances w_d and commands r, and z the so-called "error" signals which are to be minimized in some sense to meet the control objectives. As shown in (3.103) the closed-loop transfer function from w to z is given by the linear fractional transformation

$$z = F_l(P, K)w \qquad (9.27)$$

where

$$F_l(P, K) = P_{11} + P_{12}K(I - P_{22}K)^{-1}P_{21} \qquad (9.28)$$

$\mathcal{H}_2$ and $\mathcal{H}_\infty$ control involve the minimization of the $\mathcal{H}_2$ and $\mathcal{H}_\infty$ norms of $F_l(P, K)$ respectively. We will consider each of them in turn.

First some remarks about the algorithms used to solve such problems. The most general, widely available and widely used algorithms for $\mathcal{H}_2$ and $\mathcal{H}_\infty$ control problems are based on the state-space solutions in Glover and Doyle (1988) and Doyle et al. (1989). It is worth mentioning again that the similarities between $\mathcal{H}_2$ and $\mathcal{H}_\infty$ theory are most clearly evident in the aforementioned algorithms. For example, both $\mathcal{H}_2$ and $\mathcal{H}_\infty$ require the solutions to two Riccati equations, they both give controllers of state-dimension equal to that of the generalized plant P, and they both exhibit a separation structure in the controller already seen in LQG control. An algorithm for $\mathcal{H}_\infty$ control problems is summarized in Section 9.3.4.

The following assumptions are typically made in $\mathcal{H}_2$ and $\mathcal{H}_\infty$ problems:

(A1) (A, B_2, C_2) is stabilizable and detectable.

(A2) D_{12} and D_{21} have full rank.

(A3) $\begin{bmatrix} A - j\omega I & B_2 \\ C_1 & D_{12} \end{bmatrix}$ has full column rank for all ω.

(A4) $\begin{bmatrix} A - j\omega I & B_1 \\ C_2 & D_{21} \end{bmatrix}$ has full row rank for all ω.

(A5) $D_{11} = 0$ and $D_{22} = 0$.

Assumption (A1) is required for the existence of stabilizing controllers K, and assumption (A2) is sufficient to ensure the controllers are proper and hence realizable. Assumptions (A3) and (A4) ensure that the optimal controller does not try to cancel poles or zeros on the imaginary axis which would result in closed-loop instability. Assumption (A5) is conventional in $\mathcal{H}_2$ control. $D_{11} = 0$ makes P_{11} strictly proper. Recall that $\mathcal{H}_2$ is the set of strictly proper stable transfer functions. $D_{22} = 0$ makes P_{22} strictly proper and simplifies the formulas in the $\mathcal{H}_2$ algorithms. In $\mathcal{H}_\infty$, neither $D_{11} = 0$, nor $D_{22} = 0$, is required but they do significantly simplify the algorithm formulas. If they are not zero, an equivalent $\mathcal{H}_\infty$ problem can be constructed in which they are; see (Safonov et al., 1989) and (Green and Limebeer, 1995). For simplicity, it is also sometimes assumed that D_{12} and D_{21} are given by

(A6) $D_{12} = \begin{bmatrix} 0 \\ I \end{bmatrix}$ and $D_{21} = \begin{bmatrix} 0 & I \end{bmatrix}$.

This can be achieved, without loss of generality, by a scaling of u and v and a unitary transformation of w and z; see for example Maciejowski (1989). In addition, for simplicity of exposition, the following additional assumptions are sometimes made

(A7) $D_{12}^T C_1 = 0$ and $B_1 D_{21}^T = 0$.

(A8) (A, B_1) is stabilizable and (A, C_1) is detectable.

Assumption (A7) is common in $\mathcal{H}_2$ control e.g. in LQG where there are no cross terms in the cost function ($D_{12}^T C_1 = 0$), and the process noise and measurement noise are uncorrelated ($B_1 D_{21}^T = 0$). Notice that if (A7) holds then (A3) and (A4) may be replaced by (A8).

Whilst the above assumptions may appear daunting, most sensibly posed control problems will meet them. Therefore, if the software (e.g. μ-tools or the Robust Control toolbox of MATLAB) complains, then it probably means that your control problem is not well formulated and you should think again.

Lastly, it should be said that $\mathcal{H}_\infty$ algorithms, in general, find a sub-optimal controller. That is, for a specified γ a stabilizing controller is found for which $\|F_l(P, K)\|_\infty < \gamma$. If an optimal controller is required then the algorithm can be used iteratively, reducing γ until the minimum is reached within a given tolerance. In general, to find an optimal $\mathcal{H}_\infty$ controller is numerically and theoretically complicated. This contrasts significantly with $\mathcal{H}_2$ theory, in which the optimal controller is unique and can be found from the solution of just two Riccati equations.

9.3.2 $\mathcal{H}_2$ optimal control

The standard $\mathcal{H}_2$ optimal control problem is to find a stabilizing controller K which minimizes

$$\| F(s) \|_2 = \sqrt{\frac{1}{2\pi} \int_{-\infty}^{\infty} F(j\omega) F(j\omega)^T d\omega}; \quad F \triangleq F_l(P, K) \tag{9.29}$$

For a particular problem the generalized plant P will include the plant model, the interconnection structure, and the designer specified weighting functions. This is illustrated for the LQG problem in the next subsection.

As discussed in Section 4.10.1 and noted in Tables A.1 and A.2 on page 525, the $\mathcal{H}_2$ norm can be given different deterministic interpretations. It also has the following stochastic interpretation. Suppose in the general control configuration that the exogenous input w is white noise of unit intensity. That is

$$E\left\{ w(t)w(\tau)^T \right\} = I\delta(t - \tau) \tag{9.30}$$

The expected power in the error signal z is then given by

$$\begin{aligned}
E &\left\{ \lim_{T \to \infty} \frac{1}{2T} \int_{-T}^{T} z(t)^T z(t) dt \right\} \tag{9.31} \\
&= \operatorname{tr} E\left\{ z(t)z(t)^T \right\} \\
&= \frac{1}{2\pi} \int_{-\infty}^{\infty} F(j\omega) F(j\omega)^T d\omega \\
&\qquad \qquad \text{(by Parseval's Theorem)} \\
&= \|F\|_2^2 = \|F_l(P, K)\|_2^2 \tag{9.32}
\end{aligned}$$

Thus, by minimizing the $\mathcal{H}_2$ norm, the output (or error) power of the generalized system, due to a unit intensity white noise input, is minimized; we are minimizing the root-mean-square (rms) value of z.

9.3.3 LQG: a special $\mathcal{H}_2$ optimal controller

An important special case of $\mathcal{H}_2$ optimal control is the LQG problem described in subsection 9.2.1. For the stochastic system

$$\dot{x} = Ax + Bu + w_d \tag{9.33}$$
$$y = Cx + w_n \tag{9.34}$$

where

$$E\left\{ \begin{bmatrix} w_d(t) \\ w_n(t) \end{bmatrix} [w_d(\tau)^T \quad w_n(\tau)^T] \right\} = \begin{bmatrix} W & 0 \\ 0 & V \end{bmatrix} \delta(t - \tau) \tag{9.35}$$

The LQG problem is to find $u = K(s)y$ such that

$$J = E\left\{\lim_{T\to\infty} \frac{1}{T} \int_0^T [x^T Q x + u^T R u]\, dt\right\} \tag{9.36}$$

is minimized with $Q = Q^T \geq 0$ and $R = R^T > 0$.

This problem can be cast as an $\mathcal{H}_2$ optimization in the general framework in the following manner. Define an error signal z as

$$z = \begin{bmatrix} Q^{\frac{1}{2}} & 0 \\ 0 & R^{\frac{1}{2}} \end{bmatrix} \begin{bmatrix} x \\ u \end{bmatrix} \tag{9.37}$$

and represent the stochastic inputs w_d, w_n as

$$\begin{bmatrix} w_d \\ w_n \end{bmatrix} = \begin{bmatrix} W^{\frac{1}{2}} & 0 \\ 0 & V^{\frac{1}{2}} \end{bmatrix} w \tag{9.38}$$

where w is a white noise process of unit intensity. Then the LQG cost function is

$$J = E\left\{\lim_{T\to\infty} \frac{1}{T} \int_0^T z(t)^T z(t) dt\right\} = \|F_l(P, K)\|_2^2 \tag{9.39}$$

where

$$z(s) = F_l(P, K) w(s) \tag{9.40}$$

and the generalized plant P is given by

$$P = \begin{bmatrix} P_{11} & P_{12} \\ P_{21} & P_{22} \end{bmatrix} \overset{s}{=} \left[\begin{array}{c|ccc} A & W^{\frac{1}{2}} & 0 & B \\ \hline Q^{\frac{1}{2}} & 0 & 0 & 0 \\ 0 & 0 & 0 & R^{\frac{1}{2}} \\ \hline C & 0 & V^{\frac{1}{2}} & 0 \end{array}\right] \tag{9.41}$$

The above formulation of the LQG problem is illustrated in the general setting in Figure 9.9. With the standard assumptions for the LQG problem, application of the general $\mathcal{H}_2$ formulas (Doyle et al., 1989) to this formulation gives the familiar LQG optimal controller as in (9.17).

9.3.4 $\mathcal{H}_\infty$ optimal control

With reference to the general control configuration of Figure 9.8, the standard $\mathcal{H}_\infty$ optimal control problem is to find all stabilizing controllers K which minimize

$$\|F_l(P, K)\|_\infty = \max_\omega \bar\sigma(F_l(P, K)(j\omega)) \tag{9.42}$$

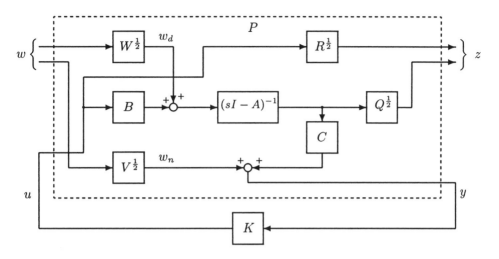

Figure 9.9: The LQG problem formulated in the general control configuration

As discussed in Section 4.10.2 the $\mathcal{H}_\infty$ norm has several interpretations in terms of performance. One is that it minimizes the peak of the maximum singular value of $F_l(P(j\omega), K(j\omega))$. It also has a time domain interpretation as the induced (worst-case) 2-norm. Let $z = F_l(P, K)w$, then

$$\|F_l(P, K)\|_\infty = \max_{w(t)\neq 0} \frac{\|z(t)\|_2}{\|w(t)\|_2} \tag{9.43}$$

where $\|z(t)\|_2 = \sqrt{\int_0^\infty \sum_i |z_i(t)|^2 dt}$ is the 2-norm of the vector signal.

In practice, it is usually not necessary to obtain an optimal controller for the $\mathcal{H}_\infty$ problem, and it is often computationally (and theoretically) simpler to design a sub-optimal one (i.e. one close to the optimal ones in the sense of the $\mathcal{H}_\infty$ norm). Let γ_{min} be the minimum value of $\|F_l(P, K)\|_\infty$ over all stabilizing controllers K. Then the $\mathcal{H}_\infty$ sub-optimal control problem is: given a $\gamma > \gamma_{min}$, find all stabilizing controllers K such that

$$\|F_l(P, K)\|_\infty < \gamma$$

This can be solved efficiently using the algorithm of Doyle et al. (1989), and by reducing γ iteratively, an optimal solution is approached. The algorithm is summarized below with all the simplifying assumptions.

General $\mathcal{H}_\infty$ algorithm. For the general control configuration of Figure 9.8 described by equations (9.24)-(9.26), with assumptions (A1) to (A8) in Section 9.3.1, there exists a stabilizing controller $K(s)$ such that $\|F_l(P, K)\|_\infty < \gamma$ if and only if

(i) $X_\infty \geq 0$ is a solution to the algebraic Riccati equation

$$A^T X_\infty + X_\infty A + C_1^T C_1 + X_\infty(\gamma^{-2} B_1 B_1^T - B_2 B_2^T)X_\infty = 0 \tag{9.44}$$

such that Re $\lambda_i \left[A + (\gamma^{-2}B_1B_1^T - B_2B_2^T)X_\infty \right] < 0, \forall i$; and

(ii) $Y_\infty \geq 0$ is a solution to the algebraic Riccati equation

$$AY_\infty + Y_\infty A^T + B_1 B_1^T + Y_\infty (\gamma^{-2}C_1^T C_1 - C_2^T C_2) Y_\infty = 0 \qquad (9.45)$$

such that Re $\lambda_i \left[A + Y_\infty(\gamma^{-2}C_1^T C_1 - C_2^T C_2) \right] < 0, \forall i$; and

(iii) $\rho(X_\infty Y_\infty) < \gamma^2$

All such controllers are then given by $K = F_l(K_c, Q)$ where

$$K_c(s) \overset{s}{=} \left[\begin{array}{c|cc} A_\infty & -Z_\infty L_\infty & Z_\infty B_2 \\ \hline F_\infty & 0 & I \\ -C_2 & I & 0 \end{array} \right] \qquad (9.46)$$

$$F_\infty = -B_2^T X_\infty, \quad L_\infty = -Y_\infty C_2^T, \quad Z_\infty = (I - \gamma^{-2}Y_\infty X_\infty)^{-1} \qquad (9.47)$$

$$A_\infty = A + \gamma^{-2}B_1 B_1^T X_\infty + B_2 F_\infty + Z_\infty L_\infty C_2 \qquad (9.48)$$

and $Q(s)$ is any stable proper transfer function such that $\|Q\|_\infty < \gamma$. For $Q(s) = 0$, we get

$$K(s) = K_{c_{11}}(s) = -Z_\infty L_\infty (sI - A_\infty)^{-1} F_\infty \qquad (9.49)$$

This is called the "central" controller and has the same number of states as the generalized plant $P(s)$. The central controller can be separated into a state estimator (observer) of the form

$$\dot{\hat{x}} = A\hat{x} + B_1 \underbrace{\gamma^{-2}B_1^T X_\infty \hat{x}}_{\widehat{w}_{\text{worst}}} + B_2 u + Z_\infty L_\infty (C_2 \hat{x} - y) \qquad (9.50)$$

and a state feedback

$$u = F_\infty \hat{x} \qquad (9.51)$$

Upon comparing the observer in (9.50) with the Kalman filter in (9.14) we see that it contains an additional term $B_1 \widehat{w}_{\text{worst}}$, where $\widehat{w}_{\text{worst}}$ can be interpreted as an estimate of the worst-case disturbance (exogenous input). Note that for the special case of $\mathcal{H}_\infty$ loop shaping this extra term is not present. This is discussed in Section 9.4.4.

γ-**iteration.** If we desire a controller that achieves γ_{min}, to within a specified tolerance, then we can perform a bisection on γ until its value is sufficiently accurate. The above result provides a test for each value of γ to determine whether it is less than γ_{min} or greater than γ_{min}.

Given all the assumptions (A1) to (A8) the above is the most simple form of the general $\mathcal{H}_\infty$ algorithm. For the more general situation, where some of the assumptions are relaxed, the reader is referred to the original source (Glover and Doyle, 1988). In practice, we would expect a user to have access to commercial software such as MATLAB and its toolboxes.

In Section 2.7, we distinguished between two methodologies for $\mathcal{H}_\infty$ controller design: the transfer function shaping approach and the signal-based approach. In the former, $\mathcal{H}_\infty$ optimization is used to shape the singular values of specified transfer functions over frequency. The maximum singular values are relatively easy to shape by forcing them to lie below user defined bounds, thereby ensuring desirable bandwidths and roll-off rates. In the signal-based approach, we seek to minimize the energy in certain error signals given a set of exogenous input signals. The latter might include the outputs of perturbations representing uncertainty, as well as the usual disturbances, noise, and command signals. Both of these two approaches will be considered again in the remainder of this section. In each case we will examine a particular problem and formulate it in the general control configuration.

A difficulty that sometimes arises with $\mathcal{H}_\infty$ control is the selection of weights such that the $\mathcal{H}_\infty$ optimal controller provides a good trade-off between conflicting objectives in various frequency ranges. Thus, for practical designs it is sometimes recommended to perform only a few iterations of the $\mathcal{H}_\infty$ algorithm. The justification for this is that the initial design, after one iteration, is similar to an $\mathcal{H}_2$ design which does trade-off over various frequency ranges. Therefore stopping the iterations before the optimal value is achieved gives the design an $\mathcal{H}_2$ flavour which may be desirable.

9.3.5   Mixed-sensitivity $\mathcal{H}_\infty$ control

Mixed-sensitivity is the name given to transfer function shaping problems in which the sensitivity function $S = (I + GK)^{-1}$ is shaped along with one or more other closed-loop transfer functions such as KS or the complementary sensitivity function $T = I - S$. Earlier in this chapter, by examining a typical one degree-of-freedom configuration, Figure 9.1, we saw quite clearly the importance of S, KS, and T.

Suppose, therefore, that we have a regulation problem in which we want to reject a disturbance d entering at the plant output and it is assumed that the measurement noise is relatively insignificant. Tracking is not an issue and therefore for this problem it makes sense to shape the closed-loop transfer functions S and KS in a one degree-of-freedom setting. Recall that S is the transfer function between d and the output, and KS the transfer function between d and the control signals. It is important to include KS as a mechanism for limiting the size and bandwidth of the controller, and hence the control energy used. The size of KS is also important for robust stability with respect to uncertainty modelled as additive plant perturbations.

The disturbance d is typically a low frequency signal, and therefore it will be successfully rejected if the maximum singular value of S is made small over the same low frequencies. To do this we could select a scalar low pass filter $w_1(s)$ with a bandwidth equal to that of the disturbance, and then find a stabilizing controller that minimizes $\|w_1 S\|_\infty$. This cost function alone is not very practical. It focuses on just one closed-loop transfer function and for plants without right-half plane zeros the optimal controller has infinite gains. In the presence of a nonminimum phase zero, the stability requirement will indirectly limit the controller gains, but it is far more useful in practice to minimize

$$\left\| \begin{bmatrix} w_1 S \\ w_2 KS \end{bmatrix} \right\|_\infty \tag{9.52}$$

where $w_2(s)$ is a scalar high pass filter with a crossover frequency approximately equal to that of the desired closed-loop bandwidth.

In general, the scalar weighting functions $w_1(s)$ and $w_2(s)$ can be replaced by matrices $W_1(s)$ and $W_2(s)$. This can be useful for systems with channels of quite different bandwidths when diagonal weights are recommended, but anything more complicated is usually not worth the effort.

Remark. Note we have here outlined an alternative way of selecting the weights from that in Example 2.11 and Section 3.4.6. There $W_1 = W_P$ was selected with a crossover frequency equal to that of the desired closed-loop bandwidth and $W_2 = W_u$ was selected as a constant, usually $W_u = I$.

To see how this mixed-sensitivity problem can be formulated in the general setting, we can imagine the disturbance d as a single exogenous input and define an error signal $z = \begin{bmatrix} z_1^T & z_2^T \end{bmatrix}^T$, where $z_1 = W_1 y$ and $z_2 = -W_2 u$, as illustrated in Figure 9.10. It is not difficult from Figure 9.10 to show that $z_1 = W_1 S w$ and

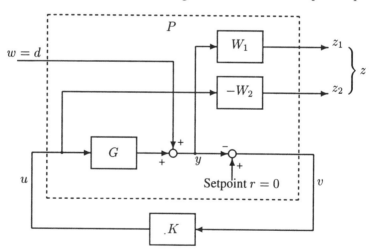

Figure 9.10: S/KS mixed-sensitivity optimization in standard form (regulation)

$z_2 = W_2 KS w$ as required, and to determine the elements of the generalized plant P as

$$P_{11} = \begin{bmatrix} W_1 \\ 0 \end{bmatrix} \quad P_{12} = \begin{bmatrix} W_1 G \\ -W_2 \end{bmatrix}$$
$$P_{21} = -I \quad P_{22} = -G \tag{9.53}$$

where the partitioning is such that

$$\begin{bmatrix} z_1 \\ z_2 \\ \hline v \end{bmatrix} = \begin{bmatrix} P_{11} & P_{12} \\ P_{21} & P_{22} \end{bmatrix} \begin{bmatrix} w \\ u \end{bmatrix} \qquad (9.54)$$

and

$$F_l(P, K) = \begin{bmatrix} W_1 S \\ W_2 K S \end{bmatrix} \qquad (9.55)$$

Another interpretation can be put on the S/KS mixed-sensitivity optimization as shown in the standard control configuration of Figure 9.11. Here we consider a tracking problem. The exogenous input is a reference command r, and the error signals are $z_1 = -W_1 e = W_1(r - y)$ and $z_2 = W_2 u$. As in the regulation problem of Figure 9.10, we have in this tracking problem $z_1 = W_1 S w$ and $z_2 = W_2 K S w$. An

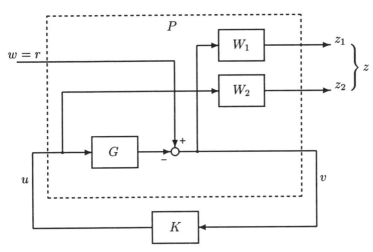

Figure 9.11: S/KS mixed-sensitivity minimization in standard form (tracking)

example of the use of S/KS mixed sensitivity minimization is given in Chapter 12, where it is used to design a rotorcraft control law. In this helicopter problem, you will see that the exogenous input w is passed through a weight W_3 before it impinges on the system. W_3 is chosen to weight the input signal and not directly to shape S or KS. This signal-based approach to weight selection is the topic of the next sub-section.

Another useful mixed sensitivity optimization problem, again in a one degree-of-freedom setting, is to find a stabilizing controller which minimizes

$$\left\| \begin{bmatrix} W_1 S \\ W_2 T \end{bmatrix} \right\|_\infty \qquad (9.56)$$

The ability to shape T is desirable for tracking problems and noise attenuation. It is also important for robust stability with respect to multiplicative perturbations at the plant output. The S/T mixed-sensitivity minimization problem can be put into the standard control configuration as shown in Figure 9.12. The elements of the

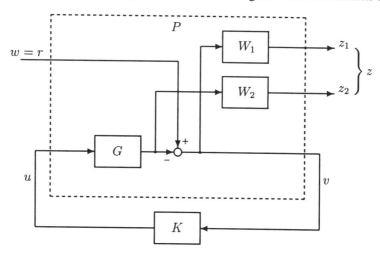

Figure 9.12: S/T mixed-sensitivity optimization in standard form

corresponding generalized plant P are

$$
\begin{aligned}
P_{11} &= \begin{bmatrix} W_1 \\ 0 \end{bmatrix} & P_{12} &= \begin{bmatrix} -W_1 G \\ W_2 G \end{bmatrix} \\
P_{21} &= I & P_{22} &= -G
\end{aligned}
\tag{9.57}
$$

Exercise 9.5 *For the cost function*

$$
\left\| \begin{bmatrix} W_1 S \\ W_2 T \\ W_3 K S \end{bmatrix} \right\|_\infty
\tag{9.58}
$$

formulate a standard problem, draw the corresponding control configuration and give expressions for the generalized plant P.

The shaping of closed-loop transfer functions as described above with the "stacked" cost functions becomes difficult with more than two functions. With two, the process is relatively easy. The bandwidth requirements on each are usually complementary and simple, stable, low-pass and high-pass filters are sufficient to carry out the required shaping and trade-offs. We stress that the weights W_i in mixed-sensitivity $\mathcal{H}_\infty$ optimal control must all be stable. If they are not, assumption (A1) in Section 9.3.1 is not satisfied, and the general $\mathcal{H}_\infty$ algorithm is not applicable. Therefore if we wish, for example, to emphasize the minimization of S at low frequencies by weighting with a term including integral action, we would have to approximate $\frac{1}{s}$ by $\frac{1}{s+\epsilon}$, where $\epsilon \ll 1$. This is exactly what was done in Example 2.11. Similarly one might be interested in weighting KS with a non-proper weight to ensure that K is small outside the system bandwidth. But the standard assumptions preclude such a weight. The trick here is to replace a non-proper term such as $(1 + \tau_1 s)$ by $(1 + \tau_1 s)/(1 + \tau_2 s)$ where $\tau_2 \ll \tau_1$. A useful discussion of the tricks involved in using "unstable" and "non-proper" weights in $\mathcal{H}_\infty$ control can be found in Meinsma (1995).

For more complex problems, information might be given about several exogenous signals in addition to a variety of signals to be minimized and classes of plant perturbations to be robust against. In which case, the mixed-sensitivity approach is not general enough and we are forced to look at more advanced techniques such as the signal-based approach considered next.

9.3.6 Signal-based $\mathcal{H}_\infty$ control

The signal-based approach to controller design is very general and is appropriate for multivariable problems in which several objectives must be taken into account simultaneously. In this approach, we define the plant and possibly the model uncertainty, we define the class of external signals affecting the system and we define the norm of the error signals we want to keep small. The focus of attention has moved to the size of signals and away from the size and bandwidth of selected closed-loop transfer functions.

Weights are used to describe the expected or known frequency content of exogenous signals and the desired frequency content of error signals. Weights are also used if a perturbation is used to model uncertainty, as in Figure 9.13, where G is the nominal model, W is a weighting function that captures the relative model fidelity over frequency, and Δ represents unmodelled dynamics usually normalized via W so that $\|\Delta\|_\infty < 1$; see Chapter 8 for more details. As in mixed-sensitivity $\mathcal{H}_\infty$ control, the weights in signal-based $\mathcal{H}_\infty$ control need to be stable and proper for the general $\mathcal{H}_\infty$ algorithm to be applicable.

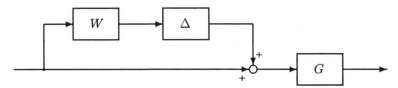

Figure 9.13: Multiplicative dynamic uncertainty model

LQG control is a simple example of the signal-based approach, in which the exogenous signals are assumed to be stochastic (or alternatively impulses in a

deterministic setting) and the error signals are measured in terms of the 2-norm. As
we have already seen, the weights Q and R are constant, but LQG can be generalized
to include frequency dependent weights on the signals leading to what is sometimes
called Wiener-Hopf design, or simply $\mathcal{H}_2$ control.

When we consider a system's response to persistent sinusoidal signals of varying
frequency, or when we consider the induced 2-norm between the exogenous input
signals and the error signals, we are required to minimize the $\mathcal{H}_\infty$ norm. In the
absence of model uncertainty, there does not appear to be an overwhelming case
for using the $\mathcal{H}_\infty$ norm rather than the more traditional $\mathcal{H}_2$ norm. However, when
uncertainty is addressed, as it always should be, $\mathcal{H}_\infty$ is clearly the more natural
approach using component uncertainty models as in Figure 9.13.

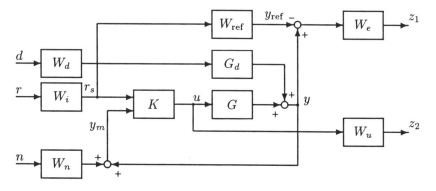

Figure 9.14: A signal-based $\mathcal{H}_\infty$ control problem

A typical problem using the signal-based approach to $\mathcal{H}_\infty$ control is illustrated in
the interconnection diagram of Figure 9.14. G and G_d are nominal models of the plant
and disturbance dynamics, and K is the controller to be designed. The weights W_d,
W_i and W_n may be constant or dynamic and describe the relative importance and/or
frequency content of the disturbances, set points, and noise signals. The weight W_{ref}
is a desired closed-loop transfer function between the weighted set point r_s and the
actual output y. The weights W_e and W_u reflect the desired frequency content of the
error $(y - y_{\text{ref}})$ and the control signals u, respectively. The problem can be cast as a
standard $\mathcal{H}_\infty$ optimization in the general control configuration by defining

$$w = \begin{bmatrix} d \\ r \\ n \end{bmatrix} \quad z = \begin{bmatrix} z_1 \\ z_2 \end{bmatrix}$$

$$v = \begin{bmatrix} r_s \\ y_m \end{bmatrix} \quad u = u \tag{9.59}$$

in the general setting of Figure 9.8.

Suppose we now introduce a multiplicative dynamic uncertainty model at the input to the plant as shown in Figure 9.15. The problem we now want to solve is: find a stabilizing controller K such that the $\mathcal{H}_\infty$ norm of the transfer function between w and z is less than 1 for all Δ, where $\|\Delta\|_\infty < 1$. We have assumed in this statement that the signal weights have normalized the 2-norm of the exogenous input signals to unity. This problem is a non-standard $\mathcal{H}_\infty$ optimization. It is a robust performance problem for which the μ-synthesis procedure, outlined in Chapter 8, can be applied. Mathematically, we require the structured singular value

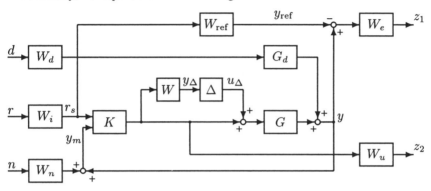

Figure 9.15: An $\mathcal{H}_\infty$ robust performance problem

$$\mu(M(j\omega)) < 1, \forall \omega \tag{9.60}$$

where M is the transfer function matrix between

$$\begin{bmatrix} d \\ r \\ n \\ \delta \end{bmatrix} \quad \text{and} \quad \begin{bmatrix} z_1 \\ z_2 \\ \epsilon \end{bmatrix} \tag{9.61}$$

and the associated block diagonal perturbation has 2 blocks: a fictitious performance block between $\begin{bmatrix} d^T & r^T & n^T \end{bmatrix}^T$ and $\begin{bmatrix} z_1^T & z_2^T \end{bmatrix}^T$, and an uncertainty block Δ between u_Δ and y_Δ. Whilst the structured singular value is a useful analysis tool for assessing designs, μ-synthesis is sometimes difficult to use and often too complex for the practical problem at hand. In its full generality, the μ-synthesis problem is not yet solved mathematically; where solutions exist the controllers tend to be of very high order; the algorithms may not always converge and design problems are sometimes difficult to formulate directly.

For many industrial control problems, a design procedure is required which offers more flexibility than mixed-sensitivity $\mathcal{H}_\infty$ control, but is not as complicated as μ-synthesis. For simplicity, it should be based on classical loop-shaping ideas and it should not be limited in its applications like LTR procedures. In the next section, we present such a controller design procedure.

9.4 $\mathcal{H}_\infty$ loop-shaping design

The loop-shaping design procedure described in this section is based on $\mathcal{H}_\infty$ robust stabilization combined with classical loop shaping, as proposed by McFarlane and Glover (1990). It is essentially a two stage design process. First, the open-loop plant is augmented by pre and post-compensators to give a desired shape to the singular values of the open-loop frequency response. Then the resulting shaped plant is robustly stabilized with respect to coprime factor uncertainty using $\mathcal{H}_\infty$ optimization. An important advantage is that no problem-dependent uncertainty modelling, or weight selection, is required in this second step.

We will begin the section with a description of the $\mathcal{H}_\infty$ robust stabilization problem (Glover and McFarlane, 1989). This is a particularly nice problem because it does not require γ-iteration for its solution, and explicit formulas for the corresponding controllers are available. The formulas are relatively simple and so will be presented in full.

Following this, a step by step procedure for $\mathcal{H}_\infty$ loop-shaping design is presented. This systematic procedure has its origin in the Ph.D. thesis of Hyde (1991) and has since been successfully applied to several industrial problems. The procedure synthesizes what is in effect a single degree-of-freedom controller. This can be a limitation if there are stringent requirements on command following. However, as shown by Limebeer et al. (1993), the procedure can be extended by introducing a second degree-of-freedom in the controller and formulating a standard $\mathcal{H}_\infty$ optimization problem which allows one to trade off robust stabilization against closed-loop model-matching. We will describe this two degrees-of-freedom extension and further show that such controllers have a special observer-based structure which can be taken advantage of in controller implementation.

9.4.1   Robust stabilization

For multivariable systems, classical gain and phase margins are unreliable indicators of robust stability when defined for each channel (or loop), taken one at a time, because simultaneous perturbations in more than one loop are not then catered for. More general perturbations like $\text{diag}\{k_i\}$ and $\text{diag}\{e^{j\theta_i}\}$, as discussed in section 9.2.2, are required to capture the uncertainty, but even these are limited. It is now common practice, as seen in Chapter 8, to model uncertainty by norm-bounded dynamic matrix perturbations. Robustness levels can then be quantified in terms of the maximum singular values of various closed-loop transfer functions.

The associated robustness tests, for a single perturbation, as described in Chapter 8, require the perturbation to be stable. This restricts the plant and perturbed plant models to have the same number of unstable poles. To overcome this, two stable perturbations can be used, one on each of the factors in a coprime factorization of the plant, as shown in Section 8.6.2. Although this uncertainty description seems unrealistic and less intuitive than the others, it is in fact quite general, and for our purposes it leads to a very useful $\mathcal{H}_\infty$ robust stabilization problem. Before presenting the problem, we will first recall the uncertainty model given in (8.62).

We will consider the stabilization of a plant G which has a normalized left coprime factorization (as discussed in Section 4.1.5)

$$G = M^{-1}N \qquad (9.62)$$

where we have dropped the subscript from M and N for simplicity. A perturbed plant

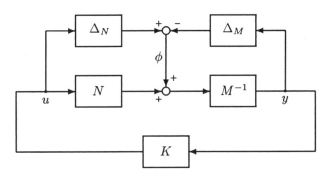

Figure 9.16: $\mathcal{H}_\infty$ robust stabilization problem

model G_p can then be written as

$$G_p = (M + \Delta_M)^{-1}(N + \Delta_N) \qquad (9.63)$$

where Δ_M, Δ_N are stable unknown transfer functions which represent the uncertainty in the nominal plant model G. The objective of robust stabilization it to stabilize not only the nominal model G, but a family of perturbed plants defined by

$$G_p = \{(M + \Delta_M)^{-1}(N + \Delta_N) : \| [\ \Delta_N \quad \Delta_M\ ] \|_\infty < \epsilon\} \qquad (9.64)$$

where $\epsilon > 0$ is then the stability margin. To maximize this stability margin is the problem of robust stabilization of normalized coprime factor plant descriptions as introduced and solved by Glover and McFarlane (1989).

For the perturbed feedback system of Figure 9.16, as already derived in (8.64), the stability property is robust if and only if the nominal feedback system is stable and

$$\gamma \triangleq \left\| \begin{bmatrix} K \\ I \end{bmatrix} (I - GK)^{-1} M^{-1} \right\|_\infty \leq \frac{1}{\epsilon} \qquad (9.65)$$

Notice that γ is the $\mathcal{H}_\infty$ norm from ϕ to $\begin{bmatrix} u \\ y \end{bmatrix}$ and $(I - GK)^{-1}$ is the sensitivity function for this positive feedback arrangement.

The lowest achievable value of γ and the corresponding maximum stability margin ϵ are given by Glover and McFarlane (1989) as

$$\gamma_{\min} = \epsilon_{max}^{-1} = \left\{ 1 - \| [N \quad M] \|_H^2 \right\}^{-\frac{1}{2}} = (1 + \rho(XZ))^{\frac{1}{2}} \qquad (9.66)$$

where $\| \cdot \|_H$ denotes Hankel norm, ρ denotes the spectral radius (maximum eigenvalue), and for a minimal state-space realization (A, B, C, D) of G, Z is the unique positive definite solution to the algebraic Riccati equation

$$(A - BS^{-1}D^T C)Z + Z(A - BS^{-1}D^T C)^T - ZC^T R^{-1}CZ + BS^{-1}B^T = 0 \quad (9.67)$$

where

$$R = I + DD^T, \quad S = I + D^T D$$

and X is the unique positive definite solution of the following algebraic Riccati equation

$$(A - BS^{-1}D^T C)^T X + X(A - BS^{-1}D^T C) - XBS^{-1}B^T X + C^T R^{-1}C = 0 \qquad (9.68)$$

Notice that the formulas simplify considerably for a strictly proper plant, i.e. when $D = 0$.

A controller (the "central" controller in McFarlane and Glover (1990)) which guarantees that

$$\left\| \begin{bmatrix} K \\ I \end{bmatrix} (I - GK)^{-1}M^{-1} \right\|_\infty \leq \gamma \qquad (9.69)$$

for a specified $\gamma > \gamma_{min}$, is given by

$$K \stackrel{s}{=} \left[\begin{array}{c|c} A + BF + \gamma^2 (L^T)^{-1}ZC^T(C + DF) & \gamma^2 (L^T)^{-1}ZC^T \\ \hline B^T X & -D^T \end{array} \right] \quad (9.70)$$

$$F = -S^{-1}(D^T C + B^T X) \qquad (9.71)$$

$$L = (1 - \gamma^2)I + XZ. \qquad (9.72)$$

The MATLAB function `coprimeunc`, listed in Table 9.2, can be used to generate the controller in (9.70). It is important to emphasize that since we can compute γ_{min} from (9.66) we get an explicit solution by solving just two Riccati equations (`aresolv`) and avoid the γ-iteration needed to solve the general $\mathcal{H}_\infty$ problem.

Remark 1 An example of the use of `coprimeunc` is given in Example 9.3 below.

Remark 2 Notice that, if $\gamma = \gamma_{min}$ in (9.70), then $L = -\rho(XZ)I + XZ$, which is singular, and thus (9.70) cannot be implemented. If for some unusual reason the truly optimal controller is required, then this problem can be resolved using a descriptor system approach, the details of which can be found in Safonov et al. (1989).

Table 9.2: MATLAB function to generate the $\mathcal{H}_\infty$ controller in (9.70)

```
% Uses the Robust Control or Mu toolbox
function [Ac,Bc,Cc,Dc,gammin]=coprimeunc(a,b,c,d,gamrel)
%
% Finds the controller which optimally ''robustifies'' a given shaped plant
% in terms of tolerating maximum coprime uncertainty.
%
% INPUTS:
% a,b,c,d: State-space description of (shaped) plant.
% gamrel: gamma used is gamrel*gammin  (typical gamrel=1.1)
%
% OUTPUTS:
% Ac,Bc,Cc,Dc: "Robustifying" controller (positive feedback).
%
S=eye(size(d'*d))+d'*d;
R=eye(size(d*d'))+d*d';
A1=a-b*inv(S)*d'*c;
Q1=c'*inv(R)*c;
R1=b*inv(S)*b';
[x1,x2,eig,xerr,wellposed,X] = aresolv(A1,Q1,R1);
% Alt. Mu toolbox:
%[x1,x2, fail, reig_min] = ric_schr([A1 -R1; -Q1 -A1']); X = x2/x1;
[x1,x2,eig,xerr,wellposed,Z] = aresolv(A1',R1,Q1);
% Alt. Mu toolbox:
%[x1, x2, fail, reig_min] = ric_schr([A1' -Q1; -R1 -A1]); Z = x2/x1;
% Optimal gamma:
gammin=sqrt(1+max(eig(X*Z)))
% Use higher gamma.....
gam = gamrel*gammin;
L=(1-gam*gam)*eye(size(X*Z)) + X*Z;
F=-inv(S)*(d'*c+b'*X);
Ac=a+b*F+gam*gam*inv(L')*Z*c'*(c+d*F);
Bc=gam*gam*inv(L')*Z*c';
Cc=b'*X;
Dc=-d';
```

Remark 3 Alternatively, from Glover and McFarlane (1989), all controllers achieving $\gamma = \gamma_{min}$ are given by $K = UV^{-1}$, where U and V are stable, (U, V) is a right coprime factorization of K, and U, V satisfy

$$\left\| \left[\begin{array}{c} -N^* \\ M^* \end{array} \right] + \left[\begin{array}{c} U \\ V \end{array} \right] \right\|_\infty = \| [N \; M] \|_H \qquad (9.73)$$

The determination of U and V is a Nehari extension problem: that is, a problem in which an unstable transfer function $R(s)$ is approximated by a stable transfer function $Q(s)$, such that $\| R + Q \|_\infty$ is minimized, the minimum being $\| R^* \|_H$. A solution to this problem is given in Glover (1984).

Exercise 9.6 *Formulate the $\mathcal{H}_\infty$ robust stabilization problem in the general control configuration of Figure 9.8, and determine a transfer function expression and a state-space realization for the generalized plant P.*

9.4.2   A systematic $\mathcal{H}_\infty$ loop-shaping design procedure

Robust stabilization alone is not much use in practice because the designer is not able to specify any performance requirements. To do this McFarlane and Glover (1990) proposed pre- and post-compensating the plant to shape the open-loop singular values prior to robust stabilization of the "shaped" plant.

If W_1 and W_2 are the pre- and post-compensators respectively, then the shaped plant G_s is given by

$$G_s = W_2 G W_1 \qquad (9.74)$$

as shown in Figure 9.17. The controller K_s is synthesized by solving the robust

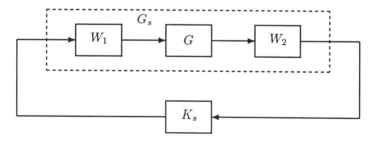

Figure 9.17: The shaped plant and controller

stabilization problem of section 9.4.1 for the shaped plant G_s with a normalized left coprime factorization $G_s = M_s^{-1} N_s$. The feedback controller for the plant G is then $K = W_1 K_s W_2$. The above procedure contains all the essential ingredients of classical loop shaping, and can easily be implemented using the formulas already presented and reliable algorithms in, for example, MATLAB.

We first present a simple SISO example, where $W_2 = 1$ and we select W_1 to get acceptable disturbance rejection. We will afterwards present a systematic procedure for selecting the weights W_1 and W_2.

Example 9.3 Glover-McFarlane $\mathcal{H}_\infty$ loop shaping for the disturbance process. *Consider*

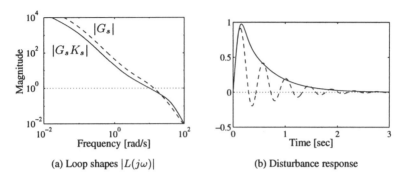

(a) Loop shapes $|L(j\omega)|$ (b) Disturbance response

Figure 9.18: Glover-McFarlane loop-shaping design for the disturbance process. Dashed line: Initial "shaped" design, G_s. Solid line: "robustified" design, $G_s K_s$

the disturbance process in (2.54) which was studied in detail in Chapter 2,

$$G(s) = \frac{200}{10s + 1}\frac{1}{(0.05s + 1)^2}, \quad G_d(s) = \frac{100}{10s + 1} \tag{9.75}$$

We want as good disturbance rejection as possible, and the gain crossover frequency w_c for the final design should be about 10 rad/s.

In Example 2.8 we argued that for acceptable disturbance rejection with minimum input usage, the loop shape ("shaped plant") $|G_s| = |GW_1|$ should be similar to $|G_d|$, so $|W_1| = |G^{-1}G_d|$ is desired. Then after neglecting the high-frequency dynamics in $G(s)$ this yields an initial weight $W_1 = 0.5$. To improve the performance at low frequencies we add integral action, and we also add a phase-advance term $s + 2$ to reduce the slope for L from -2 at lower frequencies to about -1 at crossover. Finally, to make the response a little faster we multiply the gain by a factor 2 to get the weight

$$W_1 = \frac{s + 2}{s} \tag{9.76}$$

This yields a shaped plant $G_s = GW_1$ with a gain crossover frequency of 13.7 rad/s, and the magnitude of $G_s(j\omega)$ is shown by the dashed line in Figure 9.18(a). The response to a unit step in the disturbance response is shown by the dashed line in Figure 9.18(b), and, as may expected, the response with the "controller" $K = W_1$ is too oscillatory.

We now "robustify" this design so that the shaped plant tolerates as much $\mathcal{H}_\infty$ coprime factor uncertainty as possible. This may be done with the MATLAB μ-toolbox using the command `ncfsyn` *or with the MATLAB Robust Control toolbox using the function* `coprimeunc` *given in Table 9.2:*

```
[Ac,Bc,Cc,Dc,gammin]=coprimeunc(A,B,C,D,gamrel)
```

Here the shaped plant $G_s = GW_1$ *has state-space matrices* A, B, C *and* D, *and the function returns the "robustifying" positive feedback controller* K_s *with state-space matrices* Ac, Bc, Cc *and* Dc. *In general,* K_s *has the same number of poles (states) as* G_s. gamrel *is the value of* γ *relative to* $\gamma_{\min}$, *and was in our case selected as 1.1. The returned variable* gammin *($\gamma_{\min}$) is the inverse of the magnitude of coprime uncertainty we can tolerate before we get instability. We want* $\gamma_{\min} \geq 1$ *as small as possible, and we usually require that* $\gamma_{\min}$ *is less than 4, corresponding to 25% allowed coprime uncertainty.*

By applying this to our example we get $\gamma_{\min} = 2.34$ *and an overall controller* $K = W_1 K_s$ *with 5 states (G_s, and thus K_s, has 4 states, and W_1 has 1 state). The corresponding loop shape* $|G_s K_s|$ *is shown by the solid line in Figure 9.18(a). We see that the change in the loop shape is small, and we note with interest that the slope around crossover is somewhat gentler. This translates into better margins: the gain margin (GM) is improved from 1.62 (for* G_s) *to 3.48 (for* $G_s K_s$), *and the phase margin (PM) is improved from 13.2° to 51.5°. The gain crossover frequency* w_c *is reduced slightly from 13.7 to 10.3 rad/s. The corresponding disturbance response is shown in Figure 9.18(b) and is seen to be much improved.*

Remark. The response with the controller $K = W_1 K_s$ is quite similar to that of the loop-shaping controller $K_3(s)$ designed in Chapter 2 (see curves L_3 and y_3 in Figure 2.21). The response for reference tracking with controller $K = W_1 K_s$ is not shown; it is also very similar to that with K_3 (see Figure 2.23), but it has a slightly smaller overshoot of 21% rather than 24%. To reduce this overshoot we would need to use a two degrees-of-freedom controller.

Exercise 9.7 *Design an* $\mathcal{H}_\infty$ *loop-shaping controller for the disturbance process in (9.75) using the weight* W_1 *in (9.76), i.e. generate plots corresponding to those in Figure 9.18. Next, repeat the design with* $W_1 = 2(s+3)/s$ *(which results in an initial* G_s *which would yield closed-loop instability with* $K_c = 1$). *Compute the gain and phase margins and compare the disturbance and reference responses. In both cases find* ω_c *and use (2.37) to compute the maximum delay that can be tolerated in the plant before instability arises.*

Skill is required in the selection of the weights (pre- and post-compensators W_1 and W_2), but experience on real applications has shown that robust controllers can be designed with relatively little effort by following a few simple rules. An excellent illustration of this is given in the thesis of Hyde (1991) who worked with Glover on the robust control of VSTOL (vertical and/or short take-off and landing) aircraft. Their work culminated in a successful flight test of $\mathcal{H}_\infty$ loop-shaping control laws implemented on a Harrier jump-jet research vehicle at the UK Defence Research Agency (DRA), Bedford in 1993. The $\mathcal{H}_\infty$ loop-shaping procedure has also been extensively studied and worked on by Postlethwaite and Walker (1992) in their work on advanced control of high performance helicopters, also for the UK DRA at Bedford. This application is discussed in detail in the helicopter case study in Section 12.2.

Based on these, and other studies, it is recommended that the following systematic procedure is followed when using $\mathcal{H}_\infty$ loop-shaping design:

1. Scale the plant outputs and inputs. This is very important for most design procedures and is sometimes forgotten. In general, scaling improves the conditioning of

the design problem, it enables meaningful analysis to be made of the robustness properties of the feedback system in the frequency domain, and for loop-shaping it can simplify the selection of weights. There are a variety of methods available including normalization with respect to the magnitude of the maximum or average value of the signal in question. Scaling with respect to maximum values is important if the controllability analysis of earlier chapters is to be used. However, if one is to go straight to a design the following variation has proved useful in practice:

(a) The outputs are scaled such that equal magnitudes of cross-coupling into each of the outputs is equally undesirable.

(b) Each input is scaled by a given percentage (say 10%) of its expected range of operation. That is, the inputs are scaled to reflect the relative actuator capabilities. An example of this type of scaling is given in the aero-engine case study of Chapter 12.

2. Order the inputs and outputs so that the plant is as diagonal as possible. The relative gain array can be useful here. The purpose of this pseudo-diagonalization is to ease the design of the pre- and post-compensators which, for simplicity, will be chosen to be diagonal.

Next, we discuss the selection of weights to obtain the shaped plant $G_s = W_2 G W_1$ where

$$W_1 = W_p W_a W_g \tag{9.77}$$

3. Select the elements of diagonal pre- and post-compensators W_p and W_2 so that the singular values of $W_2 G W_p$ are desirable. This would normally mean high gain at low frequencies, roll-off rates of approximately 20 dB/decade (a slope of about -1) at the desired bandwidth(s), with higher rates at high frequencies. Some trial and error is involved here. W_2 is usually chosen as a constant, reflecting the relative importance of the outputs to be controlled and the other measurements being fed back to the controller. For example, if there are feedback measurements of two outputs to be controlled and a velocity signal, then W_2 might be chosen to be diag$[1, 1, 0.1]$, where 0.1 is in the velocity signal channel. W_p contains the dynamic shaping. Integral action, for low frequency performance; phase-advance for reducing the roll-off rates at crossover; and phase-lag to increase the roll-off rates at high frequencies should all be placed in W_p if desired. The weights should be chosen so that no unstable hidden modes are created in G_s.

4. *Optional*: Align the singular values at a desired bandwidth using a further constant weight W_a cascaded with W_p. This is effectively a constant decoupler and should not be used if the plant is ill-conditioned in terms of large RGA elements (see Section 6.10.4). The align algorithm of Kouvaritakis (1974) which has been implemented in the MATLAB Multivariable Frequency-Domain Toolbox is recommended.

5. *Optional*: Introduce an additional gain matrix W_g cascaded with W_a to provide control over actuator usage. W_g is diagonal and is adjusted so that actuator rate limits are not exceeded for reference demands and typical disturbances on the scaled plant outputs. This requires some trial and error.

6. Robustly stabilize the shaped plant $G_s = W_2 G W_1$, where $W_1 = W_p W_a W_g$, using the formulas of the previous section. First, calculate the maximum stability margin $\epsilon_{\max} = 1/\gamma_{\min}$. If the margin is too small, $\epsilon_{\max} < 0.25$, then go back to step 4 and modify the weights. Otherwise, select $\gamma > \gamma_{\min}$, by about 10%, and synthesize a suboptimal controller using equation (9.70). There is usually no advantage to be gained by using the optimal controller. When $\epsilon_{\max} > 0.25$ (respectively $\gamma_{\min} < 4$) the design is usually successful. In this case, at least 25% coprime factor uncertainty is allowed, and we also find that the shape of the open-loop singular values will not have changed much after robust stabilization. A small value of $\epsilon_{\max}$ indicates that the chosen singular value loop-shapes are incompatible with robust stability requirements. That the loop-shapes do not change much following robust stabilization if γ is small (ϵ large), is justified theoretically in McFarlane and Glover (1990).

7. Analyze the design and if all the specifications are not met make further modifications to the weights.

8. Implement the controller. The configuration shown in Figure 9.19 has been found useful when compared with the conventional set up in Figure 9.1. This is because

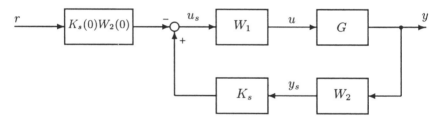

Figure 9.19: A practical implementation of the loop-shaping controller

the references do not directly excite the dynamics of K_s, which can result in large amounts of overshoot (classical derivative kick). The constant prefilter ensures a steady-state gain of 1 between r and y, assuming integral action in W_1 or G.

We will conclude this subsection with a summary of the advantages offered by the above $\mathcal{H}_\infty$ loop-shaping design procedure:

- It is relatively easy to use, being based on classical loop-shaping ideas.
- There exists a closed formula for the $\mathcal{H}_\infty$ optimal cost $\gamma_{\min}$, which in turn corresponds to a maximum stability margin $\epsilon_{\max} = 1/\gamma_{\min}$.
- No γ-iteration is required in the solution.
- Except for special systems, ones with all-pass factors, there are no pole-zero

cancellations between the plant and controller (Sefton and Glover, 1990; Tsai et al., 1992). Pole-zeros cancellations are common in many $\mathcal{H}_\infty$ control problems and are a problem when the plant has lightly damped modes.

Exercise 9.8 *First a definition and some useful properties.*

Definition: A stable transfer function matrix $H(s)$ is inner if $H^*H = I$, and co-inner if $HH^* = I$. The operator H^* is defined as $H^*(s) = H^T(-s)$.

Properties: The $\mathcal{H}_\infty$ norm is invariant under right multiplication by a co-inner function and under left multiplication by an inner function.

Equipped with the above definition and properties, show for the shaped $G_s = M_s^{-1}N_s$, that the matrix $[\,M_s \quad N_s\,]$ is co-inner and hence that the $\mathcal{H}_\infty$ loop-shaping cost function

$$\left\| \begin{bmatrix} K_s \\ I \end{bmatrix} (I - G_s K_s)^{-1} M_s^{-1} \right\|_\infty \tag{9.78}$$

is equivalent to

$$\left\| \begin{bmatrix} K_s S_s & K_s S_s G_s \\ S_s & S_s G_s \end{bmatrix} \right\|_\infty \tag{9.79}$$

where $S_s = (I - G_s K_s)^{-1}$. This shows that the problem of finding a stabilizing controller to minimise the 4-block cost function (9.79) has an exact solution.

Whilst it is highly desirable, from a computational point of view, to have exact solutions for $\mathcal{H}_\infty$ optimization problems, such problems are rare. We are fortunate that the above robust stabilization problem is also one of great practical significance.

9.4.3   Two degrees-of-freedom controllers

Many control design problems possess two degrees-of-freedom: on the one hand, measurement or feedback signals and on the other, commands or references. Sometimes, one degree-of-freedom is left out of the design, and the controller is driven (for example) by an error signal i.e. the difference between a command and the output. But in cases where stringent time-domain specifications are set on the output response, a one degree-of-freedom structure may not be sufficient. A general two degrees-of-freedom feedback control scheme is depicted in Figure 9.20. The commands and feedbacks enter the controller separately and are independently processed.

The $\mathcal{H}_\infty$ loop-shaping design procedure of McFarlane and Glover is a one degree-of-freedom design, although as we showed in Figure 9.19 a simple constant prefilter can easily be implemented for steady-state accuracy. For many tracking problems, however, this will not be sufficient and a dynamic two degrees-of-freedom design is required. In Hoyle et al. (1991) and Limebeer et al. (1993) a two degrees-of-freedom extension of the Glover-McFarlane procedure was proposed to enhance the model-matching properties of the closed-loop. With this the feedback part of the controller is designed to meet robust stability and disturbance rejection requirements in a manner similar to the one degree-of-freedom loop-shaping design procedure except that only

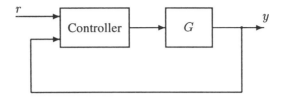

Figure 9.20: General two degrees-of-freedom feedback control scheme

a pre-compensator weight W is used. It is assumed that the measured outputs and the outputs to be controlled are the same although this assumption can be removed as shown later. An additional prefilter part of the controller is then introduced to force the response of the closed-loop system to follow that of a specified model, T_{ref}, often called the reference model. Both parts of the controller are synthesized by solving the design problem illustrated in Figure 9.21.

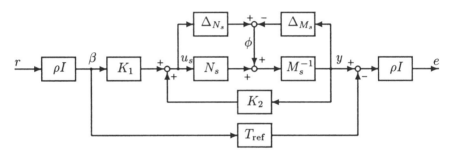

Figure 9.21: Two degrees-of-freedom $\mathcal{H}_\infty$ loop-shaping design problem

The design problem is to find the stabilizing controller $K = \begin{bmatrix} K_1 & K_2 \end{bmatrix}$ for the shaped plant $G_s = GW_1$, with a normalized coprime factorization $G_s = M_s^{-1}N_s$, which minimizes the $\mathcal{H}_\infty$ norm of the transfer function between the signals $\begin{bmatrix} r^T & \phi^T \end{bmatrix}^T$ and $\begin{bmatrix} u_s^T & y^T & e^T \end{bmatrix}^T$ as defined in Figure 9.21. The problem is easily cast into the general control configuration and solved suboptimally using standard methods and γ-iteration. We will show this later.

The control signal to the shaped plant u_s is given by

$$u_s = \begin{bmatrix} K_1 & K_2 \end{bmatrix} \begin{bmatrix} \beta \\ y \end{bmatrix} \tag{9.80}$$

where K_1 is the prefilter, K_2 is the feedback controller, β is the scaled reference, and y is the measured output. The purpose of the prefilter is to ensure that

$$\|(I - G_s K_2)^{-1} G_s K_1 - T_{\text{ref}}\|_\infty \leq \gamma \rho^{-2} \tag{9.81}$$

T_{ref} is the desired closed-loop transfer function selected by the designer to introduce time-domain specifications (desired response characteristics) into the design process; and ρ is a scalar parameter that the designer can increase to place more emphasis on model matching in the optimization at the expense of robustness.

From Figure 9.21 and a little bit of algebra, we have that

$$
\begin{bmatrix} u_s \\ y \\ e \end{bmatrix} = \begin{bmatrix} \rho(I - K_2 G_s)^{-1} K_1 & K_2(I - G_s K_2)^{-1} M_s^{-1} \\ \rho(I - G_s K_2)^{-1} G_s K_1 & (I - G_s K_2)^{-1} M_s^{-1} \\ \rho^2 \left[(I - G_s K_2)^{-1} G_s K_1 - T_{\text{ref}} \right] & \rho(I - G_s K_2)^{-1} M_s^{-1} \end{bmatrix} \begin{bmatrix} r \\ \phi \end{bmatrix}
$$

$$(9.82)$$

In the optimization, the $\mathcal{H}_\infty$ norm of this block matrix transfer function is minimized.

Notice that the (1,2) and (2,2) blocks taken together are associated with robust stabilization and the (3,1) block corresponds to model-matching. In addition, the (1,1) and (2,1) blocks help to limit actuator usage and the (3,3) block is linked to the performance of the loop. For $\rho = 0$, the problem reverts to minimizing the $\mathcal{H}_\infty$ norm of the transfer function between ϕ and $\begin{bmatrix} u_s^T & y^T \end{bmatrix}^T$, namely, the robust stabilization problem, and the two degrees-of-freedom controller reduces to an ordinary $\mathcal{H}_\infty$ loop-shaping controller.

To put the two degrees-of-freedom design problem into the standard control configuration, we can define a generalized plant P by

$$
\begin{bmatrix} u_s \\ y \\ e \\ \hline \beta \\ y \end{bmatrix} = \begin{bmatrix} P_{11} & P_{12} \\ P_{21} & P_{22} \end{bmatrix} \begin{bmatrix} r \\ \phi \\ \hline u_s \end{bmatrix}
$$

$$(9.83)$$

$$
= \begin{bmatrix} 0 & 0 & I \\ 0 & M_s^{-1} & G_s \\ -\rho^2 T_{\text{ref}} & \rho M_s^{-1} & \rho G_s \\ \hline \rho I & 0 & 0 \\ 0 & M_s^{-1} & G_s \end{bmatrix} \begin{bmatrix} r \\ \phi \\ \hline u_s \end{bmatrix}
$$

$$(9.84)$$

Further, if the shaped plant G_s and the desired stable closed-loop transfer function T_{ref} have the following state-space realizations

$$
G_s \overset{s}{=} \left[\begin{array}{c|c} A_s & B_s \\ \hline C_s & D_s \end{array} \right]
$$

$$(9.85)$$

$$
T_{\text{ref}} \overset{s}{=} \left[\begin{array}{c|c} A_r & B_r \\ \hline C_r & D_r \end{array} \right]
$$

$$(9.86)$$

then P may be realized by

$$
\left[
\begin{array}{cccc|c}
A_s & 0 & 0 & (B_s D_s^T + Z_s C_s^T) R_s^{-1/2} & B_s \\
0 & A_r & B_r & 0 & 0 \\
\hline
0 & 0 & 0 & 0 & I \\
C_s & 0 & 0 & R_s^{1/2} & D_s \\
\rho C_s & -\rho^2 C_r & -\rho^2 D_r & \rho R_s^{1/2} & \rho D_s \\
\hline
0 & 0 & \rho I & 0 & 0 \\
C_s & 0 & 0 & R_s^{1/2} & D_s
\end{array}
\right]
\tag{9.87}
$$

and used in standard $\mathcal{H}_\infty$ algorithms (Doyle et al., 1989) to synthesize the controller K. Note that $R_s = I + D_s D_s^T$, and Z_s is the unique positive definite solution to the generalized Riccati equation (9.67) for G_s. MATLAB commands to synthesize the controller are given in Table 9.3.

Remark 1 We stress that we here aim to minimize the $\mathcal{H}_\infty$ norm of the entire transfer function in (9.82). An alternative problem would be to minimize the $\mathcal{H}_\infty$ norm form r to e subject to an upper bound on $\| [\, \Delta_{N_s} \quad \Delta_{M_s} \,] \|_\infty$. This problem would involve the structured singular value, and the optimal controller could be obtained from solving a series of $\mathcal{H}_\infty$ optimization problems using DK-iteration; see Section 8.12.

Remark 2 Extra measurements. In some cases, a designer has more plant outputs available as measurements than can (or even need) to be controlled. These extra measurements can often make the design problem easier (e.g. velocity feedback) and therefore when beneficial should be used by the feedback controller K_2. This can be accommodated in the two degrees-of-freedom design procedure by introducing an output selection matrix W_o. This matrix selects from the output measurements y only those which are to be controlled and hence included in the model-matching part of the optimization. In Figure 9.21, W_o is introduced between y and the summing junction. In the optimization problem, only the equation for the error e is affected, and in the realization (9.87) for P one simply replaces ρC_s by $\rho W_o C_s$ and $\rho R_s^{1/2}$ by $\rho W_o R_s^{1/2}$ in the fifth row. For example, if there are four feedback measurements and only the first three are to be controlled then

$$
W_o = \begin{bmatrix} 1 & 0 & 0 & 0 \\ 0 & 1 & 0 & 0 \\ 0 & 0 & 1 & 0 \end{bmatrix}
\tag{9.88}
$$

Remark 3 Steady-state gain matching. The command signals r can be scaled by a constant matrix W_i to make the closed-loop transfer function from r to the controlled outputs $W_o y$ match the desired model T_{ref} exactly at steady-state. This is not guaranteed by the optimization which aims to minimize the ∞-norm of the error. The required scaling is given by

$$
W_i \triangleq \left[W_o (I - G_s(0) K_2(0))^{-1} G_s(0) K_1(0) \right]^{-1} T_{\text{ref}}(0)
\tag{9.89}
$$

Recall that $W_o = I$ if there are no extra feedback measurements beyond those that are to be controlled. The resulting controller is $K = \begin{bmatrix} K_1 W_i & K_2 \end{bmatrix}$.

Table 9.3: MATLAB commands to synthesize the $\mathcal{H}_\infty$ 2-DOF controller in (9.80)

```
% Uses MATLAB mu toolbox
%
% INPUTS: Shaped plant Gs
%         Reference model Tref
%
% OUTPUT: Two degrees-of-freedom controller K
%
% Coprime factorization of Gs
%
[As,Bs,Cs,Ds] = unpck(Gs);
[Ar,Br,Cr,Dr] = unpck(Tref);
[nr,nr] = size(Ar); [lr,mr] = size(Dr);
[ns,ns] = size(As); [ls,ms] = size(Ds);
Rs = eye(ls)+Ds*Ds.'; Ss = eye(ms)+Ds'*Ds;
A1 = (As - Bs*inv(Ss)*Ds'*Cs);
R1 = Cs'*inv(Rs)*Cs; Q1 = Bs*inv(Ss)*Bs';
[Z1, Z2, fail, reig_min] = ric_schr([A1' -R1; -Q1 -A1]); Zs = Z2/Z1;
% Alt. Robust Control toolbox:
% [Z1,Z2,eig,zerr,wellposed,Zs] = aresolv(A1',Q1,R1);
%
% Choose rho=1 (Designer's choice) and
% build the generalized plant P in (9.87)
%
rho=1;
A = daug(As,Ar);
B1 = [zeros(ns,mr) ((Bs*Ds')+(Zs*Cs'))*inv(sqrt(Rs));
       Br zeros(nr,ls)];
B2 = [Bs;zeros(nr,ms)];
C1 = [zeros(ms,ns+nr);Cs zeros(ls,nr);rho*Cs -rho*rho*Cr];
C2 = [zeros(mr,ns+nr);Cs zeros(ls,nr)];
D11 = [zeros(ms,mr+ls);zeros(ls,mr) sqrt(Rs);-rho*rho*Dr rho*sqrt(Rs)];
D12 = [eye(ms);Ds;rho*Ds];
D21 = [rho*eye(mr) zeros(mr,ls);zeros(ls,mr) sqrt(Rs)];
D22 = [zeros(mr,ms);Ds];
B = [B1 B2]; C = [C1;C2]; D = [D11 D12;D21 D22];
P = pck(A,B,C,D);
% Alternative: Use sysic to generate P from Figure 9.21
% but may get extra states, since states from Gs may enter twice.
%
% Gamma iterations to obtain H-infinity controller
%
[l1,m2] = size(D12); [l2,m1] = size(D21);
nmeas = l2; ncon = m2; gmin = 1; gmax = 5; gtol = 0.01;
[K, Gnclp, gam] = hinfsyn(P, nmeas, ncon, gmin, gmax, gtol);
% Alt. Robust toolbox, use command: hinfopt
%
```

We will conclude this subsection with a summary of the main steps required to synthesize a two degrees-of-freedom $\mathcal{H}_\infty$ loop-shaping controller.

1. Design a one degree-of-freedom $\mathcal{H}_\infty$ loop-shaping controller using the procedure of Subsection 9.4.2, but without a post-compensator weight W_2. Hence W_1.
2. Select a desired closed-loop transfer function T_{ref} between the commands and controlled outputs.
3. Set the scalar parameter ρ to a small value greater than 1; something in the range 1 to 3 will usually suffice.
4. For the shaped plant $G_s = GW_1$, the desired response T_{ref}, and the scalar parameter ρ, solve the standard $\mathcal{H}_\infty$ optimization problem defined by P in (9.87) to a specified tolerance to get $K = \begin{bmatrix} K_1 & K_2 \end{bmatrix}$. Remember to include W_o in the problem formulation if extra feedback measurements are to be used.
5. Replace the prefilter K_1, by $K_1 W_i$ to give exact model-matching at steady-state.
6. Analyse and, if required, redesign making adjustments to ρ and possibly W_1 and T_{ref}.

The final two degrees-of-freedom $\mathcal{H}_\infty$ loop-shaping controller is illustrated in Figure 9.22

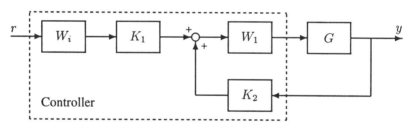

Figure 9.22: Two degrees-of-freedom $\mathcal{H}_\infty$ loop-shaping controller

9.4.4   Observer-based structure for $\mathcal{H}_\infty$ loop-shaping controllers

$\mathcal{H}_\infty$ designs exhibit a separation structure in the controller. As seen from (9.50) and (9.51) the controller has an observer/state feedback structure, but the observer is non-standard, having a disturbance term (a "worst" disturbance) entering the observer state equations. For $\mathcal{H}_\infty$ loop-shaping controllers, whether of the one or two degrees-of-freedom variety, this extra term is not present. The clear structure of $\mathcal{H}_\infty$ loop-shaping controllers has several advantages:

• It is helpful in describing a controller's function, especially to one's managers or clients who may not be familiar with advanced control.
• It lends itself to implementation in a gain-scheduled scheme, as shown by Hyde and Glover (1993).

- It offers computational savings in digital implementations and some multi-mode switching schemes, as shown in (Samar, 1995).

We will present the controller equations, for both one and two degrees-of-freedom $\mathcal{H}_\infty$ loop-shaping designs. For simplicity we will assume the shaped plant is strictly proper, with a stabilizable and detectable state-space realization

$$G_s \stackrel{s}{=} \left[\begin{array}{c|c} A_s & B_s \\ \hline C_s & 0 \end{array} \right] \tag{9.90}$$

In which case, as shown in (Sefton and Glover, 1990), the single degree-of-freedom $\mathcal{H}_\infty$ loop-shaping controller can be realized as an observer for the shaped plant plus a state-feedback control law. The equations are

$$\dot{\hat{x}}_s = A_s \hat{x}_s + H_s (C_s \hat{x}_s - y_s) + B_s u_s \tag{9.91}$$

$$u_s = K_s \hat{x}_s \tag{9.92}$$

where $\hat{x}_s$ is the observer state, u_s and y_s are respectively the input and output of the shaped plant, and

$$H_s = -Z_s C_s^T \tag{9.93}$$

$$K_s = -B_s^T \left[I - \gamma^{-2} I - \gamma^{-2} X_s Z_s \right]^{-1} X_s \tag{9.94}$$

where Z_s and X_s are the appropriate solutions to the generalized algebraic Riccati equations for G_s given in (9.67) and (9.68).

In Figure 9.23, an implementation of an observer-based $\mathcal{H}_\infty$ loop-shaping controller is shown in block diagram form. The same structure was used by Hyde and Glover (1993) in their VSTOL design which was scheduled as a function of aircraft forward speed.

Walker (1996) has shown that the two degrees-of-freedom $\mathcal{H}_\infty$ loop-shaping controller also has an observer-based structure. He considers a stabilizable and detectable plant

$$G_s \stackrel{s}{=} \left[\begin{array}{c|c} A_s & B_s \\ \hline C_s & 0 \end{array} \right] \tag{9.95}$$

and a desired closed-loop transfer function

$$T_{\text{ref}} \stackrel{s}{=} \left[\begin{array}{c|c} A_r & B_r \\ \hline C_r & 0 \end{array} \right] \tag{9.96}$$

in which case the generalized plant $P(s)$ in (9.87) simplifies to

$$P \stackrel{s}{=} \left[\begin{array}{cc|cc|c} A_s & 0 & 0 & Z_s C_s^T & B_s \\ 0 & A_r & B_r & 0 & 0 \\ \hline 0 & 0 & 0 & 0 & I \\ C_s & 0 & 0 & I & 0 \\ \rho C_s & -\rho^2 C_r & 0 & \rho I & 0 \\ \hline 0 & 0 & \rho I & 0 & 0 \\ C_s & 0 & 0 & I & 0 \end{array} \right] \stackrel{\triangle}{=} \left[\begin{array}{c|cc} A & B_1 & B_2 \\ \hline C_1 & D_{11} & D_{12} \\ C_2 & D_{21} & D_{22} \end{array} \right] \tag{9.97}$$

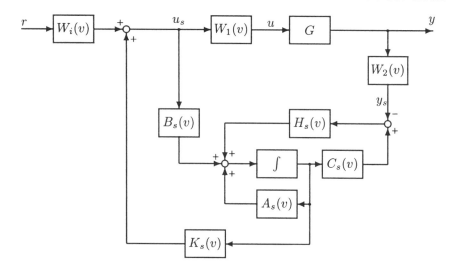

Figure 9.23: An implementation of an $\mathcal{H}_\infty$ loop-shaping controller for use when gain scheduling against a variable v

Walker then shows that a stabilizing controller $K = \begin{bmatrix} K_1 & K_2 \end{bmatrix}$ satisfying $\|F_l(P,K)\|_\infty < \gamma$ exists if, and only if,

(i) $\gamma > \sqrt{1 + \rho^2}$, and

(ii) $X_\infty \geq 0$ is a solution to the algebraic Riccati equation

$$X_\infty A + A^T X_\infty + C_1^T C_1 - \bar{F}^T(\bar{D}^T \bar{J} \bar{D})\bar{F} = 0 \qquad (9.98)$$

such that $\mathrm{Re}\,\lambda_i\left[A + B\bar{F}\right] < 0\ \forall i$, where

$$\bar{F} = (\bar{D}^T \bar{J} \bar{D})^{-1}(\bar{D}^T \bar{J} C + B^T X_\infty) \qquad (9.99)$$

$$\bar{D} = \begin{bmatrix} D_{11} & D_{12} \\ I_w & 0 \end{bmatrix} \qquad (9.100)$$

$$\bar{J} = \begin{bmatrix} I_z & 0 \\ 0 & -\gamma^2 I_w \end{bmatrix} \qquad (9.101)$$

where I_z and I_w are unit matrices of dimensions equal to those of the error signal z, and exogenous input w, respectively, in the standard configuration.

Notice that this $\mathcal{H}_\infty$ controller depends on the solution to just one algebraic Riccati equation, not two. This is a characteristic of the two degrees-of-freedom $\mathcal{H}_\infty$ loop-shaping controller (Hoyle et al., 1991).

Walker further shows that if (i) and (ii) are satisfied, then a stabilizing controller $K(s)$ satisfying $\|F_l(P, K)\|_\infty < \gamma$ has the following equations:

$$\dot{\hat{x}}_s = A_s \hat{x}_s + H_s(C_s \hat{x}_s - y_s) + B_s u_s \tag{9.102}$$

$$\dot{x}_r = A_r x_r + B_r r \tag{9.103}$$

$$u_s = -B_s^T X_{\infty 11} \hat{x}_s - B_s^T X_{\infty 12} x_r \tag{9.104}$$

where $X_{\infty 11}$ and $X_{\infty 12}$ are elements of

$$X_\infty = \begin{bmatrix} X_{\infty 11} & X_{\infty 12} \\ X_{\infty 21} & X_{\infty 22} \end{bmatrix} \tag{9.105}$$

which has been partitioned conformally with

$$A = \begin{bmatrix} A_s & 0 \\ 0 & A_r \end{bmatrix} \tag{9.106}$$

and H_s is as in (9.93).

The structure of this controller is shown in Figure 9.24, where the state-feedback gain matrices F_s and F_r are defined by

$$F_s \triangleq B_s^T X_{\infty 11} \qquad F_r \triangleq B_s^T X_{\infty 12} \tag{9.107}$$

The controller consists of a state observer for the shaped plant G_s, a model of the desired closed-loop transfer function T_{ref} (without C_r) and a state-feedback control law that uses both the observer and reference-model states.

As in the one degree-of-freedom case, this observer-based structure is useful in gain-scheduling. The reference-model part of the controller is also nice because it is often the same at different design operating points and so may not need to be changed at all during a scheduled operation of the controller. Likewise, parts of the observer may not change; for example, if the weight $W_1(s)$ is the same at all the design operating points. Therefore whilst the structure of the controller is comforting in the familiarity of its parts, it also has some significant advantages when it comes to implementation.

9.4.5 Implementation issues

Discrete-time controllers. For implementation purposes, discrete-time controllers are usually required. These can be obtained from a continuous-time design using a bilinear transformation from the s-domain to the z-domain, but there can be advantages in being able to design directly in discrete-time. In Samar (1995) and Postlethwaite et al. (1995), observer-based state-space equations are derived directly in discrete-time for the two degrees-of-freedom $\mathcal{H}_\infty$ loop-shaping controller and successfully applied to an aero engine. This application was on a real engine, a Spey

engine, which is a Rolls Royce 2-spool reheated turbofan housed at the UK Defence
Research Agency, Pyestock. As this was a real application, a number of important
implementation issues needed to be addressed. Although these are outside the general
scope of this book, they will be briefly mentioned now.

Anti-windup. In $\mathcal{H}_\infty$ loop-shaping the pre-compensator weight W_1 would
normally include integral action in order to reject low frequency disturbances acting
on the system. However, in the case of actuator saturation the integrators continue
to integrate their input and hence cause windup problems. An anti-windup scheme is
therefore required on the weighting function W_1. The approach we recommend is to
implement the weight W_1 in its *self-conditioned* or *Hanus* form. Let the weight W_1
have a realization

$$W_1 \stackrel{s}{=} \left[\begin{array}{c|c} A_w & B_w \\ \hline C_w & D_w \end{array} \right] \tag{9.108}$$

and let u be the input to the plant actuators and u_s the input to the shaped plant. Then
$u = W_1 u_s$. When implemented in Hanus form, the expression for u becomes (Hanus
et al., 1987)

$$u = \left[\begin{array}{c|cc} A_w - B_w D_w^{-1} C_w & 0 & B_w D_w^{-1} \\ \hline C_w & D_w & 0 \end{array} \right] \left[\begin{array}{c} u_s \\ u_a \end{array} \right] \tag{9.109}$$

where u_a is the actual plant input, that is the measurement at the output of the
actuators which therefore contains information about possible actuator saturation.

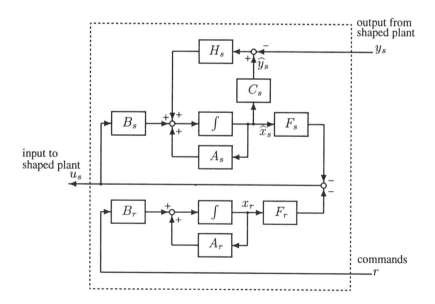

Figure 9.24: Structure of the two degrees-of-freedom $\mathcal{H}_\infty$ loop-shaping controller

The situation is illustrated in Figure 9.25, where the actuators are each modelled by a unit gain and a saturation. The Hanus form prevents windup by keeping the states of W_1 consistent with the actual plant input at all times. When there is no saturation $u_a = u$, the dynamics of W_1 remain unaffected and (9.109) simplifies to (9.108). But when $u_a \neq u$ the dynamics are inverted and driven by u_a so that the states remain consistent with the actual plant input u_a. Notice that such an implementation requires W_1 to be invertible and minimum phase.

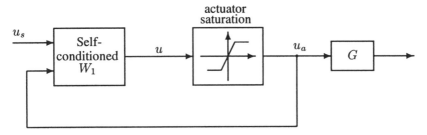

Figure 9.25: Self-conditioned weight W_1

Exercise 9.9 *Show that the Hanus form of the weight W_1 in (9.109) simplifies to (9.108) when there is no saturation i.e. when $u_a = u$.*

Bumpless transfer. In the aero-engine application, a multi-mode switched controller was designed. This consisted of three controllers, each designed for a different set of engine output variables, which were switched between depending on the most significant outputs at any given time. To ensure smooth transition from one controller to another - bumpless transfer - it was found useful to condition the reference models and the observers in each of the controllers. Thus when on-line, the observer state evolves according to an equation of the form (9.102) but when off-line the state equation becomes

$$\dot{\hat{x}}_s = A_s \hat{x}_s + H_s (C_s \hat{x}_s - y_s) + B_s u_{as} \qquad (9.110)$$

where u_{as} is the actual input to the shaped plant governed by the on-line controller. The reference model with state feedback given by (9.103) and (9.104) is not invertible and therefore cannot be self-conditioned. However, in discrete-time the optimal control also has a feed-through term from r which gives a reference model that can be inverted. Consequently, in the aero-engine example the reference models for the three controllers were each conditioned so that the inputs to the shaped plant from the off-line controller followed the actual shaped plant input u_{as} given by the on-line controller.

Satisfactory solutions to implementation issues such as those discussed above are crucial if advanced control methods are to gain wider acceptance in industry. We have tried to demonstrate here that the observer-based structure of the $\mathcal{H}_\infty$ loop-shaping controller is helpful in this regard.

9.5 Conclusion

We have described several methods and techniques for controller design, but our emphasis has been on $\mathcal{H}_\infty$ loop shaping which is easy to apply and in our experience works very well in practice. It combines classical loop-shaping ideas (familiar to most practising engineers) with an effective method for robustly stabilizing the feedback loop. For complex problems, such as unstable plants with multiple gain crossover frequencies, it may not be easy to decide on a desired loop shape. In which case, we would suggest doing an initial LQG design (with simple weights) and using the resulting loop shape as a reasonable one to aim for in $\mathcal{H}_\infty$ loop shaping.

An alternative to $\mathcal{H}_\infty$ loop shaping is a standard $\mathcal{H}_\infty$ design with a "stacked" cost function such as in S/KS mixed-sensitivity optimization. In this approach, $\mathcal{H}_\infty$ optimization is used to shape two or sometimes three closed-loop transfer functions. However, with more functions the shaping becomes increasingly difficult for the designer.

In other design situations where there are several performance objectives (e.g. on signals, model following and model uncertainty), it may be more appropriate to follow a signal-based $\mathcal{H}_2$ or $\mathcal{H}_\infty$ approach. But again the problem formulations become so complex that the designer has little direct influence on the design.

After a design, the resulting controller should be analyzed with respect to robustness and tested by nonlinear simulation. For the former, we recommend μ-analysis as discussed in Chapter 8, and if the design is not robust, then the weights will need modifying in a redesign. Sometimes one might consider synthesizing a μ-optimal controller, but this complexity is rarely necessary in practice. Moreover, one should be careful about combining controller synthesis and analysis into a single step. The following quote from Rosenbrock (1974) illustrates the dilemma:

> In synthesis the designer specifies in detail the properties which his system must have, to the point where there is only one possible solution. ... The act of specifying the requirements in detail implies the final solution, yet has to be done in ignorance of this solution, which can then turn out to be unsuitable in ways that were not foreseen.

Therefore, control system design usually proceeds iteratively through the steps of modelling, control structure design, controllability analysis, performance and robustness weights selection, controller synthesis, control system analysis and nonlinear simulation. Rosenbrock (1974) makes the following observation:

> Solutions are constrained by so many requirements that it is virtually impossible to list them all. The designer finds himself threading a maze of such requirements, attempting to reconcile conflicting demands of cost, performance, easy maintenance, and so on. A good design usually has strong aesthetic appeal to those who are competent in the subject.

10

CONTROL STRUCTURE DESIGN

Most (if not all) available control theories assume that a control structure is given at the outset. They therefore fail to answer some basic questions which a control engineer regularly meets in practice. Which variables should be controlled, which variables should be measured, which inputs should be manipulated, and which links should be made between them? The objective of this chapter is to describe the main issues involved in control structure design and to present some of the available quantitative methods, for example, for decentralized control.

10.1 Introduction

Control structure design was considered by Foss (1973) in his paper entitled "Critique of process control theory" where he concluded by challenging the control theoreticians of the day to close the gap between theory and applications in this important area. Later Morari et al. (1980) presented an overview of control structure design, hierarchical control and multilevel optimization in their paper "Studies in the synthesis of control structure for chemical processes", but the gap still remained, and still does to some extent today.

Control structure design is clearly important in the chemical process industry because of the complexity of these plants, but the same issues are relevant in most other areas of control where we have large-scale systems. For example, in the late 1980s Carl Nett (Nett, 1989; Nett and Minto, 1989) gave a number of lectures based on his experience on aero-engine control at General Electric, under the title "A quantitative approach to the selection and partitioning of measurements and manipulations for the control of complex systems". He noted that increases in controller complexity unnecessarily outpaces increases in plant complexity, and that the objective should be to

> ... minimize control system complexity subject to the achievement of accuracy specifications in the face of uncertainty.

In Chapter 1 (page 1), we described the typical steps taken in the process of designing a control system. Steps 4, 5, 6 and 7 are associated with the following tasks of *control structure design*:

1. The selection of controlled outputs (a set of variables which are to be controlled to achieve a set of specific objectives; see section 10.3).
2. The selection of manipulations and measurements (sets of variables which can be manipulated and measured for control purposes; see section 10.4).
3. The selection of a *control configuration* (a structure interconnecting measurements/commands and manipulated variables; see sections 10.6, 10.7 and 10.8).
4. The selection of a *controller type* (control law specification, e.g. PID-controller, decoupler, LQG, etc).

The distinction between the words control *structure* and control *configuration* may seem minor, but note that it is significant within the context of this book. The *control structure* (or control strategy) refers to all structural decisions included in the design of a control system. On the other hand, the *control configuration* refers only to the structuring (decomposition) of the controller K itself (also called the measurement/manipulation partitioning or input/output pairing). Control configuration issues are discussed in more detail in Section 10.6.

The selection of controlled outputs, manipulations and measurements (tasks 1 and 2 combined) is sometimes called *input/output selection*.

Ideally, the tasks involved in designing a complete control system are performed sequentially; first a "top-down" selection of controlled outputs, measurements and inputs (with little regard to the configuration of the controller K) and then a "bottom-up" design of the control system (in which the selection of the control configuration is the most important decision). However, in practice the tasks are closely related in that one decision directly influences the others, so the procedure may involve iteration.

One important reason for decomposing the control system into a specific *control configuration* is that it may allow for simple tuning of the subcontrollers without the need for a detailed plant model describing the dynamics and interactions in the process. Multivariable centralized controllers may always outperform decomposed (decentralized) controllers, but this performance gain must be traded off against the cost of obtaining and maintaining a sufficiently detailed plant model.

The number of possible control structures shows a combinatorial growth, so for most systems a careful evaluation of all alternative control structures is impractical. Fortunately, we can often from physical insight obtain a reasonable choice of controlled outputs, measurements and manipulated inputs. In other cases, simple controllability measures as presented in Chapters 5 and 6 may be used for quickly evaluating or screening alternative control structures.

Some discussion on control structure design in the process industry is given by Morari (1982), Shinskey (1988), Stephanopoulos (1989) and Balchen and Mumme (1988). A survey on control structure design is given by van de Wal and de Jager

(1995). The reader is referred to Chapter 5 (page 160) for an overview of the literature on input-output controllability analysis.

10.2   Optimization and control

The selection of controlled outputs involves selecting the variables y to be controlled at given reference values, $y \approx r$. Here the reference value r is set at some higher layer in the control hierarchy. Thus, the selection of controlled outputs (for the control layer) is usually intimately related to the hierarchical structuring of the control system which is often divided into two layers:

- *optimization layer* — computes the desired reference commands r
- *control layer* — implements these commands to achieve $y \approx r$.

Additional layers are possible, as is illustrated in Figure 10.1 which shows a typical control hierarchy for a complete chemical plant. Here the control layer is subdivided into two layers: *supervisory control* ("advanced control") and *regulatory control* ("base control"). We have also included a scheduling layer above the optimization. In general, the information flow in such a control hierarchy is based on the higher layer sending reference values (setpoints) to the layer below, and the lower layer reporting back any problems in achieving this, see Figure 10.2(b).

The optimization tends to be performed *open-loop* with limited use of feedback. On the other hand, the control layer is mainly based on *feedback* information. The optimization is often based on nonlinear steady-state models, whereas we often use linear dynamic models in the control layer (as we do throughout the book).

There is usually a time scale separation with faster lower layers as indicated in Figure 10.1. This means that the setpoints, as viewed from a given layer in the hierarchy, are updated only periodically. Between these updates, when the setpoints are constant, it is important that the system remains reasonably close to its optimum. This observation is the basis for Section 10.3 which deals with selecting outputs for the control layer.

From a theoretical point of view, the optimal coordination of the inputs and thus the optimal performance is obtained with a *centralized optimizing controller*, which combines the two layers of optimization and control; see Figure 10.2(c). All control actions in such an ideal control system would be perfectly coordinated and the control system would use on-line dynamic optimization based on a nonlinear dynamic model of the complete plant instead of, for example, infrequent steady-state optimization. However, this solution is normally not used for a number of reasons; including the cost of modelling, the difficulty of controller design, maintenance and modification, robustness problems, operator acceptance, and the lack of computing power.

As noted above we may also decompose the control layer, and from now on when we talk about control configurations, hierarchical decomposition and decentralization, we generally refer to the control layer.

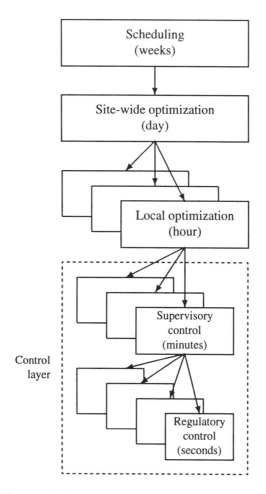

Figure 10.1: Typical control system hierarchy in a chemical plant

Mesarovic (1970) reviews some ideas related to on-line multi-layer structures applied to large-scale industrial complexes. However, according to Lunze (1992), multilayer structures, although often used in practice, lack a formal analytical treatment. Nevertheless, in the next section we provide some ideas on how to select objectives (controlled outputs) for the control layer, such that the overall goal is satisfied.

Remark 1 In accordance with Lunze (1992) we have purposely used the word *layer* rather than *level* for the hierarchical decomposition of the control system. The difference is that in a *multilevel* system all units contribute to satisfying the same goal, whereas in a *multilayer* system the different units have different objectives (which preferably contribute to the overall goal). Multilevel systems have been studied in connection with the solution of optimization problems.

Remark 2 The tasks within any layer can be performed by humans (e.g. manual control), and the interaction and task sharing between the automatic control system and the human operators are very important in most cases, e.g. an aircraft pilot. However, these issues are outside the scope of this book.

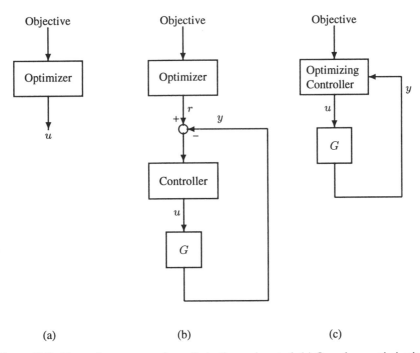

(a)   (b)   (c)

Figure 10.2: Alternative structures for optimization and control. (a) **Open-loop optimization.** (b) Closed-loop implementation with separate control layer. (c) Integrated optimization and control.

10.3   Selection of controlled outputs

A *controlled output* is an output variable (usually measured) with an associated control objective (usually a reference value). In many cases, it is clear from a physical understanding of the process what the controlled outputs should be. For example, if we consider heating or cooling a room, then we should select room temperature as the controlled output y. In other cases it is less obvious because each control objective may not be associated with a measured output variable. Then the controlled outputs y are selected to achieve the *overall system goal*, and may not appear to be important variables in themselves.

Example 10.1 Cake baking. *To get an idea of the issues involved in output selection let us consider the process of baking a cake. The overall goal is to make a cake which is well baked inside and with a nice exterior. The manipulated input for achieving this is the heat input, $u = Q$, (and we will assume that the duration of the baking is fixed, e.g. at 15 minutes). Now, if we had never baked a cake before, and if we were to construct the stove ourselves, we might consider directly manipulating the heat input to the stove, possibly with a watt-meter measurement. However, this* open-loop *implementation would not work well, as the optimal heat input depends strongly on the particular oven we use, and the operation is also sensitive to disturbances, for example, from opening the oven door or whatever else might be in the oven. Therefore, in practice we look up the optimal oven temperature in a cook book, and use a* closed-loop *implementation where a thermostat is used to keep the temperature y at its predetermined value T.*

The (a) open-loop and (b) closed-loop implementations of the cake baking process are illustrated in Figure 10.2. In (b) the "optimizer" is the cook book which has a pre-computed table of the optimal temperature profile. The reference value r for temperature is then sent down to the control layer which consists of a simple feedback controller (the thermostat).

Recall that the title of this section is selection of controlled outputs. In the cake baking process we select *oven temperature* as the controlled output y in the control layer. It is interesting to note that controlling the oven temperature in itself has no direct relation to the overall goal of making a well-baked cake. So why do we select the oven temperature as a controlled output? We now want to outline an approach for answering questions of this kind.

In the following, we let y denote the selected controlled outputs in the control layer. Note that this may also include directly using the inputs (open-loop implementation) by selecting $y = u$. Two distinct questions arise:

1. What variables y should be selected as the controlled variables?
2. What is the optimal reference value (y_{opt}) for these variables?

The second problem is one of dynamic optimization and is extensively studied. Here we want to gain some insight into the first problem. We make the following *assumptions*:

(a) The overall goal can be quantified in terms of a scalar cost function J which we want to minimize.

(b) For a given disturbance d, there exists an optimal value $u_{opt}(d)$ and corresponding value $y_{opt}(d)$ which minimizes the cost function J.

(c) The reference values r for the controlled outputs y should be constant, i.e. r should be independent of the disturbances d. Typically, some average value is selected, e.g. $r = y_{opt}(\bar{d})$

For example, in the cake baking process we may assign to each cake a number P on a scale from 0 to 10, based on cake quality. A perfectly baked cake achieves $P = 10$, and an acceptably baked cake achieves $P > 6$ (a completely burned cake may correspond to $P = 1$). In another case P could be the operating profit. In both cases we can select $J = -P$, and the overall goal of the control system is then to minimize J.

The system behaviour is a function of the independent variables u and d, so we may write $J = J(u, d)$. For a given disturbance d the optimal value of the cost function is

$$J_{opt}(d) \triangleq J(u_{opt}, d) = \min_u J(u, d) \tag{10.1}$$

Ideally, we want $u = u_{opt}$. However, this will not be achieved in practice, and we select controlled outputs y such that:

- *The input u (generated by feedback to achieve $y \approx r$) should be close to the optimal input $u_{opt}(d)$.*

Note that we have assumed that r is independent of d. The above statement is obvious, but it is nevertheless very useful. The following development aims at quantifying the statement.

What happens if $u \neq u_{opt}$? Obviously, we then have a loss which can be quantified by $J - J_{opt}$, and a reasonable objective for selecting controlled outputs y is to minimize the worst-case loss

$$\text{Worst} - \text{case loss} : \Phi \triangleq \max_{d \in \mathcal{D}} |J(u, d) - J(u_{opt}, d)| \tag{10.2}$$

Here $\mathcal{D}$ is the set of possible disturbances. As "disturbances" we should also include changes in operating point and model uncertainty.

To obtain some insight into the problem of minimizing the loss Φ, let us consider the term $J(u, d) - J_{opt}(d)$ in (10.2), where d is a fixed (generally non-zero) disturbance. We make the following additional assumptions:

(d) The cost function J is smooth, or more precisely twice differentiable.

(e) The optimization problem is unconstrained. If it is optimal to keep some variable at a constraint, then we assume that this is implemented and consider the remaining unconstrained problem.

(f) The dynamics of the problem can be neglected, that is, we consider the steady-state control and optimization.

For a fixed d we may then express $J(u, d)$ in terms of a Taylor series expansion of u around the optimal point. We get

$$J(u, d) = J_{\text{opt}} + \underbrace{\left(\frac{\partial J}{\partial u}\right)^T_{\text{opt}} (u - u_{\text{opt}})}_{=0} +$$

$$\frac{1}{2}(u - u_{\text{opt}})^T \left(\frac{\partial^2 J}{\partial u^2}\right)_{\text{opt}} (u - u_{\text{opt}}) + \cdots \quad (10.3)$$

where we have neglected terms of third order and higher (which assumes that we are reasonably close to the optimum). Also, note here that u_{opt} is a function of d so we should formally write $u_{\text{opt}}(d)$. The second term on the right hand side in (10.3) is zero at the optimal point for an unconstrained problem.

Equation (10.3) quantifies how $u - u_{\text{opt}}$ affects the cost function. To study how this relates to output selection we use a linearized model of the plant, which for a fixed d becomes $y - y_{\text{opt}} = G(u - u_{\text{opt}})$ where G is the steady-state gain matrix. If G is invertible we then get

$$u - u_{\text{opt}} = G^{-1}(y - y_{\text{opt}}) \quad (10.4)$$

If G is not invertible we may use the pseudo-inverse $G^\dagger$ which results in the smallest possible $\|u - u_{\text{opt}}\|_2$ for a given $y - y_{\text{opt}}$, and

$$J - J_{\text{opt}} \approx \frac{1}{2} \left(G^{-1}(y - y_{\text{opt}})\right)^T \left(\frac{\partial^2 J}{\partial u^2}\right)_{\text{opt}} G^{-1}(y - y_{\text{opt}}) \quad (10.5)$$

where the term $(\partial^2 J/\partial u^2)_{\text{opt}}$ is independent of y. Obviously, we want to select the controlled outputs such that $y - y_{\text{opt}}$ is small. However, this is not possible in practice. To see this, write

$$y - y_{\text{opt}} = y - r + r - y_{\text{opt}} = e + e_{\text{opt}} \quad (10.6)$$

First, we have an optimization error $e_{\text{opt}} \triangleq r - y_{\text{opt}}$, because the algorithm (e.g. a cook book) pre-computes a desired r which is different from the optimal y_{opt}. In addition, we have a control error $e = y - r$ because the control layer is not perfect, for example due to poor control performance or an incorrect measurement or estimate (steady-state bias) of y. If the control itself is perfect then $e = n$ (the measurement noise). In most cases the errors e and e_{opt} can be assumed independent.

From (10.5) and (10.6), we conclude that we should select the controlled outputs y such that:

1. G^{-1} is small (i.e. G is large); the choice of y should be such that the inputs have a large effect on y.

2. $e_{\text{opt}} = r - y_{\text{opt}}(d)$ is small; the choice of y should be such that its optimal value $y_{\text{opt}}(d)$ depends only weakly on the disturbances and other changes.

3. $e = y - r$ is small; the choice of y should be such that it is easy to keep the control error e small.

Note that $\bar{\sigma}(G^{-1}) = 1/\underline{\sigma}(G)$, and so we want the smallest singular value of the steady-state gain matrix to be large (but recall that singular values depend on scaling as is discussed below). The desire to have $\underline{\sigma}(G)$ large is consistent with our intuition that we should ensure that the controlled outputs are independent of each other. Also note that the desire to have $\underline{\sigma}(G)$ large (and preferably as large as possible) is here not related to the issue of input constraints, which was discussed in Section 6.9.

Example 10.1 Cake baking, continued. *Let us return to our initial question: why select the oven temperature as a controlled output? We have two alternatives: a closed-loop implementation with $y = T$ (the oven temperature) and an open-loop implementation with $y = u = Q$ (the heat input). From experience, we know that the optimal oven temperature T_{opt} is largely independent of disturbances and is almost the same for any oven. This means that we may always specify the same oven temperature, say $T_r = 190°C$, as obtained from the cook book. On the other hand, the optimal heat input Q_{opt} depends strongly on the heat loss, the size of the oven, etc, and may vary between, say 100W and 5000W. A cook book would then need to list a different value of Q_r for each kind of oven and would in addition need some correction factor depending on the room temperature, how often the oven door is opened, etc. Therefore, we find that it is much easier to keep $e_{opt} = T - T_{opt}$ [°C] small than to keep $Q_r - Q_{opt}$ [W] small.*

There may also be some difference in the control error $e = y - r$, but this is mainly because most ovens are designed to be operated with temperature feedback. Specifically, with a thermostat we can easily keep T close to its desired value. On the other hand, with most ovens it is difficult to accurately implement the desired heat input Q, because the knob on the stove is only a crude indicator of the actual heat input (this difference in control error could have been eliminated by implementing a watt-meter).

To use $\underline{\sigma}(G)$ to select controlled outputs, we see from (10.5) that we should first scale the outputs such that the expected magnitude of $y_i - y_{i_{opt}}$ is similar in magnitude for each output, and scale the inputs such that the effect of a given deviation $u_j - u_{j_{opt}}$ on the cost function J is similar for each input (such that $\left(\partial^2 J/\partial u^2\right)_{opt}$ is close to a constant times a unitary matrix). We must also assume that the variations in $y_i - y_{i_{opt}}$ are uncorrelated, or more precisely, we must assume:

(g) The "worst-case" combination of output deviations $y_i - y_{i_{opt}}$, corresponding to the direction of $\underline{\sigma}(G)$, can occur in practice.

Procedure for selecting controlled outputs. The use of the minimum singular value to select controlled outputs may be summarized in the following procedure:

1. From a (nonlinear) model compute the optimal parameters (inputs and outputs) for various conditions (disturbances, operating points). (This yields a "look-up" table of optimal parameter values as a function of the operating conditions.)
2. From this data obtain for each candidate output the variation in its optimal value, $v_i = (y_{i_{opt,max}} - y_{i_{opt,min}})/2$.

3. Scale the candidate outputs such that for each output the sum of the magnitudes of v_i and the control error (e.g. measurement noise) is similar (e.g. about 1).
4. Scale the inputs such that a unit deviation in each input from its optimal value has the same effect on the cost function J.
5. Select as candidates those sets of controlled outputs which correspond to a large value of $\underline{\sigma}(G)$. G is the transfer function for the effect of the scaled inputs on the scaled outputs.

The aero-engine application in Chapter 12 provides a nice illustration of output selection. There the overall goal is to operate the engine optimally in terms of fuel consumption, while at the same time staying safely away from instability. The optimization layer is a look-up table, which gives the optimal parameters for the engine at various operating points. Since the engine at steady-state has three degrees-of-freedom we need to specify three variables to keep the engine approximately at the optimal point, and five alternative sets of three outputs are given. The outputs are scaled as outlined above, and a good output set is then one with a large value of $\underline{\sigma}(G)$, provided we can also achieve good dynamic control performance.

Remark 1 In the above procedure for selecting controlled outputs, based on maximizing $\underline{\sigma}(G)$, the variation in $y_{\mathrm{opt}}(d)$ with d (which should be small) enters into the scaling of the outputs.

Remark 2 A more exact procedure, which may be used if the optimal outputs are correlated such that assumption (g) does not hold, is the following: by solving the nonlinear equations, evaluate directly the cost function J for various disturbances d and control errors e, assuming $y = r + e$ where r is kept constant at the optimal value for the nominal disturbance. The set of controlled outputs with smallest average or worst-case value of J is then preferred. This approach is usually more time consuming because the solution of the nonlinear equations must be repeated for each candidate set of controlled outputs.

Measurement selection for indirect control

The above ideas also apply for the case where the overall goal is to keep some variable z at a given value (setpoint) z_r, e.g. $J = \|z - z_r\|$. We assume we cannot measure z, and instead we attempt to achieve our goal by controlling y at some fixed value r, e.g. $r = y_{\mathrm{opt}}(\bar{d})$ where $\bar{d} = 0$ if we use deviation variables. For small changes we may assume linearity and write

$$z = G_z u + G_{zd} d \tag{10.7}$$

for which $z = z_r$ is obtained with $u = u_{\mathrm{opt}}(d)$ where $u_{\mathrm{opt}}(d) = G_z^{-1}(z_r - G_{zd}d)$. Furthermore, $y = Gu + G_d d$, so the optimal output is

$$y_{\mathrm{opt}}(d) = \underbrace{GG_z^{-1}z_r}_{r} + (G_d - GG_z^{-1}G_{zd})d \tag{10.8}$$

where one criterion for selecting outputs y is that $y_{\mathrm{opt}}(d)$ depends weakly on d.

Although the above procedure for selecting controlled outputs applies also in this simple case, it is easier to consider $J = \|z - z_r\|$ directly. Using $u = G^{-1}(y - G_d d)$ in (10.7) and introducing $y = r + e$, where $r = GG_z^{-1}z_r$ (a constant) and e is the control error (e.g. due to measurement error), we find

$$z - z_r = \underbrace{(G_{zd} - G_z G^{-1} G_d)}_{P_d} d + \underbrace{G_z G^{-1}}_{P_r} e \qquad (10.9)$$

To minimize $\|z - z_r\|$ we should therefore select controlled outputs such that $\|P_d d\|$ and $\|P_r e\|$ are small. Note that P_d depends on the scaling of disturbances d and "primary" outputs z (and are independent of the scaling of inputs u and selected outputs y, at least for square plants). The magnitude of the control error e depends on the choice of outputs. Based on (10.9) a procedure for selecting controlled outputs may be suggested.

Procedure for selecting controlled outputs for indirect control. Scale the disturbances d to be of magnitude 1 (as usual), and scale the outputs y so that the expected control error e (measurement noise) is of magnitude 1 for each output (this is different from the output scaling used in step 3 in our minimum singular value procedure). Then to minimize J we should select sets of controlled outputs which:

$$\text{Minimize } \| [P_d \quad P_r] \| \qquad (10.10)$$

Remark 1 The choice of norm in (10.10) depends on the scaling, but the choice is usually of secondary importance. The maximum singular value arises if $\|d\|_2 \le 1$ and $\|e\|_2 \le 1$, and we want to minimize $\|z - z_r\|_2$.

Remark 2 The above procedure does not require assumption (g) on uncorrelated variations in the optimal values of $y_i - y_{i_{\mathrm{opt}}}$.

Remark 3 Of course, for the choice $y = z$ we have that $y_{\mathrm{opt}} = z_r$ is independent of d and the matrix P_d in (10.9) and (10.10) is zero. However, P_r is still non-zero.

Remark 4 In some cases this measurement selection problem involves a trade-off between wanting $\|P_d\|$ small (wanting a strong correlation between measured outputs y and "primary" outputs z) and wanting $\|P_r\|$ small (wanting the effect of control errors (measurement noise) to be small).

For example, this is the case in a distillation column when we use temperatures inside the column (y) for indirect control of the product compositions (z). For a high-purity separation, we cannot place the measurement too close to the column end due to sensitivity to measurement error ($\|P_r\|$ becomes large), and we cannot place it too far from the column end due to sensitivity to disturbances ($\|P_d\|$ becomes large).

Remark 5 Indirect control is related to the idea of *inferential control* which is commonly used in the process industry. However, in inferential control the idea is usually to use the measurement of y to estimate (infer) z and then to control this estimate rather than controlling

y directly, e.g. see Stephanopoulos (1989). However, there is no universal agreement on these terms, and Marlin (1995) uses the term inferential control to mean indirect control as discussed here.

Remark 6 The problem of indirect control is closely related to that of *cascade control* discussed in Section 10.7.2 (where u is u_2, z is y_1, y is y_2, G is G_{22}, etc.); see (10.32), where we introduce P_d as the partial disturbance gain matrix. The main difference is that in cascade control we also measure and control z (y_1) in an outer loop. In this case we want $\| [\, P_d \quad P_r \,] \|$ small at high frequencies beyond the bandwidth of the outer loop involving z (y_1).

Remark 7 One might say that (10.5) and the resulting procedure for output selection, generalizes the use of P_d and P_r from the case of indirect control to the more general case of minimizing some cost function J.

Summary. Generally, the optimal values of all variables will change with time during operation (due to disturbances and other changes). For practical reasons, we have considered a hierarchical strategy where the optimization is performed only periodically. The question is then: which variables (*controlled outputs*) should be kept constant (between each optimization)? Essentially, we found that we should select variables y for which the variation in optimal value and control error is small compared to their controllable range (the range y may reach by varying the input u). This is hardly a big surprise, but it is nevertheless useful and provides the basis for our procedure for selecting controlled outputs.

The objective of the control layer is then to keep the controlled outputs at their reference values (which are computed by the optimization layer). The controlled outputs are often measured, but we may also estimate their values based on other measured variables. We may also use other measurements to improve the control of the controlled outputs, for example, by use of cascade control. Thus, the selection of controlled and measured outputs are two separate issues, although the two decisions are obviously related. The measurement selection problem is briefly discussed in the next section. In subsection 10.5 we discuss the relative gain array of the "big" transfer matrix (with all candidate outputs included), which is a useful screening tool for selecting controlled outputs.

10.4   Selection of manipulations and measurements

In some cases there are a large number of candidate measurements and/or manipulations. The need for control has three origins

- to stabilize an unstable plant,
- to reject disturbances,
- to track reference changes,

and the selection of manipulations and measurements is intimately related to these.

For measurements, the rule is to select those which have a strong relationship with the controlled outputs, or which may quickly detect a major disturbance and which together with manipulations can be used for local disturbance rejection. The issue of observability should also be considered.

The selected manipulations should have a large effect on the controlled outputs, and should be located "close" (in terms of dynamic response) to the outputs and measurements. If the plant is unstable, then the manipulations must be selected such that the unstable modes are state controllable, and the measurements must be selected such that the unstable modes are state observable.

For a more formal analysis we may consider the model $y_{\text{all}} = G_{\text{all}} u_{\text{all}} + G_{d\text{all}} d$. Here

- y_{all} = all candidate outputs (measurements)
- u_{all} = all candidate inputs (manipulations)

The model for a particular combination of inputs and outputs is then $y = Gu + G_d d$ where

$$G = S_O G_{\text{all}} S_I; \quad G_d = S_O G_{d\text{all}} \tag{10.11}$$

Here S_O is a non-square input "selection" matrix with a 1 and otherwise 0's in each row, and S_I is a non-square output "selection" matrix with a 1 and otherwise 0's in each column. For example, with $S_O = I$ all outputs are selected, and with $S_O = \begin{bmatrix} 0 & I \end{bmatrix}$ output 1 has *not* been selected.

To evaluate the alternative combinations, one may, based on G and G_d, perform an input-output controllability analysis as outlined in Chapter 6 for each combination (e.g, consider the minimum singular value, RHP-zeros, interactions, etc). At least this may be useful for eliminating some alternatives. A more involved approach, based on analyzing achievable robust performance by neglecting causality, is outlined by Lee et al. (1995). This approach is more involved both in terms of computation time and in the effort required to define the robust performance objective. An even more involved (and exact) approach would be to synthesize controllers for optimal robust performance for each candidate combination.

However, the number of combinations has a combinatorial growth, so even a simple input-output controllability analysis becomes very time-consuming if there are many alternatives. For a plant where we want to select m from M candidate manipulations, and l from L candidate measurements, the number of possibilities is

$$\binom{L}{l}\binom{M}{m} = \frac{L!}{l!(L-l)!} \frac{M!}{m!(M-m)!} \tag{10.12}$$

A few examples: for $m = l = 1$ and $M = L = 2$ the number of possibilities is 4; for $m = l = 2$ and $M = L = 4$ it is 36; for $m = l = 5$ and $M = L = 10$ it is 63504; and for $m = M, l = 5$ and $L = 100$ (selecting 5 measurements out of 100 possible) there are 75287520 possible combinations.

Remark. The number of possibilities is much larger if we consider *all* possible combinations with 1 to M inputs and 1 to L outputs. The number is (Nett, 1989): $\sum_{m=1}^{M} \sum_{l=1}^{L} \binom{L}{l} \binom{M}{m}$. For example, with $M = L = 2$ there are 4+2+2+1=9 candidates (4 structures with one input and one output, 2 structures with two inputs and one output, 2 structures with one input and two outputs, and 1 structure with two inputs and two outputs).

One way of avoiding this combinatorial problem is to base the selection directly on the "big" models G_{all} and G_{dall}. For example, one may consider the singular value decomposition and relative gain array of G_{all} as discussed in the next section. This rather crude analysis may be used, together with physical insight, rules of thumb and simple controllability measures, to perform a pre-screening in order to reduce the possibilities to a manageable number. These candidate combinations can then be analyzed more carefully.

10.5   RGA for non-square plant

A simple but effective screening tool for selecting inputs and outputs, which avoids the combinatorial problem just mentioned, is the relative gain array (RGA) of the "big" transfer matrix G_{all} with all candidate inputs and outputs included, $\Lambda = G_{\text{all}} \times G_{\text{all}}^{\dagger T}$.

Essentially, for the case of many candidate manipulations (inputs) one may consider not using those manipulations corresponding to columns in the RGA where the sum of the elements is much smaller than 1 (Cao, 1995). Similarly, for the case of many candidate measured outputs (or controlled outputs) one may consider not using those outputs corresponding to rows in the RGA where the sum of the elements is much smaller than 1.

To see this, write the singular value decomposition of G_{all} as

$$G_{\text{all}} = U\Sigma V^H = U_r \Sigma_r V_r^H \tag{10.13}$$

where Σ_r consists only of the $r = \text{rank}(G)$ non-zero singular values, U_r consists of the r first columns of U, and V_r consists of the r first columns of V. Thus, V_r consists of the input directions with a non-zero effect on the outputs, and U_r consists of the output directions we can affect (reach) by use of the inputs.

Let $e_j = \begin{bmatrix} 0 & \cdots & 0 & 1 & 0 & \cdots & 0 \end{bmatrix}^T$ be a unit vector with a 1 in position j and 0's elsewhere. Then the j'th input is $u_j = e_j^T u$. Define e_i in a similar way such that the i'th output is $y_i = e_i^T y$. We then have that $e_j^T V_r$ yields the projection of a unit input u_j onto the effective input space of G, and we follow Cao (1995) and define

$$\text{Projection for input } j = \|e_j^T V_r\|_2 \tag{10.14}$$

which is a number between 0 and 1. Similarly, $e_i^T U_r$ yields the projection of a unit output y_i onto the effective (reachable) output space of G, and we define

$$\text{Projection for output } i = \|e_i^T U_r\|_2 \tag{10.15}$$

which is a number between 0 and 1. The following theorem links the input and output (measurement) projection to the column and row sums of the RGA.

Theorem 10.1 (RGA and input and output projections.) *The i'th row sum of the RGA is equal to the square of the i'th output projection, and the j'th column sum of the RGA is equal to the square of the j'th input projection, i.e.*

$$\sum_{j=1}^{m} \lambda_{ij} = \|e_i^T U_r\|_2^2; \quad \sum_{i=1}^{l} \lambda_{ij} = \|e_j^T V_r\|_2^2 \qquad (10.16)$$

Proof: See Appendix A.4.2. $\qquad\qquad\qquad\qquad\qquad\qquad\qquad\qquad\qquad\square$

The RGA is a useful screening tool because it need only be computed once. It includes all the alternative inputs and/or outputs and thus avoids the combinatorial problem. From (10.16) we see that the row and column sums of the RGA provide a useful way of interpreting the information available in the singular vectors. For the case of extra inputs the RGA-values depend on the input scaling, and for extra outputs on the output scaling. The variables must therefore be scaled prior to the analysis. Note that for a non-singular G the RGA is scaling independent and each row and column sums to 1, i.e. the projection of any input or output is 1.

Example 10.2 *Cao (1995) considers the selection of manipulations in a chemical process for the hydrodealkylation of toluene (HDA process). The plant has 5 controlled outputs and 13 candidate manipulations. At steady-state*

$$G_{all}^T(0) = \begin{bmatrix} 0.7878 & 1.1489 & 2.6640 & -3.0928 & -0.0703 \\ 0.6055 & 0.8814 & -0.1079 & -2.3769 & -0.0540 \\ 1.4722 & -5.0025 & -1.3279 & 8.8609 & 0.1824 \\ -1.5477 & -0.1083 & -0.0872 & 0.7539 & -0.0551 \\ 2.5653 & 6.9433 & 2.2032 & -1.5170 & 8.7714 \\ 1.4459 & 7.6959 & -0.9927 & -8.1797 & -0.2565 \\ 0.0000 & 0.0000 & 0.0000 & 0.0000 & 0.0000 \\ 0.1097 & -0.7272 & -0.1991 & 1.2574 & 0.0217 \\ 0.3485 & -2.9909 & -0.8223 & 5.2178 & 0.0853 \\ -1.5899 & -0.9647 & -0.3648 & 1.1514 & -8.5365 \\ 0.0000 & 0.0002 & -0.5397 & -0.0001 & 0.0000 \\ -0.0323 & -0.1351 & 0.0164 & 0.1451 & 0.0041 \\ -0.0443 & -0.1859 & 0.0212 & 0.1951 & 0.0054 \end{bmatrix}$$

and the corresponding RGA-matrix and column-sums Λ_Σ^T are

$$
\Lambda^T = \begin{bmatrix}
0.1275 & -0.0755 & \mathbf{0.5907} & 0.1215 & 0.0034 \\
0.0656 & -0.0523 & 0.0030 & 0.1294 & 0.0002 \\
0.2780 & 0.0044 & 0.0463 & \mathbf{0.4055} & -0.0060 \\
\mathbf{0.3684} & -0.0081 & 0.0009 & 0.0383 & -0.0018 \\
-0.0599 & \mathbf{0.9017} & 0.2079 & -0.1459 & 0.0443 \\
0.1683 & 0.4042 & 0.1359 & 0.1376 & 0.0089 \\
0.0000 & 0.0000 & 0.0000 & 0.0000 & 0.0000 \\
0.0014 & -0.0017 & 0.0013 & 0.0099 & 0.0000 \\
0.0129 & -0.0451 & 0.0230 & 0.1873 & -0.0005 \\
0.0374 & -0.1277 & -0.0359 & 0.1163 & \mathbf{0.9516} \\
0.0000 & 0.0000 & 0.0268 & 0.0000 & 0.0000 \\
0.0001 & 0.0001 & 0.0000 & 0.0001 & 0.0000 \\
0.0002 & 0.0002 & 0.0001 & 0.0001 & 0.0000
\end{bmatrix}, \quad
\Lambda_\Sigma^T = \begin{bmatrix}
0.77 \\
0.15 \\
0.73 \\
0.40 \\
0.95 \\
0.85 \\
0.00 \\
0.01 \\
0.18 \\
0.94 \\
0.03 \\
0.00 \\
0.00
\end{bmatrix}
$$

There exist $\binom{13}{5} = 1287$ combinations with 5 inputs and 5 outputs, and $\binom{13}{6} = 1716$ combinations with 6 inputs and 5 outputs. The RGA may be useful in providing an initial screening.

We find from the column sums of the steady-state RGA-matrix given in Λ_Σ^T that the five inputs with the largest projections onto the input space of G_{all} are 5, 10, 6, 1, and 3 (in that order). For this selection $\underline{\sigma}(G) = 1.73$ whereas $\underline{\sigma}(G_{\mathrm{all}}) = 2.45$ with all inputs included. This shows that we have not lost much gain in the low-gain direction by using only 5 of the 13 inputs. Of course, for control purposes one must also consider higher frequencies up to crossover. The main difference at higher frequency is that input 7 (which has no steady-state effect) is effective, whereas input 1 is less effective. This may lead one to a control structure with six inputs where input u_7 is used at high frequencies and input u_1 at low frequencies.

However, there are a large number of other factors that determine controllability, such as RHP-zeros, sensitivity to uncertainty, and these must be taken into account when making the final selection.

The following example shows that although the RGA is an efficient screening tool, it must be used with some caution.

Example 10.3 *Consider a plant with 2 inputs and 4 candidate outputs of which we want to select 2. We have:*

$$
G_{\mathrm{all}} = \begin{bmatrix} 10 & 10 \\ 10 & 9 \\ 2 & 1 \\ 2 & 1 \end{bmatrix}, \quad
\Lambda = \begin{bmatrix} -2.57 & 3.27 \\ 1.96 & -1.43 \\ 0.80 & -0.42 \\ 0.80 & -0.42 \end{bmatrix}
$$

The four row sums of the RGA-matrix are 0.70, 0.53, 0.38 and 0.38. To maximize the output projection we should select outputs 1 and 2. However, this yields a plant $G_1 = \begin{bmatrix} 10 & 10 \\ 10 & 9 \end{bmatrix}$ which is ill-conditioned with large RGA-elements, $\Lambda(G_1) = \begin{bmatrix} -9 & 10 \\ 10 & -9 \end{bmatrix}$, and is likely to be difficult to control. On the other hand, selecting outputs 1 and 3 yields $G_2 = \begin{bmatrix} 10 & 10 \\ 2 & 1 \end{bmatrix}$ which *is well-conditioned with* $\Lambda(G_2) = \begin{bmatrix} -1 & 2 \\ 2 & -1 \end{bmatrix}$. *For comparison, the minimum singular values are:* $\underline{\sigma}(G_{\mathrm{all}}) = 1.05$, $\underline{\sigma}(G_1) = 0.51$, *and* $\underline{\sigma}(G_2) = 0.70$.

We discuss in Section 10.7.2 below the selection of extra measurements for use in a cascade control system.

10.6 Control configuration elements

We now assume that the measurements, manipulations and controlled outputs are fixed. The available synthesis theories presented in this book result in a multivariable controller K which connects all available measurements/commands (v) with all available manipulations (u),

$$u = Kv \tag{10.17}$$

However, such a controller may not be desirable. By control configuration selection we mean the partitioning of measurements/commands and manipulations within the control layer. More specifically, we define

Control configuration. *The restrictions imposed on the overall controller K by decomposing it into a set of local controllers (subcontrollers, units, elements, blocks) with predetermined links and with a possibly predetermined design sequence where subcontrollers are designed locally.*

In a conventional feedback system a typical restriction on K is to use a one degree-of-freedom controller (so that we have the same controller for r and $-y$). Obviously, this limits the achievable performance compared to that of a two degrees of freedom controller. In other cases we may use a two degrees-of-freedom controller, but we may impose the restriction that the feedback part of the controller (K_y) is first designed locally for disturbance rejection, and then the prefilter (K_r) is designed for command tracking. In general this will limit the achievable performance compared to a simultaneous design (see also the remark on page 105). Similar arguments apply to other cascade schemes.

Some elements used to build up a specific control configuration are:

- Cascade controllers
- Decentralized controllers
- Feedforward elements
- Decoupling elements
- Selectors

These are discussed in more detail below, and in the context of the process industry in Shinskey (1988) and Balchen and Mumme (1988). First, some definitions:

Decentralized control *is when the control system consists of independent feedback controllers which interconnect a subset of the output measurements/commands* *with a subset of the manipulated inputs. These subsets should not be used by any other controller.*

This definition of decentralized control is consistent with its use by the control community. In decentralized control we may rearrange the ordering of measurements/commands and manipulated inputs such that the feedback part of the overall controller K in (10.17) has a fixed block-diagonal structure.

Cascade control *is when the output from one controller is the input to another. This is broader than the conventional definition of cascade control which is that the output from one controller is the reference command (setpoint) to another.*

Feedforward elements *link measured disturbances and manipulated inputs.*

Decoupling elements *link one set of manipulated inputs ("measurements") with another set of manipulated inputs. They are used to improve the performance of decentralized control systems, and are often viewed as feedforward elements (although this is not correct when we view the control system as a whole) where the "measured disturbance" is the manipulated input computed by another decentralized controller.*

Selectors *are used to select for control, depending on the conditions of the system, a subset of the manipulated inputs or a subset of the outputs.*

In addition to restrictions on the structure of K, we may impose restrictions on the way, or rather in which *sequence*, the subcontrollers are designed. For most decomposed control systems we design the controllers sequentially, starting with the "fast" or "inner" or "lower-layer" control loops in the control hierarchy. In particular, this is relevant for cascade control systems, and it is sometimes also used in the design of decentralized control systems.

The choice of control configuration leads to two different ways of partitioning the control system:

- *Vertical decomposition.* This usually results from a sequential design of the control system, e.g. based on cascading (series interconnecting) the controllers in a hierarchical manner.
- *Horizontal decomposition.* This usually involves a set of independent decentralized controllers.

Remark 1 Sequential design of a decentralized controller results in a control system which is decomposed both horizontally (since K is diagonal) as well as vertically (since controllers at higher layers are tuned with lower-layer controllers in place).

Remark 2 Of course, a performance loss is inevitable if we decompose the control system. For example, for a hierarchical decentralized control system, if we select a poor configuration at the lower (base) control layer, then this may pose fundamental limitations on the achievable performance which cannot be overcome by advanced controller designs at higher layers. These limitations imposed by the lower-layer controllers may include RHP-zeros (see the aero-engine case study in Chapter 12) or strong interactions (see the distillation case study in Chapter 12 where the LV-configuration yields large RGA-elements at low frequencies).

In this section, we discuss cascade controllers and selectors, and give some justification for using such "suboptimal" configurations rather than directly designing the overall controller K. Later, in Section 10.7, we discuss in more detail the hierarchical decomposition, including cascade control, partially controlled systems and sequential controller design. Finally, in Section 10.8 we consider decentralized diagonal control.

10.6.1 Cascade control systems

We want to illustrate how a control system which is decomposed into subcontrollers can be used to solve multivariable control problems. For simplicity, we here use single-input single-output (SISO) controllers of the form

$$u_i = K_i(s)(r_i - y_i) \qquad (10.18)$$

where $K_i(s)$ is a scalar. Note that whenever we close a SISO control loop we lose the corresponding input, u_i, as a degree of freedom, but at the same time the reference, r_i, becomes a new degree of freedom.

It may look like it is not possible to handle non-square systems with SISO controllers. However, since the input to the controller in (10.18) is a reference minus a measurement, we can cascade controllers to make use of extra measurements or extra inputs. A *cascade control structure* results when either of the following two situations arise:

- The reference r_i is an output from another controller (typically used for the case of an extra measurement y_i), see Figure 10.3(a). This is *conventional cascade control*.

- The "measurement" y_i is an output from another controller (typically used for the case of an extra manipulated input u_j; e.g. in Figure 10.3(b) where u_2 is the "measurement" for controller K_1). This cascade scheme is referred to as *input resetting*.

10.6.2 Cascade control: Extra measurements

In many cases we make use of extra measurements y_2 (*secondary outputs*) to provide local disturbance rejection and linearization, or to reduce the effect of measurement noise. For example, velocity feedback is frequently used in mechanical systems, and local flow cascades are used in process systems. Let u be the manipulated input, y_1 the controlled output (with an associated control objective r_1) and y_2 the extra measurement.

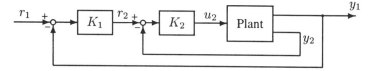

(a) Extra measurements y_2 (conventional cascade control)

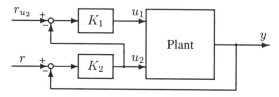

(b) Extra inputs u_2 (input resetting)

Figure 10.3: Cascade implementations

Centralized (parallel) implementation. A centralized implementation $u = K(r - y)$, where K is a 2-input-1-output controller, may be written

$$u = K_{11}(s)(r_1 - y_1) + K_{21}(s)(r_2 - y_2) \tag{10.19}$$

where in most cases $r_2 = 0$ (since we do not have a degree of freedom to control y_2).

Cascade implementation (conventional cascade control). To obtain an implementation with two SISO controllers we may cascade the controllers as illustrated in Figure 10.3(a):

$$r_2 = K_1(s)(r_1 - y_1), \tag{10.20}$$

$$u_2 = K_2(s)(r_2 - y_2), \; r_2 = \widehat{u}_1 \tag{10.21}$$

Note that the output r_2 from the slower *primary* controller K_1 is not a manipulated plant input, but rather the reference input to the faster *secondary* (or slave) controller K_2. For example, cascades based on measuring the actual manipulated variable (in which case $y_2 = u_m$) are commonly used to reduce uncertainty and nonlinearity at the plant input.

With $r_2 = 0$ in (10.19) the relationship between the centralized and cascade implementation is $K_{11} = K_2 K_1$ and $K_{21} = K_2$.

An advantage with the cascade implementation is that it more clearly decouples the design of the two controllers. It also shows more clearly that r_2 is not a degree-of-freedom at higher layers in the control system. Finally, it allows for integral action in both loops (whereas usually only K_{11} should have integral action in (10.19)).

On the other hand, a centralized implementation is better suited for direct multivariable synthesis; see the velocity feedback for the helicopter case study in Section 12.2.

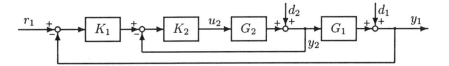

Figure 10.4: Common case of cascade control where the primary output y_1 depends directly on the extra measurement y_2.

Remark. Consider conventional cascade control in Figure 10.3(a). In the general case y_1 and y_2 are not directly related to each other, and this is sometimes referred to as *parallel cascade control*. However, it is common to encounter the situation in Figure 10.4 where y_1 depends directly on y_2. This is a special case of Figure 10.3(a) with "Plant"= $\begin{bmatrix} G_1 G_2 \\ G_2 \end{bmatrix}$, and it is considered further in Example 10.1.

Exercise 10.1 Conventional cascade control. *With reference to the special (but common) case of conventional cascade control shown in Figure 10.4, Morari and Zafiriou (1989) conclude that the use of extra measurements is useful under the following circumstances:*

(a) The disturbance d_2 is significant and G_1 is non-minimum phase.

(b) The plant G_2 has considerable uncertainty associated with it – e.g. a poorly known nonlinear behaviour – and the inner loop serves to remove the uncertainty.

In terms of design they recommended that K_2 is first designed to minimize the effect of d_2 on y_1 (with $K_1 = 0$) and then K_1 is designed to minimize the effect of d_1 on y_1. We want to derive conclusions (a) and (b) from an input-output controllability analysis, and also, (c) explain why we may choose to use cascade control if we want to use simple controllers (even with $d_2 = 0$).

Outline of solution: *(a) Note that if G_1 is minimum phase, then the input-output controllability of G_2 and $G_1 G_2$ are in theory the same, and for rejecting d_2 there is no fundamental advantage in measuring y_1 rather than y_2. (b) The inner loop $L_2 = G_2 K_2$ removes the uncertainty if it is sufficiently fast (high gain feedback) and yields a transfer function $(I + L_2)^{-1} L_2$ close to I at frequencies where K_1 is active. (c) In most cases, such as when PID controllers are used, the practical bandwidth is limited by the frequency w_u where the phase of the plant is $-180°$ (see section 5.12), so an inner cascade loop may yield faster control (for rejecting d_1 and tracking r_1) if the phase of G_2 is less than that of $G_1 G_2$.*

Exercise 10.2 *To illustrate the benefit of using inner cascades for high-order plants, case (c) in the above example, consider Figure 10.4 and let*

$$G_1 = \frac{1}{(s+1)^2}, \quad G_2 = \frac{1}{s+1}$$

We use a fast proportional controller $K_2 = 25$ in the inner loop, whereas a somewhat slower PID-controller is used in the outer loop,

$$K_1(s) = K_c \frac{(s+1)^2}{s(0.1s+1)}, \quad K_c = 5$$

Sketch the closed-loop response. What is the bandwidth for each of the two loops?

Compare this with the case where we only measure y_1, so $G = G_1 G_2$, and use a PID-controller $K(s)$ with the same dynamics as $K_1(s)$ but with a smaller value of K_c. What is the achievable bandwidth? Find a reasonable value for K_c (starting with $K_c = 1$) and sketch the closed-loop response (you will see that it is about a factor 5 slower without the inner cascade).

10.6.3   Cascade control: Extra inputs

In some cases we have more manipulated inputs than controlled outputs. These may be used to improve control performance. Consider a plant with a single controlled output y and two manipulated inputs u_1 and u_2. Sometimes u_2 is an extra input which can be used to improve the fast (transient) control of y, but if it does not have sufficient power or is too costly to use for long-term control, then after a while it is reset to some desired value ("ideal resting value").

Centralized (parallel) implementation. A centralized implementation $u = K(r - y)$ where K is a 1-input 2-output controller, may be written

$$u_1 = K_{11}(s)(r - y), \quad u_2 = K_{21}(s)(r - y) \tag{10.22}$$

Here two inputs are used to control one output, so to get a unique steady-state for the inputs u_1 and u_2, K_{11} has integral control but K_{21} does not. Then $u_2(t)$ will only be used for transient (fast) control and will return to zero (or more precisely to its desired value r_{u_2}) as $t \to \infty$.

Cascade implementation (input resetting). To obtain an implementation with two SISO controllers we may cascade the controllers as shown in Figure 10.3(b). We again let input u_2 take care of the fast control and u_1 of the long-term control. The fast control loop is then

$$u_2 = K_2(s)(r - y) \tag{10.23}$$

The objective of the other slower controller is then to use input u_1 to reset input u_2 to its desired value r_{u_2}:

$$u_1 = K_1(s)(r_{u_2} - y_1), \quad y_1 = u_2 \tag{10.24}$$

and we see that the output from the fast controller K_2 is the "measurement" for the slow controller K_1.

With $r_{u_2} = 0$ the relationship between the centralized and cascade implementation is $K_{11} = -K_1 K_2$ and $K_{21} = K_2$.

The cascade implementation again has the advantage of decoupling the design of the two controllers. It also shows more clearly that r_{u_2}, the reference for u_2, may be used as a degree-of-freedom at higher layers in the control system. Finally, we can have integral action in both K_1 and K_2, but note that the gain of K_1 should be negative (if effects of u_1 and u_2 on y are both positive).

Remark 1 Typically, the controllers in a cascade system are tuned one at a time starting with the fast loop. For example, for the control system in Figure 10.5 we would probably tune the three controllers in the order K_2 (inner cascade using fast input), K_3 (input resetting using slower input), and K_1 (final adjustment of y_1).

Remark 2 In process control, the cascade implementation of input resetting is sometimes referred to as *valve position control*, because the extra input u_2, usually a valve, is reset to a desired position by the outer cascade.

Exercise 10.3 *Draw the block diagrams for the two centralized (parallel) implementations corresponding to Figure 10.3.*

Exercise 10.4 *Derive the closed-loop transfer functions for the effect of r on y, u_1 and u_2 for the cascade input resetting scheme in Figure 10.3(b). As an example use $G = [G_{11} \quad G_{12}] = [1 \quad 1]$ and use integral action in both controllers, $K_1 = -1/s$ and $K_2 = 10/s$. Show that input u_2 is reset at steady-state.*

Example 10.4 **Two layers of cascade control.** *Consider the system in Figure 10.5 with two manipulated inputs (u_2 and u_3), one controlled output (y_1 which should be close to r_1) and two measured variables (y_1 and y_2). Input u_2 has a more direct effect on y_1 than does input u_3 (there is a large delay in $G_3(s)$). Input u_2 should only be used for transient control as it is desirable that it remains close to $r_3 = r_{u_2}$. The extra measurement y_2 is closer than y_1 to the input u_2 and may be useful for detecting disturbances (not shown) affecting G_1.*

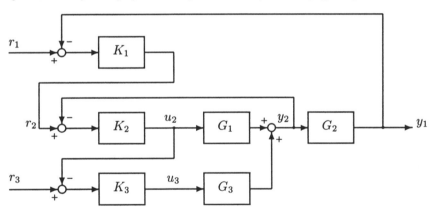

Figure 10.5: Control configuration with two layers of cascade control.

In Figure 10.5 controllers K_1 and K_2 are cascaded in a conventional manner, whereas controllers K_2 and K_3 are cascaded to achieve input resetting. The corresponding equations are

$$\widehat{u}_1 = K_1(s)(r_1 - y_1) \tag{10.25}$$

$$u_2 = K_2(s)(r_2 - y_2), \quad r_2 = \widehat{u}_1 \tag{10.26}$$

$$u_3 = K_3(s)(r_3 - y_3), \quad y_3 = u_2 \tag{10.27}$$

Controller K_1 controls the primary output y_1 at its reference r_1 by adjusting the "input" $\widehat{u}_1$, which is the reference value for y_2. Controller K_2 controls the secondary output y_2 using input

u_2. Finally, controller K_3 manipulates u_3 slowly in order to reset input u_2 to its desired value r_3.

Exercise 10.5 Process control application. *A practical case of a control system like the one in Figure 10.5 is in the use of a pre-heater to keep the reactor temperature y_1 at a given value r_1. In this case y_2 may be the outlet temperature from the pre-heater, u_2 the bypass flow (which should be reset to r_3, say 10% of the total flow), and u_3 the flow of heating medium (steam). Make a process flowsheet with instrumentation lines (not a block diagram) for this heater/reactor process.*

10.6.4 Extra inputs and outputs (local feedback)

In many cases performance may be improved with local feedback loops involving extra manipulated inputs and extra measurements. However, the improvement must be traded off against the cost of the extra actuator, measurement and control system. An example where local feedback is required to counteract the effect of high-order lags is given for a neutralization process in Figure 5.24 on page 209. The use of local feedback is also discussed by Horowitz (1991).

10.6.5 Selectors

Split-range control for extra inputs. We assumed above that the extra input is used to improve dynamic performance. Another situation is when input constraints make it necessary to add a manipulated input. In this case the control range is often split such that, for example, u_1 is used for control when $y \in [y_{\min}, y_1]$, and u_2 is used when $y \in [y_1, y_{\max}]$.

Selectors for too few inputs. A completely different situation occurs if there are too few inputs. Consider the case with one input (u) and several outputs $(y_1, y_2, \ldots)$. In this case, we cannot control all the outputs independently, so we either need to control all the outputs in some average manner, or we need to make a choice about which outputs are the most important to control. Selectors or logic switches are often used for the latter. *Auctioneering selectors* are used to decide to control one of several similar outputs. For example, this may be used to adjust the heat input (u) to keep the maximum temperature $(\max_i y_i)$ in a fired heater below some value. *Override selectors* are used when several controllers compute the input value, and we select the smallest (or largest) as the input. For example, this is used in a heater where the heat input (u) normally controls temperature (y_1), except when the pressure (y_2) is too large and pressure control takes over.

10.6.6 Why use cascade and decentralized control?

As is evident from Figure 10.5(a), decomposed control configurations can easily become quite complex and difficult to maintain and understand. It may therefore be both simpler and better in terms of control performance to set up the controller design problem as an optimization problem and let the computer do the job, resulting in a centralized multivariable controller as used in other chapters of this book.

If this is the case, why is cascade and decentralized control used in practice? There are a number of reasons, but the most important one is probably the cost associated with obtaining good plant models, which are a prerequisite for applying multivariable control. On the other hand, with cascade and decentralized control each controller is usually tuned one at a time with a minimum of modelling effort, sometimes even *on-line* by selecting only a few parameters (e.g, the gain and integral time constant of a PI-controller). *A fundamental reason for applying cascade and decentralized control is thus to save on modelling effort.* Since cascade and decentralized control systems depend more strongly on feedback rather than models as their source of information, it is usually more important (relative to centralized multivariable control) that the fast control loops be tuned to respond quickly.

Other advantages of cascade and decentralized control include the following: they are often easier to understand by operators, they reduce the need for control links and allow for decentralized implementation, their tuning parameters have a direct and "localized" effect, and they tend to be less sensitive to uncertainty, for example, in the input channels. The issue of simplified implementation and reduced computation load is also important in many applications, but is becoming less relevant as the cost of computing power is reduced.

Based on the above discussion, the main challenge is to find a *control configuration* which allows the (sub)controllers to be tuned independently based on a minimum of model information (the pairing problem). For industrial problems, the number of possible pairings is usually very high, but in most cases physical insight and simple tools, such as the RGA, can be helpful in reducing the number of alternatives to a manageable number. To be able to tune the controllers independently, we must require that the loops interact only to a limited extent. For example, one desirable property is that the steady-state gain from u_i to y_i in an "inner" loop (which has already been tuned), does not change too much as outer loops are closed. For decentralized diagonal control the RGA is a useful tool for addressing this pairing problem.

Why do we need a theory for cascade and decentralized control? We just argued that the main advantage of decentralized control was its saving on the modelling effort, but any theoretical treatment of decentralized control requires a plant model. This seems to be a contradiction. However, even though we may *not* want to use a model to tune the controllers, we may still want to use a model to decide on a control structure and to decide on whether acceptable control with a decentralized configuration is possible. The modelling effort in this case is less, because the model may be of a more "generic" nature and does not need to be modified for each particular application.

10.7   Hierarchical and partial control

A hierarchical control system results when we design the subcontrollers in a sequential manner, usually starting with the fast loops ("bottom-up"). This means that the controller at some higher layer in the hierarchy is designed based on a partially controlled plant. In this section we derive transfer functions for partial control, and provide some guidelines for designing hierarchical control systems.

10.7.1   Partial control

Partial control involves controlling only a subset of the outputs for which there is a control objective. We divide the outputs into two classes:

- y_1 – (temporarily) uncontrolled output (for which there is an associated control objective)
- y_2 – (locally) measured and controlled output

We also subdivide the available manipulated inputs in a similar manner:

- u_2 – inputs used for controlling y_2
- u_1 – remaining inputs (which *may* be used for controlling y_1)

We have inserted the word *temporarily* above, since y_1 is normally a controlled output at some higher layer in the hierarchy. However, we here consider the partially controlled system as it appears after having implemented only a local control system where u_2 is used to control y_2. In most of the development that follows we assume that the outputs y_2 are tightly controlled.

Four applications of partial control are:

1. *Sequential design of decentralized controllers.* The outputs y (which include y_1 and y_2) all have an associated control objective, and we use a hierarchical control system. We first design a controller K_2 to control the subset y_2. With this controller K_2 in place (a partially controlled system), we may then design a controller K_1 for the remaining outputs.
2. *Sequential design of conventional cascade control.* The outputs y_2 are additional measured ("secondary") variables which are not important variables in themselves. The reason for controlling y_2 is to improve the control of y_1. The references r_2 are used as degrees of freedom for controlling y_1 so the set u_1 is often empty.
3. *"True" partial control.* The outputs y (which include y_1 and y_2) all have an associated control objective, and we consider whether by controlling only the subset y_2; we can indirectly achieve acceptable control of y_1, that is, the outputs y_1 remain uncontrolled and the set u_1 remains unused.
4. *Indirect control.* The outputs y_1 have an associated control objective, but they are not measured. Instead, we aim at indirectly controlling y_1 by controlling the

"secondary" measured variables y_2 (which have no associated control objective). The references r_2 are used as degrees of freedom and the set u_1 is empty. This is similar to cascade control, but there is no "outer" loop involving y_1. Indirect control was discussed in Section 10.3.

The following table shows more clearly the difference between the four applications of partial control. In all cases there is a control objective associated with y_1 and a measurement of y_2.

	Measurement of y_1 ?	Control objective for y_2 ?
Sequential decentralized control	Yes	Yes
Sequential cascade control	Yes	No
"True" partial control	No	Yes
Indirect control	No	No

The four problems are closely related, and in all cases we want the effect of the disturbances on y_1 to be small, when y_2 is controlled. Let us derive the transfer functions for y_1 when y_2 is controlled. One difficulty is that this requires a separate analysis for each choice of y_2 and u_2, and the number of alternatives has a combinatorial growth as illustrated by (10.12).

By partitioning the inputs and outputs, the overall model $y = Gu$ may be written

$$y_1 = G_{11}u_1 + G_{12}u_2 + G_{d1}d \tag{10.28}$$

$$y_2 = G_{21}u_1 + G_{22}u_2 + G_{d2}d \tag{10.29}$$

Assume now that feedback control $u_2 = K_2(r_2 - y_2)$ is used for the "secondary" subsystem involving u_2 and y_2. By eliminating u_2 and y_2, we then get the following model for the resulting partially controlled system:

$$\begin{aligned} y_1 &= \left(G_{11} - G_{12}K_2(I + G_{22}K_2)^{-1}G_{21}\right)u_1 + \\ &\quad \left(G_{d1} - G_{12}K_2(I + G_{22}K_2)^{-1}G_{d2}\right)d + \\ &\quad G_{12}K_2(I + G_{22}K_2)^{-1}r_2 \end{aligned} \tag{10.30}$$

Remark. (10.30) may be rewritten in terms of linear fractional transformations. For example, the transfer function from u_1 to y_1 is

$$F_l(G, -K_2) = G_{11} - G_{12}K_2(I + G_{22}K_2)^{-1}G_{21} \tag{10.31}$$

Perfect control of y_2. In some cases we can assume that the control of y_2 is very fast compared to the control of y_1, so that we effectively have that y_2 is perfectly controlled when considering the control of y_1. To obtain the model we may formally let $K_2 \to \infty$ in (10.30), but it is better to set $y_2 = r_2$ and solve for u_2 in (10.29) to get

$$u_2 = -G_{22}^{-1}G_{d2}d - G_{22}^{-1}G_{21}u_1 + G_{22}^{-1}r_2$$

We have here assumed that G_{22} is square and invertible, otherwise we can get the least-square solution by replacing G_{22}^{-1} by the pseudo-inverse, $G_{22}^{\dagger}$. On substituting this into (10.28) we get

$$y_1 = \underbrace{(G_{11} - G_{12}G_{22}^{-1}G_{21})}_{\triangleq P_u} u_1 + \underbrace{(G_{d1} - G_{12}G_{22}^{-1}G_{d2})}_{\triangleq P_d} d + \underbrace{G_{12}G_{22}^{-1}}_{\triangleq P_r} r_2 \quad (10.32)$$

where P_d is called the *partial disturbance gain*, which is the disturbance gain for a system under perfect partial control, and P_u is the effect of u_1 on y_1 with y_2 perfectly controlled. In many cases the set u_1 is empty (there are no extra inputs). The advantage of the model (10.32) over (10.30) is that it is independent of K_2, but we stress that it only applies at frequencies where y_2 is tightly controlled.

Remark. Relationships similar to those given in (10.32) have been derived by many authors, e.g. see the work of Manousiouthakis et al. (1986) on block relative gains and the work of Haggblom and Waller (1988) on distillation control configurations.

10.7.2 Hierarchical control and sequential design

A hierarchical control system arises when we apply a sequential design procedure to a cascade or decentralized control system.

The idea is to first implement a local *lower-layer* (or inner) control system for controlling the outputs y_2. Next, with this lower-layer control system in place, we design a controller K_1 to control y_1. The appropriate model for designing K_1 is given by (10.30) (for the general case) or (10.32) (for the case when we can assume y_2 perfectly controlled).

The objectives for this hierarchical decomposition may vary:

1. To allow for simple or even on-line tuning of the lower-layer control system (K_2).
2. To allow the use of longer sampling intervals for the higher layers (K_1).
3. To allow simple models when designing the higher-layer control system (K_1). The high-frequency dynamics of the models of the partially controlled plant (e.g. P_u and P_r) may be simplified if K_1 is mainly effective at lower frequencies.
4. To "stabilize"[1] the plant using a lower-layer control system (K_2) such that it is amenable to manual control.

The latter is the case in many process control applications where we first close a number of faster "regulatory" loops in order to "stabilize" the plant. The higher layer control system (K_1) is then used mainly for optimization purposes, and is not required to operate the plant.

Based on these objectives, Hovd and Skogestad (1993) proposed some criteria for selecting u_2 and y_2 for use in the lower-layer control system:

---

[1] The terms "stabilize" and "unstable" as used by process operators may not refer to a plant that is unstable in a mathematical sense, but rather to a plant that is *sensitive* to disturbances and which is difficult to control manually.

1. The lower layer must quickly implement the setpoints computed by the higher layers, that is, the input-output controllability of the subsystem involving use of u_2 to control y_2 should be good (consider G_{22} and G_{d2}).
2. The control of y_2 using u_2 should provide local disturbance rejection, that is, it should minimize the effect of disturbances on y_1 (consider P_d for y_2 tightly controlled).
3. The control of y_2 using u_2 should not impose unnecessary control limitations on the remaining control problem which involves using u_1 and/or r_2 to control y_1. By "unnecessary" we mean limitations (RHP-zeros, ill-conditioning, etc.) that did not exist in the original overall problem involving u and y. Consider the controllability of P_u for y_2 tightly controlled, which should not be much worse than that of G.

These three criteria are important for selecting control configurations for distillation columns as is discussed in the next example.

Example 10.5 **Control configurations for distillation columns.** *The overall control problem for the distillation column in Figure 10.6 has 5 inputs*

$$u = [\, L \quad V \quad D \quad B \quad V_T \,]^T$$

(these are all flows: reflux L, boilup V, distillate D, bottom flow B, overhead vapour V_T) and 5 outputs

$$y = [\, y_D \quad x_B \quad M_D \quad M_B \quad p \,]^T$$

(these are compositions and inventories: top composition y_D, bottom composition x_B, condenser holdup M_D, reboiler holdup M_B, pressure p) see Figure 10.6. This problem usually has no inherent control limitations caused by RHP-zeros, but the plant has poles in or close to the origin and needs to be stabilized. In addition, for high-purity separations the 5×5 RGA-matrix may have some large elements. Another complication is that composition measurements are often expensive and unreliable.

In most cases, the distillation column is first stabilized by closing three decentralized SISO loops for level and pressure so

$$y_2 = [\, M_D \quad M_B \quad p \,]^T$$

and the remaining outputs are

$$y_1 = [\, y_D \quad x_B \,]^T$$

The three SISO loops for controlling y_2 usually interact weakly and may be tuned independently of each other. However, since each level (tank) has an inlet and two outlet flows, there exists many possible choices for u_2 (and thus for u_1). By convention, each choice ("configuration") is named by the inputs u_1 left for composition control.

For example, the "LV-configuration" used in many examples in this book refers to a partially controlled system where we use

$$u_1 = [\, L \quad V \,]^T$$

to control y_1 (and we assume that there is a control system in place which uses $u_2 = [\, D \quad B \quad V_T \,]^T$ to control y_2). The LV-configuration is good from the point of view that

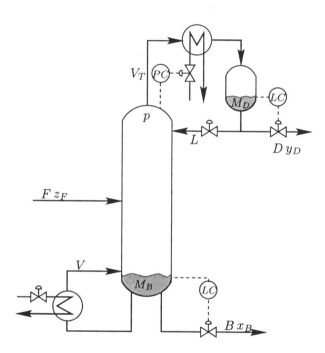

Figure 10.6: Typical distillation column controlled with the LV-configuration

control of y_1 using u_1 is nearly independent of the tuning of the controller K_2 involving y_2 and u_2. However, the problem of controlling y_1 by u_1 ("plant" P_u) is often strongly interactive with large steady-state RGA-elements in P_u.

Another configuration is the DV-configuration where

$$u_1 = [\, D \quad V \,]^T$$

and thus $u_2 = [\, L \quad B \quad V_T \,]^T$. In this case, the steady-state interactions from u_1 to y_1 are generally much less, and P_u has small RGA-elements. But the model in (10.30) depends strongly on K_2 (i.e. on the tuning of the level loops), and a slow level loop for M_D may introduce unfavourable dynamics for the response from u_1 to y_1.

There are also many other possible configurations (choices for the two inputs in u_1); with five inputs there are 10 alternative configurations. Furthermore, one often allows for the possibility of using ratios between flows, e.g. L/D, as possible degrees of freedom in u_1, and this sharply increases the number of alternatives.

Expressions which directly relate the models for various configurations, e.g. relationships between P_u^{LV}, P_d^{LV} and P_u^{DV}, P_d^{DV} etc., are given in Haggblom and Waller (1988) and Skogestad and Morari (1987a). However, it may be simpler to start from the overall 5×5 model G, and derive the models for the configurations using (10.30) or (10.32), see also the MATLAB file on page 491.

To select a good distillation control configuration, one should first consider the problem of controlling levels and pressure (y_2). This eliminates a few alternatives, so the final choice is based on the 2×2 composition control problem (y_1). If y_2 is tightly controlled then none of the configurations seem to yield RHP-zeros in P_u. Important issues to consider then are disturbance sensitivity (the partial disturbance gain P_d should be small) and the interactions (the RGA-elements of P_u). These issues are discussed by, for example, Waller et al. (1988) and Skogestad et al. (1990). Another important issue is that it is often not desirable to have tight level loops and some configurations, like the DV-configuration mentioned above, are sensitive to the tuning of K_2. Then the expressions for P_u and P_d, which are used in the references mentioned above, may not apply. This is further discussed in Skogestad (1992).

Because of the problems of interactions and the high cost of composition measurements, we often find in practice that only one of the two product compositions is controlled ("true" partial control). This is discussed in detail in Example 10.7 below. Another common solution is to make use of additional temperature measurements from the column, where their reference values are set by a composition controller in a cascade manner.

In summary, the overall 5×5 distillation control problem is solved by first designing a 3×3 controller K_2 for levels and pressure, and then designing a 2×2 controller K_1 for the composition control. This is then a case of (block) decentralized control where the controller blocks K_1 and K_2 are designed sequentially (in addition, the blocks K_1 and K_2 may themselves be decentralized).

Sequential design is also used for the design of cascade control systems. This is discussed next.

Sequential design of cascade control systems

Consider the conventional cascade control system in Figure 10.3(a), where we have additional "secondary" measurements y_2 with no associated control objective, and the objective is to improve the control of the primary outputs y_1 by locally controlling y_2. The idea is that this should reduce the effect of disturbances and uncertainty on y_1.

From (10.32), it follows that we should select secondary measurements y_2 (and inputs u_2) such that $\|P_d\|$ is small and at least smaller than $\|G_{d1}\|$. In particular, these arguments apply at higher frequencies. Furthermore, it should be easy to control y_1 by using as degrees of freedom the references r_2 (for the secondary outputs) or the unused inputs u_1. More precisely, we want the input-output controllability of the "plant" $[\,P_u \quad P_r\,]$ (or P_r if the set u_1 is empty) with disturbance model P_d, to be better than that of the plant $[\,G_{11} \quad G_{12}\,]$ (or G_{12}) with disturbance model G_{d1}.

Remark. Most of the arguments given in Section 10.2, for the separation into an optimization and a control layer, and in Section 10.3, for the selection of controlled outputs, apply to cascade control if the term "optimization layer" is replaced by "primary controller", and "control layer" is replaced by "secondary controller".

Exercise 10.6 *The block diagram in Figure 10.4 shows a cascade control system where the primary output y_1 depends directly on the extra measurement y_2, so $G_{12} = G_1 G_2$, $G_{22} = G_2$, $G_{d1} = [\,I \quad G_1\,]$ and $G_{d2} = [\,0 \quad I\,]$. Show that $P_d = [\,I \quad 0\,]$ and $P_r = G_1$ and discuss the result. Note that P_r is the "new" plant as it appears with the inner loop closed.*

10.7.3 "True" partial control

We here consider the case where we attempt to leave a set of primary outputs y_1 uncontrolled. This "true" partial control may be possible in cases where the outputs are correlated such that controlling the outputs y_2 indirectly gives acceptable control of y_1. One justification for partial control is that measurements, actuators and control links cost money, and we therefore prefer control schemes with as few control loops as possible.

To analyze the feasibility of partial control, consider the effect of disturbances on the *uncontrolled* output(s) y_1 as given by (10.32). Suppose all variables have been scaled as discussed in Section 1.4. Then we have that:

- *A set of outputs y_1 may be left uncontrolled only if the effects of all disturbances on y_1, as expressed by the elements in the corresponding partial disturbance gain matrix P_d, are less than 1 in magnitude at all frequencies.*

Therefore, to evaluate the feasibility of partial control one must for each choice of controlled outputs (y_2) and corresponding inputs (u_2), rearrange the system as in (10.28) and (10.29) and compute P_d using (10.32).

There may also be changes in r_2 (of magnitude R_2) which may be regarded as disturbances on the uncontrolled outputs y_1. From (10.32) then, we also have that:

- *A set of outputs y_1 may be left uncontrolled only if the effects of all reference changes in the controlled outputs (y_2) on y_1, as expressed by the elements in the matrix $G_{12}G_{22}^{-1}R_2$, are less than 1 in magnitude at all frequencies.*

One uncontrolled output and one unused input. "True" partial control is often considered if we have an $m \times m$ plant $G(s)$ where acceptable control of all m outputs is difficult, and we consider leaving input u_j unused and output y_i uncontrolled. In this case, instead of rearranging y into $\begin{bmatrix} y_1 \\ y_2 \end{bmatrix}$ and u into $\begin{bmatrix} u_1 \\ u_2 \end{bmatrix}$ for each candidate control configuration, we may directly evaluate the partial disturbance gain based on the overall model $y = Gu + G_d d$. The effect of a disturbance d_k on the uncontrolled output y_i is

$$P_{d_k} = \left(\frac{\partial y_i}{\partial d_k} \right)_{u_j=0,\, y_{l\neq i}=0} = \frac{[G^{-1}G_d]_{jk}}{[G^{-1}]_{ji}} \tag{10.33}$$

where "$u_j = 0, y_{l\neq i} = 0$" means that input u_j is constant and the remaining outputs $y_{l\neq i}$ are constant.

Proof of (10.33): The proof is from (Skogestad and Wolff, 1992). Rewrite $y = Gu + [G_d]_k d_k$ as $u = G^{-1}y - [G^{-1}]_k G_d d$. Set $y_l = 0$ for all $\neq i$. Then $u_j = [G^{-1}]_{ji} y_i - [G^{-1}G_d]_{jk} d_k$ and by setting $u_j = 0$ we find $y_i/d_k = [G^{-1}G_d]_{jk}/[G^{-1}]_{ji}$.   □

From (10.33) we derive direct insight into how to select the uncontrolled output and unused input:

1. Select the unused input u_j such that the j'th row in $G^{-1}G_d$ has small elements. That is, keep the input constant (unused) if its desired change is small.
2. Select the uncontrolled output y_i and unused input u_j such that the ji'th element in G^{-1} is large. That is, keep an output uncontrolled if it is insensitive to changes in the unused input with the other outputs controlled.

Example 10.6 *Consider the FCC process in Exercise 6.16 with*

$$G(0) = \begin{bmatrix} 16.8 & 30.5 & 4.30 \\ -16.7 & 31.0 & -1.41 \\ 1.27 & 54.1 & 5.40 \end{bmatrix}, \quad G^{-1}(0) = \begin{bmatrix} 0.09 & 0.02 & -0.06 \\ 0.03 & 0.03 & -0.02 \\ -0.34 & -0.32 & 0.38 \end{bmatrix}$$

where we want to leave one input unused and one output uncontrolled. From the second rule, since all elements in the third row of G^{-1} are large, it seems reasonable to let input u_3 be unused, as is done in Exercise 6.16. (The outputs are mainly selected to avoid the presence of RHP-zeros, see Exercise 6.16).

We next consider a 2×2 distillation process where it is difficult to control both outputs independently due to strong interactions, and we leave one output (y_1) uncontrolled. To improve the performance of y_1 we also consider the use of feedforward control where u_1 is adjusted based on measuring the disturbance (but we need no measurement of y_1).

Example 10.7 Partial and feedforward control of 2×2 **distillation process.** *Consider a distillation process with 2 inputs (reflux L and boilup V), 2 outputs (product compositions y_D and x_B) and 2 disturbances (feed flowrate F and feed composition z_F). We assume that changes in the reference (r_1 and r_2) are infrequent and they will not be considered. At steady-state ($s = 0$) we have*

$$G = \begin{bmatrix} 87.8 & -86.4 \\ 108.2 & -109.6 \end{bmatrix}, \ G_d = \begin{bmatrix} 7.88 & 8.81 \\ 11.72 & 11.19 \end{bmatrix}, \ G^{-1}G_d = \begin{bmatrix} -0.54 & -0.005 \\ -0.64 & -0.107 \end{bmatrix}$$
$$(10.34)$$

Since the row elements in $G^{-1}G_d$ are similar in magnitude as are also the elements of G^{-1} (between 0.3 and 0.4), the rules following (10.33) do not clearly favour any particular partial control scheme. This is confirmed by the values of P_d, which are seen to be quite similar for the four candidate partial control schemes:

$$P_{d1}^{2,2} = \begin{bmatrix} -1.36 \\ -0.011 \end{bmatrix}^T, \ P_{d1}^{2,1} = \begin{bmatrix} -1.63 \\ -0.27 \end{bmatrix}^T, \ P_{d2}^{1,2} = \begin{bmatrix} 1.72 \\ 0.014 \end{bmatrix}^T, \ P_{d2}^{1,1} = \begin{bmatrix} 2.00 \\ 0.33 \end{bmatrix}^T$$

The superscripts denote the controlled output and corresponding input. Importantly, in all four cases, the magnitudes of the elements in P_d are much smaller than in G_d, so control of one output significantly reduces the effect of the disturbances on the uncontrolled output. In particular, this is the case for disturbance 2, for which the gain is reduced from about 10 to 0.33 and less.

Let us consider in more detail scheme 1 which has the smallest disturbance sensitivity for the uncontrolled output ($P_{d1}^{2,2}$). This scheme corresponds to controlling output y_2 (the bottom composition) using u_2 (the boilup V) and with y_1 (the top composition) uncontrolled. We use a dynamic model which includes liquid flow dynamics; the model is given in Section 12.4. Frequency-dependent plots of G_d and P_d show that the conclusion at steady state also applies at higher frequencies. This is illustrated in Figure 10.7, where we show for the uncontrolled output y_1 and the worst disturbance d_1 both the open-loop disturbance gain (G_{d11}, Curve 1) and the partial disturbance gain ($P_{d11}^{2,2}$, Curve 2). For disturbance d_2 the partial disturbance gain (not shown) remains below 1 at all frequencies.

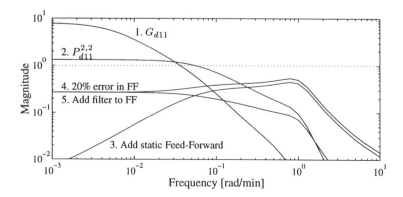

Figure 10.7: Effect of disturbance 1 on output 1 for distillation column example

The partial disturbance gain for disturbance d_1 (the feed flowrate F) is somewhat above 1 at low frequencies ($P_d(0) = -1.36$), so let us next consider how we may reduce its effect on y_1. One approach is to reduce the disturbance itself, for example, by installing a buffer tank (as in pH-example in Chapter 5.16.3). However, a buffer tank has no effect at steady-state, so it does not help in this case.

Another approach is to install a feedforward controller based on measuring d_1 and adjusting u_1 (the reflux L) which is so far unused. In practice, this is easily implemented as a ratio controller which keeps L/F constant. This eliminates the steady-state effect of d_1 on y_1 (provided the other control loop is closed). In terms of our linear model, the mathematical equivalence of this ratio controller is to use $u_1 = -0.54d_1$, where -0.54 is the $1,1$-element in $-G^{-1}G_d$. The effect of the disturbance after including this static feedforward controller is shown as curve 3 in Figure 10.7. However, due to measurement error we cannot achieve perfect feedforward control, so let us assume the error is 20%, and use $u_1 = -1.2 \cdot 0.54d_1$. The steady-state effect of the disturbance is then $P_d(0)(1 - 1.2) = 1.36 \cdot 0.2 = 0.27$, which is still acceptable. But, as seen from the frequency-dependent plot (curve 4), the effect is above 0.5 at higher frequencies, which may not be desirable. The reason for this undesirable peak is that the feedforward controller, which is purely static, reacts too fast, and in fact makes the response worse at higher frequencies (as seen when comparing curves 3 and 4 with curve 2). To avoid this we filter the feedforward action with a time constant of 3 min resulting in the following feedforward controller:

$$u_1 = -\frac{0.54}{3s+1}d_1 \qquad (10.35)$$

To be realistic we again assume an error of 20%. The resulting effect of the disturbance on the uncontrolled output is shown by curve 5, and we see that the effect is now less than 0.27 at all frequencies, so the performance is acceptable.

Remark. *In the example there are four alternative partial control schemes with quite similar disturbance sensitivity for the uncontrolled output. To decide on the best scheme, we should also perform a controllability analysis of the feedback properties of the four 1×1 problems. Performing such an analysis, we find that schemes 1 (the one chosen) and 4 are preferable, because the input in these two cases has a more direct effect on the output, and with less phase lag.*

In conclusion, for this example it is difficult to control both outputs simultaneously using feedback control due to strong interactions. However, we can almost achieve acceptable control of both outputs by leaving y_1 uncontrolled. The effect of the most difficult disturbance on y_1 can be further reduced using a simple feedforward controller (10.35) from disturbance d_1 to u_1.

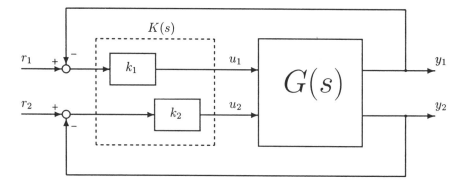

Figure 10.8: Decentralized diagonal control of a 2×2 plant

10.8 Decentralized feedback control

In this section, $G(s)$ is a square plant which is to be controlled using a diagonal controller (see Figure 10.8)

$$
K(s) = \mathrm{diag}\{k_i(s)\} = \begin{bmatrix} k_1(s) & & & \\ & k_2(s) & & \\ & & \ddots & \\ & & & k_m(s) \end{bmatrix} \tag{10.36}
$$

This is the problem of decentralized diagonal feedback control. The design of decentralized control systems involves two steps:

1. The choice of pairings (control configuration selection)
2. The design (tuning) of each controller, $k_i(s)$.

The optimal solution to this problem is very difficult mathematically, because the optimal controller is in general of infinite order and may be non-unique; we do not address it in this book. The reader is referred to the literature (e.g. Sourlas and Manousiouthakis, 1995) for more details. Rather we aim at providing simple tools for pairing selections (step 1) and for analyzing the achievable performance (controllability) of diagonally controlled plants (which may assist in step 2).

Remark. The treatment in this section may be generalized to block-diagonal controllers by, for example, introducing tools such as the block relative gain array of Manousiouthakis et al. (1986).

Notation for decentralized diagonal control. $G(s)$ denotes a square $m \times m$ plant with elements g_{ij}. $G^{ij}(s)$ denotes the remaining $(m-1) \times (m-1)$ plant obtained by removing row i and column j in $G(s)$. With a particular choice of pairing we can rearrange the columns or rows of $G(s)$ such that the paired elements are along the diagonal of $G(s)$. We then have that the controller $K(s)$ is diagonal (diag$\{k_i\}$), and we also introduce

$$\widetilde{G} \triangleq \text{diag}\{g_{ii}\} = \begin{bmatrix} g_{11} & & & \\ & g_{22} & & \\ & & \ddots & \\ & & & g_{mm} \end{bmatrix} \tag{10.37}$$

as the matrix consisting of the diagonal elements of G. The loop transfer function in loop i is denoted $L_i = g_{ii}k_i$, which is also equal to the i'th diagonal element of $L = GK$.

The magnitude of the off-diagonal elements in G (the interactions) relative to its diagonal elements are given by the matrix

$$E \triangleq (G - \widetilde{G})\widetilde{G}^{-1} \tag{10.38}$$

A very important relationship for decentralized control is given by the following factorization of the return difference operator:

$$\underbrace{(I + GK)}_{\text{overall}} = \underbrace{(I + E\widetilde{T})}_{\text{interactions}} \underbrace{(I + \widetilde{G}K)}_{\text{individual loops}} \tag{10.39}$$

or equivalently in terms of the sensitivity function $S = (I + GK)^{-1}$,

$$S = \widetilde{S}(I + E\widetilde{T})^{-1} \tag{10.40}$$

Here

$$\widetilde{S} \triangleq (I + \widetilde{G}K)^{-1} = \text{diag}\{\frac{1}{1 + g_{ii}k_i}\} \quad \text{and} \quad \widetilde{T} = I - \widetilde{S} \tag{10.41}$$

contain the sensitivity and complementary sensitivity functions for the individual loops. Note that $\widetilde{S}$ is *not* equal to the matrix of diagonal elements of S. (10.40) follows from (A.139) with $G = \widetilde{G}$ and $G' = G$. The reader is encouraged to confirm that (10.40) is correct, because most of the important results for stability and performance using decentralized control may be derived from this expression. An alternative factorization which follows from (A.140) is

$$S = (I + \widetilde{S}(\Gamma - I))^{-1}\widetilde{S}\Gamma \tag{10.42}$$

where Γ is the Performance Relative Gain Array (PRGA),

$$\Gamma(s) \triangleq \widetilde{G}(s)G^{-1}(s) \tag{10.43}$$

which is a scaled inverse of the plant. Note that $E = \Gamma^{-1} - I$. At frequencies where feedback is effective ($\widetilde{S} \approx 0$), (10.42) yields $S \approx \widetilde{S}\Gamma$ which shows that Γ is important

when evaluating performance with decentralized control. The diagonal elements of
the PRGA-matrix are equal to the diagonal elements of the RGA, $\gamma_{ii} = \lambda_{ii}$, and this
is the reason for its name. Note that the off-diagonal elements of the PRGA depend
on the relative scaling on the outputs, whereas the RGA is scaling independent. On
the other hand, the PRGA measures also one-way interaction, whereas the RGA only
measures two-way interaction.

We also will make use of the related Closed-Loop Disturbance Gain (CLDG)
matrix, defined as

$$\widetilde{G}_d(s) \triangleq \Gamma(s)G_d(s) = \widetilde{G}(s)G^{-1}(s)G_d(s) \tag{10.44}$$

The CLDG depends on both output and disturbance scaling.

10.8.1   RGA as interaction measure for decentralized control

We here follow Bristol (1966), and show that the RGA provides a measure of
the interactions caused by decentralized diagonal control. Let u_j and y_i denote a
particular input and output for the multivariable plant $G(s)$, and assume that our task
is to use u_j to control y_i. Bristol argued that there will be two extreme cases:

- Other loops open: All other inputs are constant, i.e. $u_k = 0, \forall k \neq j$.
- Other loops closed: All other outputs are constant, i.e. $y_k = 0, \forall k \neq i$.

In the latter case, it is assumed that the other loops are closed with perfect control.
Perfect control is only possible at steady-state, but it is a good approximation at
frequencies within the bandwidth of each loop. We now evaluate the effect $\partial y_i/\partial u_j$
of "our" given input u_j on "our" given output y_i for the two extreme cases. We get

$$\text{Other loops open:} \qquad \left(\frac{\partial y_i}{\partial u_j}\right)_{u_k=0, k\neq j} = g_{ij} \tag{10.45}$$

$$\text{Other loops closed:} \qquad \left(\frac{\partial y_i}{\partial u_j}\right)_{y_k=0, k\neq i} \triangleq \widehat{g}_{ij} \tag{10.46}$$

Here $g_{ij} = [G]_{ij}$ is the ij'th element of G, whereas $\widehat{g}_{ij}$ is the inverse of the ji'th
element of G^{-1}

$$\widehat{g}_{ij} = 1/[G^{-1}]_{ji} \tag{10.47}$$

To derive (10.47) note that

$$y = Gu \quad \Rightarrow \quad \left(\frac{\partial y_i}{\partial u_j}\right)_{u_k=0, k\neq j} = [G]_{ij} \tag{10.48}$$

and interchange the roles of G and G^{-1}, of u and y, and of i and j to get

$$u = G^{-1}y \quad \Rightarrow \quad \left(\frac{\partial u_j}{\partial y_i}\right)_{y_k=0, k\neq i} = [G^{-1}]_{ji} \tag{10.49}$$

and (10.47) follows. Bristol argued that the ratio between the gains in (10.45) and (10.46), corresponding to the two extreme cases, is a useful measure of interactions, and he introduced the term, ij'th relative gain defined as

$$\lambda_{ij} \triangleq \frac{g_{ij}}{\widehat{g}_{ij}} = [G]_{ij}[G^{-1}]_{ji} \qquad (10.50)$$

The Relative Gain Array (RGA) is the corresponding matrix of relative gains. From (10.50) we get $\Lambda(G) = G \times (G^{-1})^T$ where $\times$ denotes element-by-element multiplication (the Schur product). This is identical to our definition of the RGA-matrix in (3.69).

Clearly, we would like to pair variables u_j and y_i so that λ_{ij} is close to 1, because this means that the gain from u_j to y_i is unaffected by closing the other loops. On the other hand, a pairing corresponding to $\lambda_{ij}(0) < 0$ is clearly undesirable, because it means that the steady-state gain in "our" given loop changes sign when the other loops are closed. A more exact statement is given later in Theorem 10.4.

10.8.2 Stability of decentralized control systems

Consider a square plant with single-loop controllers. For a 2×2 plant there are two alternative pairings, a 3×3 plant offers 6, a 4×4 plant 24, and an $m \times m$ plant has $m!$ alternatives. Thus, tools are needed which are capable of quickly *eliminating* inappropriate control structures. In this section we provide two useful rules for pairing inputs and outputs:

1. To avoid instability caused by interactions in the crossover region one should *prefer* pairings for which the RGA-matrix in this frequency range is close to identity.
2. To avoid instability caused by interactions at low frequencies one should *avoid* pairings with negative steady-state RGA elements.

Sufficient conditions for stability

For decentralized diagonal control, it is desirable that the system can be tuned and operated one loop at a time. Assume therefore that G is stable and each individual loop is stable by itself ($\widetilde{S}$ and $\widetilde{T}$ are stable). Then from the factorization $S = \widetilde{S}(I + E\widetilde{T})^{-1}$ in (10.39) and Lemma A.5, which is a special case of the generalized Nyquist theorem, it follows that the overall system is stable (S is stable) if and only if $\det(I + E\widetilde{T}(s))$ does not encircle the origin as s traverses the Nyquist D-contour. From the spectral radius stability condition in (4.107) we then have that the overall system is stable if

$$\rho(E\widetilde{T}(j\omega)) < 1, \forall\omega \qquad (10.51)$$

This sufficient condition for overall stability can, as discussed by Grosdidier and Morari (1986), be used to obtain a number of even *weaker* stability conditions. We will consider three approaches.

1. We use the fact that $\rho(E\widetilde{T}) \leq \|E\widetilde{T}\| \leq \|E\| \|\widetilde{T}\|$ for any matrix norm, see (A.116). For example, we may use any induced norm, $\|\cdot\|_{ip}$, such as the maximum singular value ($p = 2$), maximum column ($p = 1$) or maximum row sum ($p = \infty$). A sufficient condition for overall stability is then

$$\|\widetilde{T}\|_{ip} = \max_i |\tilde{t}_i| < 1/\|E\|_{ip} \qquad (10.52)$$

2. A better (less conservative) approach is to split up $\rho(E\widetilde{T})$ using the structured singular value. From (8.92) we have $\rho(E\widetilde{T}) \leq \mu(E)\bar{\sigma}(T)$ and from (10.51) we get the following theorem (as first derived by Grosdidier and Morari, 1986):

Theorem 10.2 *Assume G is stable and that the individual loops are stable ($\widetilde{T}$ is stable). Then the entire system is closed-loop stable (T is stable) if*

$$\bar{\sigma}(\widetilde{T}) = \max_i |\tilde{t}_i| < 1/\mu(E) \quad \forall\omega \qquad (10.53)$$

Here $\mu(E)$ is the structured singular value interaction measure. It is desirable to have $\mu(E)$ small. Note that $\mu(E)$ is computed with respect to the diagonal structure of $\widetilde{T}$ (a diagonal matrix), where we may view $\widetilde{T}$ as the "design uncertainty".

Definition 10.1 *A plant is (generalized) diagonally dominant at frequencies where $\mu(E(j\omega)) < 1$.*

From (10.53) we can then allow $\bar{\sigma}(\widetilde{T}) \geq 1$ (tight control) at such frequencies and still be guaranteed closed-loop stability.

3. A third approach is to use Gershgorin's theorem, see page 502. From (10.51) we then derive the following sufficient condition for overall stability:

$$|\tilde{t}_i| < |g_{ii}|/\sum_{j\neq i} |g_{ij}| \quad \forall i, \forall\omega \qquad (10.54)$$

or alternatively, in terms of the columns,

$$|\tilde{t}_i| < |g_{ii}|/\sum_{j\neq i} |g_{ji}| \quad \forall i, \forall\omega \qquad (10.55)$$

Remark 1 We stress again that condition (10.53) is always less conservative than (10.52). This follows since: (1) $1/\mu(E)$ is *defined* to be the tightest upper bound on $\bar{\sigma}(\widetilde{T})$ which guarantees $\rho(E\widetilde{T}) \leq 1$ (see also (8.79) with $\Delta = \widetilde{T}$), and (2) $\widetilde{T}$ is diagonal so that $\bar{\sigma}(\widetilde{T}) = \|\widetilde{T}\|_{ip} = \max_i |\tilde{t}|_i$. In other words, $\mu(E) \leq \|E\|_{ip}$. Since $\mu(E) = \mu(DED^{-1})$, see (8.84) where D in this case is diagonal, it then follows from (A.127) that $\mu(E) \leq \rho(|E|)$, where $\rho(|E|)$ is the Perron root. (Note that these upper bounds on $\mu(E)$ are not general properties of the structured singular value).

Remark 2 On the other hand, we cannot say that (10.53) is always less conservative than (10.54) and (10.55). It is true that the *smallest* of the $i = 1, \ldots m$ upper bounds in (10.54) or (10.55) is always smaller (more restrictive) than $1/\mu(E)$ in (10.53). However, (10.53) imposes the *same* bound on $|\tilde{t}_i|$ for each loop, whereas (10.54) and (10.55) give *individual* bounds, some of which may be less restrictive than $1/\mu(E)$.

Remark 3 Another definition of generalized diagonal dominance is that $\rho(|E|) < 1$, where $\rho(|E|)$ is the Perron root; see (A.127). However, since as noted above $\mu(E) \leq \rho(|E|)$, it is better (less restrictive) to use $\mu(E)$ to define diagonal dominance.

Remark 4 Conditions (10.51)-(10.55) are only sufficient for stability, so we may in some cases get closed-loop stability even if they are violated; see also Remark 5 on page 440.

Remark 5 Condition (10.53) and the use of $\mu(E)$ for (nominal) stability of the decentralized control system can be generalized to include robust stability and robust performance; see equations (31a-b) in Skogestad and Morari (1989).

We now want to show that for closed-loop stability it is desirable to select pairings such that the RGA is close to the identity matrix in the crossover region. The next simple theorem, which applies to a triangular plant, will enable us to do this:

Theorem 10.3 *Suppose the plant $G(s)$ is stable. If the RGA-matrix $\Lambda(G) = I \;\forall \omega$ then stability of each of the individual loops implies stability of the entire system.*

Proof: From the definition of the RGA it follows that $\Lambda(G) = I$ can only arise from a triangular $G(s)$ or from $G(s)$-matrices that can be made triangular by interchanging rows and columns in such a way that the diagonal elements remain the same but in a different order (the pairings remain the same). A plant with a "triangularized" transfer matrix (as described above) controlled by a diagonal controller has only *one-way coupling* and will always yield a stable system provided the individual loops are stable. Mathematically, $E = (G - \tilde{G})\tilde{G}^{-1}$ can be made triangular, and since the diagonal elements of E are zero, it follows that all eigenvalues of $E\tilde{T}$ are zero, so $\rho(E\tilde{T}) = 0$ and (10.51) is satisfied.   □

RGA at crossover frequencies. In most cases, it is sufficient for overall stability to require that $G(j\omega)$ is close to triangular (or $\Lambda(G) \approx I$) at crossover frequencies. To see this, assume that $\tilde{S}$ is stable, and that $\tilde{S}\Gamma(s) = \tilde{S}\tilde{G}(s)G(s)^{-1}$ is stable and has no RHP-zeros (which is always satisfied if both G and $\tilde{G}$ are stable and have no RHP-zeros). Then from (10.42) the overall system is stable (S is stable) if and only if $(I + \tilde{S}(\Gamma - I))^{-1}$ is stable. Here $\tilde{S}(\Gamma - I)$ is stable, so from the spectral radius stability condition in (4.107) the overall system is stable if

$$\rho(\tilde{S}(\Gamma - I)(j\omega)) < 1, \quad \forall \omega \tag{10.56}$$

At low frequencies, this condition is usually satisfied because $\tilde{S}$ is small. At higher frequencies, where the elements in $\tilde{S} = \text{diag}\{\tilde{s}_i\}$ approach and possibly exceed 1 in magnitude, (10.56) may be satisfied if $G(j\omega)$ is close to triangular. This is because $\Gamma - I$ and thus $\tilde{S}(\Gamma - I)$ are then close to triangular, with diagonal elements close to zero, so the eigenvalues of $\widetilde{S}(\Gamma - I)(j\omega)$ are close to zero, (10.56) is satisfied and we have stability of S. This conclusion also holds for plants with RHP-zeros provided they are located beyond the crossover frequency range. In summary, we have established the following rule.

> **Pairing Rule 1.** *For stability of diagonal decentralized control select pairings for which $G(j\omega)$ is close to triangular, i.e. $\Lambda(G(j\omega)) \approx I$, at frequencies around crossover.*

This rule establishes the RGA in the crossover region as a useful tool for selecting input-output pairs and below we also establish the usefulness of the RGA at steady-state.

Necessary steady-state conditions for stability

A desirable property of a decentralized control system is that it has *integrity*, i.e. the closed-loop system should remain stable as subsystem controllers are brought in and out of service. Mathematically, the system possesses integrity if it remains stable when the controller K is replaced by EK where $E = \text{diag}\{\epsilon_i\}$ and the ϵ_i take on the values of 0 or 1, i.e. $\epsilon_i \in \{0, 1\}$.

An even stronger requirement is that the system remains stable as the gain in various loops are reduced (detuned) by an arbitrary factor, i.e. $\epsilon_i \in [0, 1]$. We introduce the idea of decentralized integral controllability (DIC) which is concerned with whether it is *possible* to detune a decentralized controller when there is integral action in each loop.

Definition 10.2 Decentralized Integral Controllability (DIC). *The plant $G(s)$ (corresponding to a given pairing) is DIC if there exists a decentralized controller with integral action in each loop, such that the feedback system is stable and such that each individual loop may be detuned independently by a factor ϵ_i ($0 \leq \epsilon_i \leq 1$) without introducing instability.*

Remark 1 DIC was introduced by Skogestad and Morari (1988b). A detailed survey of conditions for DIC and other related properties is given by Campo and Morari (1994).

Remark 2 Unstable plants are not DIC. The reason is that with all $\epsilon_i = 0$ we are left with the uncontrolled plant G, and the system will be (internally) unstable if $G(s)$ is unstable.

Remark 3 For $\epsilon_i = 0$ we assume that the integrator of the corresponding SISO controller has been removed, otherwise the integrator would yield internal instability.

Note that DIC considers the *existence* of a controller, so it depends only on the plant G and the chosen pairings. The steady-state RGA provides a very useful tool to test for DIC, as is clear from the following result which was first proved by Grosdidier et al. (1985):

Theorem 10.4 Steady-state RGA *Consider a stable square plant G and a diagonal controller K with integral action in all elements, and assume that the loop transfer function GK is strictly proper. If a pairing of outputs and manipulated inputs corresponds to a negative steady-state relative gain, then the closed-loop system has at least one of the following properties:*
(a) The overall closed-loop system is unstable.
(b) The loop with the negative relative gain is unstable by itself.
(c) The closed-loop system is unstable if the loop with the negative relative gain is opened (broken).

Proof: The theorem may be proved by setting $\widetilde{T} = I$ in (10.39) and applying the generalized Nyquist stability condition. Alternatively, we can use Theorem 6.7 on page 245 and select $G' = \text{diag}\{g_{ii}, G^{ii}\}$. Since $\det G' = g_{ii} \det G^{ii}$ and from (A.77) $\lambda_{ii} = \frac{g_{ii} \det G^{ii}}{\det G}$ we have $\det G'/\det G = \lambda_{ii}$ and Theorem 10.4 follows. □

Each of the three possible instabilities in Theorem 10.4 resulting from pairing on a negative value of $\lambda_{ij}(0)$ is undesirable. The worst case is (a) when the overall system is unstable, but situation (c) is also highly undesirable as it will imply instability if the loop with the negative relative gain somehow becomes inactive, for example, due to input saturation. Situation (b) is unacceptable if the loop in question is intended to be operated by itself e.g. it may be at the "bottom" of the control hierarchy, or all the other loops may become inactive, due to input saturation.

Theorem 10.4 can then be summarized in the following rule:

Pairing Rule 2. *Avoid pairings which correspond to negative values of $\lambda_{ij}(0)$.*

Consider a given pairing and assume we have reordered the inputs and outputs such that G has the paired elements along its diagonal. Then Theorem 10.4 implies that

$$A\ (reordered)\ plant\ G(s)\ is\ DIC\ only\ if\ \lambda_{ii}(0) \geq 0\ for\ all\ i. \tag{10.57}$$

The RGA is a very efficient tool because it does not have to be recomputed for each possible choice of pairing. This follows since any permutation of the rows and columns of G results in the same permutation in the RGA of G. To achieve DIC one has to pair on a positive RGA(0)-element in each row and column, and therefore one can often eliminate many alternative pairings by a simple glance at the RGA-matrix. This is illustrated by the following example.

Example 10.8 *Consider a 3×3 plant with*

$$G(0) = \begin{bmatrix} 10.2 & 5.6 & 1.4 \\ 15.5 & -8.4 & -0.7 \\ 18.1 & 0.4 & 1.8 \end{bmatrix}, \quad \Lambda(0) = \begin{bmatrix} 0.96 & \mathbf{1.45} & -1.41 \\ \mathbf{0.94} & -0.37 & 0.43 \\ -0.90 & -0.07 & \mathbf{1.98} \end{bmatrix} \tag{10.58}$$

For a 3×3 plant there are 6 alternative pairings, but from the steady-state RGA we see that there is only one positive element in column 2 ($\lambda_{12} = 1.45$), and only one positive element in row 3 ($\lambda_{33} = 1.98$), and therefore there is only one possible pairing with all RGA-elements positive ($u_1 \leftrightarrow y_2$, $u_2 \leftrightarrow y_1$, $u_3 \leftrightarrow y_3$). Thus, if we require DIC we can from a quick glance at the steady-state RGA eliminate five of the six alternative pairings.

Exercise 10.7 *Assume that the* 4×4 *matrix in (A.82) represents the steady-state model of a plant. Show that* 20 *of the* 24 *possible pairings can be eliminated by requiring DIC.*

Remarks on DIC and RGA.

1. For 2×2 and 3×3 plants we have even tighter conditions for DIC than (10.57). For 2×2 plants (Skogestad and Morari, 1988b)

$$ \text{DIC} \quad \Leftrightarrow \quad \lambda_{11}(0) > 0 \tag{10.59} $$

For 3×3 plants with positive diagonal RGA-elements of $G(0)$ and of $G^{ii}(0), i = 1, 3$ (its three principal submatrices) we have (Yu and Fan, 1990)

$$ \text{DIC} \quad \Leftrightarrow \quad \sqrt{\lambda_{11}(0)} + \sqrt{\lambda_{22}(0)} + \sqrt{\lambda_{33}(0)} \geq 1 \tag{10.60} $$

(Strictly speaking, as pointed out by Campo and Morari (1994), we do not have equivalence for the case when $\sqrt{\lambda_{11}(0)} + \sqrt{\lambda_{22}(0)} + \sqrt{\lambda_{33}(0)}$ is identical to 1, but this has little practical significance).

2. One cannot expect tight conditions for DIC in terms of the RGA for 4×4 systems or higher. The reason is that the RGA essentially only considers "corner values", $\epsilon_i = 0$ or $\epsilon_i = 1$ (integrity), for the detuning factor in each loop in the definition of DIC. This is clear from the fact that $\lambda_{ii} = \frac{g_{ii} \det G^{ii}}{\det G}$, where G corresponds to $\epsilon_i = 1$ for all i, g_{ii} corresponds to $\epsilon_i = 1$ with the other $\epsilon_k = 0$, and G^{ii} corresponds to $\epsilon_i = 0$ with the other $\epsilon_k = 1$.

3. **Determinant conditions for integrity (DIC).** The following condition is concerned with whether it is possible to design a decentralized controller for the plant such that the system possesses integrity, which is a prerequisite for having DIC:
Assume without loss of generality that the signs of the rows or columns of G have been adjusted such that all diagonal elements of G are positive, i.e. $g_{ii}(0) \geq 0$. Then one may compute the determinant of $G(0)$ and all its principal submatrices (obtained by deleting rows and corresponding columns in $G(0)$), which should all have the same sign for DIC.
This determinant condition follows by applying Theorem 6.7 to all possible combinations of $\epsilon_i = 0$ or 1 as illustrated in the proof of Theorem 10.4, and is equivalent to requiring that the so-called Niederlinski indices,

$$ N_I = \det G(0)/\Pi_i g_{ii}(0) \tag{10.61} $$

of $G(0)$ and its principal submatrices are all positive. Actually, this yields more information than the RGA, because in the RGA the terms are combined into $\lambda_{ii} = \frac{g_{ii} \det G^{ii}}{\det G}$ so we may have cases where two negative determinants result in a positive RGA-element. Nevertheless, the RGA is usually the preferred tool because it does not have to be recomputed for each pairing.

4. DIC is also closely related to D-stability, see papers by Yu and Fan (1990) and Campo and Morari (1994). The theory of D-stability provides necessary and sufficient conditions except in a few special cases, such as when the determinant of one or more of the submatrices is zero.

5. If we assume that the controllers have integral action, then $T(0) = I$, and we can derive from (10.53) that a sufficient condition for DIC is that G is generalized diagonally dominant at steady-state, that is,

$$ \mu(E(0)) < 1 $$

This is proved by Braatz (1993, p.154). However, the requirement is only sufficient for DIC and therefore cannot be used to eliminate designs. Specifically, for a 2×2 system it is easy to show (Grosdidier and Morari, 1986) that $\mu(E(0)) < 1$ is equivalent to $\lambda_{11}(0) > 0.5$, which is conservative when compared with the necessary and sufficient condition $\lambda_{11}(0) > 0$ in (10.59).

6. If the plant has $j\omega$-axis poles, e.g. integrators, it is recommended that, prior to the RGA-analysis, these are moved slightly into the LHP (e.g. by using very low-gain feedback). This will have no practical significance for the subsequent analysis.

7. Since Theorem 6.7 applies to unstable plants, we may also easily extend Theorem 10.4 to unstable plants (and in this case one may actually desire to pair on a negative RGA-element). This is shown in Hovd and Skogestad (1994a). Alternatively, one may first implement a stabilizing controller and then analyze the partially controlled system as if it were the plant $G(s)$.

8. The above results only address stability. Performance is analyzed in Section 10.8.4.

10.8.3   The RGA and right-half plane zeros

Bristol (1966) claimed that negative values of $\lambda_{ii}(0)$ implied the presence of RHP-zeros. This is not quite true, and the correct statement is (Hovd and Skogestad, 1992):

Theorem 10.5 *Consider a transfer function matrix with stable elements and no zeros or poles at $s = 0$. Assume $\lim_{s \to \infty} \lambda_{ij}(s)$ is finite and different from zero. If $\lambda_{ij}(j\infty)$ and $\lambda_{ij}(0)$ have different signs then at least one of the following must be true:*
a) The element $g_{ij}(s)$ has a RHP-zero.
b) The overall plant $G(s)$ has a RHP-zero.
c) The subsystem with input j and output i removed, $G^{ij}(s)$, has a RHP-zero.

Any such zero may be detrimental for decentralized control. In most cases the pairings are chosen such that $\lambda_{ii}(\infty)$ is positive (usually close to 1, see Pairing rule 1) which then confirms Bristol's claim that a negative $\lambda_{ii}(0)$ implies there is a RHP-zero in some subsystem.

Example 10.9 *Consider a plant*

$$G(s) = \frac{1}{5s + 1} \left(\begin{array}{cc} s + 1 & s + 4 \\ 1 & 2 \end{array} \right) \tag{10.62}$$

We find that $\lambda_{11}(\infty) = 2$ and $\lambda_{11}(0) = -1$ have different signs. Since none of the diagonal elements have RHP-zeros we conclude from Theorem 10.5 that $G(s)$ must have a RHP-zero. This is indeed true and $G(s)$ has a zero at $s = 2$.

10.8.4   Performance of decentralized control systems

In the following, we consider performance in terms of the control error

$$e = y - r = Gu + G_d d - r \tag{10.63}$$

Suppose the system has been scaled as outlined in Section 1.4, such that at each frequency:

1. Each disturbance is less than 1 in magnitude, $|d_k| < 1$.
2. Each reference change is less than the corresponding diagonal element in R, $|r_j| < R_j$.
3. For each output the acceptable control error is less than 1, $|e_i| < 1$.

For SISO systems, we found in Section 5.10 that in terms of scaled variables we must at all frequencies require

$$|1 + L| > |G_d| \quad \text{and} \quad |1 + L| > |R| \tag{10.64}$$

for acceptable disturbance rejection and command tracking, respectively. Note that L, G_d and R are all scalars in this case. For decentralized control these requirements may be directly generalized by introducing the PRGA-matrix, $\Gamma = \widetilde{G}G^{-1}$, in (10.43) and the CLDG-matrix, $\widetilde{G}_d = \Gamma G_d$, in (10.44). These generalizations will be presented and discussed next, and then subsequently proved.

Single disturbance. Consider a single disturbance, in which case G_d is a vector, and let g_{di} denote the i'th element of G_d. Let $L_i = g_{ii}k_i$ denote the loop transfer function in loop i. Consider frequencies where feedback is effective so $\widetilde{S}\Gamma$ is small (and (10.67) is valid). Then for acceptable disturbance rejection ($|e_i| < 1$) we must with decentralized control require for each loop i,

$$|1 + L_i| > |\widetilde{g}_{di}| \quad \forall i \tag{10.65}$$

which is the same as the SISO-condition (5.52) except that G_d is replaced by the CLDG, $\widetilde{g}_{di}$. In words, $\widetilde{g}_{di}$ gives the "apparent" disturbance gain as seen from loop i when the system is controlled using decentralized control.

Single reference change. Similarly, consider a change in reference for output j of magnitude R_j. Consider frequencies where feedback is effective (and (10.67) is valid). Then for acceptable reference tracking ($|e_i| < 1$) we must require for each loop i

$$|1 + L_i| > |\gamma_{ij}| \cdot |R_j| \quad \forall i \tag{10.66}$$

which is the same as the SISO-condition except for the PRGA-factor, $|\gamma_{ij}|$. In other words, when the other loops are closed the response in loop i gets slower by a factor $|\gamma_{ii}|$. Consequently, for *performance* it is desirable to have *small* elements in Γ, at least at frequencies where feedback is effective. However, at frequencies close to crossover, stability is the main issue, and since the diagonal elements of the PRGA and RGA are equal, we usually prefer to have γ_{ii} close to 1 (recall pairing rule 1 on page 438).

Proofs of (10.65) and (10.66): At frequencies where feedback is effective, $\widetilde{S}$ is small, so

$$I + \widetilde{S}(\Gamma - I) \approx I \tag{10.67}$$

and from (10.42) we have

$$S \approx \widetilde{S}\Gamma \qquad (10.68)$$

The closed-loop response then becomes

$$e = SG_d d - Sr \approx \widetilde{S}\widetilde{G}_d d - \widetilde{S}\Gamma r \qquad (10.69)$$

and the response in output i to a single disturbance d_k and a single reference change r_j is

$$e_i \approx \widetilde{s}_i \widetilde{g}_{dik} d_k - \widetilde{s}_i \gamma_{ik} r_k \qquad (10.70)$$

where $\widetilde{s}_i = 1/(1 + g_{ii}k_i)$ is the sensitivity function for loop i by itself. Thus, to achieve $|e_i| < 1$ for $|d_k| = 1$ we must require $|\widetilde{s}_i \widetilde{g}_{dik}| < 1$ and (10.65) follows. Similarly, to achieve $|e_i| < 1$ for $|r_j| = |R_j|$ we must require $|s_i \gamma_{ik} R_j| < 1$ and (10.66) follows. Also note that $|s_i \gamma_{ik}| < 1$ will imply that assumption (10.67) is valid. Since R usually has all of its elements larger than 1, in most cases (10.67) will be automatically satisfied if (10.66) is satisfied, so we normally need not check assumption (10.67). □

Remark 1 (10.68) may also be derived from (10.40) by assuming $\widetilde{T} \approx I$ which yields $(I + E\widetilde{T})^{-1} \approx (I + E)^{-1} = \Gamma$.

Remark 2 Consider a particular disturbance with model g_d. Its effect on output i with no control is g_{di}, and the ratio between $\widetilde{g}_{di}$ (the CLDG) and g_{di} is the *relative disturbance gain* (RDG) (β_i) of Stanley et al. (1985) (see also Skogestad and Morari (1987b)):

$$\beta_i \triangleq \widetilde{g}_{di}/g_{di} = [\widetilde{G}G^{-1}g_d]_i/[g_d]_i \qquad (10.71)$$

Thus β_i, which is scaling independent, gives the *change* in the effect of the disturbance caused by decentralized control. It is desirable to have β_i small, as this means that the interactions are such that they reduce the apparent effect of the disturbance, such that one does not need high gains $|L_i|$ in the individual loops.

10.8.5 Summary: Controllability analysis for decentralized control

When considering decentralized diagonal control of a plant, one should first check that the plant is controllable with any controller. If the plant is unstable, then as usual the unstable modes must be controllable and observable. In addition, the unstable modes must not be *decentralized fixed modes*, otherwise the plant cannot be stabilized with a diagonal controller (Lunze, 1992). For example, this is the case for a triangular plant if the unstable mode appears only in the off-diagonal elements.

The next step is to compute the RGA-matrix as a function of frequency, and to determine if one can find a good set of input-output pairs bearing in mind the following:

1. Prefer pairings which have the RGA-matrix close to identity at frequencies around crossover, i.e. the RGA-number $\|\Lambda(j\omega) - I\|$ should be small. This rule is to ensure that interactions from other loops do not cause instability as discussed following (10.56).

2. Avoid a pairing ij with negative steady-state RGA elements, $\lambda_{ij}(G(0))$.
3. Prefer a pairing ij where g_{ij} puts minimal restrictions on the achievable bandwidth. Specifically, the frequency ω_{uij} where $\angle g_{ij}(j\omega_{uij}) = -180°$ should be as large as possible.

   This rule favours pairing on variables "close to each other", which makes it easier to satisfy (10.65) and (10.66) physically while at the same time achieving stability. It is also consistent with the desire that $\Lambda(j\omega)$ is close to I.

When a reasonable choice of pairings has been made, one should rearrange G to have the paired elements along the diagonal and perform a controllability analysis.

4. Compute the CLDG and PRGA, and plot these as a function of frequency.
5. For systems with many loops, it is best to perform the analysis one loop at the time, that is, for each loop i, plot $|\tilde{g}_{dik}|$ for each disturbance k and plot $|\gamma_{ij}|$ for each reference j (assuming here for simplicity that each reference is of unit magnitude). For performance, we need $|1 + L_i|$ to be larger than each of these

$$\text{Performance}: \quad |1 + L_i| > \max_{k,j}\{|\tilde{g}_{dik}|, |\gamma_{ij}|\} \qquad (10.72)$$

To achieve stability of the individual loops one must analyze $g_{ii}(s)$ to ensure that the bandwidth required by (10.72) is achievable. Note that RHP-zeros in the diagonal elements may limit achievable decentralized control, whereas they may not pose any problems for a multivariable controller. Since with decentralized control we usually want to use simple controllers, the achievable bandwidth in each loop will be limited by the frequency where $\angle g_{ii}$ is $-180°$ (recall Section 5.12).

6. As already mentioned one may check for constraints by considering the elements of $G^{-1}G_d$ and making sure that they do not exceed one in magnitude within the frequency range where control is needed. Equivalently, one may for each loop i plot $|g_{ii}|$, and the requirement is then that

$$\text{To avoid input constraints}: \quad |g_{ii}| > |\tilde{g}_{dik}|, \quad \forall k \qquad (10.73)$$

at frequencies where $|\tilde{g}_{dik}|$ is larger than 1 (this follows since $\tilde{G}_d = \tilde{G}G^{-1}G_d$). This provides a direct generalization of the requirement $|G| > |G_d|$ for SISO systems. The advantage of (10.73) compared to using $G^{-1}G_d$ is that we can limit ourselves to frequencies where control is needed to reject the disturbance (where $|\tilde{g}_{dik}| > 1$).

If the plant is not controllable, then one may consider another choice of pairings and go back to Step 4. If one still cannot find any pairings which are controllable, then one should consider multivariable control.

7. If the chosen pairing *is* controllable then the analysis based on (10.72) tells us directly how large $|L_i| = |g_{ii}k_i|$ must be, and can be used as a basis for designing the controller $k_i(s)$ for loop i.

Remark. In some cases, pairings which violate the above rules may be chosen. For example, one may even choose to pair on elements with $g_{ii} = 0$ which yield $\lambda_{ii} = 0$. One then relies on the interactions to achieve the desired performance as loop i by itself has no effect. An example of this is in distillation control when the LV-configuration is *not* used, see Example 10.5.

Example 10.10 Application to distillation process. *In order to demonstrate the use of the frequency dependent RGA and CLDG for evaluation of expected diagonal control performance, we consider again the distillation process used in Example 10.7. The LV configuration is used, that is, the manipulated inputs are reflux L (u_1) and boilup V (u_2). Outputs are the product compositions y_D (y_1) and x_B (y_2). Disturbances in feed flowrate F (d_1) and feed composition z_F (d_2), are included in the model. The disturbances and outputs have been scaled such that a magnitude of 1 corresponds to a change in F of 20%, a change in z_F of 20%, and a change in x_B and y_D of 0.01 mole fraction units. The 5 state dynamic model is given in Section 12.4.*

Initial controllability analysis. $G(s)$ *is stable and has no RHP-zeros. The plant and RGA-matrix at steady-state are*

$$G(0) = \begin{bmatrix} 87.8 & -86.4 \\ 108.2 & -109.6 \end{bmatrix}, \quad \Lambda(0) = \begin{bmatrix} 35.1 & -34.1 \\ -34.1 & 35.1 \end{bmatrix} \tag{10.74}$$

The RGA-elements are much larger than 1 and indicate a plant that is fundamentally difficult to control. Fortunately, the flow dynamics partially decouple the response at higher frequencies, and we find that $\Lambda(j\omega) \approx I$ at frequencies above about 0.5 rad/min. Therefore if we can achieve sufficiently fast control, the large steady-state RGA-elements may be less of a problem.

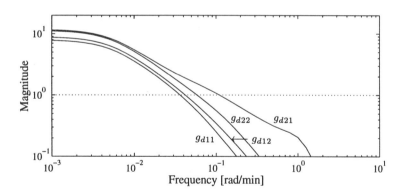

Figure 10.9: Disturbance gains $|g_{dik}|$, for effect of disturbance k on output i

The steady-state effect of the two disturbances is given by

$$G_d(0) = \begin{bmatrix} 7.88 & 8.81 \\ 11.72 & 11.19 \end{bmatrix} \tag{10.75}$$

and the magnitudes of the elements in $G_d(j\omega)$ are plotted as a function of frequency in Figure 10.9. From this plot the two disturbances seem to be equally difficult to reject with

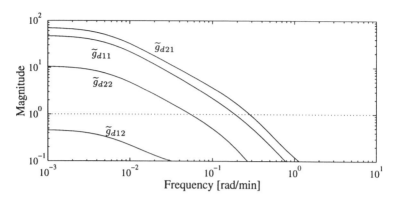

Figure 10.10: Closed-loop disturbance gains, $|\widetilde{g}_{dik}|$, for effect of disturbance k on output i

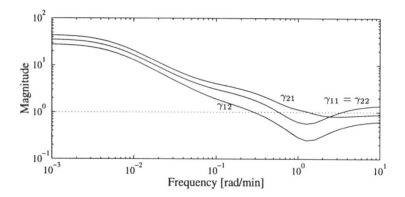

Figure 10.11: PRGA-elements $|\gamma_{ij}|$ for effect of reference j on output i

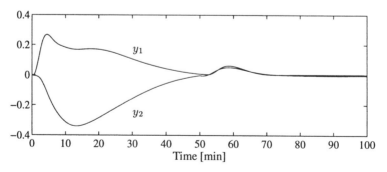

Figure 10.12: Decentralized PI-control. Responses to a unit step in d_1 at $t = 0$ and a unit step in d_2 at $t = 50$ min

magnitudes larger than 1 up to a frequency of about 0.1 rad/min. We conclude that control is needed up to 0.1 rad/min. The magnitude of the elements in $G^{-1}G_d(j\omega)$ (not shown) are all less than 1 at all frequencies (at least up to 10 rad/min), and so it will be assumed that input constraints pose no problem.

Choice of pairings. *The selection of u_1 to control y_1 and u_2 to control y_2, corresponds to pairing on positive elements of $\Lambda(0)$ and $\Lambda(j\omega) \approx I$ at high frequencies. This seems sensible, and is used in the following.*

Analysis of decentralized control. *The elements in the CLDG and PRGA matrices are shown as functions of frequency in Figures 10.10 and 10.11. At steady-state we have*

$$\Gamma(0) = \begin{bmatrix} 35.1 & -27.6 \\ -43.2 & 35.1 \end{bmatrix}, \quad \widetilde{G}_d(0) = \Gamma(0)G_d(0) = \begin{bmatrix} -47.7 & -0.40 \\ 70.5 & 11.7 \end{bmatrix} \quad (10.76)$$

In this particular case the off-diagonal elements of RGA (Λ) and PRGA (Γ) are quite similar. We note that $\widetilde{G}_d(0)$ is very different from $G_d(0)$, and this also holds at higher frequencies. For disturbance 1 (first column in $\widetilde{G}_d$) we find that the interactions increase the apparent effect of the disturbance, whereas they reduce the effect of disturbance 2, at least on output 1.

We now consider one loop at a time to find the required bandwidth. For loop 1 (output 1) we consider γ_{11} and γ_{12} for references, and $\widetilde{g}_{d11}$ and $\widetilde{g}_{d12}$ for disturbances. Disturbance 1 is the most difficult, and we need $|1 + L_1| > |\widetilde{g}_{d11}|$ at frequencies where $|\widetilde{g}_{d11}|$ is larger than 1, which is up to about 0.2 rad/min. The magnitude of the PRGA-elements are somewhat smaller than $|\widetilde{g}_{d11}|$ (at least at low frequencies), so reference tracking will be achieved if we can reject disturbance 1. From $\widetilde{g}_{d12}$ we see that disturbance 2 has almost no effect on output 1 under feedback control.

Also, for loop 2 we find that disturbance 1 is the most difficult, and from $\widetilde{g}_{d12}$ we require a loop gain larger than 1 up to about 0.3 rad/min. A bandwidth of about 0.2 to 0.3 rad/min in each loop, is required for rejecting disturbance 1, and should be achievable in practice.

Observed control performance. *To check the validity of the above results we designed two single-loop PI controllers:*

$$k_1(s) = 0.261\frac{1 + 3.76s}{3.76s}; \quad k_2(s) = -0.375\frac{1 + 3.31s}{3.31s} \quad (10.77)$$

The loop gains, $L_i = g_{ii}k_i$, with these controllers are larger than the closed-loop disturbance gains, $|\delta_{ik}|$, at frequencies up to crossover. Closed-loop simulations with these controllers are shown in Figure 10.12. The simulations confirm that disturbance 2 is more easily rejected than disturbance 1.

In summary, there is an excellent agreement between the controllability analysis and the simulations, as has also been confirmed by a number of other examples.

10.8.6   Sequential design of decentralized controllers

The results presented in this section on decentralized control are most useful for the case when the local controllers $k_i(s)$ are designed *independently*, that is, each controller is designed locally and then all the loops are closed. As discussed above, one problem with this is that the interactions may cause the overall system (T) to be unstable, even though the local loops $(\widetilde{T})$ are stable. This will not happen if the plant is diagonally dominant, such that we satisfy, for example, $\bar{\sigma}(\widetilde{T}) < 1/\mu(E)$ in (10.53).

The stability problem is avoided if the controllers are designed *sequentially* as is commonly done in practice when, for example, the bandwidths of the loops are quite different. In this case the outer loops are tuned with the inner (fast) loops in place, and each step may be considered as a SISO control problem. In particular, overall stability is determined by m SISO stability conditions. However, the issue of performance is more complicated because the closing of a loop may cause "disturbances" (interactions) into a previously designed loop. The engineer must then go back and redesign a loop that has been designed earlier. Thus sequential design may involve many iterations; see Hovd and Skogestad (1994b). The performance bounds in (10.72) are useful for determining the required bandwidth in each loop and may thus suggest a suitable sequence in which to design the controllers.

Although the analysis and derivations given in this section apply when we design the controllers sequentially, it is often useful, after having designed a lower-layer controller (the inner fast loops), to redo the analysis based on the model of the partially controlled system using (10.30) or (10.32). For example, this is usually done for distillation columns, where we base the analysis of the composition control problem on a 2×2 model of the partially controlled 5×5 plant, see Examples 10.5 and 10.10.

10.8.7 Conclusions on decentralized control

In this section, we have derived a number of conditions for the stability, e.g. (10.53) and (10.57), and performance, e.g. (10.65) and (10.66), of decentralized control systems. The conditions may be useful in determining appropriate pairings of inputs and outputs and the sequence in which the decentralized controllers should be designed. Recall, however, that in many practical cases decentralized controllers are tuned based on local models or even on-line. The conditions/bounds are also useful in an input-output controllability analysis for determining the viability of decentralized control.

Some exercises which include a controllability analysis of decentralized control are given at the end of Chapter 6.

10.9 Conclusion

The issue of control structure design is very important in applications, but it has received relatively little attention in the control community during the last 40 years. In this chapter, we have discussed the issues involved, and we have provided some ideas and tools. There is clearly a need for better tools and theory in this area.

11

MODEL REDUCTION

This chapter describes methods for reducing the order of a plant or controller model. We place considerable emphasis on reduced order models obtained by residualizing the less controllable and observable states of a balanced realization. We also present the more familiar methods of balanced truncation and optimal Hankel norm approximation.

11.1   Introduction

Modern controller design methods such as $\mathcal{H}_\infty$ and LQG, produce controllers of order at least equal to that of the plant, and usually higher because of the inclusion of weights. These control laws may be too complex with regards to practical implementation and simpler designs are then sought. For this purpose, one can either reduce the order of the plant model prior to controller design, or reduce the controller in the final stage, or both.

The central problem we address is: given a high-order linear time-invariant stable model G, find a low-order approximation G_a such that the infinity ($\mathcal{H}_\infty$ or $\mathcal{L}_\infty$) norm of the difference, $\|G - G_a\|_\infty$, is small. By model order, we mean the dimension of the state vector in a minimal realization. This is sometimes called the McMillan degree.

So far in this book we have only been interested in the infinity ($\mathcal{H}_\infty$) norm of stable systems. But the error $G - G_a$ may be unstable and the definition of the infinity norm needs to be extended to unstable systems. $\mathcal{L}_\infty$ defines the set of rational functions which have no poles on the imaginary axis, it includes $\mathcal{H}_\infty$, and its norm (like $\mathcal{H}_\infty$) is given by $\|G\|_\infty = \sup_w \bar{\sigma}\left(G(jw)\right)$.

We will describe three main methods for tackling this problem: balanced truncation, balanced residualization and optimal Hankel norm approximation. Each method gives a stable approximation and a guaranteed bound on the error in the approximation. We will further show how the methods can be employed to reduce the order of an *unstable* model G. All these methods start from a special state-space realization of G referred to as balanced. We will describe this realization, but first we will show how the techniques of truncation and residualization can be used to remove the high frequency or fast modes of a state-space realization.

11.2   Truncation and residualization

Let (A, B, C, D) be a minimal realization of a stable system $G(s)$, and partition the state vector x, of dimension n, into $\begin{bmatrix} x_1 \\ x_2 \end{bmatrix}$ where x_2 is the vector of $n - k$ states which we wish to remove. With appropriate partitioning of A, B and C, the state-space equations become

$$\begin{aligned} \dot{x}_1 &= A_{11}x_1 + A_{12}x_2 + B_1 u \\ \dot{x}_2 &= A_{21}x_1 + A_{22}x_2 + B_2 u \\ y &= C_1 x_1 + C_2 x_2 + D u \end{aligned} \qquad (11.1)$$

11.2.1   Truncation

A k-th order truncation of the realization $G \overset{s}{=} (A, B, C, D)$ is given by $G_a \overset{s}{=} (A_{11}, B_1, C_1, D)$. The truncated model G_a is equal to G at infinite frequency, $G(\infty) = G_a(\infty) = D$, but apart from this there is little that can be said in the general case about the relationship between G and G_a. If, however, A is in Jordan form then it is easy to order the states so that x_2 corresponds to high frequency or fast modes. This is discussed next.

Modal truncation.   For simplicity, assume that A has been diagonalized so that

$$A = \begin{bmatrix} \lambda_1 & 0 & \cdots & 0 \\ 0 & \lambda_2 & \cdots & 0 \\ \vdots & \vdots & \ddots & \vdots \\ 0 & 0 & \cdots & \lambda_n \end{bmatrix} \quad B = \begin{bmatrix} b_1^T \\ b_2^T \\ \vdots \\ b_n^T \end{bmatrix} \quad C = \begin{bmatrix} c_1 & c_2 & \cdots & c_n \end{bmatrix} \qquad (11.2)$$

Then, if the λ_i are ordered so that $|\lambda_1| < |\lambda_2| < \cdots$, the fastest modes are removed from the model after truncation. The difference between G and G_a following a k-th order model truncation is given by

$$G - G_a = \sum_{i=k+1}^{n} \frac{c_i b_i^T}{s - \lambda_i} \qquad (11.3)$$

and therefore

$$\|G - G_a\|_\infty \le \sum_{i=k+1}^{n} \frac{\bar{\sigma}(c_i b_i^T)}{|Re(\lambda_i)|} \qquad (11.4)$$

It is interesting to note that the error depends on the residues $c_i b_i^T$ as well as the λ_i. The distance of λ_i from the imaginary axis is therefore not a reliable indicator of whether the associated mode should be included in the reduced order model or not.

An advantage of modal truncation is that the poles of the truncated model are a subset of the poles of the original model and therefore retain any physical interpretation they might have, e.g. the phugoid mode in aircraft dynamics.

11.2.2   Residualization

In truncation, we discard all the states and dynamics associated with x_2. Suppose that instead of this we simply set $\dot{x}_2 = 0$, i.e. we *residualize* x_2, in the state-space equations. One can then solve for x_2 in terms of x_1 and u, and back substitution of x_2, then gives

$$\dot{x}_1 = (A_{11} - A_{12}A_{22}^{-1}A_{21})x_1 + (B_1 - A_{12}A_{22}^{-1}B_2)u \qquad (11.5)$$

$$y = (C_1 - C_2A_{22}^{-1}A_{21})x_1 + (D - C_2A_{22}^{-1}B_2)u \qquad (11.6)$$

Let us assume A_{22} is invertible and define

$$A_r \triangleq A_{11} - A_{12}A_{22}^{-1}A_{21} \qquad (11.7)$$

$$B_r \triangleq B_1 - A_{12}A_{22}^{-1}B_2 \qquad (11.8)$$

$$C_r \triangleq C_1 - C_2A_{22}^{-1}A_{21} \qquad (11.9)$$

$$D_r \triangleq D - C_2A_{22}^{-1}B_2 \qquad (11.10)$$

The reduced order model $G_a(s) \overset{s}{=} (A_r, B_r, C_r, D_r)$ is called a residualization of $G(s) \overset{s}{=} (A, B, C, D)$. Usually (A, B, C, D) will have been put into Jordan form, with the eigenvalues ordered so that x_2 contains the fast modes. Model reduction by residualization is then equivalent to singular perturbational approximation, where the derivatives of the fastest states are allowed to approach zero with some parameter ϵ. An important property of residualization is that it preserves the steady-state gain of the system, $G_a(0) = G(0)$. This should be no surprise since the residualization process sets derivatives to zero, which are zero anyway at steady-state. But it is in stark contrast to truncation which retains the system behaviour at infinite frequency. This contrast between truncation and residualization follows from the simple bilinear relationship $s \to \frac{1}{s}$ which relates the two (e.g. Liu and Anderson, 1989)

It is clear from the discussion above that truncation is to be preferred when accuracy is required at high frequencies, whereas residualization is better for low frequency modelling.

Both methods depend to a large extent on the original realization and we have suggested the use of the Jordan form. A better realization, with many useful properties, is the balanced realization which will be considered next.

11.3   Balanced realizations

In words only: A *balanced realization* is an asymptotically stable minimal realization in which the controllability and observability Gramians are equal and diagonal.

More formally: Let (A, B, C, D) be a minimal realization of a stable, rational transfer function $G(s)$, then (A, B, C, D) is called *balanced* if the solutions to the following Lyapunov equations

$$AP + PA^T + BB^T = 0 \qquad (11.11)$$
$$A^T Q + QA + C^T C = 0 \qquad (11.12)$$

are $P = Q = diag(\sigma_1, \sigma_2, \ldots, \sigma_n) \triangleq \Sigma$, where $\sigma_1 \geq \sigma_2 \geq \ldots \geq \sigma_n > 0$. P and Q are the controllability and observability Gramians, also defined by

$$P \triangleq \int_0^\infty e^{At} BB^T e^{A^T t} dt \qquad (11.13)$$

$$Q \triangleq \int_0^\infty e^{A^T t} C^T C e^{At} dt \qquad (11.14)$$

Σ is therefore simply referred to as the Gramian of $G(s)$. The σ_i are the ordered Hankel singular values of $G(s)$, more generally defined as $\sigma_i \triangleq \lambda_i^{\frac{1}{2}}(PQ), i = 1, \ldots, n$. Notice that $\sigma_1 = \|G\|_H$, the Hankel norm of $G(s)$.

Any minimal realization of a stable transfer function can be balanced by a simple state similarity transformation, and routines for doing this are available in MATLAB. For further details on computing balanced realizations, see Laub et al. (1987). Note that balancing does not depend on D.

So what is so special about a balanced realization? In a balanced realization the value of each σ_i is associated with a state x_i of the balanced system. And the size of σ_i is a relative measure of the contribution that x_i makes to the input-output behaviour of the system; also see the discussion on page 156. Therefore if $\sigma_1 \gg \sigma_2$, then the state x_1 affects the input-output behaviour much more than x_2, or indeed any other state because of the ordering of the σ_i. After balancing a system, each state is just as controllable as it is observable, and a measure of a state's joint observability and controllability is given by its associated Hankel singular value. This property is fundamental to the model reduction methods in the remainder of this chapter which work by removing states having little effect on the system's input-output behaviour.

11.4 Balanced truncation and balanced residualization

Let the balanced realization (A, B, C, D) of $G(s)$ and the corresponding Σ be partitioned compatibly as

$$A = \begin{bmatrix} A_{11} & A_{12} \\ A_{21} & A_{22} \end{bmatrix}, \quad B = \begin{bmatrix} B_1 \\ B_2 \end{bmatrix}, \quad C = \begin{bmatrix} C_1 & C_2 \end{bmatrix} \qquad (11.15)$$

$$\Sigma = \begin{bmatrix} \Sigma_1 & 0 \\ 0 & \Sigma_2 \end{bmatrix} \qquad (11.16)$$

where $\Sigma_1 = diag(\sigma_1, \sigma_2, \ldots, \sigma_k)$, $\Sigma_2 = diag(\sigma_{k+1}, \sigma_{k+2}, \ldots, \sigma_n)$ and $\sigma_k > \sigma_{k+1}$.

Balanced truncation. The reduced order model given by (A_{11}, B_1, C_1, D) is called a *balanced truncation* of the full order system $G(s)$. This idea of balancing the system and then discarding the states corresponding to small Hankel singular values was first introduced by Moore (1981). A balanced truncation is also a balanced realization (Pernebo and Silverman, 1982), and the infinity norm of the error between $G(s)$ and the reduced order system is bounded by twice the sum of the last $n - k$ Hankel singular values, i.e. twice the trace of Σ_2 or simply "twice the sum of the tail" (Glover, 1984; Enns, 1984). For the case of repeated Hankel singular values, Glover (1984) shows that each repeated Hankel singular value is to be counted only once in calculating the sum.

A precise statement of the bound on the approximation error is given in Theorem 11.1 below.

Useful algorithms that compute balanced truncations without first computing a balanced realization have been developed by Tombs and Postlethwaite (1987) and Safonov and Chiang (1989). These still require the computation of the observability and controllability Gramians, which can be a problem if the system to be reduced is of very high order. In such cases the technique of Jaimoukha et al. (1992), based on computing approximate solutions to Lyapunov equations, is recommended.

Balanced residualization. In balanced truncation above, we discarded the least controllable and observable states corresponding to Σ_2. In balanced residualization, we simply set to zero the derivatives of all these states. The method was introduced by Fernando and Nicholson (1982) who called it a singular perturbational approximation of a balanced system. The resulting balanced residualization of $G(s)$ is (A_r, B_r, C_r, D_r) as given by the formulas (11.7)–(11.10).

Liu and Anderson (1989) have shown that balanced residualization enjoys the same error bound as balanced truncation. An alternative derivation of the error bound, more in the style of Glover (1984), is given by Samar et al. (1995). A precise statement of the error bound is given in the following theorem.

Theorem 11.1 *Let $G(s)$ be a stable rational transfer function with Hankel singular values $\sigma_1 > \sigma_2 > \ldots > \sigma_N$ where each σ_i has multiplicity r_i and let $G_a^k(s)$ be obtained by truncating or residualizing the balanced realization of $G(s)$ to the first $(r_1 + r_2 + \ldots + r_k)$ states. Then*

$$\|G(s) - G_a^k(s)\|_\infty \leq 2(\sigma_{k+1} + \sigma_{k+2} + \ldots + \sigma_N). \qquad (11.17)$$

The following two exercises are to emphasize that (i) balanced truncation preserves the steady state-gain of the system and (ii) balanced residualization is related to balanced truncation by the bilinear transformation $s \to s^{-1}$.

Exercise 11.1 *The steady-state gain of a full order balanced system (A, B, C, D) is $D - CA^{-1}B$. Show, by algebraic manipulation, that this is also equal to $D_r - C_r A_r^{-1} B_r$, the steady-state gain of the balanced residualization given by (11.7)–(11.10).*

Exercise 11.2 *Let $G(s)$ have a balanced realization $\left[\begin{array}{c|c} A & B \\ \hline C & D \end{array}\right]$, then*

$$\left[\begin{array}{c|c} A^{-1} & A^{-1}B \\ \hline -CA^{-1} & D - CA^{-1}B \end{array}\right]$$

is a balanced realization of $H(s) \triangleq G(s^{-1})$, and the Gramians of the two realizations are the same.

1. *Write down an expression for a balanced truncation $H_t(s)$ of $H(s)$.*
2. *Apply the reverse transformation $s^{-1} \rightarrow s$ to $H_t(s)$, and hence show that $G_r(s) \triangleq H_t(s^{-1})$ is a balanced residualization of $G(s)$ as defined by (11.7)–(11.10).*

11.5 Optimal Hankel norm approximation

In this approach to model reduction, the problem that is directly addressed is the following: given a stable model $G(s)$ of order (McMillan degree) n, find a reduced order model $G_h^k(s)$ of degree k such that the Hankel norm of the approximation error, $\|G(s) - G_h^k(s)\|_H$, is minimized.

The Hankel norm of any stable transfer function $E(s)$ is defined as

$$\|E(s)\|_H \triangleq \rho^{\frac{1}{2}}(PQ) \tag{11.18}$$

where P and Q are the controllability and observability Gramians of $E(s)$. It is also the maximum Hankel singular value of $E(s)$. So in the optimization we seek an error which is in some sense closest to being completely unobservable and completely uncontrollable, which seems sensible. A more detailed discussion of the Hankel norm was given in Section 4.10.4.

The Hankel norm approximation problem has been considered by many but especially Glover (1984). In Glover (1984) a complete treatment of the problem is given, including a closed-form optimal solution and a bound on the infinity norm of the approximation error. The infinity norm bound is of particular interest because it is better than that for balanced truncation and residualization.

The theorem below gives a particular construction for optimal Hankel norm approximations of square stable transfer functions.

Theorem 11.2 *Let $G(s)$ be a stable, square, transfer function $G(s)$ with Hankel singular values $\sigma_1 \geq \sigma_2 \geq \cdots \geq \sigma_k \geq \sigma_{k+1} = \sigma_{k+2} = \cdots = \sigma_{k+l} > \sigma_{k+l+1} \geq \cdots \geq \sigma_n > 0$, then an optimal Hankel norm approximation of order k, $G_h^k(s)$, can be constructed as follows.*

Let (A, B, C, D) be a balanced realization of $G(s)$ with the Hankel singular values reordered so that the Gramian matrix is

$$\begin{aligned} \Sigma &= diag\,(\sigma_1, \sigma_2, \cdots, \sigma_k, \sigma_{k+l+1}, \cdots, \sigma_n, \sigma_{k+1}, \cdots, \sigma_{k+l}) \quad (11.19) \\ &\triangleq diag\,(\Sigma_1, \sigma_{k+1}I) \end{aligned}$$

Partition (A, B, C, D) to conform with Σ:

$$A = \begin{bmatrix} A_{11} & A_{12} \\ A_{21} & A_{22} \end{bmatrix} \quad B = \begin{bmatrix} B_1 \\ B_2 \end{bmatrix} \quad C = [C_1 \quad C_2] \tag{11.20}$$

Define $(\widehat{A}, \widehat{B}, \widehat{C}, \widehat{D})$ by

$$\widehat{A} \triangleq \Gamma^{-1} \left(\sigma_{k+1}^2 A_{11}^T + \Sigma_1 A_{11} \Sigma_1 - \sigma_{k+1} C_1^T U B_1^T \right) \tag{11.21}$$

$$\widehat{B} \triangleq \Gamma^{-1} \left(\Sigma_1 B_1 + \sigma_{k+1} C_1^T U \right) \tag{11.22}$$

$$\widehat{C} \triangleq C_1 \Sigma_1 + \sigma_{k+1} U B_1^T \tag{11.23}$$

$$\widehat{D} \triangleq D - \sigma_{k+1} U \tag{11.24}$$

where U is a unitary matrix satisfying

$$B_2 = -C_2^T U \tag{11.25}$$

and

$$\Gamma \triangleq \Sigma_1^2 - \sigma_{k+1}^2 I \tag{11.26}$$

The matrix $\widehat{A}$ has k "stable" eigenvalues (in the open left-half plane); the remaining ones are in the open right-half plane. Then

$$G_h^k(s) + F(s) \overset{s}{=} \left[\begin{array}{c|c} \widehat{A} & \widehat{B} \\ \hline \widehat{C} & \widehat{D} \end{array} \right] \tag{11.27}$$

where $G_h^k(s)$ is a stable optimal Hankel norm approximation of order k, and $F(s)$ is an anti-stable (all poles in the open right-half plane) transfer function of order $n - k - l$. The Hankel norm of the error between G and the optimal approximation G_h^k is equal to the $(k+1)$'th Hankel singular value of G:

$$\|G - G_h^k\|_H = \sigma_{k+1}(G) \tag{11.28}$$

Remark 1 The $k+1$'th Hankel singular value is generally not repeated, but the possibility is included in the theory for completeness.

Remark 2 The order k of the approximation can either be selected directly, or indirectly by choosing the "cut-off" value σ_k for the included Hankel singular values. In the latter case, one often looks for large "gaps" in the relative magnitude, σ_k / σ_{k+1}.

Remark 3 There is an infinite number of unitary matrices U satisfying (11.25); one choice is $U = -C_2(B_2^T)^\dagger$.

Remark 4 If $\sigma_{k+1} = \sigma_n$, i.e. only the smallest Hankel singular value is deleted, then $F = 0$, otherwise $(\widehat{A}, \widehat{B}, \widehat{C}, \widehat{D})$ has a non-zero anti-stable part and G_h^k has to be separated from F.

Remark 5 For non-square systems, an optimal Hankel norm approximation can be obtained by first augmenting $G(s)$ with zero to form a square system. For example, if $G(s)$ is flat, define $\bar{G}(s) \triangleq \begin{bmatrix} G(s) \\ 0 \end{bmatrix}$ which is square, and let $\bar{G}_h(s) = \begin{bmatrix} G_1(s) \\ G_2(s) \end{bmatrix}$ be a k-th order optimal Hankel norm approximation of $\bar{G}(s)$ such that $\|\bar{G}(s) - \bar{G}_h(s)\|_H = \sigma_{k+1}\left(\bar{G}(s)\right)$. Then

$$\sigma_{k+1}\left(G(s)\right) \leq \|G - G_1\|_H \leq \|\bar{G} - \bar{G}_h\|_H = \sigma_{k+1}(\bar{G}) = \sigma_{k+1}(G)$$

Consequently, this implies that $\|G - G_1\|_H = \sigma_{k+1}(G)$ and $G_1(s)$ is an optimal Hankel norm approximation of $G(s)$.

Remark 6 The Hankel norm of a system does not depend on the D-matrix in the system's state-space realization. The choice of the D-matrix in G_h^k is therefore arbitrary except when $F = 0$, in which case it is equal to $\widehat{D}$.

Remark 7 The infinity norm does depend on the D-matrix, and therefore the D-matrix of G_h^k can be chosen to reduce the infinity norm of the approximation error (without changing the Hankel norm). Glover (1984) showed that through a particular choice of D, called D_o, the following bound could be obtained:

$$\|G - G_h^k - D_o\|_\infty \leq \sigma_{k+1} + \delta \tag{11.29}$$

where

$$\delta \triangleq \|F - D_o\|_\infty \leq \sum_{i=1}^{n-k-l} \sigma_i\left(F(-s)\right) \leq \sum_{i=1}^{n-k-l} \sigma_{i+k+l}\left(G(s)\right) \tag{11.30}$$

This results in an infinity norm bound on the approximation error, $\delta \leq \sigma_{k+1} + \sigma_{k+l+1} + \cdots + \sigma_n$, which is equal to the "sum of the tail" or less since the Hankel singular value σ_{k+1}, which may be repeated, is only included once. Recall that the bound for the error in balanced truncation and balanced residualization is *twice* the "sum of the tail".

11.6 Two practical examples

In this section, we make comparisons between the three main model reduction techniques presented by applying them to two practical examples. The first example is on the reduction of a plant model and the second considers the reduction of a two degrees-of-freedom controller. Our presentation is similar to that in Samar et al. (1995).

11.6.1 Reduction of a gas turbine aero-engine model

For the first example, we consider the reduction of a model of a Rolls Royce Spey gas turbine engine. This engine will be considered again in Chapter 12. The model has 3 inputs, 3 outputs, and 15 states. Inputs to the engine are fuel flow, variable nozzle area and an inlet guide vane with a variable angle setting. The outputs to be controlled are the high pressure compressor's spool speed, the ratio of the high pressure compressor's outlet pressure to engine inlet pressure, and the low pressure compressor's exit Mach number measurement. The model describes the engine at 87% of maximum thrust with sea-level static conditions. The Hankel singular values for the 15 state model are listed in Table 11.1 below. Recall that the $\mathcal{L}_\infty$ error bounds after reduction are "twice the sum of the tail" for balanced residualization and balanced truncation and the "sum of the tail" for optimal Hankel norm approximation. Based on this we decided to reduce the model to 6 states.

1) 2.0005e+01	6) 6.2964e-01	11) 1.3621e-02
2) 4.0464e+00	7) 1.6689e-01	12) 3.9967e-03
3) 2.7546e+00	8) 9.3407e-02	13) 1.1789e-03
4) 1.7635e+00	9) 2.2193e-02	14) 3.2410e-04
5) 1.2965e+00	10) 1.5669e-02	15) 3.3073e-05

Table 11.1: Hankel singular values of the gas turbine aero-engine model

Figure 11.1 shows the singular values (not Hankel singular values) of the reduced and full order models plotted against frequency for the residualized, truncated and optimal Hankel norm approximated cases respectively. The D matrix used for optimal

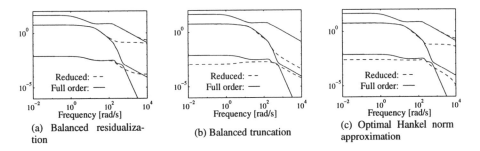

(a) Balanced residualization

(b) Balanced truncation

(c) Optimal Hankel norm approximation

Figure 11.1: Singular values for model reductions of the aero-engine from 15 to 6 states

Hankel norm approximation is such that the error bound given in (11.29) is met. It can be seen that the residualized system matches perfectly at steady-state. The singular values of the error system $(G - G_a)$, for each of the three approximations are shown in Figure 11.2(a). The infinity norm of the error system is computed to be 0.295 for balanced residualization and occurs at 208 rad/s; the corresponding error norms for balanced truncation and optimal Hankel norm approximation are 0.324 and 0.179 occurring at 169 rad/sec and 248 rad/sec, respectively. The theoretical upper bounds for these error norms are 0.635 (twice the sum of the tail) for residualization

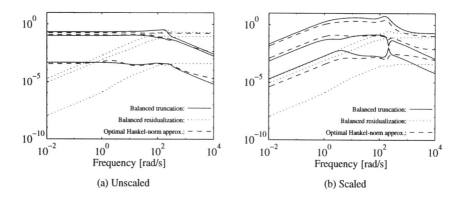

(a) Unscaled   (b) Scaled

Figure 11.2: Singular values for the scaled and unscaled error systems

and truncation, and 0.187 (using (11.29)) for optimal Hankel norm approximation respectively. It should be noted that the plant under consideration is desired to have a closed-loop bandwidth of around 10 rad/sec. The error around this frequency, therefore, should be as small as possible for good controller design. Figure 11.2(a) shows that the error for balanced residualization is the smallest in this frequency range.

Steady-state gain preservation. It is sometimes desirable to have the steady-state gain of the reduced plant model the same as the full order model. For example, this is the case if we want to use the model for feedforward control. The truncated and optimal Hankel norm approximated systems do not preserve the steady-state gain and have to be scaled, i.e. the model approximation G_a is replaced by $G_a W_s$, where $W_s = G_a(0)^{-1}G(0)$, G being the full order model. The scaled system no longer enjoys the bounds guaranteed by these methods and $\|G - G_a W_s\|_\infty$ can be quite large as is shown in Figure 11.2(b). Note that the residualized system does not need scaling, and the error system for this case has been shown again only for ease of comparison. The infinity norms of these errors are computed and are found to degrade to 5.71 (at 151 rad/sec) for the scaled truncated system and 2.61 (at 168.5 rad/sec) for the scaled optimal Hankel norm approximated system. The truncated and Hankel norm approximated systems are clearly worse after scaling since the errors in the critical frequency range around crossover become large despite the improvement at steady-state. Hence residualization is to be preferred over these other techniques whenever good low frequency matching is desired.

Impulse and step responses from the second input to all the outputs for the three reduced systems (with the truncated and optimal Hankel norm approximated systems scaled) are shown in Figures 11.3 and 11.4 respectively. The responses for the other inputs were found to be similar. The simulations confirm that the residualized model's

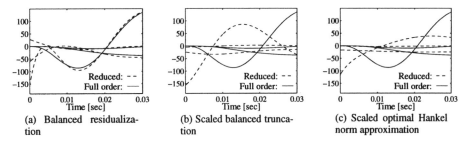

Figure 11.3: Aero-engine: Impulse responses (2nd input)

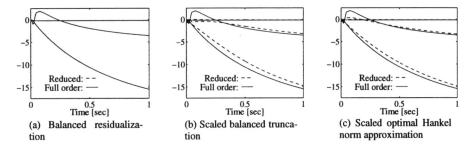

Figure 11.4: Aero-engine: Step responses (2nd input)

response is closer to the full order model's response.

11.6.2 Reduction of an aero-engine controller

We now consider reduction of a two degrees-of-freedom H_∞ loop-shaping controller. The plant for which the controller is designed is the full order gas turbine engine model described in example 11.6.1 above.

A robust controller was designed using the procedure outlined in Section 9.4.3; see Figure 9.21 which describes the design problem. $T_{\text{ref}}(s)$ is the desired closed-loop transfer function, ρ is a design parameter, $G_s = M_s^{-1} N_s$ is the shaped plant and $(\Delta_{N_s}, \Delta_{M_s})$ are perturbations on the normalized coprime factors representing uncertainty. We denote the actual closed-loop transfer function (from β to y) by $T_{y\beta}$.

The controller $K = [K_1 \; K_2]$, which excludes the loop-shaping weight W_1 (which includes 3 integral action states), has 6 inputs (because of the two degrees-of-freedom structure), 3 outputs, and 24 states. It has not been scaled (i.e. the steady-state value of $T_{y\beta}$ has not been matched to that of T_{ref} by scaling the prefilter). It is reduced to 7 states in each of the cases that follow.

Let us first compare the magnitude of $T_{y\beta}$ with that of the specified model T_{ref}. By magnitude, we mean singular values. These are shown in Figure 11.5(a). The infinity

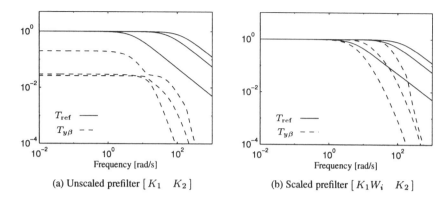

(a) Unscaled prefilter $[\, K_1 \quad K_2 \,]$ (b) Scaled prefilter $[\, K_1 W_i \quad K_2 \,]$

Figure 11.5: Singular values of T_{ref} and $T_{y\beta}$

norm of the difference $T_{y\beta} - T_{\mathrm{ref}}$ is computed to be 0.974 and occurs at 8.5 rad/sec. Note that we have $\rho = 1$ and the γ achieved in the $\mathcal{H}_\infty$ optimization is 2.32, so that $\|T_{y\beta} - T_{\mathrm{ref}}\|_\infty \leq \gamma\rho^{-2}$ as required; see (9.81). The prefilter is now scaled so that $T_{y\beta}$ matches T_{ref} exactly at steady-state, i.e. we replace K_1 by $K_1 W_i$ where $W_i = T_{y\beta}(0)^{-1}T_{\mathrm{ref}}(0)$. It is argued by Hoyle et al. (1991) that this scaling produces better model matching at all frequencies, because the $\mathcal{H}_\infty$ optimization process has already given $T_{y\beta}$ the same magnitude frequency response shape as the model T_{ref}. The scaled transfer function is shown in Figure 11.5(b), and the infinity norm of the difference $(T_{y\beta} - T_{\mathrm{ref}})$ computed to be 1.44 (at 46 rad/sec). It can be seen that this scaling has not degraded the infinity norm of the error significantly as was claimed by Hoyle et al. (1991). To ensure perfect steady-state tracking the controller is always scaled in this way. We are now in a position to discuss ways of reducing the controller. We will look at the following two approaches:

1. The scaled controller $[\, K_1 W_i \quad K_2 \,]$ is reduced. A balanced residualization of this controller preserves the controller's steady-state gain and would not need to be scaled again. Reductions via truncation and optimal Hankel norm approximation techniques, however, lose the steady-state gain. The prefilters of these reduced controllers would therefore need to be rescaled to match $T_{\mathrm{ref}}(0)$.
2. The full order controller $[\, K_1 \quad K_2 \,]$ is directly reduced without first scaling the prefilter. In which case, scaling is done after reduction.

We now consider the first approach. A balanced residualization of $[\, K_1 W_i \quad K_2 \,]$ is obtained. The theoretical upper bound on the infinity norm of the error (twice the sum of the tail) is 0.698, i.e.

$$\| \, K_1 W_i - (K_1 W_i)_a \quad K_2 - K_{2a} \, \|_\infty \leq 0.698 \qquad (11.31)$$

where the subscript a refers to the low order approximation. The actual error norm is computed to be 0.365. $T_{y\beta}$ for this residualization is computed and its magnitude plotted in Figure 11.6(a). The infinity norm of the difference $(T_{y\beta} - T_{\text{ref}})$ is computed

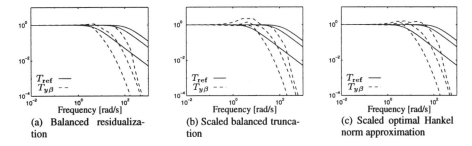

(a) Balanced residualization

(b) Scaled balanced truncation

(c) Scaled optimal Hankel
norm approximation

Figure 11.6: Singular values of T_{ref} and $T_{y\beta}$ for reduced $[\, K_1 W_i \quad K_2 \,]$

to be 1.44 (at 43 rad/sec). This value is very close to that obtained with the full order controller $[\, K_1 W_i \quad K_2 \,]$, and so the closed-loop response of the system with this reduced controller is expected to be very close to that with the full order controller. Next $[\, K_1 W_i \quad K_2 \,]$ is reduced via balanced truncation. The bound given by (11.31) still holds. The steady-state gain, however, falls below the adjusted level, and the prefilter of the truncated controller is thus scaled. The bound given by (11.31) can no longer be guaranteed for the prefilter (it is in fact found to degrade to 3.66), but it holds for $K_2 - K_{2a}$. Singular values of T_{ref} and $T_{y\beta}$ for the scaled truncated controller are shown in Figure 11.6(b). The infinity norm of the difference is computed to be 1.44 and this maximum occurs at 46 rad/sec. Finally $[\, K_1 W_i \quad K_2 \,]$ is reduced by optimal Hankel norm approximation. The following error bound is theoretically guaranteed:

$$\| K_1 W_i - (K_1 W_i)_a \quad K_2 - K_{2a} \|_\infty \leq 0.189 \qquad (11.32)$$

Again the reduced prefilter needs to be scaled and the above bound can no longer be guaranteed; it actually degrades to 1.87. Magnitude plots of $T_{y\beta}$ and T_{ref} are shown in Figure 11.6(c), and the infinity norm of the difference is computed to be 1.43 and occurs at 43 rad/sec.

It has been observed that both balanced truncation and optimal Hankel norm approximation cause a lowering of the system steady-state gain. In the process of adjustment of these steady-state gains, the infinity norm error bounds are destroyed. In the case of our two degrees-of-freedom controller, where the prefilter has been optimized to give closed-loop responses within a tolerance of a chosen ideal model, large deviations may be incurred. Closed-loop responses for the three reduced controllers discussed above are shown in Figures 11.7, 11.8 and 11.9.

It is seen that the residualized controller performs much closer to the full order controller and exhibits better performance in terms of interactions and overshoots. It may not be possible to use the other two reduced controllers if the deviation from

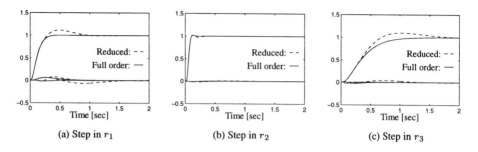

Figure 11.7: Closed-loop step responses: $[\,K_1 W_i \quad K_2\,]$ balanced residualized

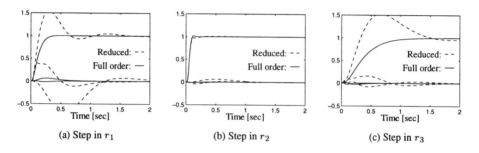

Figure 11.8: Closed-loop step responses: $[\,K_1 W_i \quad K_2\,]$ balanced truncated

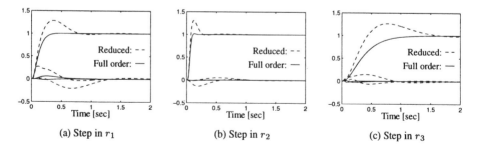

Figure 11.9: Closed-loop step responses: $[\,K_1 W_i \quad K_2\,]$ optimal Hankel norm approximated and rescaled

the specified model becomes larger than the allowable tolerance, in which case the number of states by which the controller is reduced would probably have to be reduced. It should also be noted from (11.31) and (11.32) that the guaranteed bound for $K_2 - K_{2a}$ is lowest for optimal Hankel norm approximation.

Let us now consider the second approach. The controller $[\,K_1 \quad K_2\,]$ obtained from the H_∞ optimization algorithm is reduced directly. The theoretical upper bound on the error for balanced residualization and truncation is

$$\|\,K_1 - K_{1a} \quad K_2 - K_{2a}\,\|_\infty \leq 0.165 \qquad (11.33)$$

The residualized controller retains the steady-state gain of $[\,K_1 \quad K_2\,]$. It is therefore scaled with the same W_i as was required for scaling the prefilter of the full order controller. Singular values of T_{ref} and $T_{y\beta}$ for this reduced controller are shown in Figure 11.10(a), and the infinity norm of the difference was computed to be 1.50 at 44 rad/sec. $[\,K_1 \quad K_2\,]$ is next truncated. The steady-state gain of the truncated controller

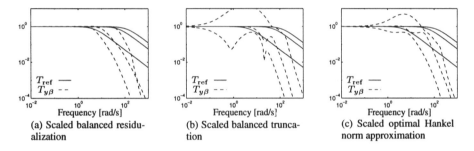

(a) Scaled balanced residualization   (b) Scaled balanced truncation   (c) Scaled optimal Hankel norm approximation

Figure 11.10: Singular values of T_{ref} and $T_{y\beta}$ for reduced $[\,K_1 \quad K_2\,]$

is lower than that of $[\,K_1 \quad K_2\,]$, and it turns out that this has the effect of reducing the steady-state gain of $T_{y\beta}$. Note that the steady-state gain of $T_{y\beta}$ is already less than that of T_{ref} (Figure 11.5). Thus in scaling the prefilter of the truncated controller, the steady-state gain has to be pulled up from a lower level as compared with the previous (residualized) case. This causes greater degradation at other frequencies. The infinity norm of $(T_{y\beta} - T_{\mathrm{ref}})$ in this case is computed to be 25.3 and occurs at 3.4 rad/sec (see Figure 11.10(b)). Finally $[\,K_1 \quad K_2\,]$ is reduced by optimal Hankel norm approximation. The theoretical bound given in (11.29) is computed and found to be 0.037, i.e. we have

$$\|\,K_1 - K_{1a} \quad K_2 - K_{2a}\,\|_\infty \leq 0.037 \qquad (11.34)$$

The steady-state gain falls once more in the reduction process, and again a larger scaling is required. Singular value plots for $T_{y\beta}$ and T_{ref} are shown in Figure 11.10(c). $\|T_{y\beta} - T_{\mathrm{ref}}\|_\infty$ is computed to be 4.5 and occurs at 5.1 rad/sec.

Some closed-loop step response simulations are shown in Figures 11.11, 11.12 and 11.13. It can be seen that the truncated and Hankel norm approximated systems have deteriorated to an unacceptable level. Only the residualized system maintains an acceptable level of performance.

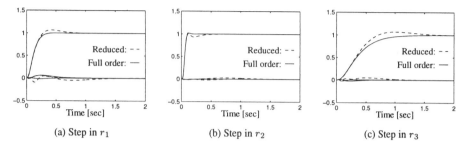

(a) Step in r_1 (b) Step in r_2 (c) Step in r_3

Figure 11.11: Closed-loop step responses: $[\,K_1 \quad K_2\,]$ balanced residualized and scaled

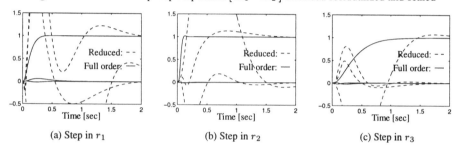

(a) Step in r_1 (b) Step in r_2 (c) Step in r_3

Figure 11.12: Closed-loop step responses: $[\,K_1 \quad K_2\,]$ balanced truncated and scaled

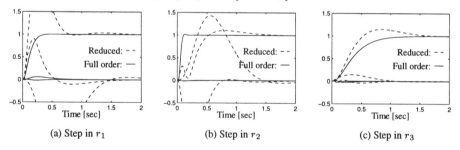

(a) Step in r_1 (b) Step in r_2 (c) Step in r_3

Figure 11.13: Closed-loop step responses: $[\,K_1 \quad K_2\,]$ optimal Hankel norm approximated and scaled

We have seen that the first approach yields better model matching, though at the expense of a larger infinity norm bound on $K_2 - K_{2a}$ (compare (11.31) and (11.33), or (11.32) and (11.34)). We have also seen how the scaling of the prefilter in the first approach gives poorer performance for the truncated and optimal Hankel norm approximated controllers, relative to the residualized one.

In the second case, all the reduced controllers need to be scaled, but a "larger" scaling is required for the truncated and optimal Hankel norm approximated controllers. There appears to be no formal proof of this observation. It is, however, intuitive in the sense that controllers reduced by these two methods yield poorer model matching at steady-state as compared with that achieved by the full order controller. A larger scaling is therefore required for them than is required by the full order or residualized controllers. In any case, this larger scaling gives poorer model matching at other frequencies, and only the residualized controller's performance is deemed acceptable.

11.7 Reduction of unstable models

Balanced truncation, balanced residualization and optimal Hankel norm approximation only apply to stable models. In this section we will briefly present two approaches for reducing the order of an unstable model.

11.7.1 Stable part model reduction

Enns (1984) and Glover (1984) proposed that the unstable model could first be decomposed into its stable and anti-stable parts. Namely

$$G(s) = G_u(s) + G_s(s) \tag{11.35}$$

where $G_u(s)$ has all its poles in the closed right-half plane and $G_s(s)$ has all its poles in the open left-half plane. Balanced truncation, balanced residualization or optimal Hankel norm approximation can then be applied to the stable part $G_s(s)$ to find a reduced order approximation $G_{sa}(s)$. This is then added to the anti-stable part to give

$$G_a(s) = G_u(s) + G_{sa}(s) \tag{11.36}$$

as an approximation to the full order model $G(s)$.

11.7.2 Coprime factor model reduction

The coprime factors of a transfer function $G(s)$ are stable, and therefore we could reduce the order of these factors using balanced truncation, balanced residualization or optimal Hankel norm approximation, as proposed in the following scheme (McFarlane and Glover, 1990).

- Let $G(s) = M^{-1}(s)N(s)$, where $M(s)$ and $N(s)$ are stable left-coprime factors of $G(s)$.
- Approximate $[N \ M]$ of degree n by $[N_a \ M_a]$ of degree $k < n$, using balanced truncation, balanced residualization or optimal Hankel norm approximation.

- Realize the reduced order transfer function $G_a(s)$, of degree k, by $G_a(s) = M_a^{-1} N_a$.

A dual procedure could be written down based on a right coprime factorization of $G(s)$.

For related work in this area, we refer the reader to (Anderson and Liu, 1989; Meyer, 1987). In particular, Meyer (1987) has derived the following result:

Theorem 11.3 *Let* (N, M) *be a normalized left-coprime factorization of* $G(s)$ *of degree n. Let* $[N_a \ M_a]$ *be a degree k balanced truncation of* $[N \ M]$ *which has Hankel singular values* $\sigma_1 \geq \sigma_2 \geq \ldots \geq \sigma_k > \sigma_{k+1} \geq \cdots \geq \sigma_n > 0$. *Then* (N_a, M_a) *is a normalized left-coprime factorization of* $G_a = M_a^{-1} N_a$, *and* $[N_a \ M_a]$ *has Hankel singular values* $\sigma_1, \sigma_2, \ldots, \sigma_k$.

Exercise 11.3 *Is Theorem 11.3 true, if we replace balanced truncation by balanced residualization?*

11.8   Model reduction using MATLAB

The commands in Table 11.2 from the MATLAB μ-toolbox may be used to perform model reduction for stable systems. For an unstable system the commands in Table 11.3 may be used.

Table 11.2: MATLAB commands for model reduction of a stable system

```
% Uses the Mu-toolbox
sysd=strans(sys);                % order states in Jordan form according to speed
syst=strunc(sysd,k);             % then: truncate leaving k states in syst.
sysr=sresid(sysd,k);             % or: residualize leaving k states in sysr.
%
[sysb,hsig]=sysbal(sys);         % obtain balanced realization.
sysbt=strunc(sysb,k);            % then: balanced truncation leaving k states.
sysbr=sresid(sysb,k);            % or: balanced residualization.
sysh=hankmr(sysb,hsig,k,'d');    % or: optimal Hankel norm approximation
```

Table 11.3: MATLAB commands for model reduction of an unstable system

```
% Uses the Mu-toolbox
[syss,sysu]=sdecomp(sys);        % decompose into stable and unstable part.
sys1=sresid(sysbal(syss),ks);    % balanced residualization of stable part.
sys1=hankmr(sysbal(syss),ks,'d');% or: Hankel norm approx. of stable part.
syssbr=madd(sys1,sysu);          % realize reduced-order system.
%
[nlcf,hsig,nrcf]=sncfbal(sys);   % balanced realization of coprime factors.
nrcfr=sresid(nrcf,k);            % residualization of coprime factors.
syscbr=cf2sys(nrcfr);            % realize reduced-order system.
```

Alternatively, the command [ar,br,cr,dr]=ohklmr(a,b,c,d,1,k) in the MATLAB Robust Control Toolbox finds directly the optimal Hankel norm approximation of an unstable plant based on first decomposing the system into the stable and unstable parts. It avoids the sometimes numerically ill-conditioned step of first finding a balanced realization.

11.9 Conclusion

We have presented and compared three main methods for model reduction based on balanced realizations: balanced truncation, balanced residualization and optimal Hankel norm approximation.

Residualization, unlike truncation and optimal Hankel norm approximation, preserves the steady-state gain of the system, and, like truncation, it is simple and computationally inexpensive. It is observed that truncation and optimal Hankel norm approximation perform better at high frequencies, whereas residualization performs better at low and medium frequencies, i.e. up to the critical frequencies. Thus for plant model reduction, where models are not accurate at high frequencies to start with, residualization would seem to be a better option. Further, if the steady-state gains are to be kept unchanged, truncated and optimal Hankel norm approximated systems require scaling, which may result in large errors. In such a case, too, residualization would be a preferred choice.

Frequency weighted model reduction has been the subject of numerous papers over the past few years. The idea is to emphasize frequency ranges where better matching is required. This, however, has been observed to have the effect of producing larger errors (greater mismatching) at other frequencies (Anderson, 1986; Enns, 1984). In order to get good steady-state matching, a relatively large weight would have to be used at steady-state, which would cause poorer matching elsewhere. The choice of weights is not straightforward, and an error bound is available only for weighted Hankel norm approximation. The computation of the bound is also not as easy as in the unweighted case (Anderson and Liu, 1989). Balanced residualization can in this context, be seen as a reduction scheme with implicit low and medium frequency weighting.

For controller reduction, we have shown in a two degrees-of-freedom example, the importance of scaling and steady-state gain matching.

In general, steady-state gain matching may not be crucial, but the matching should usually be good near the desired closed-loop bandwidth. Balanced residualization has been seen to perform close to the full order system in this frequency range. Good approximation at high frequencies may also sometimes be desired. In such a case, using truncation or optimal Hankel norm approximation with appropriate frequency weightings may yield better results.

12

CASE STUDIES

In this chapter, we present three case studies which illustrate a number of important practical issues, namely: weights selection in $\mathcal{H}_\infty$ mixed-sensitivity design, disturbance rejection, output selection, two degrees-of-freedom $\mathcal{H}_\infty$ loop-shaping design, ill-conditioned plants, μ analysis and μ synthesis.

12.1 Introduction

The complete design process for an industrial control system will normally include the following steps:

1. *Plant modelling:* to determine a mathematical model of the plant either from experimental data using identification techniques, or from physical equations describing the plant dynamics, or a combination of these.
2. *Plant input-output controllability analysis:* to discover what closed-loop performance can be expected and what inherent limitations there are to 'good' control, and to assist in deciding upon an initial control structure and may be an initial selection of performance weights.
3. *Control structure design:* to decide on which variables to be manipulated and measured and which links should be made between them.
4. *Controller design:* to formulate a mathematical design problem which captures the engineering design problem and to synthesize a corresponding controller.
5. *Control system analysis:* to assess the control system by analysis and simulation against the performance specifications or the designer's expectations.
6. *Controller implementation:* to implement the controller, almost certainly in software for computer control, taking care to address important issues such as anti-windup and bumpless transfer.
7. *Control system commissioning:* to bring the controller on-line, to carry out on-site testing and to implement any required modifications before certifying that the controlled plant is fully operational.

In this book we have focused on steps 2, 3, 4 and 5, and in this chapter we will present three case studies which demonstrate many of the ideas and practical techniques which can be used in these steps. The case studies are not meant to produce the 'best' controller for the application considered but rather are used here to illustrate a particular technique from the book.

In case study 1, a helicopter control law is designed for the rejection of atmospheric turbulence. The gust disturbance is modelled as an extra input to an S/KS $\mathcal{H}_\infty$ mixed-sensitivity design problem. Results from nonlinear simulations indicate significant improvement over a standard S/KS design. For more information on the applicability of $\mathcal{H}_\infty$ control to advanced helicopter flight, the reader is referred to Walker and Postlethwaite (1996) who describe the design and ground-based piloted simulation testing of a high performance helicopter flight control system.

Case study 2 illustrates the application and usefulness of the two degrees-of-freedom $\mathcal{H}_\infty$ loop-shaping approach by applying it to the design of a robust controller for a high performance areo-engine. Nonlinear simulation results are shown. Efficient and effective tools for control structure design (input-output selection) are also described and applied to this problem. This design work on the aero-engine has been further developed and forms the basis of a multi-mode controller which has been implemented and successfully tested on a Rolls-Royce Spey engine test facility at the UK Defence Research Agency, Pyestock (Samar, 1995).

The final case study is concerned with the control of an idealized distillation column. A very simple plant model is used, but it is sufficient to illustrate the difficulties of controlling ill-conditioned plants and the adverse effects of model uncertainty. The structured singular value μ is seen to be a powerful tool for robustness analysis.

Case studies 1, 2 and 3 are based on papers by Postlethwaite et al. (1994), Samar and Postlethwaite (1994), and Skogestad et al. (1988), respectively.

12.2   Helicopter control

This case study is used to illustrate how weights can be selected in $\mathcal{H}_\infty$ mixed-sensitivity design, and how this design problem can be modified to improve disturbance rejection properties.

12.2.1   Problem description

In this case study, we consider the design of a controller to reduce the effects of atmospheric turbulence on helicopters. The reduction of the effects of gusts is very important in reducing a pilot's workload, and enables aggressive manoeuvers to be carried out in poor weather conditions. Also, as a consequence of decreased buffeting, the airframe and component lives are lengthened and passenger comfort is increased.

The design of rotorcraft flight control systems, for robust stability and performance, has been studied over a number of years using a variety of methods including: H_∞ optimization (Yue and Postlethwaite, 1990; Postlethwaite and Walker, 1992); eigenstructure assignment (Manness and Murray-Smith, 1992; Samblancatt et al., 1990); sliding mode control (Foster et al., 1993); and H_2 design (Takahashi, 1993). The H_∞ controller designs have been particularly successful (Walker et al., 1993), and have proved themselves in piloted simulations. These designs have used frequency information about the disturbances to limit the system sensitivity but in general there has been no explicit consideration of the effects of atmospheric turbulence. Therefore by incorporating practical knowledge about the disturbance characteristics, and how they affect the real helicopter, improvements to the overall performance should be possible. We will demonstrate this below.

The nonlinear helicopter model we will use for simulation purposes was developed at the Defence Research Agency (DRA), Bedford (Padfield, 1981) and is known as the Rationalized Helicopter Model (RHM). A turbulence generator module has recently been included in the RHM and this enables controller designs to be tested on-line for their disturbance rejection properties. It should be noted that the model of the gusts affects the helicopter equations in a complicated fashion and is self contained in the code of the RHM. For design purposes we will imagine that the gusts affect the model in a much simpler manner.

We will begin by repeating the design of Yue and Postlethwaite (1990) which used an $S/KS\,\mathcal{H}_\infty$ mixed sensitivity problem formulation without explicitly considering atmospheric turbulence. We will then, for the purposes of design, represent gusts as a perturbation in the velocity states of the helicopter model and include this disturbance as an extra input to the S/KS design problem. The resulting controller is seen to be substantially better at rejecting atmospheric turbulence than the earlier standard S/KS design.

12.2.2 The helicopter model

The aircraft model used in our work is representative of the Westland Lynx, a twin-engined multi-purpose military helicopter, approximately 9000 lbs gross weight, with a four-blade semi-rigid main rotor. The unaugmented aircraft is unstable, and exhibits many of the cross-couplings characteristic of a single main-rotor helicopter. In addition to the basic rigid body, engine and actuator components, the model also includes second order rotor flapping and coning modes for off-line use. The model has the advantage that essentially the same code can be used for a real-time piloted simulation as for a workstation-based off-line handling qualities assessment.

The equations governing the motion of the helicopter are complex and difficult to formulate with high levels of precision. For example, the rotor dynamics are particularly difficult to model. A robust design methodology is therefore essential for high performance helicopter control. The starting point for this study was to obtain an eighth-order differential equation modelling the small-perturbation rigid motion

State	Description
θ	Pitch attitude
ϕ	Roll attitude
p	Roll rate (body-axis)
q	Pitch rate (body-axis)
ξ	Yaw rate
v_x	Forward velocity
v_y	Lateral velocity
v_z	Vertical velocity

Table 12.1: Helicopter state vector

of the aircraft about hover. The corresponding state-space model is

$$\dot{x} = Ax + Bu \qquad (12.1)$$
$$y = Cx \qquad (12.2)$$

where the matrices A, B and C for the appropriately scaled system are available over the Internet as described in the preface. The 8 state rigid body vector x is given in the Table 12.1. The outputs consist of four controlled outputs

- Heave velocity $\dot{H}$
- Pitch attitude θ
- Roll attitude ϕ $\left.\right\} y_1$
- Heading rate $\dot{\psi}$

together with two additional (body-axis) measurements

- Roll rate p
- Pitch rate q $\left.\right\} y_2$

The controller (or pilot in manual control) generates four blade angle demands which are effectively the helicopter inputs, since the actuators (which are typically modelled as first order lags) are modelled as unity gains in this study. The blade angles are

- main rotor collective
- longitudinal cyclic
- lateral cyclic $\left.\right\} u$
- tail rotor collective

The action of each of these blade angles can be briefly described as follows. The main rotor collective changes all the blades of the main rotor by an equal amount and so roughly speaking controls lift. The longitudinal and lateral cyclic inputs change the main rotor blade angles differently thereby tilting the lift vector to give longitudinal and lateral motion, respectively. The tail rotor is used to balance the torque generated by the main rotor, and so stops the helicopter spinning around; it is also used to give lateral motion. This description, which assumes the helicopter inputs and outputs are decoupled, is useful to get a feeling of how a helicopter works but the dynamics are actually highly coupled. They are also unstable, and about some operating points exhibit non-minimum phase characteristics.

We are interested in the design of *full-authority controllers*, which means that the controller has total control over the blade angles of the main and tail rotors, and is interposed between the pilot and the actuation system. It is normal in conventional helicopters for the controller to have only limited authority leaving the pilot to close the loop for much of the time (manual control). With a full-authority controller, the pilot merely provides the reference commands.

One degree-of-freedom controllers as shown in Figure 12.1 are to be designed. Notice that in the standard one degree-of-freedom configuration the pilot reference

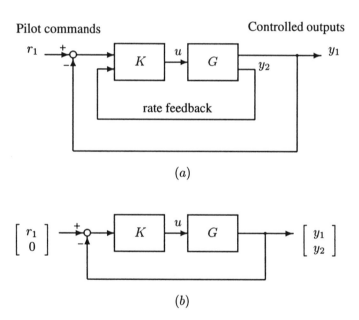

(a)

(b)

Figure 12.1: Helicopter control structure (a) as implemented, (b) in the standard one degree-of-freedom configuration

commands r_1 are augmented by a zero vector because of the rate feedback signals. These zeros indicate that there are no *a priori* performance specifications on $y_2 = [p \quad q]^T$.

12.2.3 $\mathcal{H}_\infty$ mixed-sensitivity design

We will consider the $\mathcal{H}_\infty$ mixed-sensitivity design problem illustrated in Figure 12.2. It can be viewed as a tracking problem as previously discussed in Chapter 9 (see

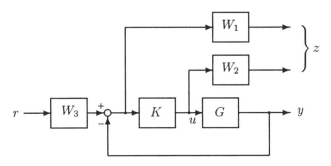

Figure 12.2: S/KS mixed-sensitivity minimization

Figure 9.11), but with an additional weight W_3. W_1 and W_2 are selected as loop-shaping weights whereas W_3 is signal-based. The optimization problem is to find a stabilizing controller K to minimize the cost function

$$\left\| \left[\begin{array}{c} W_1 S W_3 \\ W_2 K S W_3 \end{array} \right] \right\|_\infty \tag{12.3}$$

This cost was also considered by Yue and Postlethwaite (1990) in the context of helicopter control. Their controller was successfully tested on a piloted flight simulator at DRA Bedford and so we propose to use the same weights here. The design weights W_1, W_2 and W_3 were selected as

$$
\begin{align}
W_1 &= diag\left\{ 0.5\frac{s+12}{s+0.012},\ 0.89\frac{s+2.81}{s+0.005},\ 0.89\frac{s+2.81}{s+0.005}, \right. \nonumber \\
&\qquad \left. 0.5\frac{s+10}{s+0.01},\ \frac{2s}{(s+4)(s+4.5)},\ \frac{2s}{(s+4)(s+4.5)} \right\} \tag{12.4}
\end{align}
$$

$$W_2 = 0.5\frac{s+0.0001}{s+10}I_4 \tag{12.5}$$

$$W_3 = diag\{1,\ 1,\ 1,\ 1,\ 0.1,\ 0.1\} \tag{12.6}$$

The reasoning behind these selections of Yue and Postlethwaite (1990) is summarized below.

Selection of $W_1(s)$: For good tracking accuracy in each of the controlled outputs the sensitivity function is required to be small. This suggests forcing integral action into the controller by selecting an s^{-1} shape in the weights associated with the controlled outputs. It was not thought necessary to have exactly zero steady-state errors and therefore these weights were given a finite gain of 500 at low frequencies. (Notice that a pure integrator cannot be included in W_1 anyway, since the standard $\mathcal{H}_\infty$ optimal control problem would not then be well posed in the sense that the corresponding generalized plant P could not then be stabilized by the feedback controller K). In tuning W_1 it was found that a finite attenuation at high frequencies was useful in reducing overshoot. Therefore, high-gain low-pass filters were used in the primary channels to give accurate tracking up to about 6 rad/s. The presence of unmodelled rotor dynamics around 10 rad/s limits the bandwidth of W_1. With four inputs to the helicopter, we can only expect to independently control four outputs. Because of the rate feedback measurements the sensitivity function S is a six by six matrix and therefore two of its singular values (corresponding to p and q) are always close to one across all frequencies. All that can be done in these channels is to improve the disturbance rejection properties around crossover, 4 to 7 rad/s, and this was achieved using second-order band-pass filters in the rate channels of W_1.

Selection of $W_2(s)$: The same first-order high-pass filter is used in each channel with a corner frequency of 10 rad/s to limit input magnitudes at high frequencies and thereby limit the closed-loop bandwidth. The high frequency gain of W_2 can be increased to limit fast actuator movement. The low frequency gain of W_2 was set to approximately -100 dB to ensure that the cost function is dominated by W_1 at low frequencies.

Selection of $W_3(s)$: W_3 is a weighting on the reference input r. It is chosen to be a constant matrix with unity weighting on each of the output commands and a weighting of 0.1 on the fictitious rate demands. The reduced weighting on the rates (which are not directly controlled) enables some disturbance rejection on these outputs, without them significantly affecting the cost function. The main aim of W_3 is to force equally good tracking of each of the primary signals.

For the controller designed using the above weights, the singular value plots of S and KS are shown in Figures 12.3(a) and 12.3(b). These have the general shapes and bandwidths designed for and, as already mentioned, the controlled system performed well in piloted simulation. The effects of atmospheric turbulence will be illustrated later after designing a second controller in which disturbance rejection is explicitly included in the design problem.

12.2.4   Disturbance rejection design

In the design below we will assume that the atmospheric turbulence can be modelled as gust velocity components that perturb the helicopter's velocity states v_x, v_y and v_z by $d = \begin{bmatrix} d_1 & d_2 & d_3 \end{bmatrix}^T$ as in the following equations. The disturbed system is therefore expressed as

$$\dot{x} \;=\; Ax + A\begin{bmatrix} 0 \\ d \end{bmatrix} + Bu \tag{12.7}$$

$$y \;=\; Cx. \tag{12.8}$$

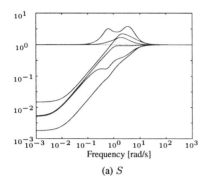 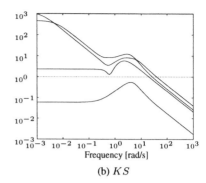

(a) S (b) KS

Figure 12.3: Singular values of S and KS (S/KS design)

Define $B_d \triangleq$ columns 6,7 and 8 of A. Then we have

$$\dot{x} \;=\; Ax + Bu + B_d d \tag{12.9}$$
$$y \;=\; Cx \tag{12.10}$$

which in transfer function terms can be expressed as

$$y = G(s)u + G_d(s)d \tag{12.11}$$

where $G(s) = C(sI - A)^{-1}B$, and $G_d(s) = C(sI - A)^{-1}B_d$. The design problem
we will solve is illustrated in Figure 12.4. The optimization problem is to find a
stabilizing controller K that minimizes the cost function

$$\left\| \begin{bmatrix} W_1 S W_3 & -W_1 S G_d W_4 \\ W_2 K S W_3 & -W_2 K S G_d W_4 \end{bmatrix} \right\|_\infty \tag{12.12}$$

which is the $\mathcal{H}_\infty$ norm of the transfer function from $\begin{bmatrix} r \\ d \end{bmatrix}$ to z. This is easily cast into
the general control configuration and solved using standard software. Notice that if
we set W_4 to zero the problem reverts to the S/KS mixed-sensitivity design of the
previous subsection. To synthesize the controller we used the same weights W_1, W_2
and W_3 as in the S/KS design, and selected $W_4 = \alpha I$, with α a scalar parameter used
to emphasize disturbance rejection. After a few iterations we finalized on $\alpha = 30$.
For this value of α, the singular value plots of S and KS, see Figures 12.5(a) and
12.5(b), are quite similar to those of the S/KS design, but as we will see in the next
subsection there is a significant improvement in the rejection of gusts. Also, since
G_d shares the same dynamics as G, and W_4 is a constant matrix, the degree of the
disturbance rejection controller is the same as that for the S/KS design.

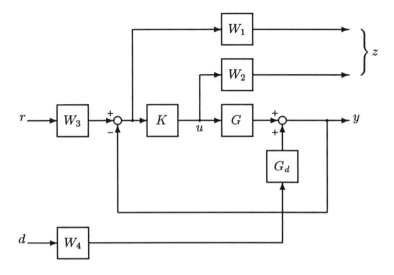

Figure 12.4: Disturbance rejection design

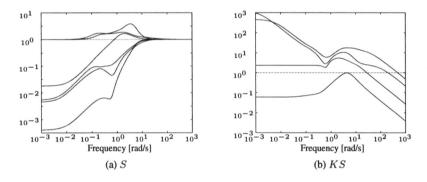

           (a) S (b) KS

Figure 12.5: Singular values of S and KS (disturbance rejection design)

12.2.5 Comparison of disturbance rejection properties of the two designs

To compare the disturbance rejection properties of the two designs we simulated both controllers on the RHM nonlinear helicopter model equipped with a statistical discrete gust model for atmospheric turbulence, (Dahl and Faulkner, 1979). With this simulation facility, gusts cannot be generated at hover and so the nonlinear model was trimmed at a forward flight speed of 20 knots (at an altitude of 100 ft), and the effect of turbulence on the four controlled outputs observed. Recall that both designs were

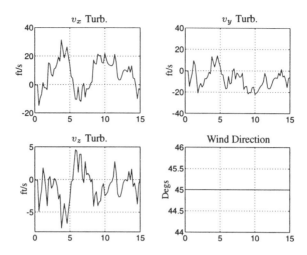

Figure 12.6: Velocity components of turbulence (time in seconds)

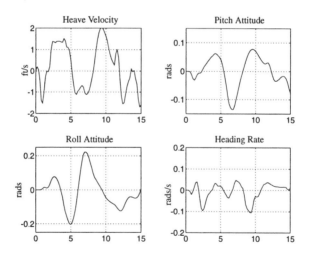

Figure 12.7: Response to turbulence of the S/KS design (time in seconds)

based on a linearized model about hover and therefore these tests at 20 knots also demonstrate the robustness of the controllers. Tests were carried out for a variety of gusts, and in all cases the disturbance rejection design was significantly better than the S/KS design.

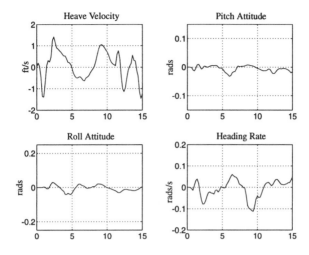

Figure 12.8: Response to turbulence of the disturbance rejection design (time in seconds)

In Figure 12.6, we show a typical gust generated by the RHM. The effects of this on the controlled outputs are shown in Figures 12.7 and 12.8 for the S/KS design and the disturbance rejection design, respectively. Compared with the S/KS design, the disturbance rejection controller practically halves the turbulence effect on heavy velocity, pitch attitude and roll attitude. The change in the effect on heading rate is small.

12.2.6   Conclusions

The two controllers designed were of the same degree and had similar frequency domain properties. But by incorporating knowledge about turbulence activity into the second design, substantial improvements in disturbance rejection were achieved. The reduction of the turbulence effects by a half in heave velocity, pitch attitude and roll attitude indicates the possibility of a significant reduction in a pilot's workload, allowing more aggressive manoeuvers to be carried out with greater precision. Passenger comfort and safety would also be increased.

The study was primarily meant to illustrate the ease with which information about disturbances can be beneficially included in controller design. The case study also demonstrated the selection of weights in $\mathcal{H}_\infty$ mixed-sensitivity design.

12.3 Aero-engine control

In this case study, we apply a variety of tools to the problem of output selection, and illustrate the application of the two degrees-of-freedom $\mathcal{H}_\infty$ loop-shaping design procedure.

12.3.1 Problem description

This case study explores the application of advanced control techniques to the problem of control structure design and robust multivariable controller design for a high performance gas turbine engine. The engine under consideration is the Spey engine which is a Rolls-Royce 2-spool reheated turbofan, used to power modern military aircraft. The engine has two compressors: a low pressure (LP) compressor or fan, and a high pressure (HP) or core compressor as shown in Figure 12.9. The

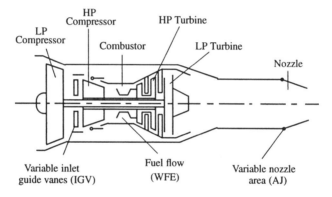

Figure 12.9: Schematic of the aero-engine

high pressure flow at the exit of the core compressor is combusted and allowed to partially expand through the HP and LP turbines which drive the two compressors. The flow finally expands to atmospheric pressure at the nozzle exit, thus producing thrust for aircraft propulsion. The efficiency of the engine and the thrust produced depends on the pressure ratios generated by the two compressors. If the pressure ratio across a compressor exceeds a certain maximum, it may no longer be able to hold the pressure head generated and the flow will tend to reverse its direction. This happens in practice, with the flow actually going negative, but it is only a momentary effect. When the back pressure has cleared itself, positive flow is re-established but, if flow conditions do not change, the pressure builds up causing flow reversal again. Thus the flow surges back and forth at high frequency, the phenomenon being referred to as *surge*. Surging causes excessive aerodynamic pulsations which are transmitted through the whole machine and must be avoided at all costs. However, for higher performance and greater efficiency the compressors must also be operated close to their surge lines. The primary aim of the control system is thus to control engine thrust whilst regulating compressor surge margins. But these engine parameters, namely thrust and the two compressor surge margins, are not directly measurable. There are, however, a number of measurements available which represent these quantities, and our first task is to choose from the available measurements, the ones that are in some sense better for control purposes. This is the problem of output selection as discussed in Chapter 10.

The next step is the design of a robust multivariable controller which provides satisfactory performance over the entire operating range of the engine. Since the aero-engine is a highly nonlinear system, it is normal for several controllers to be designed at different operating points and then to be scheduled across the flight envelope. Also in an aero-engine there are a number of parameters, apart from the ones being primarily controlled, that are to be kept within specified safety limits, e.g. the turbine blade temperature. The number of parameters to be controlled and/or limited exceeds the number of available inputs, and hence all these parameters cannot be controlled independently at the same time. The problem can be tackled by designing a number of scheduled controllers, each for a different set of output variables, which are then switched between, depending on the most significant limit at any given time. The switching is usually done by means of lowest-wins or highest-wins gates, which serve to propagate the output of the most suitable controller to the plant input. Thus, a switched gain-scheduled controller can be designed to cover the full operating range and all possible configurations. In Postlethwaite et al. (1995) a digital multi-mode scheduled controller is designed for the Spey engine under consideration here. In their study gain-scheduling was not required to meet the design specifications. Below we will describe the design of a robust controller for the primary engine outputs using the two degrees-of-freedom $\mathcal{H}_\infty$ loop-shaping approach. The same methodology was used in the design of Postlethwaite et al. (1995) which was successfully implemented and tested on the Spey engine.

12.3.2 Control structure design: output selection

The Spey engine has three inputs, namely fuel flow (WFE), a nozzle with a variable area (AJ), and inlet guide vanes with a variable angle setting (IGV):

$$u = [\text{WFE} \quad \text{AJ} \quad \text{IGV}]^T$$

In this study, there are six output measurements available,

$$y_{\text{all}} = [\text{NL} \quad \text{OPR1} \quad \text{OPR2} \quad \text{LPPR} \quad \text{LPEMN} \quad \text{NH}]^T$$

as described below. For each one of the six output measurements, a look-up table provides its desired optimal value (set point) as a function of the operating point. However, with three inputs we can only control three outputs independently so the first question we face is, which three?

Engine thrust (one of the parameters to be controlled) can be defined in terms of the LP compressor's spool speed (NL), the ratio of the HP compressor's outlet pressure to engine inlet pressure (OPR1), or the engine overall pressure ratio (OPR2). We will choose from these three measurements the one that is best for control.

• Engine thrust: Select one of NL, OPR1 and OPR2 (outputs 1, 2 and 3).

Similarly, surge margin of the LP compressor can be represented by either the LP compressor's pressure ratio (LPPR) or the LP compressor's exit Mach number measurement (LPEMN), and a selection between the two has to be made.

• Surge margin: Select one of LPPR and LPEMN (outputs 4 and 5).

In this study we will not consider control of the HP compressor's surge margin, or other configurations concerned with the limiting of engine temperatures. Our third output will be the HP compressor's spool speed (NH), which it is also important to maintain within safe limits. (NH is actually the HP spool speed made dimensionless by dividing by the square root of the total inlet temperature and scaled so that it is a percentage of the maximum spool speed at a standard temperature of 288.15°K).

• Spool speed: Select NH (output 6).

We have now subdivided the available outputs into three subsets, and decided to select one output from each subset. This gives rise to the six candidate output sets as listed in Table 12.2.

We now apply some of the tools given in Chapter 10 for tackling the output selection problem. It is emphasized at this point that a good physical understanding of the plant is very important in the context of this problem, and some measurements may have to be screened beforehand on practical grounds. A 15 state linear model of the engine (derived from a nonlinear simulation at 87% of maximum thrust) will be used in the analysis that follows. The model is available over the Internet (as described in the preface), along with actuator dynamics which result in a plant model of 18 states for controller design. The nonlinear model used in this case study was provided by the UK Defence Research Agency at Pyestock with the permission of Rolls-Royce Military Aero Engines Ltd.

Scaling. Some of the tools we will use for control structure selection are dependent on the scalings employed. Scaling the inputs and the candidate measurements therefore, is vital before comparisons are made and can also improve the conditioning of the problem for design purposes. We use the method of scaling described in Section 9.4.2. The outputs are scaled such that equal magnitudes of cross-coupling into each of the outputs are equally undesirable. We have chosen to scale the thrust-related outputs such that one unit of each scaled measurement represents 7.5% of maximum thrust. A step demand on each of these scaled outputs would thus correspond to a demand of 7.5% (of maximum) in thrust. The surge margin-related outputs are scaled so that one unit corresponds to 5% surge margin. If the controller designed provides an interaction of less than 10% between the scaled outputs (for unit reference steps), then we would have 0.75% or less change in thrust for a step demand of 5% in surge margin, and a 0.5% or less change in surge margin for a 7.5% step demand in thrust. The final output NH (which is already a scaled variable) was further scaled (divided by 2.2) so that a unit change in NH corresponds to a 2.2% change in NH. The inputs are scaled by 10% of their expected ranges of operation.

Set No.	Candidate controlled outputs	RHP zeros < 100 rad/sec	$\underline{\sigma}(G(0))$
1	NL, LPPR, NH (1, 4, 6)	none	0.060
2	OPR1, LPPR, NH (2, 4, 6)	none	0.049
3	OPR2, LPPR, NH (3, 4, 6)	30.9	0.056
4	NL, LPEMN, NH (1, 5, 6)	none	0.366
5	OPR1, LPEMN, NH (2, 5, 6)	none	0.409
6	OPR2, LPEMN, NH (3, 5, 6)	27.7	0.392

Table 12.2: RHP zeros and minimum singular value for the six candidate output sets

Steady-state model. With these scalings the steady-state model $y_{all} = G_{all}u$ (with all the candidate outputs included) and the corresponding RGA-matrix, $\Lambda = G_{all} \times G_{all}^{\dagger T}$, are given by

$$G_{all} = \begin{bmatrix} 0.696 & -0.046 & -0.001 \\ 1.076 & -0.027 & 0.004 \\ 1.385 & 0.087 & -0.002 \\ 11.036 & 0.238 & -0.017 \\ -0.064 & -0.412 & 0.000 \\ 1.474 & -0.093 & 0.983 \end{bmatrix} \quad \Lambda(G_{all}) = \begin{bmatrix} 0.009 & 0.016 & 0.000 \\ 0.016 & 0.008 & -0.000 \\ 0.006 & 0.028 & -0.000 \\ 0.971 & -0.001 & 0.002 \\ -0.003 & 0.950 & 0.000 \\ 0.002 & -0.000 & 0.998 \end{bmatrix}$$

$$(12.13)$$

and the singular value decomposition of $G_{all}(0) = U_0 \Sigma_0 V_0^H$ is

$$U_0 = \begin{bmatrix} 0.062 & 0.001 & -0.144 & -0.944 & -0.117 & -0.266 \\ 0.095 & 0.001 & -0.118 & -0.070 & -0.734 & 0.659 \\ 0.123 & -0.025 & 0.133 & -0.286 & 0.640 & 0.689 \\ 0.977 & -0.129 & -0.011 & 0.103 & -0.001 & -0.133 \\ -0.006 & 0.065 & -0.971 & 0.108 & 0.195 & 0.055 \\ 0.131 & 0.989 & 0.066 & -0.000 & 0.004 & -0.004 \end{bmatrix}$$

$$\Sigma_0 = \begin{bmatrix} 11.296 & 0 & 0 \\ 0 & 0.986 & 0 \\ 0 & 0 & 0.417 \\ 0 & 0 & 0 \\ 0 & 0 & 0 \\ 0 & 0 & 0 \end{bmatrix} \quad V_0 = \begin{bmatrix} 1.000 & -0.007 & -0.021 \\ 0.020 & -0.154 & 0.988 \\ 0.010 & 0.988 & 0.154 \end{bmatrix}$$

The RGA-matrix of G_{all}, the overall non-square gain matrix, is sometimes a useful screening tool when there are many alternatives. The six row-sums of the RGA-matrix are

$$\Lambda_\Sigma = [\,0.025 \quad 0.023 \quad 0.034 \quad 0.972 \quad 0.947 \quad 1.000\,]^T$$

and from (10.16) this indicates that we should select outputs 4, 5 and 6 (corresponding to the three largest elements) in order to maximize the projection of the selected outputs onto the space corresponding to the three non-zero singular values. However, this selection is not one of our six candidate output sets because there is no output directly related to engine thrust (outputs 1, 2 and 3).

We now proceed with a more detailed input-output controllability analysis of the six candidate output sets. In the following, $G(s)$ refers to the transfer function matrix for the effect of the three inputs on the selected three outputs.

Minimum singular value. In Chapter 10, we showed that a reasonable criterion for selecting controlled outputs y is to make $\|G^{-1}(y - y_{opt})\|$ small, in particular at steady-state. Here $y - y_{opt}$ is the deviation in y from its optimal value. At steady-state this deviation arises mainly from errors in the (look-up table) set point due to disturbances and unknown variations in the operating point. If we assume that, with the scalings given above, the magnitude $|(y - y_{opt})_i|$ is similar (close to 1) for each of the six outputs, then we should select a set of outputs such that the elements in $G^{-1}(0)$ are small, or alternatively, such that $\underline{\sigma}\,(G(0))$ is as large as possible. In Table 12.2 we have listed $\underline{\sigma}\,(G(0))$ for the six candidate output sets. We conclude that we can eliminate sets 1, 2 and 3, and consider only sets 4, 5 and 6. For these three sets we find that the value of $\bar{\sigma}(G(0))$ is between 0.366 and 0.409 which is only slightly smaller than $\bar{\sigma}(G_{\text{all}}(0)) = 0.417$.

Remark. The three eliminated sets all include output 4, LPPR. Interestingly, this output is associated with the largest element in the gain matrix $G_{\text{all}}(0)$ of 11.0, and is thus also associated with the largest singular value (as seen from the first column of U). This illustrates that the preferred choice is often not associated with $\bar{\sigma}(G)$.

Right-half plane zeros. Right-half plane (RHP) zeros limit the achievable performance of a feedback loop by limiting the open-loop gain-bandwidth product. They can be a cause of concern, particularly if they lie within the desired closed-loop bandwidth. Also, choosing different outputs for feedback control can give rise to different numbers of RHP zeros at different locations. The choice of outputs should be such that a minimum number of RHP zeros are encountered, and should be as far removed from the imaginary axis as possible.

Table 12.2 shows the RHP zeros slower than 100 rad/sec for all combinations of prospective output variables. The closed-loop bandwidth requirement for the aero-engine is approximately 10 rad/sec. RHP zeros close to this value or smaller (closer to the origin) will therefore, cause problems and should be avoided. It can be seen that the variable OPR2 introduces (relatively) slow RHP zeros. It was observed that these zeros move closer to the origin at higher thrust levels. Thus Sets 3 and 6 are unfavourable for closed-loop control. This along with the minimum singular value analysis leaves us with sets 4 and 5 for further consideration

Relative gain array (RGA). We here consider the RGAs of the candidate square transfer function matrices $G(s)$ with three outputs,

$$\Lambda(G(s)) = G(s) \times G^{-T}(s) \qquad (12.14)$$

In Section 3.6.2, it is argued that the RGA provides useful information for the analysis of input-output controllability and for the pairing of inputs and outputs. Specifically input and output variables should be paired so that the diagonal elements of the RGA are as close as possible to unity. Furthermore, if the plant has large RGA elements and an inverting controller is used, the closed-loop system will have little robustness in the face of diagonal input uncertainty. Such a perturbation is quite common due to uncertainty in the actuators. Thus we want Λ to have small elements and for diagonal dominance we want $\Lambda - I$ to be small. These two objectives can be combined in the single objective of a small RGA-number, defined as

$$\text{RGA-number} \triangleq \|\Lambda - I\|_{\text{sum}} = \sum_{i=j} | 1 - \lambda_{ij} | + \sum_{i \neq j} | \lambda_{ij} | \qquad (12.15)$$

The lower the RGA number, the more preferred is the control structure. Before calculating the RGA number over frequency we rearranged the output variables so that the steady-state RGA matrix was as close as possible to the identity matrix.

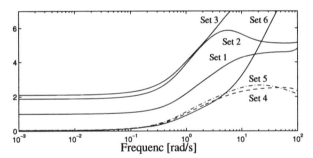

Figure 12.10: RGA numbers

The RGA numbers for the six candidate output sets are shown in Figure 12.10. As in the minimum singular value analysis above, we again see that Sets 1,2 and 3 are less favourable. Once more, sets 4 and 5 are the best but too similar to allow a decisive selection.

Hankel singular values. Notice that Sets 4 and 5 differ only in one output variable, NL in Set 4 and OPR1 in Set 5. Therefore, to select between them we next consider the Hankel singular values of the two transfer functions between the three inputs and output NL and output OPR1, respectively. Hankel singular values reflect the joint controllability and observability of the states of a balanced realization (as described in Section 11.3). Recall that the Hankel singular values are invariant under state transformations but they do depend on scaling.

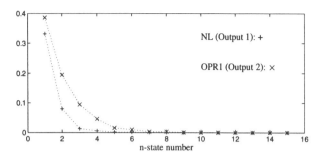

Figure 12.11: Hankel singular values

Figure 12.11 shows the Hankel singular values of the two transfer functions for outputs NL and OPR1, respectively. The Hankel singular values for OPR1 are larger, which indicates that OPR1 has better state controllability and observability properties than NL. In other words, output OPR1 contains more information about the system internal states than output NL. It therefore seems to be preferable to use OPR1 for control purposes rather than NL, and hence (in the absence of no other information) Set 5 is our final choice.

12.3.3   A two degrees-of-freedom $\mathcal{H}_\infty$ loop-shaping design

The design procedure given in Section 9.4.3 will be used to design a two degrees-of-freedom $\mathcal{H}_\infty$ loop-shaping controller for the 3-input 3-output plant G. An 18 state linear plant model G (including actuator dynamics), is available over the Internet. It is based on scaling, output selection, and input-output pairing as described below. To summarize, the selected outputs (Set 5) are

- engine inlet pressure, OPR1
- LP compressor's exit mach number measurement, LPEMN
- HP compressor's spool speed, NH

and the corresponding inputs are

- fuel flow, WFE
- nozzle area, AJ
- inlet guide vane angle, IGV

The corresponding steady-state ($s = 0$) model and RGA-matrix is

$$G = \begin{bmatrix} 1.076 & -0.027 & 0.004 \\ -0.064 & -0.412 & 0.000 \\ 1.474 & -0.093 & 0.983 \end{bmatrix}, \quad \Lambda(G) = \begin{bmatrix} 1.002 & 0.004 & -0.006 \\ 0.004 & 0.996 & -0.000 \\ -0.006 & -0.000 & 1.006 \end{bmatrix} \quad (12.16)$$

Pairing of inputs and outputs. The pairing of inputs and outputs is important because it makes the design of the prefilter easier in a two degrees-of-freedom control configuration and simplifies the selection of weights. It is of even greater importance if a decentralized control scheme is to be used, and gives insight into the working of the plant. In Chapter 10, it is argued that negative entries on the principal diagonal of the steady-state RGA should be avoided and that the outputs in G should be (re)arranged such that the RGA is close to the identity matrix. For the selected output set, we see from (12.16) that no rearranging of the outputs is needed. That is, we should pair OPR1, LPEMN and NH with WFE, AJ and IGV, respectively.

$\mathcal{H}_{\infty}$ **loop-shaping design.** We follow the design procedure given in Section 9.4.3. In steps 1 to 3 we discuss how pre- and post-compensators are selected to obtain the desired shaped plant (loop shape) $G_s = W_2GW_1$ where $W_1 = W_pW_aW_b$. In steps 4 to 6 we present the subsequent $\mathcal{H}_{\infty}$ design.

1. The singular values of the plant are shown in Figure 12.12(a) and indicate a need for extra low frequency gain to give good steady-state tracking and disturbance rejection. The precompensator weight is chosen as simple integrators, i.e. $W_p =$

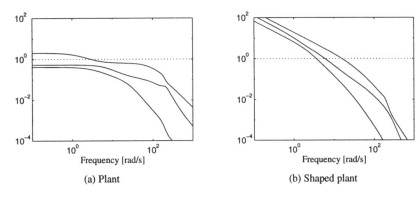

(a) Plant   (b) Shaped plant

Figure 12.12: Singular values for plant and shaped plant

   $\frac{1}{s}I_3$, and the post-compensator weight is selected as $W_2 = I_3$.
2. W_2GW_p is next aligned at 7 rad/sec. The align gain W_a (used in front of W_p) is the approximate real inverse of the shaped system at the specified frequency. The crossover is thus adjusted to 7 rad/sec in order to give a closed-loop bandwidth of approximately 10 rad/sec. Alignment should not be used if the plant is ill-conditioned with large RGA elements at the selected alignment frequency. In our case the RGA elements are small (see Figure 12.10) and hence alignment is not expected to cause problems.
3. An additional gain W_g is used in front of the align gain to give some control over actuator usage. W_g is adjusted so that the actuator rate limits are not exceeded for reference and disturbance steps on the scaled outputs. By some trial and error, W_g is chosen to be $\text{diag}(1, 2.5, 0.3)$. This indicates that the second actuator (AJ) is made to respond at higher rates whereas the third actuator (IGV) is made slower.

The shaped plant now becomes $G_s = GW_1$ where $W_1 = W_pW_aW_g$. Its singular values are shown in Figure 12.12(b).

4. γ_{min} in (9.66) for this shaped plant is found to be 2.3 which indicates that the shaped plant is compatible with robust stability.

5. ρ is set to 1 and the reference model T_{ref} is chosen as $T_{ref} = \text{diag}\{\frac{1}{0.018s+1},$ $\frac{1}{0.008s+1}, \frac{1}{0.2s+1}\}$. The third output NH is thus made slower than the other two in following reference inputs.

6. The standard $\mathcal{H}_\infty$ optimization defined by P in (9.87) is solved. γ iterations are performed and a slightly suboptimal controller achieving $\gamma = 2.9$ is obtained. Moving closer to optimality introduces very fast poles in the controller which, if the controller is to be discretized, would ask for a very high sample rate. Choosing a slightly suboptimal controller alleviates this problem and also improves on the H_2 performance. The prefilter is finally scaled to achieve perfect steady-state model matching. The controller (with the weights W_1 and W_2) has 27 states.

12.3.4 Analysis and simulation results

Step responses of the linear controlled plant model are shown in Figure 12.13. The decoupling is good with less than 10% interactions. Although not shown here

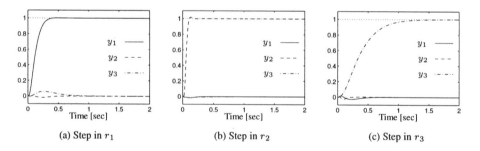

(a) Step in r_1 (b) Step in r_2 (c) Step in r_3

Figure 12.13: Reference step responses

the control inputs were analyzed and the actuator signals were found to lie within specified limits. Responses to disturbance steps on the outputs were also seen to meet the problem specifications. Notice that because there are two degrees-of-freedom in the controller structure, the reference to output and disturbance to output transfer functions can be given different bandwidths.

The robustness properties of the closed-loop system are now analyzed. Figure 12.14(a) shows the singular values of the sensitivity function. The peak value is less than 2 (actually it is $1.44 = 3.2$ dB), which is considered satisfactory. Figure 12.14(b) shows the maximum singular values of $T = (I - GW_1K_2)^{-1}GW_1K_2$ and $T_I = (I - W_1K_2G)^{-1}W_1K_2G$. Both of these have small peaks and go to zero quickly at high frequencies. From Section 9.2.2, this indicates good robustness both

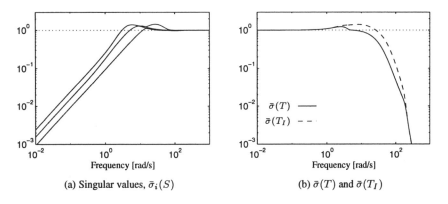

(a) Singular values, $\bar{\sigma}_i(S)$

(b) $\bar{\sigma}(T)$ and $\bar{\sigma}(T_I)$

Figure 12.14: Sensitivity and complementary sensitivity functions

with respect to multiplicative output and multiplicative input plant perturbations. Nonlinear simulation results are shown in Figure 12.15. Reference signals are

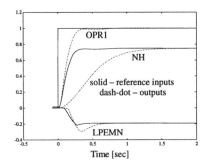

Figure 12.15: Nonlinear simulation results

given to each of the scaled outputs simultaneously. The solid lines show the references, and the dash-dot lines, the outputs. It can be seen that the controller exhibits good performance with low interactions.

12.3.5 Conclusions

The case study has demonstrated the ease with which the two degrees-of-freedom $\mathcal{H}_\infty$ loop-shaping design procedure can be applied to a complex engineering system. Some tools for control structure design have also been usefully applied to the aero-engine example. We stress that a good control structure selection is very important. It results in simpler controllers and in general, a simpler design exercise.

12.4   Distillation process

A typical distillation column is shown in Figure 10.6 on page 426. The overall 5×5 control problem is discussed in Example 10.5 (page 425) and you are advised to read this first. The commonly used LV- and DV-configurations, which are discussed below, are partially controlled systems where 3 loops for liquid level and pressure have already been closed.

For a general discussion on distillation column control, the reader is also referred to Shinskey (1984), Skogestad and Morari (1987a) and the survey paper by Skogestad (1992).

We have throughout the book studied a particular high-purity binary distillation column with 40 theoretical stages (39 trays and a reboiler) plus a total condenser. This is "column A" in Skogestad et al. (1990). The feed is an equimolar liquid mixture of two components with a relative volatility of 1.5. The pressure p is assumed constant (perfect control of p using V_T as an input). The operating variables (e.g. reflux and boilup rates) are such that we nominally have 99% purity for each product (y_D and x_B). The nominal holdups on all stages, including the reboiler and condenser, are $M_i^*/F = 0.5$ min. The liquid flow dynamics are modelled by a simple linear relationship, $L_i(t) = L_i^* + (M_i(t) - M_i^*)/\tau_L$, where $\tau_L = 0.063$ min (the same value is used on all trays). No actuator or measurement dynamics are included. This results in a model with 82 states. This distillation process is difficult to control because of strong interactions between the two product compositions. More information, including steady-state profiles along the column, is available over the internet.

This distillation process has been used as an illustrative example throughout the book, and so to avoid unnecessary repetition we will simply summarize what has been done and refer to the many exercises and examples for more details.

Remark 1 The complete linear distillation column model with 4 inputs (L, V, D, B), 4 outputs (y_D, x_B, M_D, M_B), 2 disturbances (F, z_F) and 82 states is available over the internet. The states are the mole fractions and liquid holdups on each of the 41 stages. By closing the two level loops (M_D and M_B) this model may be used to generate the model for any configuration (LV, DV, etc.). The MATLAB commands for generating the LV-, DV- and DB-configurations are given in Table 12.3.

Remark 2 A 5 state LV-model, obtained by model reducing the above model, is given on page 494. This model is also available over the internet.

12.4.1   Idealized LV-model

The following idealized model of the distillation process, originally from Skogestad et al. (1988), has been used in examples throughout the book:

$$G(s) = \frac{1}{75s+1} \begin{bmatrix} 87.8 & -86.4 \\ 108.2 & -109.6 \end{bmatrix} \tag{12.17}$$

Table 12.3: MATLAB program for generating model of various distillation configurations

```
% Uses MATLAB Mu toolbox
% G4: State-space model (4 inputs, 2 disturbances, 4 outputs, 82 states)
% Level controllers using D and B (P-controllers; bandwidth = 10 rad/min):
Kd = 10; Kb = 10;
%
% Now generate the LV-configuration from G4 using sysic:
systemnames = 'G4 Kd Kb';
inputvar = '[L(1); V(1); d(2)]';
outputvar = '[G4(1);G4(2)]';
input_to_G4 = '[L; V; Kd; Kb; d ]';
input_to_Kd = '[G4(3)]';
input_to_Kb = '[G4(4)]';
sysoutname ='Glv';
cleanupsysic='yes'; sysic;
%
% Modifications needed to generate DV-configuration:
Kl = 10; Kb = 10;
systemnames = 'G4 Kl Kb';
inputvar = '[D(1); V(1); d(2)]';
input_to_G4 = '[Kl; V; D; Kb; d ]';
input_to_Kl = '[G4(3)]';
input_to_Kb = '[G4(4)]';
sysoutname ='Gdv';
%
% Modifications needed to generate DB-configuration:
Kl = 10; Kv = 10;
systemnames = 'G4 Kl Kv';
inputvar = '[D(1); B(1); d(2)]';
input_to_G4 = '[Kl; Kv; D; B; d ]';
input_to_Kl = '[G4(3)]';
input_to_Kv = '[G4(4)]';
sysoutname ='Gdb';
```

The inputs are the reflux (L) and boilup (V), and the controlled outputs are the top and bottom product compositions (y_D and x_B). This is a very crude model of the distillation process, but it provides an excellent example of an ill-conditioned process where control is difficult, primarily due to the presence of input uncertainty.

We refer the reader to the following places in the book where the model (12.17) is used:

Example 3.6 (page 75): SVD-analysis. The singular values are plotted as a function of frequency in Figure 3.6(b) on page 76.

Example 3.7 (page 75): Discussion of the physics of the process and the interpretation of directions.

Example 3.11 (page 90): The condition number, $\gamma(G)$, is 141.7, and the 1, 1-element of the RGA, $\lambda_{11}(G)$, is 35.1 (at all frequencies).

Motivating Example No. 2 (page 93): Introduction to robustness problems with inverse-based controller using simulation with 20% input uncertainty.

Exercise 3.7 (page 96): Design of robust SVD-controller.

Exercise 3.8 (page 96): Combined input and output uncertainty for inverse-based controller.

Exercise 3.9 (page 97): Attempt to "robustify" an inverse-based design using McFarlane-Glover $\mathcal{H}_\infty$ loop-shaping procedure.

Example 6.5 (page 232): Magnitude of inputs for rejecting disturbances (in feed rate and feed composition) at steady state.

Example 6.6 (page 241): Sensitivity to input uncertainty with feedforward control (RGA).

Example 6.7 (page 242): Sensitivity to input uncertainty with inverse-based controller, sensitivity peak (RGA).

Example 6.11 (page 246): Sensitivity to element-by-element uncertainty (relevant for identification).

Example 8.1 (page 294): Coupling between uncertainty in transfer function elements.

Example in Section 8.11.3 (page 329): μ for robust performance which explains poor performance in Motivating Example No. 2.

Example in Section 8.12.4 (page 337): Design of μ-optimal controller using DK-iteration.

The model in (12.17) has also been the basis for two benchmark problems.

Original benchmark problem. The original control problem was formulated by Skogestad et al. (1988) as a bound on the weighted sensitivity with frequency-bounded input uncertainty. The optimal solution to this problem is provided by the one degree-of-freedom μ-optimal controller given in the example in Section 8.12.4 where a peak μ-value of 0.974 (remark 1 on page 1) was obtained.

CDC benchmark problem. The original problem formulation is unrealistic in that there is no bound on the input magnitudes. Furthermore, the bounds on performance and uncertainty are given in the frequency domain (in terms of weighted $\mathcal{H}_\infty$ norm), whereas many engineers feel that time domain specifications are more realistic. Limebeer (1991) therefore suggested the following CDC-specifications. The set of plants Π is defined by

$$\widetilde{G}(s) = \frac{1}{75s+1} \begin{bmatrix} 0.878 & -0.864 \\ 1.082 & -1.096 \end{bmatrix} \begin{bmatrix} k_1 e^{-\theta_1 s} & 0 \\ 0 & k_2 e^{-\theta_2 s} \end{bmatrix}$$
$$k_i \in [0.8 \quad 1.2], \quad \theta_i \in [0 \quad 1.0] \qquad (12.18)$$

In physical terms this means 20% gain uncertainty and up to 1 minute delay in each input channel. The specification is to achieve for for every plant $\widetilde{G} \in \Pi$:

S1: Closed-loop stability.

S2: For a unit step demand in channel 1 at $t = 0$ the plant output y_1 (tracking) and y_2 (interaction) should satisfy:

- $y_1(t) \geq 0.9$ for all $t \geq 30$ min
- $y_1(t) \leq 1.1$ for all t
- $0.99 \leq y_1(\infty) \leq 1.01$
- $y_1(t) \leq 0.5$ for all t
- $-0.01 \leq y_2(\infty) \leq 0.01$

The same corresponding requirements hold for a unit step demand in channel 2.

S3: $\bar{\sigma}(K_y \widetilde{S}) < 0.316$, $\forall \omega$

S4: $\bar{\sigma}(\widetilde{G} K_y) < 1$ for $\omega \geq 150$

Note that a two degrees-of-freedom controller may be used and K_y then refers to the feedback plant of the controller. In practice, specification S4 is indirectly satisfied by S3. Note that the uncertainty description $G_p = G(I + w_I \Delta_I)$ with $w_I = \frac{s+0.2}{0.5s+1}$ (as used in the examples in the book) only allows for about 0.9 minute time delay error. To get a weight $w_I(s)$ which includes the uncertainty in (12.18) we may use the procedure described in Example 7.5, equations (7.26) or (7.27) with $r_k = 0.2$ and $\theta_{max} = 1$.

Several designs have been presented which satisfy the specifications for the CDC-problem in (12.18). For example, a two degrees-of-freedom $\mathcal{H}_\infty$ loop-shaping design is given by Limebeer et al. (1993), and an extension of this by Whidborne et al. (1994). A two degrees-of-freedom μ-optimal design is presented by Lundström et al. (1996).

12.4.2 Detailed LV-model

In the book we have also used a 5 state dynamic model of the distillation process which includes liquid flow dynamics (in addition to the composition dynamics) as well as disturbances. This 5 state model was obtained from model reduction of the detailed model with 82 states. The steady-state gains for the two disturbances are given in (10.75).

The 5-state model is similar to (12.17) at low frequencies, but the model is much less interactive at higher frequencies. The physical reason for this is that the liquid flow dynamics decouple the response and make $G(j\omega)$ upper triangular at higher frequencies. The effect is illustrated in Figure 12.16 where we show the singular values and the magnitudes of the RGA-elements as functions of frequency. As a comparison, the RGA element $\lambda_{11}(G) = 35.1$ at all frequencies (and not just at steady-state) for the simplified model in (12.17). The implication is that control at crossover frequencies is easier than expected from the simplified model (12.17).

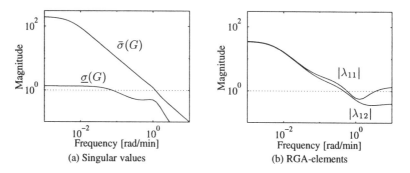

(a) Singular values   (b) RGA-elements

Figure 12.16: Detailed 5-state model of distillation column

Applications based on the 5 state model are found in:

Example 10.7 (page 430): Controllability analysis of partial control and feedforward control.

Example in Section 10.10 (page 445): Controllability analysis of decentralized control.

Details on the 5 state model. A state-space realization is

$$G(s) \stackrel{s}{=} \left[\begin{array}{c|c} A & B \\ \hline C & 0 \end{array} \right], \quad G_d(s) \stackrel{s}{=} \left[\begin{array}{c|c} A & B_d \\ \hline C & 0 \end{array} \right] \tag{12.19}$$

where

$$A = \begin{bmatrix} -.005131 & 0 & 0 & 0 & 0 \\ 0 & -.07366 & 0 & 0 & 0 \\ 0 & 0 & -.1829 & 0 & 0 \\ 0 & 0 & 0 & -.4620 & .9895 \\ 0 & 0 & 0 & -.9895 & -.4620 \end{bmatrix}, \quad B = \begin{bmatrix} -.629 & .624 \\ .055 & -.172 \\ .030 & -.108 \\ -.186 & -.139 \\ -1.23 & -.056 \end{bmatrix}$$

$$C = \begin{bmatrix} -.7223 & -.5170 & .3386 & -.1633 & .1121 \\ -.8913 & .4728 & .9876 & .8425 & .2186 \end{bmatrix}, \quad B_d = \begin{bmatrix} -0.062 & -0.067 \\ 0.131 & 0.040 \\ 0.022 & -0.106 \\ -0.188 & 0.027 \\ -0.045 & 0.014 \end{bmatrix}$$

Scaling. The model is scaled such that a magnitude of 1 corresponds to the following: 0.01 mole fraction units for each output (y_D and x_B), the nominal feed flowrate for the two inputs (L and V) and a 20% change for each disturbance (feed rate F and feed composition z_F). Notice that the steady-state gains computed with this model are slightly different from the ones used in the examples.

Remark. A similar dynamic LV-model, but with 8 states, is given by Green and Limebeer (1995), who also design an $\mathcal{H}_\infty$ loop-shaping controller.

Exercise 12.1 *Repeat the μ-optimal design based on DK-iteration in Section 8.12.4 using the model (12.19).*

12.4.3 Idealized DV-model

Finally, we have also made use of an idealized model for the DV-configuration:

$$G(s) = \frac{1}{75s + 1} \begin{bmatrix} -87.8 & 1.4 \\ -108.2 & -1.4 \end{bmatrix} \tag{12.20}$$

In this case the condition number $\gamma(G) = 70.8$ is still large, but the RGA elements are small (about 0.5).

Example 6.8 (page 243) Bounds on the sensitivity peak show that an inverse-based controller is robust with respect to diagonal input uncertainty.

Example 8.9 (page 319): μ for robust stability with a diagonal controller is computed. The difference between diagonal and full-block input uncertainty is significant.

Remark. In practice, the DV-configuration may not be as favourable as indicated by these examples, because the level controller at the top of the column is not perfect as was assumed when deriving (12.20).

12.4.4   Further distillation case studies

The full distillation model, which is available over the internet, may form the basis for several case studies (projects). These could include input-output controllability analysis, controller design, robustness analysis, and closed-loop simulation. The following cases may be considered:

1. Model with 4 inputs and 4 outputs
2. LV-configuration (studied extensively in this book)
3. DV-configuration
4. DB-configuration

The models in the latter three cases are generated from the 4×4 model by closing two level loops (see the MATLAB file in Table 12.3) to get a partially controlled plant with 2 inputs and 2 outputs (in addition to the two disturbances).

Remark 1 For the DV- and DB-configurations the resulting model depends quite strongly on the tuning of the level loops, so one may consider separately the two cases of tight level control (e.g. $K = 10$, as in Table 12.3) or loosely tuned level control (e.g. $K = 0.2$ corresponding to a time constant of 5 min). Level control tuning may also be considered as a source of uncertainty.

Remark 2 The models do not include actuator or measurement dynamics, which may also be considered as a source of uncertainty.

Remark 3 The model for the DB-configuration contains a pure integrator in one direction and the steady-state gain matrix is therefore singular.

12.5   Conclusion

The case studies in this chapter have served to demonstrate the usefulness and ease of application of many of the techniques discussed in the book. Realistic problems have been considered but the idea has been to illustrate the techniques rather than to provide "optimal" solutions.

For the helicopter problem, practice was obtained in the selection of weights in $\mathcal{H}_\infty$ mixed-sensitivity design, and it was seen how information about disturbances could easily be considered in the design problem.

In the aero-engine study, we applied a variety of tools to the problem of output selection and then designed a two degrees-of-freedom $\mathcal{H}_\infty$ loop-shaping controller.

The final case study was a collection of examples and exercises on the distillation process considered throughout the book. This served to illustrate the difficulties of controlling ill-conditioned plants and the adverse effects of model uncertainty. The structured singular value played an important role in the robustness analysis.

You should now be in a position to move straight to Appendix B, to complete a major project on your own and to sit the sample exam.

Good luck!

APPENDIX A

MATRIX THEORY AND NORMS

The topics in this Appendix are included as background material for the book, and should ideally be studied before reading Chapter 3.

After studying the Appendix the reader should feel comfortable with a range of mathematical tools including eigenvalues, eigenvectors and the singular value decomposition; the reader should appreciate the difference between various norms of vectors, matrices, signals and systems, and know how these norms can be used to measure performance.

The main references are: Strang (1976) and Horn and Johnson (1985) on matrices, and Zhou et al. (1996) on norms.

A.1 Basics

Let us start with a complex scalar

$$c = \alpha + j\beta, \quad \text{where } \alpha = \text{Re } c, \ \beta = \text{Im } c$$

To compute the magnitude $|c|$, we multiply c by its conjugate $\bar{c} \triangleq \alpha - j\beta$ and take the square root, i.e.

$$|c| = \sqrt{\bar{c}c} = \sqrt{\alpha^2 - j^2\beta^2} = \sqrt{\alpha^2 + \beta^2}$$

A complex *column vector* a with m components (elements) is written

$$a = \begin{bmatrix} a_1 \\ a_2 \\ \vdots \\ a_m \end{bmatrix}$$

where a_i is a complex scalar. a^T (the transposed) is used to denote a row vector.

Now consider a complex $l \times m$ matrix A with elements $a_{ij} = \text{Re } a_{ij} + j \text{Im } a_{ij}$. l is the number of rows (the number of "outputs" when viewed as an *operator*) and m

is the number of columns ("inputs"). Mathematically, we write $A \in \mathbb{C}^{l \times m}$ if A is a complex matrix, or $A \in \mathbb{R}^{l \times m}$ if A is a real matrix. Note that a column vector a with m elements may be viewed as an $m \times 1$ matrix.

The transpose of a matrix A is A^T (with elements a_{ji}), the conjugate is $\bar{A}$ (with elements $\operatorname{Re} a_{ij} - j \operatorname{Im} a_{ij}$), the conjugate transpose (or Hermitian adjoint) matrix is $A^H \triangleq \bar{A}^T$ (with elements $\operatorname{Re} a_{ji} - j\operatorname{Im} a_{ji}$), the trace is $\operatorname{tr} A$ (sum of diagonal elements), and the determinant is $\det A$. By definition, the inverse of a non-singular matrix A, denoted A^{-1}, satisfies $A^{-1}A = AA^{-1} = I$, and is given by

$$A^{-1} = \frac{\operatorname{adj} A}{\det A} \tag{A.1}$$

where $\operatorname{adj} A$ is the adjugate (or "classical adjoint") of A which is the transposed matrix of cofactors c_{ij} of A,

$$c_{ij} = [\operatorname{adj} A]_{ji} \triangleq (-1)^{i+j} \det A^{ij} \tag{A.2}$$

Here A^{ij} is a submatrix formed by deleting row i and column j of A. As an example, for a 2×2 matrix we have

$$A = \begin{bmatrix} a_{11} & a_{12} \\ a_{21} & a_{22} \end{bmatrix}; \quad \det A = a_{11}a_{22} - a_{12}a_{21}$$

$$A^{-1} = \frac{1}{\det A} \begin{bmatrix} a_{22} & -a_{12} \\ -a_{21} & a_{11} \end{bmatrix} \tag{A.3}$$

We also have

$$(AB)^T = B^T A^T, \quad (AB)^H = B^H A^H \tag{A.4}$$

and, assuming the inverses exist,

$$(AB)^{-1} = B^{-1} A^{-1} \tag{A.5}$$

A square matrix A is symmetric if $A^T = A$, and Hermitian if $A^H = A$. A Hermitian matrix is said to be positive definite if $x^H A x > 0$ for any non-zero vector x.

A.1.1   Some useful matrix identities

Lemma A.1 The matrix inversion lemma. *Let A_1, A_2, A_3 and A_4 be matrices with compatible dimensions such that the matrices $A_2 A_3 A_4$ and $(A_1 + A_2 A_3 A_4)$ are defined. Also assume that the inverses given below exist. Then*

$$(A_1 + A_2 A_3 A_4)^{-1} = A_1^{-1} - A_1^{-1} A_2 (A_4 A_1^{-1} A_2 + A_3^{-1})^{-1} A_4 A_1^{-1} \tag{A.6}$$

Proof: Postmultiply (or premultiply) the right hand side in (A.6) by $A_1 + A_2 A_3 A_4$. This gives the identity matrix.   □

Lemma A.2 Inverse of a partitioned matrix. *If A_{11}^{-1} and X^{-1} exist then*

$$\begin{bmatrix} A_{11} & A_{12} \\ A_{21} & A_{22} \end{bmatrix}^{-1} = \begin{bmatrix} A_{11}^{-1} + A_{11}^{-1}A_{12}X^{-1}A_{21}A_{11}^{-1} & -A_{11}^{-1}A_{12}X^{-1} \\ -X^{-1}A_{21}A_{11}^{-1} & X^{-1} \end{bmatrix} \quad (A.7)$$

where $X \triangleq A_{22} - A_{21}A_{11}^{-1}A_{12}$ is the Schur complement of A_{11} in A; also see (A.15).
Similarly if A_{22}^{-1} and Y^{-1} exist then

$$\begin{bmatrix} A_{11} & A_{12} \\ A_{21} & A_{22} \end{bmatrix}^{-1} = \begin{bmatrix} Y^{-1} & -Y^{-1}A_{12}A_{22}^{-1} \\ -A_{22}^{-1}A_{21}Y^{-1} & A_{22}^{-1} + A_{22}^{-1}A_{21}Y^{-1}A_{12}A_{22}^{-1} \end{bmatrix} \quad (A.8)$$

where $Y \triangleq A_{11} - A_{12}A_{22}^{-1}A_{21}$ is the Schur complement of A_{22} in A; also see (A.16).

A.1.2 Some determinant identities

The determinant is defined only for square matrices, so let A be an $n \times n$ matrix. The matrix is non-singular if $\det A$ is non-zero. The determinant may be defined inductively as $\det A = \sum_{i=1}^{n} a_{ij}c_{ij}$ (expansion along column j) or $\det A = \sum_{j=1}^{n} a_{ij}c_{ij}$ (expansion along row i), where c_{ij} is the ij'th cofactor given in (A.2). This inductive definition begins by defining the determinant of an 1×1 matrix (a scalar) to be the value of the scalar, i.e. $\det a = a$. We then get for a 2×2 matrix $\det A = a_{11}a_{22} - a_{12}a_{21}$ and so on. From the definition we directly get that $\det A = \det A^T$. Some other determinant identities are given below:

1. Let A_1 and A_2 be square matrices of the same dimension. Then

$$\det(A_1A_2) = \det(A_2A_1) = \det A_1 \cdot \det A_2 \quad (A.9)$$

2. Let c be a complex scalar and A an $n \times n$ matrix. Then

$$\det(cA) = c^n \det(A) \quad (A.10)$$

3. Let A be a non-singular matrix. Then

$$\det A^{-1} = 1/\det A \quad (A.11)$$

4. Let A_1 and A_2 be matrices of compatible dimensions such that both matrices A_1A_2 and A_2A_1 are square (but A_1 and A_2 need not themselves be square). Then

$$\det(I + A_1A_2) = \det(I + A_2A_1) \quad (A.12)$$

This is actually a special case of Schur's formula given in (A.14). (A.12) is useful in the field of control because it yields $\det(I + GK) = \det(I + KG)$.

5. The determinant of a triangular or block-triangular matrix is the product of the determinants of the diagonal blocks:

$$\det \begin{bmatrix} A_{11} & A_{12} \\ 0 & A_{22} \end{bmatrix} = \det \begin{bmatrix} A_{11} & 0 \\ A_{21} & A_{22} \end{bmatrix} = \det(A_{11}) \cdot \det(A_{22}) \quad (A.13)$$

6. **Schur's formula** for the determinant of a partitioned matrix:

$$\det \begin{bmatrix} A_{11} & A_{12} \\ A_{21} & A_{22} \end{bmatrix} = \det(A_{11}) \cdot \det(A_{22} - A_{21}A_{11}^{-1}A_{12})$$

$$= \det(A_{22}) \cdot \det(A_{11} - A_{12}A_{22}^{-1}A_{21}) \quad \text{(A.14)}$$

where it is assumed that A_{11} and/or A_{22} are non-singular.

Proof: Note that A has the following decomposition if A_{11} is non-singular:

$$\begin{bmatrix} A_{11} & A_{12} \\ A_{21} & A_{22} \end{bmatrix} = \begin{bmatrix} I & 0 \\ A_{21}A_{11}^{-1} & I \end{bmatrix} \begin{bmatrix} A_{11} & 0 \\ 0 & X \end{bmatrix} \begin{bmatrix} I & A_{11}^{-1}A_{12} \\ 0 & I \end{bmatrix} \quad \text{(A.15)}$$

where $X = A_{22} - A_{21}A_{11}^{-1}A_{12}$. The first part of (A.14) is proved by evaluating the determinant using (A.9) and (A.13). Similarly, if A_{22} is non-singular,

$$\begin{bmatrix} A_{11} & A_{12} \\ A_{21} & A_{22} \end{bmatrix} = \begin{bmatrix} I & A_{12}A_{22}^{-1} \\ 0 & I \end{bmatrix} \begin{bmatrix} Y & 0 \\ 0 & A_{22} \end{bmatrix} \begin{bmatrix} I & 0 \\ A_{22}^{-1}A_{21} & I \end{bmatrix} \quad \text{(A.16)}$$

where $Y = A_{11} - A_{12}A_{22}^{-1}A_{21}$, and the last part of (A.14) follows. $\square$

A.2 Eigenvalues and eigenvectors

Definition A.1 **Eigenvalues and eigenvectors.** *Let A be a square $n \times n$ matrix. The eigenvalues λ_i, $i = 1, \ldots, n$, are the n solutions to the n'th order characteristic equation*

$$\det(A - \lambda I) = 0 \quad \text{(A.17)}$$

The (right) eigenvector t_i corresponding to the eigenvalue λ_i is the nontrivial solution $(t_i \neq 0)$ to

$$(A - \lambda_i I)t_i = 0 \quad \Leftrightarrow \quad At_i = \lambda_i t_i \quad \text{(A.18)}$$

The corresponding left eigenvectors q_i satisfy

$$q_i^H(A - \lambda_i I) = 0 \quad \Leftrightarrow \quad q_i^H A = \lambda_i q_i^H \quad \text{(A.19)}$$

When we just say eigenvector *we mean the right eigenvector.*

Remark. The left eigenvectors of A are the (right) eigenvectors of A^H.

The eigenvalues are sometimes called characteristic gains. The set of eigenvalues of A is called the spectrum of A. The largest of the absolute values of the eigenvalues of A is the *spectral radius* of A, $\rho(A) \triangleq \max_i |\lambda_i(A)|$.

Note that if t is an eigenvector then so is αt for any constant α. Therefore, the eigenvectors are usually normalized to have unit length, i.e. $t_i^H t_i = 1$. An important result for eigenvectors is that *eigenvectors corresponding to distinct eigenvalues are always linearly independent.* For repeated eigenvalues, this may not always be the case, that is, not all $n \times n$ matrices have n linearly independent eigenvectors (these are the so-called "defective" matrices).

The eigenvectors may be collected as columns in the matrix T and the eigenvalues $\lambda_1, \lambda_2, \ldots, \lambda_n$ as diagonal elements in the matrix Λ:

$$T = \{t_1, t_2, \ldots, t_n\}; \quad \Lambda = \text{diag}\{\lambda_1, \lambda_2, \ldots, \lambda_n\} \tag{A.20}$$

We may then write (A.18) in the following form

$$AT = T\Lambda \tag{A.21}$$

Let us now consider using T for "diagonalization" of the matrix A which is possible when the eigenvectors are linearly independent and T^{-1} exists. This always happens if the eigenvalues are distinct, and may also happen in other cases, e.g. for $A = I$. From (A.21) we then get that the eigenvector matrix diagonalizes A in the following manner

$$\Lambda = T^{-1}AT \tag{A.22}$$

Remark. We use Λ to denote both the eigenvalue matrix and the relative gain array (RGA). Usually this does not cause problems, since λ_i (with one index) always is the eigenvalue, whereas λ_{ij} (with two indices) refers to the ij'th element in the RGA.

A.2.1   Eigenvalue properties

Let λ_i denote the eigenvalues of A in the following properties:

1. The sum of the eigenvalues of A is equal to the trace of A (sum of the diagonal elements): $\text{tr}A = \sum_i \lambda_i$.
2. The product of the eigenvalues of A is equal to the determinant of A: $\det A = \prod_i \lambda_i$.
3. The eigenvalues of an upper or lower triangular matrix are equal to the diagonal elements of the matrix.
4. For a real matrix the eigenvalues are either real, or occur in complex conjugate pairs.
5. A and A^T have the same eigenvalues (but in general different eigenvectors).
6. The inverse A^{-1} exists if and only if all eigenvalues of A are non-zero. The eigenvalues of A^{-1} are then $1/\lambda_1, \ldots, 1/\lambda_n$.
7. The matrix $A + cI$ has eigenvalues $\lambda_i + c$.
8. The matrix cA^k where k is an integer has eigenvalues $c\lambda_i^k$.
9. Consider the $l \times m$ matrix A and the $m \times l$ matrix B. Then the $l \times l$ matrix AB and the $m \times m$ matrix BA have the same non-zero eigenvalues. To be more specific assume $l > m$. Then the matrix AB has the same m eigenvalues as BA plus $l - m$ eigenvalues which are identically equal to zero.

10. Eigenvalues are invariant under similarity transformations, that is, A and DAD^{-1} have the same eigenvalues.

11. The same eigenvector matrix diagonalizes the matrix A and the matrix $(I + A)^{-1}$. *(Proof:* $T^{-1}(I + A)^{-1}T = (T^{-1}(I + A)T)^{-1} = (I + \Lambda)^{-1}$.)

12. *Gershgorin's theorem.* The eigenvalues of the $n \times n$ matrix A lie in the union of n circles in the complex plane, each with centre a_{ii} and radius $r_i = \sum_{j \neq i} |a_{ij}|$ (sum of off-diagonal elements in row i). They also lie in the union of n circles, each with centre a_{ii} and radius $r'_i = \sum_{j \neq i} |a_{ji}|$ (sum of off-diagonal elements in column i).

13. A matrix is positive definite if and only if all its eigenvalues are real and positive.

From the above properties we have, for example, that

$$\lambda_i(S) = \lambda_i((I + L)^{-1}) = \frac{1}{\lambda_i(I + L)} = \frac{1}{1 + \lambda_i(L)} \qquad (A.23)$$

In this book we are sometimes interested in the eigenvalues of a real (state) matrix, and in other cases in the eigenvalues of a complex transfer function matrix evaluated at a given frequency, e.g. $L(j\omega)$. It is important to appreciate this difference.

A.2.2 Eigenvalues of the state matrix

Consider a system described by the linear differential equations

$$\dot{x} = Ax + Bu \qquad (A.24)$$

Unless A is diagonal this is a set of coupled differential equations. For simplicity, assume that the eigenvectors of A are linearly independent and introduce the new state vector $z = T^{-1}x$, that is, $x = Tz$. We then get

$$T\dot{z} = ATz + Bu \quad \Leftrightarrow \quad \dot{z} = \Lambda z + T^{-1}Bu \qquad (A.25)$$

which is a set of uncoupled differential equations in terms of the new states $z = Tx$. The unforced solution (i.e. with $u = 0$) for each state z_i is $z_i = z_{0i}e^{\lambda_i t}$ where z_{0i} is the value of the state at $t = 0$. If λ_i is real, then we see that this mode is stable ($z_i \to 0$ as $t \to \infty$) if and only if $\lambda_i < 0$. If $\lambda_i = \text{Re}\lambda_i + j\text{Im}\lambda_i$ is complex, then we get $e^{\lambda_i t} = e^{\text{Re}\lambda_i t}(\cos(\text{Im}\lambda_i t) + j\sin(\text{Im}\lambda_i t))$ and the mode is stable ($z_i \to 0$ as $t \to \infty$) if and only if $\text{Re}\lambda_i < 0$. The fact that the new state z_i is complex is of no concern since the real physical states $x = Tz$ are of course real. Consequently, a linear system is stable if and only if all the eigenvalues of the state matrix A have real parts less than 0, that is, lie in the open left-half plane.

A.2.3 Eigenvalues of transfer functions

The eigenvalues of the loop transfer function matrix, $\lambda_i(L(j\omega))$, evaluated as a function of frequency, are sometimes called the characteristic loci, and to some extent they generalize $L(j\omega)$ for a scalar system. In Chapter 8, we make use of $\lambda_i(L)$ to study the stability of the $M\Delta$-structure where $L = M\Delta$. Even more important in this context is the spectral radius, $\rho(L) = \max_i |\lambda_i(L(j\omega))|$.

A.3 Singular Value Decomposition

Definition A.2 Unitary matrix. *A (complex) matrix U is unitary if*

$$U^H = U^{-1} \tag{A.26}$$

All the eigenvalues of a unitary matrix have absolute value equal to 1, and all its singular values (as we shall see from the definition below) are equal to 1.

Definition A.3 SVD. *Any complex $l \times m$ matrix A may be factorized into a singular value decomposition*

$$A = U\Sigma V^H \tag{A.27}$$

where the $l \times l$ matrix U and the $m \times m$ matrix V are unitary, and the $l \times m$ matrix Σ contains a diagonal matrix Σ_1 of real, non-negative singular values, σ_i, arranged in a descending order as in

$$\Sigma = \begin{bmatrix} \Sigma_1 \\ 0 \end{bmatrix}; \quad l \geq m \tag{A.28}$$

or

$$\Sigma = [\Sigma_1 \quad 0]; \quad l \leq m \tag{A.29}$$

where

$$\Sigma_1 = \text{diag}\{\sigma_1, \sigma_2, \ldots, \sigma_k\}; \quad k = \min(l, m) \tag{A.30}$$

and

$$\bar{\sigma} \equiv \sigma_1 \geq \sigma_2 \geq \ldots \geq \sigma_k \equiv \underline{\sigma} \tag{A.31}$$

The unitary matrices U and V form orthonormal bases for the column (output) space and the row (input) space of A. The column vectors of V, denoted v_i, are called right or input singular vectors and the column vectors of U, denoted u_i, are called left or output singular vectors. We define $\bar{u} \equiv u_1$, $\bar{v} \equiv v_1$, $\underline{u} \equiv u_k$ and $\underline{v} \equiv v_k$.

Note that the decomposition in (A.27) is not unique since $A = U'\Sigma V'^H$, where $U' = US$, $V' = VS^{-1}$, $S = \text{diag}\{e^{j\theta_i}\}$ and θ_i is any real number, is also an SVD of A. However, the singular values, σ_i, *are* unique.

The singular values are the positive square roots of the $k = \min(l, m)$ largest eigenvalues of both AA^H and $A^H A$. We have

$$\sigma_i(A) = \sqrt{\lambda_i(A^H A)} = \sqrt{\lambda_i(AA^H)} \tag{A.32}$$

Also, the columns of U and V are unit eigenvectors of AA^H and $A^H A$, respectively. To derive (A.32) write

$$AA^H = (U\Sigma V^H)(U\Sigma V^H)^H = (U\Sigma V^H)(V\Sigma^H U^H) = U\Sigma\Sigma^H U^H \qquad (A.33)$$

or equivalently since U is unitary and satisfies $U^H = U^{-1}$ we get

$$(AA^H)U = U\Sigma\Sigma^H \qquad (A.34)$$

We then see that U is the matrix of eigenvectors of AA^H and $\{\sigma_i^2\}$ are its eigenvalues. Similarly, we have that V is the matrix of eigenvectors of $A^H A$.

A.3.1 Rank

Definition A.4 *The **rank** of a matrix is equal to the number of non-zero singular values of the matrix. Let* $rank(A) = r$*, then the matrix A is called rank deficient if* $r < k = \min(l, m)$*, and we have singular values* $\sigma_i = 0$ *for* $i = r + 1, \ldots k$*. A rank deficient square matrix is a singular matrix (non-square matrices are always singular).*

The rank of a matrix is unchanged after left or right multiplication by a non-singular matrix. Furthermore, for an $l \times m$-matrix A and an $m \times p$-matrix B, the rank of their product AB is bounded as follows (Sylvester's inequality):

$$\text{rank}(A) + \text{rank}(B) - m \le \text{rank}(AB) \le \min(\text{rank}(A), \text{rank}(B)) \qquad (A.35)$$

A.3.2 Singular values of a 2×2 matrix

In general, the singular values must be computed numerically. For 2×2 matrices, however, an analytic expression is easily derived. Introduce

$$b \triangleq tr(A^H A) = \sum_{i,j} |a_{ij}|^2, \quad c \triangleq \det(A^H A)$$

Now the sum of the eigenvalues of a matrix is equal to its trace and the product is equal to its determinant, so

$$\lambda_1 + \lambda_2 = b, \quad \lambda_1 \cdot \lambda_2 = c$$

Upon solving for λ_1 and λ_2, and using $\sigma_i(A) = \sqrt{\lambda_i(A^H A)}$ we get

$$\bar\sigma(A) = \sqrt{\frac{b + \sqrt{b^2 - 4c}}{2}}; \quad \underline\sigma(A) = \sqrt{\frac{b - \sqrt{b^2 - 4c}}{2}} \qquad (A.36)$$

For example, for $A = \begin{bmatrix} 1 & 2 \\ 3 & 4 \end{bmatrix}$ we have $b = \sum |a_{ij}|^2 = 1 + 4 + 9 + 16 = 30$, $c = (\det A)^2 = (-2)^2 = 4$, and we find $\bar\sigma(A) = 5.465$ and $\underline\sigma(A) = 0.366$.

Note that for singular 2×2 matrices (with $\det A = 0$ and $\underline{\sigma}(A) = 0$) we get $\bar{\sigma}(A) = \sqrt{\sum |a_{ij}|^2} \triangleq \|A\|_F$ (the Frobenius norm), which is actually a special case of (A.126).

A.3.3 SVD of a matrix inverse

Since $A = U \Sigma V^H$ we get, provided the $m \times m$ A is non-singular, that

$$A^{-1} = V \Sigma^{-1} U^H \tag{A.37}$$

This is the SVD of A^{-1} but with the order of the singular values reversed. Let $j = m - i + 1$. Then it follows from (A.37) that

$$\sigma_i(A^{-1}) = 1/\sigma_j(A), \quad u_i(A^{-1}) = v_j(A), \quad v_i(A^{-1}) = u_j(A) \tag{A.38}$$

and in particular

$$\bar{\sigma}(A^{-1}) = 1/\underline{\sigma}(A) \tag{A.39}$$

A.3.4 Singular value inequalities

The singular values bound the magnitude of the eigenvalues (also see (A.116)):

$$\underline{\sigma}(A) \leq |\lambda_i(A)| \leq \bar{\sigma}(A) \tag{A.40}$$

The following is obvious from the SVD-definition:

$$\bar{\sigma}(A^H) = \bar{\sigma}(A) \quad \text{and} \quad \bar{\sigma}(A^T) = \bar{\sigma}(A) \tag{A.41}$$

The next important property is proved below (eq. A.97):

$$\bar{\sigma}(AB) \leq \bar{\sigma}(A)\bar{\sigma}(B) \tag{A.42}$$

For a non-singular A (or B) we also have a lower bound on $\bar{\sigma}(AB)$

$$\underline{\sigma}(A)\bar{\sigma}(B) \leq \bar{\sigma}(AB) \quad \text{or} \quad \bar{\sigma}(A)\underline{\sigma}(B) \leq \bar{\sigma}(AB) \tag{A.43}$$

We also have a lower bound on the minimum singular value

$$\underline{\sigma}(A)\underline{\sigma}(B) \leq \underline{\sigma}(AB) \tag{A.44}$$

For a partitioned matrix the following inequalities are useful:

$$\max\{\bar{\sigma}(A), \bar{\sigma}(B)\} \leq \bar{\sigma} \begin{bmatrix} A \\ B \end{bmatrix} \leq \sqrt{2} \max\{\bar{\sigma}(A), \bar{\sigma}(B)\} \tag{A.45}$$

$$\bar{\sigma} \begin{bmatrix} A \\ B \end{bmatrix} \leq \bar{\sigma}(A) + \bar{\sigma}(B) \tag{A.46}$$

The following equality for a block-diagonal matrix is used extensively in the book:

$$\bar{\sigma}\begin{bmatrix} A & 0 \\ 0 & B \end{bmatrix} = \max\{\bar{\sigma}(A), \bar{\sigma}(B)\} \tag{A.47}$$

Another very useful result is Fan's theorem (Horn and Johnson, 1991, p. 140 and p. 178):

$$\sigma_i(A) - \bar{\sigma}(B) \leq \sigma_i(A+B) \leq \sigma_i(A) + \bar{\sigma}(B) \tag{A.48}$$

Two special cases of (A.48) are:

$$|\bar{\sigma}(A) - \bar{\sigma}(B)| \leq \bar{\sigma}(A+B) \leq \bar{\sigma}(A) + \bar{\sigma}(B) \tag{A.49}$$

$$\underline{\sigma}(A) - \bar{\sigma}(B) \leq \underline{\sigma}(A+B) \leq \underline{\sigma}(A) + \bar{\sigma}(B) \tag{A.50}$$

(A.50) yields

$$\underline{\sigma}(A) - 1 \leq \underline{\sigma}(I+A) \leq \underline{\sigma}(A) + 1 \tag{A.51}$$

On combining (A.39) and (A.51) we get a relationship that is useful when evaluating the amplification of closed-loop systems:

$$\underline{\sigma}(A) - 1 \leq \frac{1}{\bar{\sigma}(I+A)^{-1}} \leq \underline{\sigma}(A) + 1 \tag{A.52}$$

A.3.5 SVD as a sum of rank 1 matrices

Let r denote the rank of the $l \times m$ matrix A. We may then consider the SVD as a decomposition of A into r $l \times m$ matrices, each of rank 1. We have

$$A = U\Sigma V^H = \sum_{i=1}^{r} \sigma_i u_i v_i^H \tag{A.53}$$

The remaining terms from $r+1$ to $k = \min(l, m)$ have singular values equal to 0 and give no contribution to the sum. The first and most important submatrix is given by $A_1 = \sigma_1 u_1 v_1^H$. If we now consider the residual matrix

$$A^1 = A - A_1 = A - \sigma_1 u_1 v_1^H \tag{A.54}$$

then

$$\sigma_1(A^1) = \sigma_2(A) \tag{A.55}$$

That is, the largest singular value of A^1 is equal to the second singular value of the original matrix. This shows that the direction corresponding to $\sigma_2(A)$ is the second most important direction, and so on.

A.3.6   Singularity of matrix $A + E$

From the left inequality in (A.50) we find that

$$\bar{\sigma}(E) < \underline{\sigma}(A) \quad \Rightarrow \quad \underline{\sigma}(A + E) > 0 \tag{A.56}$$

and $A + E$ is non-singular. On the other hand, there always exists an E with $\bar{\sigma}(E) = \underline{\sigma}(A)$ which makes $A + E$ singular, e.g. choose $E = -\underline{u}\,\underline{\sigma}\,\underline{v}^H$; see (A.53). Thus the smallest singular value $\underline{\sigma}(A)$ measures how near the matrix A is to being singular or rank deficient. This test is often used in numerical analysis, and it is also an important inequality in the formulation of robustness tests.

A.3.7   Economy-size SVD

Since there are only $r = \text{rank}(A) \leq \min(l, m)$ non-zero singular values, and since only the non-zero singular values contribute to the overall result, the singular value decomposition of A is sometimes written as an economy-size SVD, as follows

$$A^{l \times m} = U_r^{l \times r} \Sigma_r^{r \times r} (V_r^{m \times r})^H \tag{A.57}$$

where the matrices U_r and V_r contain only the first r columns of the matrices U and V introduced above. Here we have used the notation $A^{l \times m}$ to indicate that A is an $l \times m$ matrix. The economy-size SVD is used for computing the pseudo inverse, see (A.60).

Remark. The "economy-size SVD" presently used in MATLAB is not quite as economic as the one given in (A.57) as it uses m instead of r.

A.3.8   Pseudo-inverse (Generalized inverse)

Consider the set of linear equations

$$y = Ax \tag{A.58}$$

with a given $l \times 1$ vector y and a given $l \times m$ matrix A. A *least squares* solution to (A.58) is an $m \times 1$ vector x such that $\|x\|_2 = \sqrt{x_1^2 + x_2^2 \cdots x_m^2}$ is minimized among all vectors for which $\|y - Ax\|_2$ is minimized. The solution is given in terms of the pseudo-inverse (Moore-Penrose generalized inverse) of A:

$$x = A^\dagger y \tag{A.59}$$

The pseudo-inverse may be obtained from an SVD of $A = U\Sigma V^H$ by

$$A^\dagger = V_r \Sigma_r^{-1} U_r^H = \sum_{i=1}^{r} \frac{1}{\sigma_i(A)} v_i u_i^H \tag{A.60}$$

where r is the number of non-zero singular values of A. We have that

$$\underline{\sigma}(A) = 1/\bar{\sigma}(A^\dagger) \tag{A.61}$$

Note that $A^\dagger$ exists for any matrix A, even for a singular square matrix and a non-square matrix. The pseudo-inverse also satisfies

$$AA^\dagger A = A \quad \text{and} \quad A^\dagger AA^\dagger = A^\dagger$$

Note the following cases (where r is the rank of A):

1. $r = l = m$, i.e. A is non-singular. In this case $A^\dagger = A^{-1}$ is the inverse of the matrix.
2. $r = m \le l$, i.e. A has full column rank. This is the "conventional least squares problem" where we want to minimize $\|y - Ax\|_2$, and the solution is

$$A^\dagger = (A^H A)^{-1} A^H \tag{A.62}$$

   In this case $A^\dagger A = I$, so $A^\dagger$ is a *left inverse* of A.
3. $r = l \le m$, i.e. A has full row rank. In this case we have an infinite number of solutions to (A.58) and we seek the one that minimizes $\|x\|_2$. We get

$$A^\dagger = A^H (AA^H)^{-1} \tag{A.63}$$

   In this case $AA^\dagger = I$, so $A^\dagger$ is a *right inverse* of A.
4. $r < k = \min(l, m)$ (general case). In this case both matrices $A^H A$ and AA^H are rank deficient, and we have to use the SVD to obtain the pseudo-inverse. In this case A has neither a left nor a right inverse.

Principal component regression (PCR)

We note that the pseudo-inverse in (A.60) may be very sensitive to noise and "blow up" if the smallest non-zero singular value, σ_r is small. In the PCR method one avoids this problem by using only the $q \le r$ first singular values which can be distinguished from the noise. The PCR pseudo-inverse then becomes

$$A_{PCR}^\dagger = \sum_{i=1}^{q} \frac{1}{\sigma_i} v_i u_i^H \tag{A.64}$$

Remark. This is similar in spirit to the use of Hankel singular values for model reduction.

A.3.9 Condition number

The **condition number** of a matrix is defined in this book as the ratio

$$\gamma(A) = \sigma_1(A)/\sigma_k(A) = \bar{\sigma}(A)/\underline{\sigma}(A) \tag{A.65}$$

where $k = \min(l, m)$. A matrix with a large condition number is said to be ill-conditioned. This definition yields an infinite condition number for rank deficient matrices. For a non-singular matrix we get from (A.39)

$$\gamma(A) = \bar{\sigma}(A) \cdot \bar{\sigma}(A^{-1}) \qquad (A.66)$$

Other definitions for the condition number of a non-singular matrix are also in use, for example,

$$\gamma_p(A) = \|A\| \cdot \|A^{-1}\| \qquad (A.67)$$

where $\|A\|$ denotes any matrix norm. If we use the induced 2-norm (maximum singular value) then this yields (A.66). From (A.66) and (A.42), we get for non-singular matrices

$$\gamma(AB) \le \gamma(A)\gamma(B) \qquad (A.68)$$

The **minimized condition number** is obtained by minimizing the condition number over all possible scalings. We have

$$\gamma^*(G) \triangleq \min_{D_I, D_O} \gamma(D_O G D_I) \qquad (A.69)$$

where D_I and D_O are diagonal scaling matrices. For a 2×2 matrix, the minimized condition number is given by (Grosdidier et al., 1985) as

$$\gamma^*(G) = \|\Lambda\|_{i1} + \sqrt{\|\Lambda\|_{i1}^2 - 1} \qquad (A.70)$$

where $\|\Lambda\|_{i1}$ is the induced 1-norm (maximum column sum) of the RGA-matrix of G. Note that, $\Lambda(G) = I$ and $\gamma^*(G) = 1$ for a triangular 2×2 matrix. If we allow only scaling on one side then we get the input and output minimized condition numbers:

$$\gamma_I^*(G) \triangleq \min_{D_I} \gamma(G D_I), \quad \gamma_O^*(G) \triangleq \min_{D_O} \gamma(D_O G) \qquad (A.71)$$

Remark. To compute these minimized condition numbers we define

$$H = \begin{bmatrix} 0 & G^{-1} \\ G & 0 \end{bmatrix} \qquad (A.72)$$

Then we have, as proven by Braatz and Morari (1994):

$$\sqrt{\gamma^*(G)} = \min_{D_I, D_O} \bar{\sigma}(DHD^{-1}), \quad D = \text{diag}\{D_I, D_O\} \qquad (A.73)$$

$$\sqrt{\gamma_I^*(G)} = \min_{D_I} \bar{\sigma}(DHD^{-1}), \quad D = \text{diag}\{D_I, I\} \qquad (A.74)$$

$$\sqrt{\gamma_O^*(G)} = \min_{D_O} \bar{\sigma}(DHD^{-1}), \quad D = \text{diag}\{I, D_O\} \qquad (A.75)$$

These convex optimization problems may be solved using available software for the upper bound on the structured singular value $\mu_\Delta(H)$; see (8.87). In calculating $\mu_\Delta(H)$, we use for $\gamma^*(G)$ the structure $\Delta = \text{diag}\{\Delta_{\text{diag}}, \Delta_{\text{diag}}\}$, for $\gamma_I^*(G)$ the structure $\Delta = \text{diag}\{\Delta_{\text{diag}}, \Delta_{\text{full}}\}$, and for $\gamma_O^*(G)$ the structure $\Delta = \text{diag}\{\Delta_{\text{full}}, \Delta_{\text{diag}}\}$.

A.4   Relative Gain Array

The relative gain array (RGA) was introduced by Bristol (1966). Many of its properties were stated by Bristol, but they were not proven rigorously until the work by Grosdidier et al. (1985). Some additional properties are given in Hovd and Skogestad (1992).

The Relative Gain Array of a complex non-singular $m \times m$ matrix A, denoted RGA(A) or $\Lambda(A)$, is a complex $m \times m$ matrix defined by

$$\text{RGA}(A) \equiv \Lambda(A) \triangleq A \times (A^{-1})^T \tag{A.76}$$

where the operation $\times$ denotes element by element multiplication (Hadamard or Schur product). If A is real then $\Lambda(A)$ is also real.
Example:

$$A_1 = \begin{bmatrix} 1 & -2 \\ 3 & 4 \end{bmatrix}, \; A_1^{-1} = \begin{bmatrix} 0.4 & 0.2 \\ -0.3 & 0.1 \end{bmatrix}, \; \Lambda(A_1) = \begin{bmatrix} 0.4 & 0.6 \\ 0.6 & 0.4 \end{bmatrix}$$

A.4.1   Properties of the RGA

Most of the properties below follow directly if we write the RGA-elements in the form

$$\lambda_{ij} = a_{ij} \cdot \widehat{a}_{ji} = a_{ij} \frac{c_{ij}}{\det A} = (-1)^{i+j} \frac{a_{ij} \det A^{ij}}{\det A} \tag{A.77}$$

where $\widehat{a}_{ji}$ denotes the ji'th element of the matrix $\widehat{A} \triangleq A^{-1}$, A^{ij} denotes the matrix A with row i and column j deleted, and $c_{ij} = (-1)^{i+j} \det A^{ij}$ is the ij'th cofactor of the matrix A.

For any non-singular $m \times m$ matrix A, the following properties hold:

1. $\Lambda(A^{-1}) = \Lambda(A^T) = \Lambda(A)^T$
2. Any permutation of the rows and columns of A results in the same permutation in the RGA. That is, $\Lambda(P_1 A P_2) = P_1 \Lambda(A) P_2$ where P_1 and P_2 are permutation matrices. (A permutation matrix has a single 1 in every row and column and all other elements equal to 0.) $\Lambda(P) = P$ for any permutation matrix.
3. The sum of the elements in each row (and each column) of the RGA is 1. That is, $\sum_{i=1}^m \lambda_{ij} = 1$ and $\sum_{j=1}^m \lambda_{ij} = 1$.
4. $\Lambda(A) = I$ if and only if A is a lower or upper triangular matrix; and in particular the RGA of a diagonal matrix is the identity matrix.
5. The RGA is scaling invariant. Therefore, $\Lambda(D_1 A D_2) = \Lambda(A)$ where D_1 and D_2 are diagonal matrices.
6. The RGA is a measure of sensitivity to relative element-by-element uncertainty in the matrix. More precisely, the matrix A becomes singular if a single element in A is perturbed from a_{ij} to $a'_{ij} = a_{ij}(1 - \frac{1}{\lambda_{ij}})$.

7. The norm of the RGA is closely related to the minimized condition number, $\gamma^*(A) = \min_{D_1,D_2} \gamma(D_1 A D_2)$ where D_1 and D_2 are diagonal matrices. We have the following lower and conjectured upper bounds on $\gamma^*(A)$

$$\|\Lambda\|_m - \frac{1}{\gamma^*(A)} \le \gamma^*(A) \le \|\Lambda\|_{\text{sum}} + k(m) \tag{A.78}$$

where $k(m)$ is a constant, $\|\Lambda\|_m \triangleq 2\max\{\|\Lambda\|_{i1}, \|\Lambda\|_{i\infty}\}$, and $\|\Lambda\|_{\text{sum}} = \sum_{ij} |\lambda_{ij}|$ (the matrix norms are defined in Section A.5.2). The lower bound is proved by Nett and Manousiouthakis (1987). The upper bound is proved for 2×2 matrices with $k(2) = 0$ (Grosdidier et al., 1985), but it is only conjectured for the general case with $k(3) = 1$ and $k(4) = 2$ (Skogestad and Morari, 1987c; Nett and Manousiouthakis, 1987). Note that $\|\Lambda\|_m \le \|\Lambda\|_{\text{sum}}$ where the equality always holds in the 2×2 case. Consequently, for 2×2 matrices γ^* and $\|\Lambda\|_{\text{sum}}$ are always very close in magnitude (also see (A.70)):

$$2 \times 2 \text{ matrix}: \quad \|\Lambda\|_{\text{sum}} - \frac{1}{\gamma^*(A)} \le \gamma^*(A) \le \|\Lambda\|_{\text{sum}} \tag{A.79}$$

8. The diagonal elements of the matrix ADA^{-1} are given in terms of the corresponding row-elements of the RGA (Skogestad and Morari, 1987c). For any diagonal matrix $D = \text{diag}\{d_i\}$ we have

$$[ADA^{-1}]_{ii} = \sum_{j=1}^{m} \lambda_{ij}(A) d_j \tag{A.80}$$

$$[A^{-1}DA]_{ii} = \sum_{i=1}^{m} \lambda_{ij}(A) d_i \tag{A.81}$$

9. It follows from Property 3 that Λ always has at least one eigenvalue and one singular value equal to 1.

Proofs of some of the properties: Property 3: Since $AA^{-1} = I$ it follows that $\sum_{j=1}^{m} a_{ij}\hat{a}_{ji} = 1$. From the definition of the RGA we then have that $\sum_{j=1}^{m} \lambda_{ij} = 1$. *Property 4*: If the matrix is upper triangular then $a_{ij} = 0$ for $i > j$. It then follows that $c_{ij} = 0$ for $j > i$ and all the off-diagonal RGA-elements are zero. *Property 5*: Let $A' = D_1 A D_2$. Then $a'_{ij} = d_{1i} d_{2j} a_{ij}$ and $\hat{a}'_{ij} = \frac{1}{d_{2j}} \frac{1}{d_{1i}} \hat{a}_{ij}$ and the result follows. *Property 6*: The determinant can be evaluated by expanding it in terms of any row or column, e.g. by row i, $\det A = \sum_i (-1)^{i+j} a_{ij} \det A^{ij}$. Let A' denote A with a'_{ij} substituted for a_{ij}. By expanding the determinant of A' by row i and then using (A.77) we get

$$\det A' = \underbrace{\det A - (-1)^{i+j} \frac{a_{ij}}{\lambda_{ij}} \det A^{ij}}_{\det A} = 0$$

Property 8: The ii'th element of the matrix $B = ADA^{-1}$ is $b_{ii} = \sum_j d_j a_{ij}\hat{a}_{ji} = \sum_j d_j \lambda_{ij}$.
□   □

Example A.1

$$A_2 = \begin{bmatrix} 56 & 66 & 75 & 97 \\ 75 & 54 & 82 & 28 \\ 18 & 66 & 25 & 38 \\ 9 & 51 & 8 & 11 \end{bmatrix} ; \quad \Lambda(A_2) = \begin{bmatrix} 6.16 & -0.69 & -7.94 & 3.48 \\ -1.77 & 0.10 & 3.16 & -0.49 \\ -6.60 & 1.73 & 8.55 & -2.69 \\ 3.21 & -0.14 & -2.77 & 0.70 \end{bmatrix} \quad \text{(A.82)}$$

In this case, $\gamma(A_2) = \bar{\sigma}(A_2)/\underline{\sigma}(A_2) = 207.68/1.367 = 151.9$ and $\gamma^*(A_2) = 51.73$ (obtained numerically using (A.73)). Furthermore, $\|\Lambda\|_m = 2\max\{22.42, 19.58\} = 44.84$, and $\|\Lambda\|_{\text{sum}} = 50.19$, so (A.78) with $k(m) = k(4) = 2$ is satisfied. The matrix A_2 is non-singular and the 1, 3-element of the RGA is $\lambda_{13}(A_2) = -7.94$. Thus from Property 6 the matrix A_2 becomes singular if the 1, 3-element is perturbed from 75 to $75(1 - \frac{1}{-7.94}) = 84.45$.

A.4.2 RGA of a non-square matrix

The RGA may be generalized to a non-square $l \times m$ matrix A by use of the pseudo inverse $A^\dagger$ defined in (A.60). We have

$$\Lambda(A) = A \times (A^\dagger)^T \tag{A.83}$$

Properties 1 (transpose and inverse) and 2 (permutations) of the RGA also hold for non-square matrices, but the remaining properties do not apply in the general case. However, they partly apply if A is either of full row rank or full column rank.

1. *A has full row rank, $r = \text{rank}(A) = l$* (i.e. A has at least as many inputs as outputs, and the outputs are linearly independent). In this case $AA^\dagger = I$, and the following properties hold:

   (a) The RGA is independent of output scaling, i.e. $\Lambda(DA) = \Lambda(A)$.
   (b) The elements in each row of the RGA sum to 1, $\sum_j^m \lambda_{ij} = 1$.
   (c) The elements of column j of the RGA sum to the square of the 2-norm of the j'th row in V_r,

   $$\sum_{i=1}^l \lambda_{ij} = \|e_j^T V_r\|_2^2 \le 1 \tag{A.84}$$

   Here V_r contains the first r input singular vectors for G, and e_j is an $m \times 1$ basis vector for input u_j; $e_j = \begin{bmatrix} 0 & \cdots & 0 & 1 & 0 & \cdots & 0 \end{bmatrix}^T$ where 1 appears in position j.

   (d) The diagonal elements of $B = ADA^\dagger$ are $b_{ii} = \sum_{j=1}^m d_j a_{ij} \hat{a}_{ji} = \sum_{j=1}^m d_j \lambda_{ij}$, where $\hat{a}_{ji}$ denotes the ji'th element of $A^\dagger$ and D is any diagonal matrix.

2. *A has full column rank, $r = \text{rank}(A) = m$* (i.e. A has no more inputs than outputs, and the inputs are linearly independent). In this case $A^\dagger A = I$, and the following properties hold:

   (a) The RGA is independent of input scaling, i.e. $\Lambda(AD) = \Lambda(A)$.

(b) The elements in each column of the RGA sum to 1, $\sum_i^l \lambda_{ij} = 1$.

(c) The elements of row i of the RGA sum to the square of the 2-norm of the i'th row in U_r,

$$\sum_{i=1}^m \lambda_{ij} = \|e_i^T U_r\|_2^2 \leq 1 \qquad (A.85)$$

Here U_r contains the first r output singular vectors for G, and e_i is an $l \times 1$ basis vector for output y_i; $e_i = \begin{bmatrix} 0 & \cdots & 0 & 1 & 0 & \cdots & 0 \end{bmatrix}^T$ where 1 appears in position i.

(d) The diagonal elements of $B = A^\dagger D A$ are equal to $b_{jj} = \sum_{i=1}^l \hat{a}_{ji} d_i a_{ij} = \sum_{i=1}^l d_i \lambda_{ij}$, where $\hat{a}_{ji}$ denotes the ji'th element of $A^\dagger$ and D is any diagonal matrix.

3. *General case.* For a general square or non-square matrix which has neither full row nor full column rank, identities (A.84) and (A.85) still apply.

From this it also follows that the rank of any matrix is equal to the sum of its RGA-elements: Let the $l \times m$ matrix G have rank r, then

$$\sum_{i,j} \lambda_{ij}(G) = \text{rank}(G) = r \qquad (A.86)$$

Proofs of (A.84) and (A.85): We will prove these identities for the general case. Write the SVD of G as $G = U_r \Sigma_r V_r^H$ (this is the economy-size SVD from (A.57)) where Σ_r is invertible. We have that $g_{ij} = e_i^H U_r \Sigma_r V_r^H e_j$, $[G^\dagger]_{ji} = e_j^H V_r \Sigma_r^{-1} U_r^H e_i$, $U_r^H U_r = I_r$ and $V_r^H V_r = I_r$, where I_r denotes identity matrix of dim $r \times r$. For the row sum (A.85) we then get

$$\sum_{j=1}^m \lambda_{ij} = \sum_{j=1}^m e_i^H U_r \Sigma_r V_r^H e_j e_j^H V_r \Sigma_r^{-1} U_r^H e_i =$$

$$e_i^H U_r \Sigma_r V_r^H \underbrace{\sum_{j=1}^m e_j e_j^H}_{I_m} V_r \Sigma_r^{-1} U_r^H e_i = e_i^H U_r U_r^H e_i = \|e_i^H U_r\|_2^2$$

The result for the column sum (A.84) is proved in a similar fashion. $\qquad \square$

Remark. The extension of the RGA to non-square matrices was suggested by Chang and Yu (1990) who also stated most of its properties, although in a somewhat incomplete form. More general and precise statements are found in e.g. Cao (1995).

A.4.3 Computing the RGA with MATLAB

If G is a constant matrix then the RGA can be computed using

```
RGA = G.*pinv(G.');
```

If $G(j\omega)$ is a frequency-dependent matrix generated using the μ toolbox, e.g.

```
G=pck(A,B,C,D); omega=logspace(-2,2,41); Gw=frsp(G,omega);
```

then the RGA as a function of frequency can be computed using

```
RGAw = veval('.*',Gw,vpinv(vtp(Gw)));
```

A.5   Norms

It is useful to have a single number which gives an overall measure of the size of a vector, a matrix, a signal, or a system. For this purpose we use functions which are called norms. The most commonly used norm is the Euclidean vector norm, $\|e\|_2 = \sqrt{|e_1|^2 + |e_2|^2 + \cdots |e_m|^2}$. This is simply the distance between two points y and x, where $e_i = y_i - x_i$ is the difference in their i'th coordinates.

Definition A.5 *A norm of e (which may be a vector, matrix, signal or system) is a real number, denoted* $\|e\|$, *that satisfies the following properties:*

1. *Non-negative:* $\|e\| \geq 0$.
2. *Positive:* $\|e\| = 0 \Leftrightarrow e = 0$ *(for semi-norms we have* $\|e\| = 0 \Leftarrow e = 0$).
3. *Homogeneous:* $\|\alpha \cdot e\| = |\alpha| \cdot \|e\|$ *for all complex scalars* α.
4. *Triangle inequality:*

$$\|e_1 + e_2\| \leq \|e_1\| + \|e_2\| \tag{A.87}$$

More precisely, e is an element in a vector space V over the field $\mathbb{C}$ of complex numbers, and the properties above must be satisfied $\forall e, e_1, e_2 \in V$ and $\forall \alpha \in \mathbb{C}$.

In this book we consider the norms of four different objects (norms on four different vector spaces):

1. e is a constant vector.
2. e is a constant matrix.
3. e is a time dependent signal, $e(t)$, which at each fixed t is a constant scalar or vector.
4. e is a "system", a transfer function $G(s)$ or impulse response $g(t)$, which at each fixed s or t is a constant scalar or matrix.

Cases 1 and 2 involve *spatial* norms and the question that arises is: how do we average or sum up the channels? Cases 3 and 4 involve function norms or *temporal* norms where we want to "average" or "sum up" as a function of time or frequency. Note that the first two are finite dimensional norms, while the latter two are infinite-dimensional.

Remark. Notation for norms. The reader should be aware that the notation on norms in the literature is not consistent, and one must be careful to avoid confusion. First, in spite of the fundamental difference between spatial and temporal norms, the same notation, $\| \cdot \|$, is generally used for both of them, and we adopt this here. Second, the same notation is often used to denote entirely different norms. For example, consider the infinity-norm, $\|e\|_\infty$. If e is a constant vector, then $\|e\|_\infty$ is the largest element in the vector (we often use $\|e\|_{max}$ for this). If $e(t)$ is a scalar time signal, then $\|e(t)\|_\infty$ is the peak value of $|e(t)|$ as a function of time. If E is a constant matrix then $\|E\|_\infty$ may denote the the largest matrix element (we use $\|A\|_{max}$ for this), while other authors use $\|E\|_\infty$ to denote the largest matrix row-sum (we use $\|E\|_{i\infty}$ for this). Finally, if $E(s)$ is a stable proper system (transfer function), then $\|E\|_\infty$ is the $\mathcal{H}_\infty$ norm which is the peak value of the maximum singular value of E, $\|E(s)\|_\infty = \max_w \bar{\sigma}(E(j\omega))$ (which is how we mostly use the ∞-norm in this book).

A.5.1   Vector norms

We will consider a vector a with m elements, that is, the vector space is $V = \mathbb{C}^m$. To illustrate the different norms we will calculate each of them for the vector

$$b = \begin{bmatrix} b_1 \\ b_2 \\ b_3 \end{bmatrix} = \begin{bmatrix} 1 \\ 3 \\ -5 \end{bmatrix} \tag{A.88}$$

We will consider three norms which are special cases of the vector p-norm

$$\|a\|_p = (\sum_i |a_i|^p)^{1/p} \tag{A.89}$$

where we must have $p \geq 1$ to satisfy the triangle inequality (property 4 of a norm). Here a is a column vector with elements a_i and $|a_i|$ is the absolute value of the complex scalar a_i.

Vector 1-norm (or sum-norm). This is sometimes referred to as the "taxi-cab norm", as in two dimensions it corresponds to the distance between two places when following the "streets" (New York style). We have

$$\|a\|_1 \triangleq \sum_i |a_i| \qquad (\|b\|_1 = 1 + 3 + 5 = 9) \tag{A.90}$$

Vector 2-norm (Euclidean norm). This is the most common vector norm, and corresponds to the shortest distance between two points

$$\|a\|_2 \triangleq \sqrt{\sum_i |a_i|^2} \qquad (\|b\|_2 = \sqrt{1 + 9 + 25} = 5.916) \tag{A.91}$$

The Euclidean vector norm satisfies the property

$$a^H a = \|a\|_2^2 \tag{A.92}$$

where a^H denotes the complex conjugate transpose of the vector a.

Vector ∞ -norm (or max norm). This is the largest element magnitude in the vector. We use the notation $\|a\|_{max}$ so that

$$\|a\|_{max} \equiv \|a\|_\infty \triangleq \max_i |a_i| \qquad (\|b\|_{max} = |-5| = 5) \tag{A.93}$$

Since the various vector norms only differ by constant factors, they are often said to be *equivalent*. For example, for a vector with m elements

$$\|a\|_{\max} \leq \|a\|_2 \leq \sqrt{m}\,\|a\|_{\max} \qquad (A.94)$$

$$\|a\|_2 \leq \|a\|_1 \leq \sqrt{m}\,\|a\|_2 \qquad (A.95)$$

In Figure A.1 the differences between the vector norms are illustrated by plotting the contours for $\|a\|_p = 1$ for the case with $m = 2$.

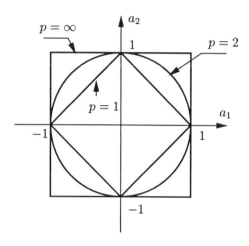

Figure A.1: Contours for the vector p-norm, $\|a\|_p = 1$ for $p = 1, 2, \infty$

A.5.2   Matrix norms

We will consider a constant $l \times m$ matrix A The matrix A may represent, for example, the frequency response, $G(j\omega)$, of a system $G(s)$ with m inputs and l outputs. For numerical illustrations we will use the following 2×2 matrix example

$$A_0 = \begin{bmatrix} 1 & 2 \\ -3 & 4 \end{bmatrix} \qquad (A.96)$$

Definition A.6 *A norm on a matrix* $\|A\|$ *is a* **matrix norm** *if, in addition to the four norm properties in Definition A.5, it also satisfies the multiplicative property (also called the consistency condition):*

$$\|AB\| \leq \|A\| \cdot \|B\| \qquad (A.97)$$

Property (A.97) is very important when combining systems, and forms the basis for the small gain theorem. Note that there exist *norms on matrices* (thus satisfying the

four properties of a norm), which are not *matrix norms* (thus *not* satisfying (A.97)).
Such norms are sometimes called *generalized matrix norms*. The only generalized
matrix norm considered in this book is the largest-element norm, $\|A\|_{\max}$.

Let us first examine three norms which are direct extensions of the definitions of
the vector p-norms.

Sum matrix norm. This is the sum of the element magnitudes

$$\|A\|_{\text{sum}} = \sum_{i,j} |a_{ij}| \qquad (\|A_0\|_{\text{sum}} = 1 + 2 + 3 + 4 = 10) \qquad \text{(A.98)}$$

Frobenius matrix norm (or Euclidean norm). This is the square root of the sum
of the squared element magnitudes

$$\|A\|_F = \sqrt{\sum_{i,j} |a_{ij}|^2} = \sqrt{\text{tr}(A^H A)} \qquad (\|A_0\|_F = \sqrt{30} = 5.477) \qquad \text{(A.99)}$$

The trace tr is the sum of the the diagonal elements, and A^H is the complex conjugate
transpose of A. The Frobenius norm is important in control because it is used for
summing up the channels, for example, when using LQG optimal control.

Max element norm. This is the largest element magnitude,

$$\|A\|_{\max} = \max_{i,j} |a_{ij}| \qquad (\|A_0\|_{\max} = 4) \qquad \text{(A.100)}$$

This norm is *not* a matrix norm as it does not satisfy (A.97). However note that
$\sqrt{lm}\,\|A\|_{\max}$ *is a matrix norm*.

The above three norms are sometimes called the $1-$, $2-$ and $\infty-$norm, respectively, but this notation is *not* used in this book to avoid confusion with the more important induced p-norms introduced next.

Induced matrix norms

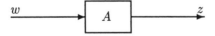

Figure A.2: Representation of (A.101)

Induced matrix norms are important because of their close relationship to signal
amplification in systems. Consider the following equation which is illustrated in
Figure A.2

$$z = Aw \qquad \text{(A.101)}$$

We may think of w as the input vector and z as the output vector and consider the
"amplification" or "gain" of the matrix A as defined by the ratio $\|z\|/\|w\|$. The

maximum gain for all possible input directions is of particular interest. This is given by the *induced norm* which is defined as

$$\|A\|_{ip} \triangleq \max_{w \neq 0} \frac{\|Aw\|_p}{\|w\|_p} \tag{A.102}$$

where $\|w\|_p = (\sum_i |w_i|^p)^{1/p}$ denotes the vector p-norm. In other words, we are looking for a direction of the vector w such that the ratio $\|z\|_p/\|w\|_p$ is maximized. Thus, the induced norm gives the largest possible "amplifying power" of the matrix. The following equivalent definition is also used

$$\|A\|_{ip} = \max_{\|w\|_p \leq 1} \|Aw\|_p = \max_{\|w\|_p = 1} \|Aw\|_p \tag{A.103}$$

For the induced 1-, 2- and ∞-norms the following identities hold:

$$\|A\|_{i1} = \max_j (\sum_i |a_{ij}|) \quad \text{"maximum column sum"} \tag{A.104}$$

$$\|A\|_{i\infty} = \max_i (\sum_j |a_{ij}|) \quad \text{"maximum row sum"} \tag{A.105}$$

$$\|A\|_{i2} = \bar{\sigma}(A) = \sqrt{\rho(A^H A)} \quad \text{"singular value or spectral norm"} \tag{A.106}$$

where the spectral radius $\rho(A) = \max_i |\lambda_i(A)|$ is the largest eigenvalue of the matrix A. Note that the induced 2-norm of a matrix is equal to the (largest) singular value, and is often called the spectral norm. For the example matrix in (A.96) we get

$$\|A_0\|_{i1} = 6; \quad \|A_0\|_{i\infty} = 7; \quad \|A_0\|_{i2} = \bar{\sigma}(A_0) = 5.117 \tag{A.107}$$

Theorem A.3 *All induced norms* $\|A\|_{ip}$ *are matrix norms and thus satisfy the multiplicative property*

$$\|AB\|_{ip} \leq \|A\|_{ip} \cdot \|B\|_{ip} \tag{A.108}$$

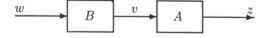

Figure A.3: Representation of (A.109)

Proof: Consider the following set of equations which is illustrated graphically in Figure A.3.

$$z = Av, \quad v = Bw \quad \Rightarrow z = ABw \tag{A.109}$$

From the definition of the induced norm we get by first introducing $v = Bw$, then multiplying the numerator and denominator by $\|v\|_p \neq 0$, and finally maximizing each term involving w and v independently, that

$$\|AB\|_{ip} \triangleq \max_{w \neq 0} \frac{\|A \overbrace{Bw}^{v}\|_p}{\|w\|_p} = \max_{w \neq 0} \frac{\|Av\|_p}{\|v\|_p} \cdot \frac{\|Bw\|_p}{\|w\|_p} \leq \max_{v \neq 0} \frac{\|Av\|_p}{\|v\|_p} \cdot \max_{w \neq 0} \frac{\|Bw\|_p}{\|w\|_p}$$

and (A.108) follows from the definition of an induced norm.   □

Implications of the multiplicative property

For matrix norms the multiplicative property $\|AB\| \leq \|A\| \cdot \|B\|$ holds for matrices A and B of any dimension as long as the product AB exists. In particular, it holds if we choose A and B as vectors. From this observation we get:

1. Choose B to be a vector, i.e $B = w$. Then for any matrix norm we have from (A.97) that

$$\|Aw\| \leq \|A\| \cdot \|w\| \tag{A.110}$$

We say that the "matrix norm $\|A\|$ is compatible with its corresponding vector norm $\|w\|$". Clearly, from (A.102) any induced matrix p-norm is compatible with its corresponding vector p-norm. Similarly, the Frobenius norm is compatible with the vector 2-norm (since when w is a vector $\|w\|_F = \|w\|_2$).

2. From (A.110) we also get for any matrix norm that

$$\|A\| \geq \max_{w \neq 0} \frac{\|Aw\|}{\|w\|} \tag{A.111}$$

Note that the induced norms are defined such that we have equality in (A.111). The property $\|A\|_F \geq \bar{\sigma}(A)$ then follows since $\|w\|_F = \|w\|_2$.

3. Choose both $A = z^H$ and $B = w$ as vectors. Then using the Frobenius norm or induced 2-norm (singular value) in (A.97) we derive the Cauchy-Schwarz inequality

$$|z^H w| \leq \|z\|_2 \cdot \|w\|_2 \tag{A.112}$$

where z and w are column vectors of the same dimension and $z^H w$ is the Euclidean inner product between the vectors z and w.

4. The inner product can also be used to define the angle ϕ between two vectors z and w

$$\phi = \cos^{-1} \left(\frac{|z^H w|}{\|z\|_2 \cdot \|w\|_2} \right) \tag{A.113}$$

Note that with this definition, ϕ is between 0° and 90°.

A.5.3   The spectral radius

The spectral radius $\rho(A)$ is the magnitude of the largest eigenvalue of the matrix A,

$$\rho(A) = \max_i |\lambda_i(A)| \qquad (A.114)$$

It is *not* a norm, as it does not satisfy norm properties 2 and 4 in Definition A.5. For example, for

$$A_1 = \begin{bmatrix} 1 & 0 \\ 10 & 1 \end{bmatrix}, \quad A_2 = \begin{bmatrix} 1 & 10 \\ 0 & 1 \end{bmatrix} \qquad (A.115)$$

we have $\rho(A_1) = 1$ and $\rho(A_2) = 1$. However, $\rho(A_1 + A_2) = 12$ and $\rho(A_1 A_2) = 101.99$, which neither satisfy the triangle inequality (property 4 of a norm) nor the multiplicative property in (A.97).

Although the spectral radius is not a norm, it provides a lower bound on any matrix norm, which can be very useful.

Theorem A.4 *For any matrix norm (and in particular for any induced norm)*

$$\rho(A) \leq \|A\| \qquad (A.116)$$

Proof: Since $\lambda_i(A)$ is an eigenvalue of A, we have that $At_i = \lambda_i t_i$ where t_i denotes the eigenvector. We get

$$|\lambda_i| \cdot \|t_i\| = \|\lambda_i t_i\| = \|At_i\| \leq \|A\| \cdot \|t_i\| \qquad (A.117)$$

(the last inequality follows from (A.110)). Thus for any matrix norm $|\lambda_i(A)| \leq \|A\|$ and since this holds for all eigenvalues the result follows. $\square$

For our example matrix in (A.96) we get $\rho(A_0) = \sqrt{10} \approx 3.162$ which is less than all the induced norms ($\|A_0\|_{i1} = 6, \|A_0\|_{i\infty} = 7, \bar{\sigma}(A_0) = 5.117$) and also less than the Frobenius norm ($\|A\|_F = 5.477$) and the sum-norm ($\|A\|_{\text{sum}} = 10$).

A simple physical interpretation of (A.116) is that the eigenvalue measures the gain of the matrix only in certain directions (given by the eigenvectors), and must therefore be less than that for a matrix norm which allows any direction and yields the maximum gain, recall (A.111).

A.5.4   Some matrix norm relationships

The various norms of the matrix A are closely related as can be seen from the following inequalities taken from Golub and van Loan (1989, p. 15) and Horn and Johnson (1985, p. 314). Let A be an $l \times m$ matrix, then

$$\bar{\sigma}(A) \leq \|A\|_F \leq \sqrt{\min(l, m)}\, \bar{\sigma}(A) \qquad (A.118)$$

$$\|A\|_{\max} \leq \bar{\sigma}(A) \leq \sqrt{lm}\, \|A\|_{\max} \qquad (A.119)$$

$$\bar{\sigma}(A) \leq \sqrt{\|A\|_{i1}\|A\|_{i\infty}} \tag{A.120}$$

$$\frac{1}{\sqrt{m}}\|A\|_{i\infty} \leq \bar{\sigma}(A) \leq \sqrt{l}\,\|A\|_{i\infty} \tag{A.121}$$

$$\frac{1}{\sqrt{l}}\|A\|_{i1} \leq \bar{\sigma}(A) \leq \sqrt{m}\,\|A\|_{i1} \tag{A.122}$$

$$\max\{\bar{\sigma}(A), \|A\|_F, \|A\|_{i1}, \|A\|_{i\infty}\} \leq \|A\|_{\text{sum}} \tag{A.123}$$

All these norms, except $\|A\|_{\max}$, are matrix norms and satisfy (A.97). The inequalities are tight; that is, there exist matrices of any size for which the equality holds. Note from (A.119) that the maximum singular value is closely related to the largest element of the matrix. Therefore, $\|A\|_{\max}$ can be used as a simple and readily available estimate of $\bar{\sigma}(A)$.

An important property of the Frobenius norm and the maximum singular value (induced 2-norm) is that they are invariant with respect to unitary transformations, i.e. for unitary matrices U_i, satisfying $U_i U_i^H = I$, we have

$$\|U_1 A U_2\|_F = \|A\|_F \tag{A.124}$$

$$\bar{\sigma}(U_1 A U_2) = \bar{\sigma}(A) \tag{A.125}$$

From an SVD of the matrix $A = U\Sigma V^H$ and (A.124), we then obtain an important relationship between the Frobenius norm and the singular values, $\sigma_i(A)$, namely

$$\|A\|_F = \sqrt{\sum_i \sigma_i^2(A)} \tag{A.126}$$

The Perron-Frobenius theorem, which applies to a square matrix A, states that

$$\min_D \|DAD^{-1}\|_{i1} = \min_D \|DAD^{-1}\|_{i\infty} = \rho(|A|) \tag{A.127}$$

where D is a diagonal "scaling" matrix, $|A|$ denotes the matrix A with all its elements replaced by their magnitudes, and $\rho(|A|) = \max_i |\lambda_i(|A|)|$ is the Perron root (Perron-Frobenius eigenvalue). The Perron root is greater than or equal to the spectral radius, $\rho(A) \leq \rho(|A|)$.

A.5.5   Matrix and vector norms in MATLAB

The following MATLAB commands are used for matrices:

$$\bar{\sigma}(A) = \|A\|_{i2} \quad \texttt{norm(A,2)} \text{ or } \texttt{max(svd(A))}$$
$$\|A\|_{i1} \quad \texttt{norm(A,1)}$$
$$\|A\|_{i\infty} \quad \texttt{norm(A,'inf')}$$
$$\|A\|_F \quad \texttt{norm(A,'fro')}$$
$$\|A\|_{\text{sum}} \quad \texttt{sum (sum(abs(A)))}$$
$$\|A\|_{\text{max}} \quad \texttt{max(max(abs(A)))} \text{ (which is not a matrix norm)}$$
$$\rho(A) \quad \texttt{max(abs(eig(A)))}$$
$$\rho(|A|) \quad \texttt{max(eig(abs(A)))}$$
$$\gamma(A) = \bar{\sigma}(A)/\underline{\sigma}(A) \quad \texttt{cond(A)}$$

For vectors:
$$\|a\|_1 \quad \texttt{norm(a,1)}$$
$$\|a\|_2 \quad \texttt{norm(a,2)}$$
$$\|a\|_{\text{max}} \quad \texttt{norm(a,'inf')}$$

A.5.6 Signal norms

We will consider the temporal norm of a time-varying (or frequency-varying) signal, $e(t)$. In contrast with spatial norms (vector and matrix norms), we find that the choice of temporal norm makes a big difference. As an example, consider Figure A.4 which shows two signals, $e_1(t)$ and $e_2(t)$. For $e_1(t)$ the infinity-norm (peak) is one, $\|e_1(t)\|_\infty = 1$, whereas since the signal does not "die out" the 2-norm is infinite, $\|e_1(t)\|_2 = \infty$. For $e_2(t)$ the opposite is true.

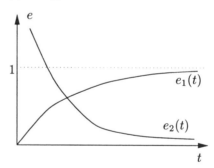

Figure A.4: Signals with entirely different 2-norms and ∞-norms.

For signals we may compute the norm in two steps:

1. "Sum up" the channels at a given time or frequency using a vector norm (for a scalar signal we simply take the absolute value).
2. "Sum up" in time or frequency using a temporal norm.

Recall from above, that the vector norms are "equivalent" in the sense that their values differ only by a constant factor. Therefore, it does not really make too much difference

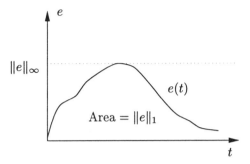

Figure A.5: Signal 1-norm and ∞-norm.

which norm we use in step 1. We normally use the same p-norm both for the vector and the signal, and thus define the temporal p-norm, $\|e(t)\|_p$, of a time-varying vector as

$$l_p \text{ norm:} \quad \|e(t)\|_p = \left(\int_{-\infty}^{\infty} \sum_i |e_i(\tau)|^p d\tau \right)^{1/p} \tag{A.128}$$

The following temporal norms of signals are commonly used:

1-norm in time (integral absolute error (IAE), see Figure A.5):

$$\|e(t)\|_1 = \int_{-\infty}^{\infty} \sum_i |e_i(\tau)| d\tau \tag{A.129}$$

2-norm in time (quadratic norm, integral square error (ISE), "energy" of signal):

$$\|e(t)\|_2 = \sqrt{\int_{-\infty}^{\infty} \sum_i |e_i(\tau)|^2 d\tau} \tag{A.130}$$

∞-norm in time (peak value in time, see Figure A.5):

$$\|e(t)\|_\infty = \max_\tau \left(\max_i |e_i(\tau)| \right) \tag{A.131}$$

In addition, we will consider the power-norm or RMS-norm (which is actually only a semi-norm since it does not satisfy norm property 2)

$$\|e(t)\|_{\text{pow}} = \lim_{T \to \infty} \sqrt{\frac{1}{2T} \int_{-T}^{T} \sum_i |e_i(\tau)|^2 d\tau} \tag{A.132}$$

Remark 1 In most cases we assume $e(t) = 0$ for $t < 0$ so the lower value for the integration may be changed to $\tau = 0$.

Remark 2 To be mathematically correct we should have used $\sup_\tau$ (least upper bound) rather than $\max_\tau$ in (A.131), since the maximum value may not actually be achieved (e.g. if it occurs for $t = \infty$).

A.5.7   Signal interpretation of various system norms

Two system norms are considered in Section 4.10. These are the $\mathcal{H}_2$ norm, $\|G(s)\|_2 = \|g(t)\|_2$ and the $\mathcal{H}_\infty$ norm, $\|G(s)\|_\infty$. The main reason for including this section is to show that there are many ways of evaluating performance in terms of signals, and to show that the $\mathcal{H}_2$ and $\mathcal{H}_\infty$ norms are useful measures in this context. This in turn will be useful in helping us to understand how to select performance weights in controller design problems. The proofs of the results in this section require a good background in functional analysis and can be found in Doyle et al. (1992) and Zhou et al. (1996).

Consider a system G with input d and output e, such that

$$e = Gd \tag{A.133}$$

For performance we may want the output signal e to be "small" for any allowed input signals d. We therefore need to specify:

1. What d's are allowed. (Which set does d belong to?)
2. What we mean by "small". (Which norm should we use for e?)

Some possible input signal sets are:

1. $d(t)$ consists of impulses, $\delta(t)$. These generate step changes in the states, which is the usual way of introducing the LQ-objective and gives rise to the $\mathcal{H}_2$ norm.
2. $d(t)$ is a white noise process with zero mean.
3. $d(t) = \sin(\omega t)$ with fixed frequency, applied from $t = -\infty$ (which corresponds to the steady-state sinusoidal response).
4. $d(t)$ is a set of sinusoids with all frequencies allowed.
5. $d(t)$ is bounded in energy, $\|d(t)\|_2 \leq 1$.
6. $d(t)$ is bounded in power, $\|d(t)\|_{\text{pow}} \leq 1$.
7. $d(t)$ is bounded in magnitude, $\|d(t)\|_\infty \leq 1$.

The first three sets of input are specific signals, whereas the latter three are classes of inputs with bounded norm. The physical problem at hand determines which of these input classes is the most reasonable.

To measure the output signal one may consider the following norms:

1. 1-norm, $\|e(t)\|_1$
2. 2-norm (energy), $\|e(t)\|_2$
3. ∞-norm (peak magnitude), $\|e(t)\|_\infty$
4. Power, $\|e(t)\|_{\text{pow}}$

Other norms are possible, but again, it is engineering issues that determine which norm is the most appropriate. We will now consider which system norms result from the definitions of input classes, and output norms, respectively. That is, we want to find the appropriate system gain to test for performance. The results for SISO systems in which $d(t)$ and $e(t)$ are scalar signals, are summarized in Tables A.1 and A.2. In these tables $G(s)$ is the transfer function and $g(t)$ is its corresponding impulse response. Note in particular the

$$\mathcal{H}_\infty \text{ norm:} \quad \|G(s)\|_\infty = \max_{d(t)} \frac{\|e(t)\|_2}{\|d(t)\|_2} \qquad (A.134)$$

and

$$l_1 \text{ norm:} \quad \|g(t)\|_1 = \max_{d(t)} \frac{\|e(t)\|_\infty}{\|d(t)\|_\infty} \qquad (A.135)$$

We see from Tables A.1 and A.2 that the $\mathcal{H}_2$ and $\mathcal{H}_\infty$ norms appear in several positions. This gives some basis for their popularity in control. In addition, the $\mathcal{H}_\infty$ norm results if we consider $d(t)$ to be the set of sinusoids with all frequencies allowed, and measure the output using the 2-norm (not shown in Tables A.1 and A.2, but discussed in Section 3.3.5). Also, the $\mathcal{H}_2$ norm results if the input is white noise and we measure the output using the 2-norm.

	$d(t) = \delta(t)$	$d(t) = \sin(\omega t)$
$\|e\|_2$	$\|G(s)\|_2$	∞ (usually)
$\|e\|_\infty$	$\|g(t)\|_\infty$	$\bar{\sigma}(G(j\omega))$
$\|e\|_{\text{pow}}$	0	$\frac{1}{\sqrt{2}}\bar{\sigma}(G(j\omega))$

Table A.1: System norms for two specific input signals and three different output norms

	$\|d\|_2$	$\|d\|_\infty$	$\|d\|_{\text{pow}}$
$\|e\|_2$	$\|G(s)\|_\infty$	∞	∞ (usually)
$\|e\|_\infty$	$\|G(s)\|_2$	$\|g(t)\|_1$	∞ (usually)
$\|e\|_{\text{pow}}$	0	$\leq \|G(s)\|_\infty$	$\|G(s)\|_\infty$

Table A.2: System norms for three sets of norm-bounded input signals and three different output norms. The entries along the diagonal are induced norms.

The results in Tables A.1 and A.2 may be generalized to MIMO systems by use of the appropriate matrix and vector norms. In particular, the induced norms along the diagonal in Table A.2 generalize if we use for the $\mathcal{H}_\infty$ norm $\|G(s)\|_\infty = \max_\omega \bar{\sigma}(G(j\omega))$, and for the l_1 norm we use $\|g(t)\|_1 = \max_i \|g_i(t)\|_1$, where $g_i(t)$ denotes row i of the impulse response matrix. The fact that the $\mathcal{H}_\infty$ norm and l_1 norm are induced norms makes them well suited for robustness analysis, for example, using the small gain theorem. The two norms are also closely related as can be seen from the following bounds for a proper scalar system

$$\|G(s)\|_\infty \leq \|g(t)\|_1 \leq (2n+1)\|G(s)\|_\infty \qquad (A.136)$$

Here n is the number of states in a minimal realization. A multivariable generalization for a strictly proper system is given by Zhou et al. (1996) as

$$\sigma_1 \leq \|G(s)\|_\infty \leq \int_0^\infty \bar\sigma(g(t))dt \leq 2\sum_{i=1}^n \sigma_i \leq 2n\|G(s)\|_\infty \qquad (A.137)$$

where σ_i is the i'th Hankel singular value of $G(s)$ and $g(t) = Ce^{At}B$ is the impulse response matrix.

A.6 Factorization of the sensitivity function

Consider two plant models, G a nominal model and G' an alternative model, and assume that the same controller is applied to both plants. Then the corresponding sensitivity functions are

$$S = (I+GK)^{-1}, \quad S' = (I+G'K)^{-1} \qquad (A.138)$$

A.6.1 Output perturbations

Assume that G' is related to G by either an output multiplicative perturbation E_O, or an inverse output multiplicative perturbation E_{iO}. Then S' can be factorized in terms of S as follows

$$S' = S(I+E_OT)^{-1}; \quad G' = (I+E_O)G \qquad (A.139)$$

$$S' = S(I-E_{iO}S)^{-1}(I-E_{iO}); \quad G' = (I-E_{iO})^{-1}G \qquad (A.140)$$

For a square plant, E_O and E_{iO} can be obtained from a given G and G' by

$$E_O = (G'-G)G^{-1}; \quad E_{iO} = (G'-G)G'^{-1} \qquad (A.141)$$

Proof of (A.139):

$$I+G'K = I+(I+E_O)GK = (I+E_O\underbrace{GK(I+GK)^{-1}}_{T})(I+GK) \quad (A.142)$$

Proof of (A.140):

$$\begin{aligned}
I+G'K &= I+(I-E_{iO})^{-1}GK = (I-E_{iO})^{-1}((I-E_{iO})+GK) \\
&= (I-E_{iO})^{-1}(I-E_{iO}\underbrace{(I+GK)^{-1}}_{S})(I+GK) \qquad (A.143)
\end{aligned}$$

Similar factorizations may be written in terms of the complementary sensitivity function (Horowitz and Shaked, 1975; Zames, 1981). For example, by writing (A.139) in the form $S = S'(I + E_O T)$ and using the fact $S - S' = T' - T$, we get

$$T' - T = S' E_O T \qquad \text{(A.144)}$$

A.6.2 Input perturbations

For a square plant, the following factorization in terms of input multiplicative uncertainty E_I is useful:

$$S' = S(I + GE_I G^{-1} T)^{-1} = SG(I + E_I T_I)^{-1} G^{-1}; \quad G' = G(I + E_I) \quad \text{(A.145)}$$

where $T_I = KG(I + KG)^{-1}$ is the input complementary sensitivity function.

Proof: Substitute $E_O = GE_I G^{-1}$ into (A.139) and use $G^{-1} T = T_I G^{-1}$. □

Alternatively, we may factor out the controller to get

$$S' = (I + TK^{-1} E_I K)^{-1} S = K^{-1}(I + T_I E_I)^{-1} KS \qquad \text{(A.146)}$$

Proof: Start from $I + G'K = I + G(I + E_I)K$ and factor out $(I + GK)$ to the left. □

A.6.3 Stability conditions

The next Lemma follows directly from the generalized Nyquist Theorem and the factorization (A.139):

Lemma A.5 *Assume that the negative feedback closed-loop system with loop transfer function $G(s)K(s)$ is stable. Suppose $G' = (I + E_O)G$, and let the number of open loop unstable poles of $G(s)K(s)$ and $G'(s)K(s)$ be P and P', respectively. Then the negative feedback closed-loop system with loop transfer function $G'(s)K(s)$ is stable if and only if*

$$\mathcal{N}(\det(I + E_O T)) = P - P' \qquad \text{(A.147)}$$

where $\mathcal{N}$ denotes the number of clockwise encirclements of the origin as s traverses the Nyquist D-contour in a clockwise direction.

Proof: Let $\mathcal{N}(f)$ denote the number of clockwise encirclements of the origin by $f(s)$ as s traverses the Nyquist D contour in a clockwise direction. For the encirclements of the product of two functions we have $\mathcal{N}(f_1 f_2) = \mathcal{N}(f_1) + \mathcal{N}(f_2)$. This together with (A.142) and the fact $\det(AB) = \det A \cdot \det B$ yields

$$\mathcal{N}(\det(I + G'K)) = \mathcal{N}(\det(I + E_O T)) + \mathcal{N}(\det(I + GK)) \qquad \text{(A.148)}$$

For stability we need from Theorem 4.7 that $\mathcal{N}(\det(I + G'K)) = -P'$, but we know that $\mathcal{N}(\det(I + GK)) = -P$ and hence Lemma A.5 follows. The Lemma is from Hovd and Skogestad (1994a); similar results, at least for stable plants, have been presented by e.g. Grosdidier and Morari (1986) and Nwokah and Perez (1991). □

In other words, (A.147) tells us that for stability $\det(I + E_O T)$ must provide the required additional number of clockwise encirclements. If (A.147) is not satisfied then the negative feedback system with $G'K$ must be unstable. We show in Theorem 6.7 how the information about what happens at $s = 0$ can be used to determine stability.

A.7 Linear fractional transformations

Linear fractional transformations (LFTs), as they are currently used in the control literature for analysis and design, were introduced by Doyle (1984). Consider a matrix P of dimension $(n_1 + n_2) \times (m_1 + m_2)$ and partition it as follows:

$$P = \begin{bmatrix} P_{11} & P_{12} \\ P_{21} & P_{22} \end{bmatrix} \tag{A.149}$$

Let the matrices Δ and K have dimensions $m_1 \times n_1$ and $m_2 \times n_2$, respectively (compatible with the upper and lower partitions of P, respectively). We adopt the following notation for the lower and upper linear fractional transformations

$$F_l(P, K) \triangleq P_{11} + P_{12} K (I - P_{22} K)^{-1} P_{21} \tag{A.150}$$

$$F_u(P, \Delta) \triangleq P_{22} + P_{21} \Delta (I - P_{11} \Delta)^{-1} P_{12} \tag{A.151}$$

where subscript l denotes lower and subscript u upper. In the following, let R denote a matrix function resulting from an LFT.

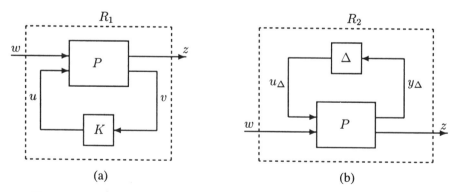

Figure A.6: (a) R_1 as lower LFT in terms of K. (b) R_2 as upper LFT in terms of Δ.

The lower fractional transformation $F_l(P, K)$ is the transfer function R_1 resulting from wrapping (positive) feedback K around the lower part of P as illustrated in

Figure A.6(a). To see this, note that the block diagram in Figure A.6(a) may be written as

$$z = P_{11}w + P_{12}u, \quad v = P_{21}w + P_{22}u, \quad u = Kv \qquad \text{(A.152)}$$

Upon eliminating v and u from these equations we get

$$z = R_1 w = F_l(P, K)w = [P_{11} + P_{12}K(I - P_{22}K)^{-1}P_{21}]w \qquad \text{(A.153)}$$

In words, R_1 is written as a lower LFT of P in terms of the parameter K. Similarly, in Figure A.6(b) we illustrate the upper LFT, $R_2 = F_u(P, \Delta)$, obtained by wrapping (positive) feedback Δ around the upper part of P.

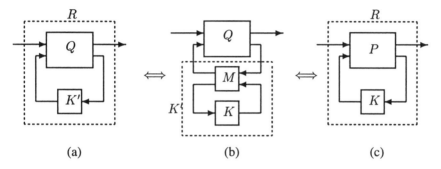

(a)   (b)   (c)

Figure A.7: An interconnection of LFTs yields an LFT

A.7.1 Interconnection of LFTs

An important property of LFTs is that any interconnection of LFTs is again an LFT. Consider Figure A.7 where R is written in terms of a lower LFT of K', which again is a lower LFT of K, and we want to express R directly as an LFT of K. We have

$$R = F_l(Q, K') \quad \text{where} \quad K' = F_l(M, K) \qquad \text{(A.154)}$$

and we want to obtain the P (in terms of Q and M) such that

$$R = F_l(P, K) \qquad \text{(A.155)}$$

We find

$$P = \begin{bmatrix} P_{11} & P_{12} \\ P_{21} & P_{22} \end{bmatrix} =$$

$$\begin{bmatrix} Q_{11} + Q_{12}M_{11}(I - Q_{22}M_{11})^{-1}Q_{21} & Q_{12}(I - M_{11}Q_{22})^{-1}M_{12} \\ M_{21}(I - Q_{22}M_{11})^{-1}Q_{21} & M_{22} + M_{21}Q_{22}(I - M_{11}Q_{22})^{-1}M_{12} \end{bmatrix}$$

$$\text{(A.156)}$$

Similar expressions apply when we use *upper* LFTs. For

$$R = F_u(M, \Delta') \quad \text{where} \quad \Delta' = F_u(Q, \Delta) \tag{A.157}$$

we get $R = F_u(P, \Delta)$ where P is given in terms of Q and M by (A.156).

A.7.2   Relationship between F_l and F_u.

F_l and F_u are obviously closely related. If we know $R = F_l(M, K)$, then we may directly obtain R in terms of an upper transformation of K by reordering M. We have

$$F_u(\widetilde{M}, K) = F_l(M, K) \tag{A.158}$$

where

$$\widetilde{M} = \begin{bmatrix} 0 & I \\ I & 0 \end{bmatrix} M \begin{bmatrix} 0 & I \\ I & 0 \end{bmatrix} \tag{A.159}$$

A.7.3   Inverse of LFTs

On the assumption that all the relevant inverses exist we have

$$(F_l(M, K))^{-1} = F_l(\widetilde{M}, K) \tag{A.160}$$

where $\widetilde{M}$ is given by

$$\widetilde{M} = \begin{bmatrix} M_{11}^{-1} & -M_{11}^{-1} M_{12} \\ M_{21} M_{11}^{-1} & M_{22} - M_{21} M_{11}^{-1} M_{12} \end{bmatrix} \tag{A.161}$$

This expression follows easily from the matrix inversion lemma in (A.6).

A.7.4   LFT in terms of the inverse parameter

Given an LFT in terms of K, it is possible to derive an equivalent LFT in terms of K^{-1}. If we assume that all the relevant inverses exist we have

$$F_l(M, K) = F_l(\widehat{M}, K^{-1}) \tag{A.162}$$

where $\widehat{M}$ is given by

$$\widehat{M} = \begin{bmatrix} M_{11} - M_{12} M_{22}^{-1} M_{21} & -M_{12} M_{22}^{-1} \\ M_{22}^{-1} M_{21} & M_{22}^{-1} \end{bmatrix} \tag{A.163}$$

This expression follows from the fact that $(I + L)^{-1} = I - L(I + L)^{-1}$ for any square matrix L.

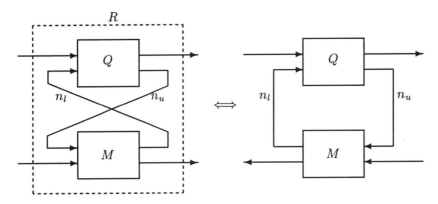

Figure A.8: Star product of Q and M, $R = \mathcal{S}(Q, M)$

A.7.5   Generalized LFT: The matrix star product

A generalization of the upper and lower LFTs above is provided by Redheffer's star product. Consider Figure A.8 where Q and M are interconnected such that the last n_u outputs from Q are the first n_u inputs of M, and the first n_l outputs from M are the last n_l inputs of Q. The corresponding partitioned matrices are

$$Q = \begin{bmatrix} Q_{11} & Q_{12} \\ Q_{21} & Q_{22} \end{bmatrix}, \quad M = \begin{bmatrix} M_{11} & M_{12} \\ M_{21} & M_{22} \end{bmatrix}$$

The overall matrix R with these interconnections closed (see Figure A.8) is called the star product, $\mathcal{S}(Q, M)$, between Q and M. We find that

$$R = \mathcal{S}(Q, M) =$$
$$\begin{bmatrix} Q_{11} + Q_{12}M_{11}(I - Q_{22}M_{11})^{-1}Q_{21} & Q_{12}(I - M_{11}Q_{22})^{-1}M_{12} \\ M_{21}(I - Q_{22}M_{11})^{-1}Q_{21} & M_{22} + M_{21}Q_{22}(I - M_{11}Q_{22})^{-1}M_{12} \end{bmatrix}$$
$$(A.164)$$

Note that $\mathcal{S}(Q, M)$ depends on the chosen partitioning of the matrices Q and M. If one of the matrices is *not* partitioned then this means that this matrix has no external inputs and outputs, and $\mathcal{S}(Q, M)$ then gives the "maximum" interconnection. For example, we have for the LFTs

$$F_l(P, K) = \mathcal{S}(P, K) \tag{A.165}$$

$$F_u(P, \Delta) = \mathcal{S}(\Delta, P) \tag{A.166}$$

The order in the last equation is not a misprint. Of course, this assumes that the dimensions of K and Δ are smaller than those of P. The corresponding command to generate (A.164) in the MATLAB μ-toolbox is

```
starp(Q,M,nu,nl)
```

where n_u and n_l are as shown in Figure A.8. If n_u and n_l are not specified then this results in a "maximum" interconnection involving the corresponding LFT in (A.165) or (A.166).

APPENDIX B

PROJECT WORK and SAMPLE EXAM

B.1 Project work

Students are encouraged to formulate their own project based on an application they are working on. Otherwise, the project is given by the instructor. In either case, a preliminary statement of the problem must be approved before starting the project; see the first item below.

A useful collection of benchmark problems for control system design is provided in Davison (1990). The helicopter, aero-engine and distillation case studies in Chapter 12, and the chemical reactor in Example 6.16, also provide the basis for several projects. These models are available over the internet.

1. *Introduction: Preliminary problem definition.*

    (i) Give a simple description of the engineering problem with the aid of one or two diagrams.

    (ii) Discuss briefly the control objectives.

    (iii) Specify the exogenous inputs (disturbances, noise, setpoints), the manipulated inputs, the measurements, and the controlled outputs (exogenous outputs).

    (iv) Describe the most important sources of model uncertainty.

    (v) What specific control problems do you expect, e.g. due to interactions, RHP-zeros, saturation, etc.

    The preliminary statement of no more than 3 pages must be handed in and approved before starting the project.

2. *Plant model.* Specify all parameters, operating conditions, etc. and obtain a linear model of the plant. Comment: You may need to consider more than one operating point.

3. *Analysis of the plant.* For example, compute the steady-state gain matrix, plot the gain elements as a function of frequency, obtain the poles and zeros (both of

the individual elements and the overall system), compute the SVD and comment on directions and the condition number, perform an RGA-analysis, a disturbance analysis, etc.. Does the analysis indicate that the plant is difficult to control?

4. *Initial controller design.* Design at least two controllers, for example, using

   (i) Decentralized control (PID).

   (ii) Centralized control (LQG, LTR, $\mathcal{H}_2$ (in principle same as LQG, but with a different way of choosing weights), $\mathcal{H}_\infty$ loop shaping, $\mathcal{H}_\infty$ mixed sensitivity, etc.).

   (iii) A decoupler combined with PI-control.

5. *Simulations.* Perform simulations in the time domain for the closed-loop system.

6. *Robustness analysis using μ.*

   (i) Choose suitable performance and uncertainty weights. Plot the weights as functions of frequency.

   (ii) State clearly how RP is defined for your problem (using block diagrams).

   (iii) Compute μ for NP, RS, and RP.

   (iv) Perform a sensitivity analysis. For example, change the weights (e.g. to make one output channel faster and another slower), move uncertainties around (e.g. from input to output), change Δ's from a diagonal to full matrix, etc.

   Comment: You may need to move back to step (a) and redefine your weights if you find out from step (c) that your original weights are unreasonable.

7. *Optional: $\mathcal{H}_\infty$ or μ-optimal controller design.* Design an $\mathcal{H}_\infty$ or μ-optimal controller and see if you can improve the response and satisfy RP. Compare simulations with previous designs.

8. *Discussion.* Discuss the main results. You should also comment on the usefulness of the project as an aid to learning and give suggestions on how the project activity might be improved.

9. *Conclusion.*

B.2   Sample exam

A Norwegian-style 5-hour exam.

Problem 1 (35%). Controllability analysis.

   Perform a controllability analysis (compute poles, zeros, RGA ($\lambda_{11}(s)$), check for constraints, discuss the use of decentralized control (pairings), etc.) for the following four plants. You can assume that the plants have been scaled properly.

   (a) 2×2 plant:

$$G(s) = \frac{1}{(s+2)(s-1.1)} \begin{bmatrix} s-1 & 1 \\ 90 & 10(s-1) \end{bmatrix} \qquad (B.1)$$

(b) SISO plant with disturbance:

$$g(s) = 200 \frac{-0.1s + 1}{(s + 10)(0.2s + 1)}; \quad g_d(s) = \frac{40}{s + 1} \qquad \text{(B.2)}$$

(c) Plant with two inputs and one output:

$$y(s) = \frac{s}{0.2s + 1} u_1 + \frac{4}{0.2s + 1} u_2 + \frac{3}{0.02s + 1} d \qquad \text{(B.3)}$$

(d) Consider the following 2×2 plant with 1 disturbance given in state-space form:

$$\dot{x}_1 = -0.1x_1 + 0.01u_1$$

$$\dot{x}_2 = -0.5x_2 + 10u_2$$

$$\dot{x}_3 = 0.25x_1 + 0.25x_2 - 0.25x_3 + 1.25d$$

$$y_1 = 0.8x_3; \quad y_2 = 0.1x_3$$

i. Construct a block diagram representation of the system with each block in the form $k/(1 + \tau s)$.
ii. Perform a controllability analysis.

Problem 2 (25%). General Control Problem Formulation.

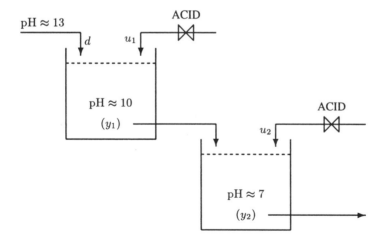

Figure B.1: Neutralization process

Consider the neutralization process in Figure B.1 where acid is added in two stages. Most of the neutralization takes place in tank 1 where a large amount of acid is used (input u_1) to obtain a pH of about 10 (measurement y_1). In tank 2 the pH is fine-tuned

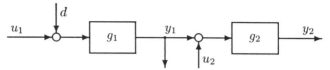

Figure B.2: Block diagram of neutralization process

to about 7 (output y_2) by using a small amount of acid (input u_2). This description is is just to give you some idea of a real process; all the information you need to solve the problem is given below.

A block diagram of the process is shown in Figure B.2. It includes one disturbance, two inputs and two measurements (y_1 and y_2). The main control objective is to keep $y_2 \approx r_2$. In addition, we would like to reset input 2 to its nominal value, that is, we want $u_2 \approx r_{u_2}$ at low frequencies. Note that there is no particular control objective for y_1.

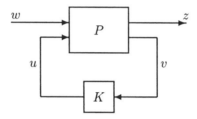

Figure B.3: General control configuration

(a) Define the general control problem, that is, find z, w, u, v and P (see Figure B.3).

(b) Define an $\mathcal{H}_\infty$ control problem based on P. Discuss briefly what you want the unweighted transfer functions from d to z to look like, and use this to say a little about how the performance weights should be selected.

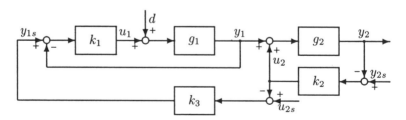

Figure B.4: Proposed control structure for neutralization process

(c) A simple practical solution based on single loops is shown in Figure B.4. Explain briefly the idea behind this control structure, and find the interconnection matrix P and the generalized controller $K = \text{diag}\{ k_1, \quad k_2, \quad k_3 \}$. Note that u and y are different in this case, while w and z are the same as in (a).

Problem 3 (40%). Various.

Give brief answers to each of the following questions:
  (a) Consider the plant

$$\dot{x}(t) = a(1 + 1.5\delta_a)x(t) + b(1 + 0.2\delta_b)u(t); \quad y = x$$

where $|\delta_a| \leq 1$ and $|\delta_b| \leq 1$. For a feedback controller $K(s)$ derive the interconnection matrix M for robust stability.

  (b) For the above case consider using the condition $\min_D \bar{\sigma}(DMD^{-1}) < 1$ to check for robust stability (RS). What is D (give as few parameters as possible)? Is the RS-condition tight in this case?

  (c) When is the condition $\rho(M\Delta) < 1$ necessary and sufficient for robust stability? Based on $\rho(M\Delta) < 1$, derive the RS-condition $\mu(M) < 1$. When is this last condition necessary and sufficient?

  (d) Let

$$G_p(s) = \begin{bmatrix} g_{11} + w_1\Delta_1 & g_{12} + w_2\Delta_2 \\ g_{21} + w_3\Delta_1 & g_{22} \end{bmatrix}, \quad |\Delta_1| \leq 1, |\Delta_2| \leq 1$$

Represent this uncertainty as $G_p = G + W_1\Delta W_2$ where Δ is diagonal. Determine the corresponding $M\Delta$-structure and derive the RS-condition.

  (e) Let

$$G_p(s) = \frac{1 - \theta s}{1 + \theta s}; \quad \theta = \theta_0(1 + w\Delta), \quad |\Delta| < 1$$

and consider the controller $K(s) = c/s$. Put this into the $M\Delta$-structure and find the RS-condition.

  (f) Show by a counterexample that in general $\bar{\sigma}(AB)$ is not equal to $\bar{\sigma}(BA)$. Under what conditions is $\mu(AB) = \mu(BA)$?

  (g) The PRGA matrix is defined as $\Gamma = G_{\text{diag}}G^{-1}$. What is its relationship to the RGA?

BIBLIOGRAPHY

Anderson, B. D. O. (1986). Weighted Hankel-norm approximation: Calculation of bounds, *Systems & Control Letters* **7**(4): 247–255.

Anderson, B. D. O. and Liu, Y. (1989). Controller reduction: Concepts and approaches, *IEEE Transactions on Automatic Control* **34**(8): 802–812.

Anderson, B. D. O. and Moore, J. B. (1989). *Optimal Control: Linear Quadratic Methods*, Prentice-Hall, Englewood Cliffs, New Jersey.

Balas, G. J., Doyle, J. C., Glover, K., Packard, A. and Smith, R. (1993). *μ-Analysis and Synthesis Toolbox User's Guide*, MathWorks, Natick, Mass.

Balchen, J. G. and Mumme, K. (1988). *Process Control. Structures and Applications*, Van Nostrand Reinhold, New York.

Bode, H. W. (1945). *Network Analysis and Feedback Amplifier Design*, D. Van Nostrand Co., New York.

Boyd, S. and Barratt, C. (1991). *Linear Controller Design — Limits of Performance*, Prentice-Hall, Englewood Cliffs.

Boyd, S. and Desoer, C. A. (1985). Subharmonic functions and performance bounds in linear time-invariant feedback systems, *IMA J. Math. Contr. and Info.* **2**: 153–170.

Boyd, S., Ghaoui, L. E., Feron, E. and Balakrishnan, V. (1994). *Linear Matrix Inequalities in System and Control Theory*, SIAM, Philadelphia.

Braatz, R. D. (1993). *Robust Loopshaping for Process Control*, PhD thesis, California Institute of Technology, Pasadena.

Braatz, R. D. and Morari, M. (1994). Minimizing the Euclidean condition number, *SIAM Journal on Control and Optimization* **32**(6): 1763–1768.

Braatz, R. D., Morari, M. and Skogestad, S. (1996). Loopshaping for robust performance, *International Journal of Robust and Nonlinear Control* **6**.

Bristol, E. H. (1966). On a new measure of interactions for multivariable process control, *IEEE Transactions on Automatic Control* **AC-11**: 133–134.

Campo, P. J. and Morari, M. (1994). Achievable closed-loop properties of systems under decentralised control: Conditions involving the steady-state gain, *IEEE Transactions on Automatic Control* **AC-39**: 932–942.

Cao, Y. (1995). *Control Structure Selection for Chemical Processes Using Input-output Controllability Analysis*, PhD thesis, University of Exeter.

Chang, J. W. and Yu, C. C. (1990). The relative gain for non-square multivariable systems, *Chem. Eng. Sci.* **45**: 1309–1323.

Chen, C. T. (1984). *Linear System Theory and Design*, Holt, Rinehart and Winston, Inc., New York.

Chen, J. (1995). Sensitivity integral relations and design trade-offs in linear multivariable feedback-systems, *IEEE Transactions on Automatic Control* **AC-40**(10): 1700–1716.

Chiang, R. Y. and Safonov, M. G. (1992). *Robust Control Toolbox User's Guide*, MathWorks,

South Natick.
Churchill, R. V., Brown, J. W. and Verhey, R. F. (1974). *Complex Variables and Applications*, McGraw-Hill, New York.
Dahl, H. J. and Faulkner, A. J. (1979). Helicopter simulation in atmospheric turbulence, *Vertica* pp. 65–78.
Daoutidis, P. and Kravaris, C. (1992). Structural evaluation of control configurations for multivariable nonlinear processes, *Chemical Engineering Science* 47: 1091–1107.
Davison, E. J. (ed.) (1990). *Benchmark Problems for Control System Design*, Report of the IFAC Theory Committee, International Federation of Automatic Control.
Desoer, C. A. and Vidyasagar, M. (1975). *Feedback Systems: Input-Output Properties*, Academic Press, New York.
Doyle, J. C. (1978). Guaranteed margins for LQG regulators, *IEEE Transactions on Automatic Control* AC-23(4): 756–757.
Doyle, J. C. (1982). Analysis of feedback systems with structured uncertainties, *IEE Proceedings, Part D* 129(6): 242–250.
Doyle, J. C. (1983). Synthesis of robust controllers and filters, *Proc. IEEE Conf. on Decision and Control*, San Antonio, Texas, pp. 109–114.
Doyle, J. C. (1984). *Lecture Notes on Advances in Multivariable Control*, ONR/Honeywell Workshop, Minneapolis, October.
Doyle, J. C. (1986). Redondo Beach lecture notes, Internal Report, Caltech, Pasadena.
Doyle, J. C., Francis, B. and Tannenbaum, A. (1992). *Feedback Control Theory*, Macmillan Publishing Company.
Doyle, J. C., Glover, K., Khargonekar, P. P. and Francis, B. A. (1989). State-space solutions to standard $\mathcal{H}_2$ and $\mathcal{H}_\infty$ control problems, *IEEE Transactions on Automatic Control* AC-34(8): 831–847.
Doyle, J. C. and Stein, G. (1981). Multivariable feedback design: Concepts for a classical/modern synthesis, *IEEE Transactions on Automatic Control* AC-26(1): 4–16.
Doyle, J. and Stein, G. (1979). Robustness with observers, *IEEE Transactions on Automatic Control* 24(4): 607–611.
Eaton, J. W. and Rawlings, J. B. (1992). Model-predictive control of chemical processes, *Chemical Engineering Science* 69(1): 3–9.
Engell, S. (1988). *Optimale Lineare Regelung*, Vol. 18 of *Fachberichte Messen, Steuern, Regeln*, Springer-Verlag, Berlin.
Enns, D. (1984). Model reduction with balanced realizations: An error bound and a frequency weighted generalization, *Proceedings of the 23rd IEEE Conference on Decision and Control*, Las Vegas, NV, USA, pp. 127–32.
Fernando, K. V. and Nicholson, H. (1982). Singular perturbational model reduction of balanced systems, *IEEE Transactions on Automatic Control* AC-27(2): 466–468.
Foss, A. S. (1973). Critique of chemical process control theory, *AIChE Journal* 19: 209–214.
Foster, N. P., Spurgeon, S. K. and Postlethwaite, I. (1993). Robust model-reference tracking control with a sliding mode applied to an act rotorcraft, *19th European Rotorcraft Forum*, Italy.
Francis, B. (1987). *A course in $\mathcal{H}_\infty$ control theory*, Lecture Notes in Control and Information Sciences, Springer-Verlag, Berlin.
Francis, B. A. and Zames, G. (1984). On $\mathcal{H}_\infty$ optimal sensitivity theory for SISO feedback systems, *IEEE Transactions on Automatic Control* AC-29(1): 9–16.
Frank, P. M. (1968a). Vollständige Vorhersage im stetigen Regelkreis mit Totzeit, Teil I, *Regelungstechnik* 16(3): 111–116.
Frank, P. M. (1968b). Vollständige Vorhersage im stetigen Regelkreis mit Totzeit, Teil II, *Regelungstechnik* 16(5): 214–218.
Freudenberg, J. S. and Looze, D. P. (1988). *Frequency Domain Properties of Scalar and*

Multivariable Feedback Systems, Vol. 104 of *Lecture Notes in Control and Information Sciences*, Springer-Verlag, Berlin.

Freudenberg, J.S. Looze, D. (1985). Right half planes poles and zeros and design tradeoffs in feedback systems, *IEEE Transactions on Automatic Control* pp. 555–565.

Gjøsæter, O. B. (1995). *Structures for Multivariable Robust Process Control*, PhD thesis, Norwegian University of Science and Technology, Trondheim.

Glover, K. (1984). All optimal Hankel-norm approximations of linear multivariable systems and their L^∞-error bounds, *International Journal of Control* **39**(6): 1115–93.

Glover, K. and Doyle, J. C. (1988). State-space formulae for all stabilizing controller that satisfy an $\mathcal{H}_\infty$ norm bound and relations to risk sensitivity, *Systems and Control Letters* **11**: 167–172.

Glover, K. and McFarlane, D. (1989). Robust stabilization of normalized coprime factor plant descriptions with $\mathcal{H}_\infty$ bounded uncertainty, *IEEE Transactions on Automatic Control* **AC-34**(8): 821–830.

Golub, G. H. and van Loan, C. F. (1989). *Matrix Computations*, John Hopkins University Press, Baltimore.

Green, M. and Limebeer, D. J. N. (1995). *Linear Robust Control*, Prentice-Hall, Englewood Cliffs.

Grosdidier, P. and Morari, M. (1986). Interaction measures for systems under decentralized control, *Automatica* **22**: 309–319.

Grosdidier, P., Morari, M. and Holt, B. R. (1985). Closed-loop properties from steady-state gain information, *Industrial and Engineering Chemistry Process Design and Development* **24**: 221–235.

Haggblom, K. E. and Waller, K. (1988). Transformations and consistency relations of distillation control structures, *AIChE Journal* **34**: 1634–1648.

Hanus, R., Kinnaert, M. and Henrotte, J. (1987). Conditioning technique, a general anti-windup and bumpless transfer method, *Automatica* **23**(6): 729–739.

Havre, K. (1995). Personal communication. Norwegian University of Science and Technology, Trondheim.

Helton, J. (1976). Operator theory and broadband matching, *In Proceedings of the 11th Annual Allerton Conference on Communications, Control and Computing* .

Holt, B. R. and Morari, M. (1985a). Design of resilient processing plants V — The effect of deadtime on dynamic resilience, *Chemical Engineering Science* **40**: 1229–1237.

Holt, B. R. and Morari, M. (1985b). Design of resilient processing plants VI — The effect of right plane zeros on dynamic resilience, *Chemical Engineering Science* **40**: 59–74.

Horn, R. A. and Johnson, C. R. (1985). *Matrix Analysis*, Cambridge University Press.

Horn, R. A. and Johnson, C. R. (1991). *Topics in Matrix Analysis*, Cambridge University Press.

Horowitz, I. M. (1963). *Synthesis of Feedback Systems*, Academic Press, London.

Horowitz, I. M. (1991). Survey of quantitative feedback theory (QFT), *International Journal of Control* **53**(2): 255–291.

Horowitz, I. M. and Shaked, U. (1975). Superiority of transfer function over state-variable methods in linear time-invariant feedback system design, *IEEE Transactions on Automatic Control* **AC-20**(1): 84–97.

Hovd, M. (1992). *Studies on Control Structure Selection and Design of Robust Decentralized and SVD Controllers*, PhD thesis, Norwegian University of Science and Technology, Trondheim.

Hovd, M., Braatz, R. D. and Skogestad, S. (1994). SVD controllers for $\mathcal{H}_2$-, $\mathcal{H}_\infty$-, and μ-optimal control, *Proc. 1994 American Control Conference*, Baltimore, pp. 1233–1237.

Hovd, M. and Skogestad, S. (1992). Simple frequency-dependent tools for control system analysis, structure selection and design, *Automatica* **28**(5): 989–996.

Hovd, M. and Skogestad, S. (1993). Procedure for regulatory control structure selection with application to the FCC process, *AIChE Journal* **39**(12): 1938–1953.

Hovd, M. and Skogestad, S. (1994a). Pairing criteria for decentralised control of unstable plants, *Industrial and Engineering Chemistry Research* **33**: 2134–2139.

Hovd, M. and Skogestad, S. (1994b). Sequential design of decentralized controllers, *Automatica* **30**(10): 1601–1607.

Hoyle, D., Hyde, R. A. and Limebeer, D. J. N. (1991). An $\mathcal{H}_\infty$ approach to two degree of freedom design, *Proceedings of the 30th IEEE Conference on Decision and Control*, Brighton, UK, pp. 1581–1585.

Hung, Y. S. and MacFarlane, A. G. J. (1982). *Multivariable Feedback: A Quasi-Classical Approach*, Vol. 40 of *Lecture Notes in Control and Information Sciences*, Springer-Verlag, Berlin.

Hyde, R. A. (1991). *The Application of Robust Control to VSTOL Aircraft*, PhD thesis, University of Cambridge.

Hyde, R. A. and Glover, K. (1993). The application of scheduled $\mathcal{H}_\infty$ controllers to a VSTOL aircraft, *IEEE Transactions on Automatic Control* **AC-38**(7): 1021–1039.

Jaimoukha, I. M., Kasenally, E. M. and Limebeer, D. J. N. (1992). Numerical solution of large scale Lyapunov equations using Krylov subspace methods, *Proceedings of the 31st IEEE Conference on Decision and Control*.

Johnson, C. R. and Shapiro, H. M. (1986). Mathematical aspects of the relative gain array $(A \circ A^{-T})$, *SIAM Journal on Algebraic and Discrete Methods* **7**(4): 627–644.

Kailath, T. (1980). *Linear Systems*, Prentice-Hall, Englewood Cliffs.

Kalman, R. (1964). When is a linear control system optimal?, *Journal of Basic Engineering — Transaction on ASME — Series D* **86**: 51–60.

Kouvaritakis, B. (1974). *Characteristic Locus Methods for Multivariable Feedback Systems Design*, PhD thesis, University of Manchester Institute of Science and Technology, UK.

Kwakernaak, H. (1969). Optimal low-sensitivity linear feedback systems, *Automatica* **5**(3): 279–286.

Kwakernaak, H. (1985). Minimax frequency-domain performance and robustness optimization of linear feedback systems, *IEEE Transactions on Automatic Control* **AC-30**(10): 994–1004.

Kwakernaak, H. (1993). Robust control and $\mathcal{H}_\infty$-optimization — Tutorial paper, *Automatica* **29**: 255–273.

Kwakernaak, H. and Sivan, R. (1972). *Linear Optimal Control Systems*, Wiley Interscience, New York.

Laub, A. J., Heath, M. T., Page, C. C. and Ward, R. C. (1987). Computation of balancing transformations and other applications of simultaneous diagonalization algorithms, *IEEE Transactions on Automatic Control* **AC-32**(2): 115–122.

Laughlin, D. L., Jordan, K. G. and Morari, M. (1986). Internal model control and process uncertainty – mapping uncertainty regions for SISO controller-design, *International Journal of Control* **44**(6): 1675–1698.

Laughlin, D. L., Rivera, D. E. and Morari, M. (1987). Smith predictor design for robust performance, *International Journal of Control* **46**(2): 477–504.

Lee, J. H., Braatz, R. D., Morari, M. and Packard, A. (1995). Screening tools for robust control structure selection, *Automatica* **31**(2): 229–235.

Limebeer, D. J. N. (1991). The specification and purpose of a controller design case study, *Proc. IEEE Conf. Decision Contr.*, Brighton, UK, pp. 1579–1580.

Limebeer, D. J. N., Kasenally, E. M. and Perkins, J. D. (1993). On the design of robust two degree of freedom controllers, *Automatica* **29**(1): 157–168.

Liu, Y. and Anderson, B. D. O. (1989). Singular perturbation approximation of balanced systems, *International Journal of Control* **50**(4): 1379–1405.

Lundström, P. (1994). *Studies on Robust Multivariable Distillation Control*, PhD thesis,

Norwegian University of Science and Technology, Trondheim.

Lundström, P., Skogestad, S. and Doyle, J. C. (1996). Two degrees of freedom controller design for an ill-conditioned plant using μ-synthesis, *To appear in IEEE Transactions on Control System Technology*.

Lunze, J. (1992). *Feedback Control of Large-Scale Systems*, Prentice-Hall, Englewood Cliffs.

MacFarlane, A. G. J. and Karcanias, N. (1976). Poles and zeros of linear multivariable systems: A survey of algebraic, geometric and complex variable theory, *International Journal of Control* **24**: 33–74.

MacFarlane, A. G. J. and Kouvaritakis, B. (1977). A design technique for linear multivariable feedback systems, *International Journal of Control* **25**: 837–874.

Maciejowski, J. M. (1989). *Multivariable Feedback Design*, Addison-Wesley, Wokingham U.K.

Manness, M. A. and Murray-Smith, D. J. (1992). Aspects of multivariable flight control law design for helicopters using eigenstructure assignment, *Journal of American Helicopter Society* pp. 18–32.

Manousiouthakis, V., Savage, R. and Arkun, Y. (1986). Synthesis of decentralized process control structures using the concept of block relative gain, *AIChE Journal* **32**: 991–1003.

Marlin, T. (1995). *Process Control*, Mc-Graw Hill.

McFarlane, D. and Glover, K. (1990). *Robust Controller Design Using Normalized Coprime Factor Plant Descriptions*, Vol. 138 of *Lecture Notes in Control and Information Sciences*, Springer-Verlag, Berlin.

McMillan, G. K. (1984). *pH Control*, Instrument Society of America, Research Triangle Park, North Carolina.

Meinsma, G. (1995). Unstable and nonproper weights in $\mathcal{H}_\infty$ control, *Automatica* **31**(1): 1655–1658.

Mesarovic, M. (1970). Multilevel systems and concepts in process control, *Proc. of the IEEE*, Vol. 58, pp. 111–125.

Meyer, D. G. (1987). *Model Reduction via Factorial Representation*, PhD thesis, Stanford University.

Moore, B. C. (1981). Principal component analysis in linear systems: controllability, observability and model reduction, *IEEE Transactions on Automatic Control* **AC-26**(1): 17–32.

Morari, M. (1982). Integrated plant control: A solution at hand or a research topic for the next decade?, *in* T. F. Edgar and D. E. Seborg (eds), *Proc. Chemical Process Control 2*, AIChE, New York, pp. 467–495.

Morari, M. (1983). Design of resilient processing plants III – a general framework for the assessment of dynamic resilience, *Chemical Engineering Science* **38**: 1881–1891.

Morari, M., Arkun, Y. and Stephanopoulos, G. (1980). Studies in the synthesis of control structures for chemical process, Part I: Formulation of the problem. Process decomposition and the classification of the control tasks. Analysis of the optimizing control structures., *AIChE Journal* **26**(2): 220–232.

Morari, M. and Zafiriou, E. (1989). *Robust Process Control*, Prentice-Hall, Englewood Cliffs.

Nett, C. N. (1986). Algebraic aspects of linear-control system stability, *IEEE Transactions on Automatic Control* **AC-31**(10): 941–949.

Nett, C. N. (1989). A quantitative approach to the selection and partitioning of measurements and manipulations for the control of complex systems, *Presentation at Caltech Control Workshop*, Pasadena, USA, Jan. 1989.

Nett, C. N. and Manousiouthakis, V. (1987). Euclidean condition and block relative gain - connections conjectures, and clarifications, *IEEE Transactions on Automatic Control* **AC-32**(5): 405–407.

Nett, C. N. and Minto, K. D. (1989). A quantitative approach to the selection and partitioning of measurements and manipulations for the control of complex systems, Copy of transparencies from talk at *American Control Conference*, Pittsburgh, Pennsylvania, June 1989.

Niemann, H. and Stoustrup, J. (1995). Special Issue on Loop Transfer Recovery, *International Journal of Robust and Nonlinear Control* 7(7): November.

Nwokah, O. D. I. and Perez, R. (1991). On multivariable stability in the gain space, *Automatica* 27(6): 975–983.

Owen, J. G. and Zames, G. (1992). Robust $\mathcal{H}_\infty$ disturbance minimization by duality, *Systems & Control Letters* 19(4): 255–263.

Packard, A. (1988). *What's New with μ*, PhD thesis, University of California, Berkeley.

Packard, A. and Doyle, J. C. (1993). The complex structured singular value, *Automatica* 29(1): 71–109.

Packard, A., Doyle, J. C. and Balas, G. (1993). Linear, multivariable robust-control with a μ-perspective, *Journal of Dynamic Systems Measurement and Control — Transactions of the ASME* 115(2B): 426–438.

Padfield, G. D. (1981). Theoretical model of helicopter flight mechanics for application to piloted simulation, *Technical Report 81048*, Defence Research Agency (formerly Royal Aircraft Establishment), UK.

Perkins, J. D. (ed.) (1992). *IFAC Workshop on Interactions Between Process Design and Process Control*, (London, Sept. 1992), Pergamon Press, Oxford.

Pernebo, L. and Silverman, L. M. (1982). Model reduction by balanced state space representation, *IEEE Transactions on Automatic Control* AC-27(2): 382–387.

Poolla, K. and Tikku, A. (1995). Robust performance against time-varying structured perturbations, *IEEE Transactions on Automatic Control* 40(9): 1589–1602.

Postlethwaite, I., Foster, N. P. and Walker, D. J. (1994). Rotorcraft control law design for rejection of atmospheric turbulence, *Proceedings of IEE Conference, Control 94*, Warwick, pp. 1284–1289.

Postlethwaite, I. and MacFarlane, A. G. J. (1979). *A Complex Variable Approach to the Analysis of Linear Multivariable Feedback Systems*, Vol. 12 of *Lecture Notes in Control and Information Sciences*, Springer-Verlag, Berlin.

Postlethwaite, I., Samar, R., Choi, B.-W. and Gu, D.-W. (1995). A digital multi-mode $\mathcal{H}_\infty$ controller for the spey turbofan engine, *3rd European Control Conference*, Rome, Italy, pp. 147–152.

Postlethwaite, I. and Walker, D. J. (1992). Advanced control of high performance rotorcraft, *Institute of Mathematics and Its Applications Conference on Aerospace Vehicle Dynamics and Control, Cranfield Institute of Technology* pp. 615–619.

Qiu, L. and Davison, E. J. (1993). Performance limitations of non-minimum phase systems in the servomechanism problem, *Automatica* 29(2): 337–349.

Rosenbrock, H. H. (1966). On the design of linear multivariable systems, *Third IFAC World Congress*. Paper 1a.

Rosenbrock, H. H. (1970). *State-space and Multivariable Theory*, Nelson, London.

Rosenbrock, H. H. (1974). *Computer-Aided Control System Design*, Academic Press, New York.

Safonov, M. G. (1982). Stability margins of diagonally perturbed multivariable feedback systems, *IEE Proceedings, Part D* 129(6): 251–256.

Safonov, M. G. and Athans, M. (1977). Gain and phase margin for multiloop LQG regulators, *IEEE Transactions on Automatic Control* AC-22(2): 173–179.

Safonov, M. G. and Chiang, R. Y. (1989). A Schur method for balanced-truncation model reduction, *IEEE Transactions on Automatic Control* AC-34: 729–733.

Safonov, M. G., Limebeer, D. J. N. and Chiang, R. Y. (1989). Simplifying the $\mathcal{H}_\infty$ theory via loop-shifting, matrix-pencil and descriptor concepts, *International Journal of Control* 50(6): 2467–2488.

Samar, R. (1995). *Robust Multi-Mode Control of High Performance Aero-Engines*, PhD thesis,

University of Leicester.

Samar, R. and Postlethwaite, I. (1994). Multivariable controller design for a high performance areo engine, *Proceedings IEE Conference Control 94*, Warwick, pp. 1312–1317.

Samar, R., Postlethwaite, I. and Gu, D.-W. (1995). Model reduction with balanced realizations, *International Journal of Control* **62**(1): 33–64.

Samblancatt, C., Apkarian, P. and Patton, R. J. (1990). Improvement of helicopter robustness and performance control law using eigenstructure techniques and $\mathcal{H}_\infty$ synthesis, *16th European Rotorcraft Forum*, Scotland. Paper No. 2.3.1.

Seborg, D. E., Edgar, T. F. and Mellichamp, D. A. (1989). *Process Dynamics and Control*, Wiley, New York.

Sefton, J. and Glover, K. (1990). Pole-zero cancellations in the general $\mathcal{H}_\infty$ problem with reference to a two block design, *Systems & Control Letters* **14**: 295–306.

Shamma, J. S. (1994). Robust stability with time-varying structured uncertainty, *IEEE Transactions on Automatic Control* **AC-39**: 714–724.

Shinskey, F. G. (1984). *Distillation Control*, 2nd edn, McGraw Hill, New York.

Shinskey, F. G. (1988). *Process Control Systems*, 3rd edn, McGraw Hill, New York.

Skogestad, S. (1992). Dynamics and control of distillation columns — a critical survey, *IFAC symposium DYCORD+'92*, Maryland, pp. 1–25.

Skogestad, S. and Havre, K. (1996). The use of RGA and condition number as robustness measures, In *Proc. European Symposium on Computer-Aided Process Engineering (ESCAPE'96)*, Rhodes, Greece, May 1996.

Skogestad, S., Lundström, P. and Jacobsen, E. (1990). Selecting the best distillation control configuration, *AIChE Journal* **36**(5): 753–764.

Skogestad, S. and Morari, M. (1987a). Control configuration selection for distillation columns, *AIChE Journal* **33**(10): 1620–1635.

Skogestad, S. and Morari, M. (1987b). Effect of disturbance directions on closed-loop performance, *Industrial and Engineering Chemistry Research* **26**: 2029–2035.

Skogestad, S. and Morari, M. (1987c). Implications of large RGA elements on control performance, *Industrial and Engineering Chemistry Research* **26**: 2323–2330.

Skogestad, S. and Morari, M. (1988a). Some new properties of the structured singular value, *IEEE Transactions on Automatic Control* **AC-33**(12): 1151–1154.

Skogestad, S. and Morari, M. (1988b). Variable selection for decentralized control, *AIChE Annual Meeting*, Washington DC. Paper 126f. Reprinted in *Modeling, Identification and Control*, 1992, Vol. **13**, No. 2, 113–125.

Skogestad, S. and Morari, M. (1989). Robust performance of decentralized control systems by independent designs, *Automatica* **25**(1): 119–125.

Skogestad, S., Morari, M. and Doyle, J. C. (1988). Robust control of ill-conditioned plants: High-purity distillation, *IEEE Transactions on Automatic Control* **AC-33**(12): 1092–1105.

Skogestad, S. and Wolff, E. A. (1992). Controllability measures for disturbance rejection, *IFAC Workshop on Interactions between Process Design and Process Control*, London, UK, pp. 23–29.

Sourlas, D. D. and Manousiouthakis, V. (1995). Best achievable decentralized performance, *IEEE Transactions on Automatic Control* **AC-40**(11): 1858–1871.

Stanley, G., Marino-Galarraga, M. and McAvoy, T. J. (1985). Shortcut operability analysis. 1. The relative disturbance gain, *Industrial and Engineering Chemistry Process Design and Development* **24**: 1181–1188.

Stein, G. and Athans, M. (1987). The LQG/LTR procedure for multivariable feedback design, *IEEE Transactions on Automatic Control* **32**(2): 105–114.

Stein, G. and Doyle, J. C. (1991). Beyond singular values and loopshapes, *AIAA J. of Guidance and Control* **14**: 5–16.

Stephanopoulos, G. (1989). *Chemical Process Control*, Prentice-Hall, Englewood Cliffs, New Jersey.

Strang, G. (1976). *Linear Algebra and Its Applications*, Academic Press, New York.

Takahashi, M. D. (1993). Synthesis and evaluation of an $\mathcal{H}_2$ control law for a hovering helicopter, *Journal of Guidance, Control and Dynamics* **16**: 579–584.

Tøffner-Clausen, S., Andersen, P., Stoustrup, J. and Niemann, H. H. (1995). A new approach to μ-synthesis for mixed perturbation sets, *Proc. of 3rd European Control Conference*, Rome, Italy, pp. 147–152.

Toker, O. and Ozbay, H. (1995). On np-hardness of the purely complex μ computation, *Proceedings of the American Control Conference*, pp. 447–451.

Tombs, M. S. and Postlethwaite, I. (1987). Truncated balanced realization of a stable nonminimal state-space system, *International Journal of Control* **46**: 1319–1330.

Tsai, M., Geddes, E. and Postlethwaite, I. (1992). Pole-zero cancellations and closed-loop properties of an $\mathcal{H}_\infty$ mixed sensitivity design problem, *Automatica* **3**: 519–530.

van de Wal, M. and de Jager, B. (1995). Control structure design: A survey, *Proc. of American Control Conference*, Seattle, pp. 225–229.

van Diggelen, F. and Glover, K. (1994a). A Hadamard weighted loop shaping design procedure for robust decoupling, *Automatica* **30**(5): 831–845.

van Diggelen, F. and Glover, K. (1994b). State-space solutions to Hadamard weighted $\mathcal{H}_\infty$ and $\mathcal{H}_2$ control-problems, *International Journal of Control* **59**(2): 357–394.

Vidyasagar, M. (1985). *Control System Synthesis: A Factorization Approach*, MIT Press, Cambridge, MA.

Walker, D. J. (1996). On the structure of a two degrees-of-freedom controller, *To appear in International Journal of Control*.

Walker, D. J. and Postlethwaite, I. (1996). Advanced helicopter flight control using two degrees-of-freedom $\mathcal{H}_\infty$ optimization, *Journal of Guidance, Control and Dynamics* **19**(2): March–April.

Walker, D. J., Postlethwaite, I., Howitt, J. and Foster, N. P. (1993). Rotorcraft flying qualities improvement using advanced control, *American Helicopter Society/NASA Conference*, San Francisco.

Waller, K. V., Häggblom, K. E., Sandelin, P. M. and Finnerman, D. H. (1988). Disturbance sensitivity of distillation control structures, *AIChE Journal* **34**: 853–858.

Wang, Z. Q., Lundström, P. and Skogestad, S. (1994). Representation of uncertain time delays in the $\mathcal{H}_\infty$ framework, *International Journal of Control* **59**(3): 627–638.

Weinmann, A. (1991). *Uncertain Models and Robust Control*, Springer-Verlag, Berlin.

Whidborne, J. F., Postlethwaite, I. and Gu, D. W. (1994). Robust controller design using $\mathcal{H}_\infty$ loop shaping and the method of inequalities, *IEEE Transactions on Control Systems technology* **2**(4): 455–461.

Willems (1970). *Stability Theory of Dynamical Systems*, Nelson.

Wolff, E. (1994). *Studies on Control of Integrated Plants*, PhD thesis, Norwegian University of Science and Technology, Trondheim.

Wonham, M. (1974). *Linear Multivariable Systems*, Springer-Verlag, Berlin.

Youla, D. C., Jabr, H. A. and Bongiorno, J. J. (1976). Modern Wiener-Hopf design of optimal controllers, part II: The multivariable case., *IEEE Transactions on Automatic Control* **AC-21**: 319–38.

Youla, D. C., Jabr, H. A. and Lu, C. N. (1974). Single-loop feedback stabilization of linear multivariable dynamical plants, *Automatica* **10**: 159–173.

Young, P. M. (1993). *Robustness with Parametric and Dynamic Uncertainties*, PhD thesis, California Institute of Technology, Pasadena.

Young, P. M. (1994). Controller design with mixed uncertainties, *Proceedings of the American Control Conference*, Baltimore, Maryland, USA, pp. 2333–2337.

Young, P. M., Newlin, M. and Doyle, J. C. (1992). Practical computation of the mixed μ problem, *Proceedings of the American Control Conference*, Chicago, pp. 2190–2194.

Yu, C. C. and Fan, M. K. H. (1990). Decentralised integral controllability and D-stability, *Chemical Engineering Science* **45**: 3299–3309.

Yu, C. C. and Luyben, W. (1986). Design of multiloop SISO controllers in multivariable processes, *Industrial and Engineering Chemistry Process Design and Development* **25**: 498–503.

Yu, C. C. and Luyben, W. L. (1987). Robustness with respect to integral controllability, *Industrial and Engineering Chemistry Research* **26**(5): 1043–1045.

Yue, A. and Postlethwaite, I. (1990). Improvement of helicopter handling qualities using $\mathcal{H}_\infty$ optimization, *IEE Proceedings-D Control Theory and Applications* **137**: 115–129.

Zafiriou, E. (ed.) (1994). *IFAC Workshop on Integration of Process Design and Control*, (Baltimore, June 1994), Pergamon Press, Oxford. See also special issue of *Computers & Chemical Engineering*, Vol. 20, No. 4, 1996.

Zames, G. (1981). Feedback and optimal sensitivity: model reference transformations, multiplicative seminorms, and approximate inverse, *IEEE Transactions on Automatic Control* **AC-26**: 301–320.

Zhou, K., Doyle, J. C. and Glover, K. (1996). *Robust and Optimal Control*, Prentice-Hall, Upper Saddle River.

Ziegler, J. G. and Nichols, N. B. (1942). Optimum settings for automatic controllers, *Transactions of the A.S.M.E.* **64**: 759–768.

Ziegler, J. G. and Nichols, N. B. (1943). Process lags in automatic-control circuits, *Transactions of the A.S.M.E.* **65**: 433–444.

INDEX

---

[1] Page numbers in *italic* refer to definitions.